SCHAUM'S OUTLINE OF

THEORY AND PROBLEMS

OF

BUSINESS STATISTICS
Second Edition

•

LEONARD J. KAZMIER, Ph.D.

Arizona State University

•

SCHAUM'S OUTLINE SERIES
McGRAW-HILL PUBLISHING COMPANY

*New York St. Louis San Francisco Auckland Bogotá Caracas
Hamburg Lisbon London Madrid Mexico Montreal New Delhi
Oklahoma City Paris San Juan São Paulo Singapore
Sydney Tokyo Toronto*

LEONARD J. KAZMIER is Professor of Decision and Information Systems at Arizona State University. He completed bachelor's and master's degrees at Wayne State University, Detroit, and earned the Ph.D. at The Ohio State University. He is the author or coauthor of books in management concepts, statistical analysis, and computer applications. A charter member of the Decision Sciences Institute, Professor Kazmier is also a member of the Academy of Management and the American Statistical Association. He has taught at Wayne State University, the University of Notre Dame, and Arizona State University.

Schaum's Outline of Theory and Problems of
BUSINESS STATISTICS

3 4 5 6 7 8 9 10 11 12 13 14 15 16 17 18 19 20 SHP SHP 8 9 2 1 0 9

ISBN 0-07-033533-8

Sponsoring Editor, John Aliano
Production Supervisor, Louise Karam
Editing Supervisor, Marthe Grice

Library of Congress Cataloging-in-Publication Data

Kazmier, Leonard J.
    Schaum's outline of theory and problems of business statistics
by Leonard J. Kazmier.—2nd ed.
        p.   cm.
    Includes index.
    ISBN 0-07-033533-8
    1. Commercial statistics. 2. Economics—Statistical methods.
3. Statistics. I. Title. II. Title: Outline of theory and
problems of business statistics. III. Title: Business statistics.
HF1017.K39 1988
519.5—dc19
                                                                87-35372
                                                                    CIP

Preface

This Outline covers the basic methods of statistical description, statistical inference, and statistical decision analysis that typically are included in introductory and intermediate-level courses in business and economic statistics.

The purpose of this book is to present the concepts and methods of statistical analysis clearly and concisely. Along these lines, verbal explanations have been minimized in favor of presenting concrete examples. Because this Outline has been written particularly for those whose primary interest is the *application* of statistical techniques, mathematical derivations are omitted.

Although the book has been developed as a supplement to textbooks in business statistics, the content is sufficiently complete to permit its use as a course text. In terms of content, both the classical and contemporary methods of statistical analysis are included, and the topics have been grouped so as to be consistent with the organization of most of the textbooks concerned with the application of statistical methods in business and economics. The solved problems at the end of each chapter include complete solutions, while the supplementary problems include check answers only. Instructors can obtain a *Solutions Manual* that contains complete solutions to the supplementary problems by writing directly to the author on stationery that includes the institutional letterhead.

I express gratitude to Minitab, Inc. for permission to incorporate Minitab computer output in this book. The output from Minitab is typical of the many available software packages in statistical analysis. It is a user-friendly system that is frequently cited in textbooks in business statistics. I am also indebted to the literary executor of the late Sir Ronald A. Fisher, F.R.S., to Dr. Frank Yates, F.R.S., and to the Longman Group, Ltd., London, for permission to adapt and reprint Tables III and IV from their book, *Statistical Tables for Biological, Agricultural and Medical Research.* Finally, special thanks go to my daughter, Marian Lamb, for her careful and cheerful proofreading and editing of the entire manuscript.

LEONARD J. KAZMIER

Contents

Chapter 1 ANALYZING BUSINESS DATA 1

    1.1 Definition of Business Statistics 1
    1.2 Descriptive and Inferential Statistics 1
    1.3 Classical Statistics and Bayesian Decision Analysis 1
    1.4 Discrete and Continuous Variables 1
    1.5 Obtaining Data Through Experiments and Surveys 2
    1.6 Methods of Random Sampling..................................... 2
    1.7 Using a Computer to Generate Random Numbers 3

Chapter 2 STATISTICAL PRESENTATIONS 7

    2.1 Frequency Distributions 7
    2.2 Class Intervals... 7
    2.3 Histograms and Frequency Polygons 8
    2.4 Frequency Curves 9
    2.5 Cumulative Frequency Distributions 10
    2.6 Relative Frequency Distributions 11
    2.7 The "And Under" Type of Frequency Distribution 11
    2.8 Bar Charts and Line Graphs.................................... 11
    2.9 Pie Charts.. 12
    2.10 Computer Output 13

Chapter 3 DESCRIBING BUSINESS DATA: MEASURES OF LOCATION 30

    3.1 Measures of Location in Data Sets 30
    3.2 The Arithmetic Mean... 30
    3.3 The Weighted Mean... 30
    3.4 The Median 31
    3.5 The Mode.. 31
    3.6 Relationship Among the Mean, Median, and Mode 32
    3.7 Quartiles, Deciles, and Percentiles.......................... 32
    3.8 The Arithmetic Mean for Grouped Data 32
    3.9 The Median for Grouped Data 33
    3.10 The Mode for Grouped Data 34
    3.11 Quartiles, Deciles, and Percentiles for Grouped Data 34
    3.12 Computer Output 35

Chapter 4 DESCRIBING BUSINESS DATA: MEASURES OF VARIABILITY 45

    4.1 Measures of Variability in Data Sets.......................... 45
    4.2 The Range 45
    4.3 Modified Ranges... 45

4.4 The Average Deviation 46

4.5 The Variance and Standard Deviation 46

4.6 Shortcut Calculations of the Variance and Standard Deviation 47

4.7 Use of the Standard Deviation 48

4.8 The Coefficient of Variation 49

4.9 Pearson's Coefficient of Skewness 49

4.10 The Range and Modified Ranges for Grouped Data 50

4.11 The Average Deviation for Grouped Data 50

4.12 The Variance and Standard Deviation for Grouped Data 51

4.13 Computer Output 52

Chapter 5 PROBABILITY **65**

5.1 Basic Definitions of Probability 65

5.2 Expressing Probability 66

5.3 Mutually Exclusive and Nonexclusive Events 67

5.4 The Rules of Addition 67

5.5 Independent Events, Dependent Events, and Conditional Probability 68

5.6 The Rules of Multiplication 69

5.7 Bayes' Theorem 71

5.8 Joint Probability Tables 72

5.9 Permutations 73

5.10 Combinations 74

Chapter 6 PROBABILITY DISTRIBUTIONS FOR DISCRETE RANDOM VARIABLES: BINOMIAL, HYPERGEOMETRIC, AND POISSON **91**

6.1 Probability Distributions for Random Variables 91

6.2 The Expected Value and Variance for a Discrete Random Variable 91

6.3 The Binomial Distribution 93

6.4 The Binomial Distribution Expressed by Proportions 94

6.5 The Hypergeometric Distribution 95

6.6 The Poisson Distribution 96

6.7 Poisson Approximation of Binomial Probabilities 97

6.8 Computer Applications 98

Chapter 7 PROBABILITY DISTRIBUTIONS FOR CONTINUOUS RANDOM VARIABLES: NORMAL AND EXPONENTIAL **110**

7.1 Continuous Random Variables 110

7.2 The Normal Probability Distribution 110

7.3 Percentile Points for Normally Distributed Variables 113

7.4 Normal Approximation of Binomial Probabilities 114

7.5 Normal Approximation of Poisson Probabilities 115

7.6 The Exponential Probability Distribution 116

7.7 Computer Applications 116

CONTENTS

Chapter *8* SAMPLING DISTRIBUTIONS AND CONFIDENCE INTERVALS FOR THE MEAN **128**

8.1 Point Estimation 128
8.2 Sampling Distribution of the Mean 128
8.3 Confidence Intervals for the Mean Using the Normal Distribution 131
8.4 Determining the Required Sample Size for Estimating the Mean 132
8.5 Student's *t* Distribution and Confidence Intervals for the Mean 132
8.6 Summary Table for Interval Estimation of the Population Mean 133
8.7 Computer Output 133

Chapter *9* OTHER CONFIDENCE INTERVALS **144**

9.1 Confidence Intervals for the Difference Between Two Population Means Using the Normal Distribution 144
9.2 Student's *t* Distribution and Confidence Intervals for the Difference Between Two Population Means 145
9.3 Confidence Intervals for the Proportion Using the Normal Distribution 146
9.4 Determining the Required Sample Size for Estimating the Proportion 146
9.5 Confidence Intervals for the Difference Between Two Population Proportions 147
9.6 The χ^2 (Chi-Square) Distribution and Confidence Intervals for the Variance and Standard Deviation 148
9.7 Computer Output 149

Chapter *10* TESTING HYPOTHESES CONCERNING THE VALUE OF THE POPULATION MEAN **157**

10.1 Basic Steps in Hypothesis Testing 157
10.2 Testing a Hypothesized Value of the Mean Using the Normal Distribution 158
10.3 Type I and Type II Errors in Hypothesis Testing 161
10.4 Determining the Required Sample Size for Testing the Mean 163
10.5 Testing a Hypothesized Value of the Mean Using Student's *t* Distribution 164
10.6 The *P*-Value Approach to Testing Null Hypotheses Concerning the Population Mean 164
10.7 The Confidence Interval Approach to Testing Null Hypotheses Concerning the Population Mean 164
10.8 Summary Table for Testing a Hypothesized Value of the Mean 165
10.9 Computer Output 165

Chapter *11* TESTING OTHER HYPOTHESES **178**

11.1 Testing the Difference Between Two Means Using the Normal Distribution 178
11.2 Testing the Difference Between Two Means Using Student's *t* Distribution ... 180
11.3 Testing the Difference Between Two Means Based on Paired Observations ... 180
11.4 Testing a Hypothesized Value of the Population Proportion Using the Binomial Distribution 182
11.5 Testing a Hypothesized Value of the Population Proportion Using the Normal Distribution 183
11.6 Determining Required Sample Size for Testing the Proportion 184

11.7 Testing the Difference Between Two Population Proportions 184
11.8 Testing a Hypothesized Value of the Variance Using the Chi-Square Distribution 185
11.9 The *F* Distribution and Testing the Difference Between Two Variances 186
11.10 Alternative Approaches to Testing Null Hypotheses 187
11.11 Computer Output 187

Chapter *12* THE CHI-SQUARE TEST **200**

12.1 The Chi-Square Test as a Hypothesis-Testing Procedure 200
12.2 Goodness of Fit Tests 200
12.3 Tests for Independence of Two Categorical Variables (Contingency Table Tests) 202
12.4 Testing Hypotheses Concerning Proportions 204
12.5 Computer Output 207

Chapter *13* ANALYSIS OF VARIANCE **222**

13.1 Basic Rationale Associated with Testing the Difference Among Several Means 222
13.2 One-Factor Completely Randomized Design (One-Way ANOVA) 223
13.3 Two-Way Analysis of Variance 224
13.4 The Randomized Block Design (Two-Way ANOVA, One Observation per Cell) 224
13.5 Two-Factor Completely Randomized Design (Two-Way ANOVA, *n* Observa-
tions per Cell) 225
13.6 Additional Considerations 226
13.7 Computer Applications 227

Chapter *14* LINEAR REGRESSION AND CORRELATION ANALYSIS **242**

14.1 Objectives and Assumptions of Regression Analysis 242
14.2 The Scatter Plot 242
14.3 The Method of Least Squares for Fitting a Regression Line 243
14.4 Residuals and Residual Plots 244
14.5 The Standard Error of Estimate 244
14.6 Inferences Concerning the Slope 245
14.7 Confidence Intervals for the Conditional Mean 245
14.8 Prediction Intervals for Individual Values of the Dependent Variable 246
14.9 Objectives and Assumptions of Correlation Analysis 246
14.10 The Coefficient of Determination 247
14.11 The Coefficient of Correlation 247
14.12 The Covariance Approach to Understanding the Correlation Coefficient 249
14.13 Significance of the Correlation Coefficient 249
14.14 Pitfalls and Limitations Associated with Regression and Correlation Analysis 249
14.15 Computer Output 250

Chapter *15* MULTIPLE REGRESSION AND CORRELATION **262**

15.1 Objectives and Assumptions of Multiple Linear Regression Analysis 262
15.2 Additional Concepts in Multiple Regression Analysis 262

CONTENTS

15.3 The Use of Indicator (Dummy) Variables 263

15.4 Residuals and Residual Plots 264

15.5 Analysis of Variance in Linear Regression Analysis 264

15.6 Objectives and Assumptions of Multiple Correlation Analysis 265

15.7 Additional Concepts in Multiple Correlation Analysis 266

15.8 Pitfalls and Limitations Associated with Multiple Regression and Correlation
Analysis 266

15.9 Computer Output 267

Chapter 16   TIME SERIES ANALYSIS AND BUSINESS FORECASTING **275**

16.1 The Classical Time Series Model 275

16.2 Trend Analysis 276

16.3 Analysis of Cyclical Variations 277

16.4 Measurement of Seasonal Variations 277

16.5 Applying Seasonal Adjustments 278

16.6 Forecasting Based on Trend and Seasonal Factors 278

16.7 Cyclical Forecasting and Business Indicators 279

16.8 Exponential Smoothing as a Forecasting Method 280

16.9 Computer Output 281

Chapter 17   INDEX NUMBERS FOR BUSINESS AND ECONOMIC DATA **291**

17.1 Introduction 291

17.2 Construction of Simple Indexes 291

17.3 Construction of Aggregate Price Indexes 291

17.4 Link Relatives 292

17.5 Shifting the Base Period 292

17.6 Splicing Two Series of Index Numbers................................ 293

17.7 The Consumer Price Index (CPI) 293

17.8 Purchasing Power and the Deflation of Time Series Values 293

17.9 Other Published Indexes 293

**Chapter 18   BAYESIAN DECISION ANALYSIS: PAYOFF TABLES AND DECISION
TREES** **302**

18.1 The Structure of Payoff Tables 302

18.2 Decision Making Based upon Probabilities Alone 303

18.3 Decision Making Based upon Economic Consequences Alone 304

18.4 Decision Making Based upon Both Probabilities and Economic Consequences:
The Expected Payoff Criterion 306

18.5 Decision Tree Analysis 307

18.6 Expected Utility as the Decision Criterion 308

**Chapter 19   BAYESIAN DECISION ANALYSIS: THE USE OF SAMPLE INFOR-
MATION** **322**

19.1 The Expected Value of Perfect Information ($EVPI$) 322

19.2 Prior and Posterior Probability Distributions.......................... 323

CONTENTS

19.3 Bayesian Posterior Analysis and the Value of Sample Information (After Sampling) 325

19.4 Preposterior Analysis: The Expected Value of Sample Information (*EVSI*) Prior to Sampling 327

19.5 Expected Net Gain from Sampling (*ENGS*) and Optimum Sample Size 329

Chapter 20 BAYESIAN DECISION ANALYSIS: APPLICATION OF THE NORMAL DISTRIBUTION 338

20.1 Introduction 338

20.2 Determining the Parameters of the Normal Prior Probability Distribution 338

20.3 Defining the Linear Payoff Functions and Determining the Best Act 340

20.4 Linear Piecewise Loss Functions and the Expected Value of Perfect Information (*EVPI*) 342

20.5 Bayesian Posterior Analysis 344

20.6 Preposterior Analysis and the Expected Value of Sample Information (*EVSI*) 346

20.7 Expected Net Gain from Sampling (*ENGS*) and Optimum Sample Size 347

20.8 Bayesian Decision Analysis vs. Classical Decision Procedures 348

Chapter 21 NONPARAMETRIC STATISTICAL TESTS 357

21.1 Scales of Measurement 357

21.2 Parametric vs. Nonparametric Statistical Methods 357

21.3 The Runs Test for Randomness 358

21.4 One Sample: The Sign Test 358

21.5 One Sample: The Wilcoxon Test 359

21.6 Two Independent Samples: The Mann–Whitney Test 360

21.7 Paired Observations: The Sign Test 360

21.8 Paired Observations: The Wilcoxon Test 361

21.9 Several Independent Samples: The Kruskal–Wallis Test 361

Appendix 1 TABLE OF RANDOM NUMBERS 372

Appendix 2 BINOMIAL PROBABILITIES 373

Appendix 3 VALUES OF $e^{-\lambda}$ 376

Appendix 4 POISSON PROBABILITIES 377

Appendix 5 PROPORTIONS OF AREA FOR THE STANDARD NORMAL DISTRIBUTION 381

Appendix 6 PROPORTIONS OF AREA FOR THE *t* DISTRIBUTION 382

Appendix 7 PROPORTIONS OF AREA FOR THE χ^2 DISTRIBUTION 383

CONTENTS

Appendix *8* VALUES OF *F* EXCEEDED WITH PROBABILITIES OF 5 AND 1 PERCENT 383

Appendix *9* UNIT NORMAL LOSS FUNCTION 387

Appendix *10* CRITICAL VALUES OF *T* IN THE WILCOXON TEST 388

INDEX 391

Chapter 1

Analyzing Business Data

1.1 DEFINITION OF BUSINESS STATISTICS

Statistics refers to the techniques by which quantitative data are collected, organized, and analyzed. The focal point of business statistics is managerial decision making.

1.2 DESCRIPTIVE AND INFERENTIAL STATISTICS

Descriptive statistics include the techniques that are used to summarize and describe numerical data. These methods can either be graphical or involve computational analysis (see Chapters 2, 3, and 4).

EXAMPLE 1. The monthly sales volume for a product during the past year can be described and made meaningful by preparing a bar chart or a line graph (as described in Section 2.8). The relative sales by month can be highlighted by calculating an index number for each month such that the deviation from 100 for any given month indicates the percentage deviation of sales in that month as compared with average monthly sales during the entire year.

Inferential statistics include those techniques by which decisions about a statistical population are made based only on a sample having been observed, or possibly, by the use of managerial judgments. Because such decisions are made under conditions of uncertainty, the use of probability concepts is required. Whereas the measured characteristics of a sample are called *sample statistics*, measured characteristics of a statistical population, or universe, are called *population parameters*. The process of measuring the characteristics of all of the members of a defined population is called a *census*. Chapters 5 through 7 are concerned with probability concepts, and most of the chapters which follow Chapter 7 are concerned with the application of these concepts in statistical inference.

EXAMPLE 2. In order to estimate the voltage required to cause an electrical device to fail, a sample of such devices can be subjected to increasingly higher voltages until each device fails. Based on these sample results, the probability of failure at various voltage levels for the other devices in the sampled population can be estimated.

1.3 CLASSICAL STATISTICS AND BAYESIAN DECISION ANALYSIS

The methods of *classical statistics* are concerned with the analysis of sampled (objective) data for the purpose of inference, with the exclusions of any personal judgments or opinions (see Chapters 8–15). *Bayesian decision analysis* incorporates the use of managerial judgments in the statistical analysis and also places special emphasis on the possible economic gains or the possible losses associated with alternative decision acts (see Chapters 18–20).

EXAMPLE 3. By the classical approach to statistical inference, the uncertain level of sales for a new product would be estimated solely on the basis of market studies carried out in a number of locations selected in accordance with the requirements of scientific sampling. By the Bayesian approach, the judgments of managers who have had experience with similar products would be obtained and used as the basis of arriving at an estimated sales volume. This subjective estimate could then be combined with objective sample data to arrive at a combined estimate of the sales volume.

1.4 DISCRETE AND CONTINUOUS VARIABLES

A *discrete variable* can have only observed values at isolated points along a scale of values. In business statistics, such data typically occur through the process of *counting*; hence, the values generally are expressed as integers (whole numbers) only. A *continuous variable* can assume a value at any fractional point along a specified interval of values. Continuous data are generated by the process of *measuring.*

EXAMPLE 4. Examples of discrete data are the number of persons per household, the units of an item in inventory, and the number of assembled components which are found to be defective. Examples of continuous data are the weight of a shipment, the length of time before the first failure of a device, and the average number of persons per household in a large community. Note that an *average number* of persons can be a fractional value and is thus a continuous variable, even though the *number* per household is a discrete variable.

1.5   OBTAINING DATA THROUGH EXPERIMENTS AND SURVEYS

One way that data can be obtained is through direct observation. A *statistical experiment* is a form of direct observation in which there is control over some or all of the factors that may influence the variable being studied.

EXAMPLE 5. Two methods of assembling a component could be compared by having one group of employees use one of the methods and a second group of employees use the other method. The members of the first group are carefully matched to the members of the second group in terms of such factors as age and experience.

In some situations, it is not possible to collect data directly but, rather, the information has to be obtained from individual respondents. A *statistical survey* is a process of collecting data by asking individuals to provide the data. The data may be obtained through such methods as the personal interview, telephone interview, or written questionnaire.

EXAMPLE 6. An analyst in a state's Department of Economic Security may need to determine what increases or decreases in the employment level are planned by business firms located in that state. A standard method by which such data can be obtained is to conduct a survey of the business firms.

1.6   METHODS OF RANDOM SAMPLING

Random sampling is a type of sampling in which every item in a population of interest, or *target population,* has a known, and usually equal, chance of being chosen for inclusion in a sample. A random sample is also called a *probability sample* or a *scientific sample.* There are four principal methods of random sampling: simple random, systematic, stratified, and cluster.

A *simple random sample* is one in which items are chosen individually from the entire target population on the basis of chance. Such chance selection is similar to the random drawing of numbers in a lottery. However, in statistical sampling a *table of random numbers* or a *random-number generator* computer program generally is used to identify the numbered items in the population that are to be selected for the sample.

EXAMPLE 7. Appendix 1 is an abbreviated table of random numbers. Suppose we wish to take a simple random sample of 10 accounts receivable from a population of 90 such accounts, with the accounts being numbered 01 to 90. We would enter the table of random numbers "blindly" by literally closing our eyes and pointing to a starting position. Then we would read the digits in groups of two in any direction to choose the accounts for our sample. Suppose we begin reading numbers (as pairs) starting from the number on line 6, column 1. The 10 account numbers for the sample would be 66, 06, 59, 94, 78, 70, 08, 67, 12, and 65. However, since there are only 90 accounts, the number 94 cannot be included. Instead, the next number (11) is included in the sample. If any of the selected numbers are repeated, they are included only once in the sample.

A *systematic sample* is a random sample in which the items are selected from the population at a uniform interval of a listed order, such as choosing every tenth account receivable for the sample. The first account of the 10 accounts to be included in the sample would be chosen randomly (perhaps by reference to a table of random numbers). A particular concern with systematic sampling is the existence of any periodic, or cyclical, factor in the population listing that could lead to a systematic error in the sample results.

EXAMPLE 8.   If every twelfth house is at a corner location in a neighborhood surveyed for adequate street lighting, a systematic sample would include a systematic bias if every twelfth household were included in the survey. In this case, either all or none of the surveyed households would be at a corner location.

In *stratified sampling* the items in the population are first classified into separate subgroups, or strata, by the researcher on the basis of one or more important characteristics. Then a simple random or systematic sample is taken separately from each stratum. Such a sampling plan can be used to ensure proportionate representation of various population subgroups in the sample. Further, the required sample size to achieve a given level of precision typically is smaller than it is with simple random sampling, thereby reducing sampling cost.

EXAMPLE 9.   In a study of student attitudes toward on-campus housing, we have reason to believe that important differences may exist between undergraduate and graduate students, and between men and women students. Therefore, a stratified sampling plan should be considered in which a simple random sample is taken separately from the four strata: male undergraduate, female undergraduate, male graduate, and female graduate.

Cluster sampling is a type of random sampling in which the population items occur naturally in subgroups. Entire subgroups, or clusters, are then randomly sampled.

EXAMPLE 10.   If an analyst in a state's Department of Economic Security needs to study the hourly wage rates being paid in a metropolitan area, it would be difficult to obtain a listing of all the wage earners in the target population. However, a listing of the *firms* in that area can be obtained much more easily. The analyst then can take a simple random sample of the identified firms, which represent *clusters* of employees, and obtain the wage rates being paid to the employees of these firms.

1.7   USING A COMPUTER TO GENERATE RANDOM NUMBERS

Computer software is widely available to generate randomly selected digits within any specified range of values. Problem 1.9 illustrates the use of such software.

Solved Problems

DESCRIPTIVE AND INFERENTIAL STATISTICS

1.1   Indicate which of the following terms or operations are concerned with a sample or sampling (S), and those which are concerned with a population (P): (*a*) Group measures called *parameters*, (*b*) use of *inferential statistics*, (*c*) taking a *census*, (*d*) judging the quality of an incoming shipment of fruit by inspecting several crates of the large number included in the shipment.

(*a*) P, (*b*) S, (*c*) P, (*d*) S

CLASSICAL STATISTICS AND BAYESIAN DECISION ANALYSIS

1.2   Indicate which of the following types of information could be used only in Bayesian decision analysis (B), and those which could be used in either classical or Bayesian analysis (CB): (*a*)

Managerial judgment about the likely level of sales for a new product, (*b*) survey results for a sample of previous customers, (*c*) financial analysts' forecasts of stock market averages.

(*a*) B, (*b*) CB, (*c*) B

DISCRETE AND CONTINUOUS VARIABLES

1.3   For the following types of values, designate discrete variables (D) and continuous variables (C): (*a*) Weight of the contents of a package of cereal, (*b*) diameter of a bearing, (*c*) number of defective items produced, (*d*) number of individuals in a geographic area who are collecting unemployment benefits, (*e*) the average number of prospective customers contacted per sales representative during the past month, (*f*) dollar amount of sales.

(*a*) C, (*b*) C, (*c*) D, (*d*) D, (*e*) C, (*f*) D (*Note:* Although monetary amounts are discrete, when the amounts are large relative to the one-cent discrete units, they generally are treated as continuous data.)

OBTAINING DATA THROUGH EXPERIMENTS AND SURVEYS

1.4   Indicate which of the following data-gathering procedures would be considered an experiment (E), and which would be considered a survey (S): (*a*) A political poll of how individuals intend to vote in an upcoming election, (*b*) customers in a shopping mall interviewed about why they shop there, (*c*) comparing two approaches to marketing an annuity policy by having each approach used in comparable geographic areas.

(*a*) S, (*b*) S, (*c*) E

1.5   In the area of statistical measurements, such as questionnaires, *reliability* refers to the consistency of the measuring instrument and *validity* refers to the accuracy of the instrument. Thus, if a questionnaire yields similar results when completed by two equivalent groups of respondents, then the questionnaire can be described as being reliable. Does the fact that an instrument is reliable thereby guarantee that it is valid?

The reliability of a measuring instrument does not guarantee that it is valid for a particular purpose. An instrument that is reliable is consistent in the repeated measurements that are produced, but the measurements may all include a common error, or bias, component. (See the next Solved Problem.)

1.6   Refer to Solved Problem 1.5, above. Can a survey instrument that is not reliable have validity for a particular purpose?

An instrument that is not reliable cannot be valid for *any* particular purpose. In the absence of reliability, there is no consistency in the results that are obtained. An analogy to a rifle range can illustrate this concept. Bullet holes that are closely clustered on a target are indicative of the reliability (consistency) in firing the rifle. In such a case the validity (accuracy) may be improved by adjusting the sights so that the bullet holes subsequently will be centered at the bull's-eye of the target. But widely dispersed bullet holes would indicate a lack of reliability, and under such a condition no adjustment in the sights can lead to a high score.

METHODS OF RANDOM SAMPLING

1.7   For the purpose of statistical inference a *representative* sample is desired. Yet, the methods of statistical inference only require that a *random* sample be obtained. Why?

There is no sampling method that can guarantee a representative sample. The best we can do is to avoid any consistent or systematic bias by the use of random (probability) sampling. While a random sample rarely will be exactly representative of the target population from which it was obtained, use of this procedure does guarantee that only chance factors underlie the amount of difference between the sample and the population.

1.8 An oil company wants to determine the factors affecting consumer choice of gasoline service stations in a test area, and therefore has obtained the names and addresses of and available personal information for all the registered car owners residing in that area. Describe how a sample of this list could be obtained using each of the four methods of random sampling described in this chapter.

For a *simple random sample*, the listed names could be numbered sequentially, and then the individuals to be sampled could be selected by using a table of random numbers. For a *systematic sample*, every *n*th (such as 5th) person on the list could be contacted, starting randomly within the first five names. For a *stratified sample*, we can classify the owners by their type of car, the value of their car, sex, or age, and then take a simple random or systematic sample from each defined stratum. For a *cluster sample*, we could choose to interview all the registered car owners residing in randomly selected blocks in the test area. Having a geographic basis, this type of cluster sample can also be called an *area sample*.

USING A COMPUTER TO GENERATE RANDOM NUMBERS

1.9 A state economic analyst wishes to obtain a simple random sample of 30 business firms from the 435 listed firms for a geographic region. The firms are identified by the sequential identification numbers 001 through 435. Use available computer software to obtain the 30 identification numbers of the firms to be included in the study.

Figure 1-1 includes the listing of the identification numbers of the 30 randomly selected firms. Note that by chance firm 94 was selected three times. Therefore, two more firms would be needed to achieve the desired sample size of 30.

```
MTB ) RANDOM 30 DIGITS, PUT IN C1;
SUBC) INTEGERS FROM 1 TO 435.
MTB ) NAME FOR C1 IS 'SAMPLE'
MTB ) PRINT 'SAMPLE'
SAMPLE
     8    56   184    94   275    78   135   172   303   296    94
    90   405   169   239   114   100    26   212   331   215    18
   141   326   290   297   201   140   248    94
```

Fig. 1-1 Minitab output. (Copyright Pennsylvania State University.)

Supplementary Problems

DESCRIPTIVE AND INFERENTIAL STATISTICS

1.10 Indicate which of the following terms or operations are concerned with a sample or sampling (S), and those which are concerned with a population (P): (*a*) Universe, (*b*) group measures called *statistics*, (*c*) application of probability concepts, (*d*) inspection of every item that is manufactured.

Ans. (*a*) P, (*b*) S, (*c*) S, (*d*) P

CLASSICAL STATISTICS AND BAYESIAN DECISION ANALYSIS

1.11 Indicate which of the following types of information could be used only in Bayesian decision analysis (B) and those which could be used in either classical or Bayesian analysis (CB): (*a*) Objective data, (*b*) sample data, (*c*) subjective data, (*d*) managerial judgments.

Ans. (*a*) CB, (*b*) CB, (*c*) B, (*d*) B

DISCRETE AND CONTINUOUS VARIABLES

1.12 For the following types of values, designate discrete variables (D) and continuous variables (C): (*a*) Number of units of an item held in stock, (*b*) ratio of current assets to current liabilities, (*c*) total tonnage shipped, (*d*) quantity shipped, in units, (*e*) volume of traffic on a toll road, (*f*) attendance at the company's annual meeting.

   Ans.   (*a*) D, (*b*) C, (*c*) C, (*d*) D, (*e*) D, (*f*) D

OBTAINING DATA THROUGH EXPERIMENTS AND SURVEYS

1.13 Indicate which of the following data-gathering procedures would be considered an experiment (E), and which would be considered a survey (S): (*a*) comparing the results of a new approach to training airline ticket agents to those of the traditional approach, (*b*) evaluating two different sets of assembly instructions for a toy by having two comparable groups of children assemble the toy using the different instructions, (*c*) having a product-evaluation magazine send subscribers a questionnaire asking them to rate the products that they have recently purchased.

   Ans.   (*a*) E, (*b*) E, (*c*) S

METHODS OF RANDOM SAMPLING

1.14 Identify whether the simple random (R) or the systematic (S) sampling method is used in the following: (*a*) Using a table of random numbers, (*b*) interviewing every one-hundredth adult entering an amusement park, randomly starting at the fifty-fifth person to enter the park.

   Ans.   (*a*) R, (*b*) S

1.15 For the following group-oriented sampling situations, identify whether the stratified (St) or the cluster (C) sampling method would be used: (*a*) Estimating the voting preferences of people who live in various neighborhoods, (*b*) studying consumer attitudes with the belief that there are important differences according to age and sex.

   Ans.   (*a*) C, (*b*) St

USING A COMPUTER TO GENERATE RANDOM NUMBERS

1.16 An auditor wishes to take a simple random sample of 50 accounts from the 5,250 accounts receivable in a large firm. The accounts are sequentially numbered from 0001 or 5250. Use available computer software to obtain a listing of the required 50 random numbers.

Chapter 2

Statistical Presentations

2.1 FREQUENCY DISTRIBUTIONS

A *frequency distribution* is a table in which possible values for a variable are grouped into classes, and the number of observed values which fall into each class is recorded. Data organized in a frequency distribution are called *grouped data*. In contrast, for *ungrouped data* every observed value of the random variable is listed.

EXAMPLE 1. A frequency distribution of weekly wages is shown in Table 2.1. Note that the amounts are reported to the nearest dollar. When a remainder that is to be rounded is "exactly 0.5" (exactly $0.50 in this case), the convention is to round to the nearest *even* number. Thus a weekly wage of $259.50 would have been rounded to $260 as part of the data-reporting process.

**Table 2.1   A Frequency Distribution of Weekly Wages
for 100 Unskilled Workers**

Weekly wage	Number of workers (f)
$240–259	7
260–279	20
280–299	33
300–319	25
320–339	11
340–359	4
Total	100

2.2   CLASS INTERVALS

For each class in a frequency distribution, the lower and upper *stated class limits* indicate the values included within the class. (See the first column of Table 2.1.) In contrast, the *exact class limits*, or *class boundaries*, are the specific points that serve to separate adjoining classes along a measurement scale for continuous variables. Exact class limits can be determined by identifying the points that are halfway between the upper and lower stated class limits, respectively, of adjoining classes. The *class interval* identifies the range of values included within a class and can be determined by subtracting the lower exact class limit from the upper exact class limit for the class. When exact limits are not identified, the class interval can be determined by subtracting the lower stated limit for a class from the lower stated limit of the adjoining next-higher class. Finally, for certain purposes the values in a class often are represented by the *class midpoint*, which can be determined by adding one-half of the class interval to the lower exact limit of the class.

EXAMPLE 2. Table 2.2 presents the exact class limits and the class midpoints for the frequency distribution in Table 2.1.

EXAMPLE 3. Calculated by the two approaches, the class interval for the first class in Table 2.2 is

$259.50 − $239.50 = $20 (subtraction of the lower exact class limit from the upper exact class limit of the class)

$260 − $240 = $20 (subtraction of the lower stated class limit of the class from the lower stated class limit of the adjoining next-higher class)

Table 2.2 Weekly Wages for 100 Unskilled Workers

Weekly wage (class limits)	Exact class limits*	Class midpoint	Number of workers
$240–259	$239.50–259.50	$249.50	7
260–279	259.50–279.50	269.50	20
280–299	279.50–299.50	289.50	33
300–319	299.50–319.50	309.50	25
320–339	319.50–339.50	329.50	11
340–359	339.50–359.50	349.50	4
			Total 100

*In general, only one additional significant digit is expressed in exact class limits as compared with stated class limits. However, because with monetary units the next more precise unit of measurement after "nearest dollar" is usually defined as "nearest cent," in this case two additional digits are expressed.

Computationally, it is generally desirable that all class intervals in a given frequency distribution be equal. A formula which can be used to determine the approximate class interval is

$$\text{Approximate interval} = \frac{\left[\begin{array}{c}\text{largest value in}\\\text{ungrouped data}\end{array}\right] - \left[\begin{array}{c}\text{smallest value in}\\\text{ungrouped data}\end{array}\right]}{\text{number of classes desired}} \qquad (2.1)$$

EXAMPLE 4. For the original, ungrouped data that were grouped in Table 2.1, suppose the highest observed wage was $358 and the lowest observed wage was $242. Given the objective of having six classes with equal class intervals,

$$\text{Approximate interval} = \frac{358 - 242}{6} = \$19.33$$

The closest convenient class size is thus $20.

For data that are distributed in a highly nonuniform way, such as annual salary data for a variety of occupations, *unequal class intervals* may be desirable. In such a case, the larger class intervals are used for the ranges of values in which there are relatively few observations.

2.3 HISTOGRAMS AND FREQUENCY POLYGONS

A *histogram* is a bar graph of a frequency distribution. As indicated in Fig. 2-1, typically the exact class limits are entered along the horizontal axis of the graph while the numbers of observations are listed along the vertical axis.

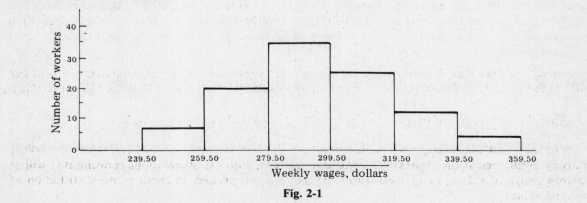

Fig. 2-1

EXAMPLE 5.   A histogram for the frequency distribution of weekly wages in Table 2.2 is shown in Fig. 2-1.

A *frequency polygon* is a line graph of a frequency distribution. As indicated in Fig. 2-2, the two axes of this graph are similar to those of the histogram except that the midpoint of each class typically is identified along the horizontal axis. The number of observations in each class is represented by a dot above the midpoint of the class, and these dots are joined by a series of line segments to form a "many-sided figure," or polygon.

EXAMPLE 6.   A frequency polygon for the distribution of weekly wages in Table 2.2 is shown in Fig. 2-2.

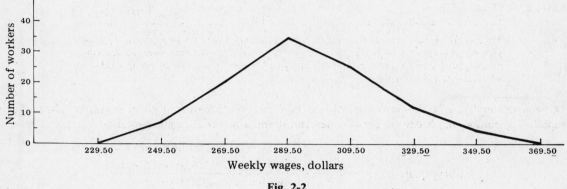

Fig. 2-2

2.4   FREQUENCY CURVES

A frequency curve is a smoothed frequency polygon.

EXAMPLE 7.   Figure 2-3 is a frequency curve for the distribution of weekly wages in Table 2.2.

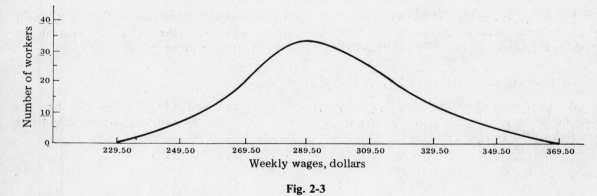

Fig. 2-3

In terms of skewness, a frequency curve can be: (1) *negatively skewed*: nonsymmetrical with the "tail" to the left; (2) *positively skewed*: nonsymmetrical with the "tail" to the right; or (3) *symmetrical*.

EXAMPLE 8.   The concept of frequency curve skewness is illustrated graphically in Fig. 2-4.

In terms of kurtosis, a frequency curve can be: (1) *platykurtic*: flat, with the observations distributed relatively evenly across the classes; (2) *leptokurtic*: peaked, with the observations concentrated within a narrow range of values; or (3) *mesokurtic*: neither flat nor peaked, in terms of the distribution of observed values.

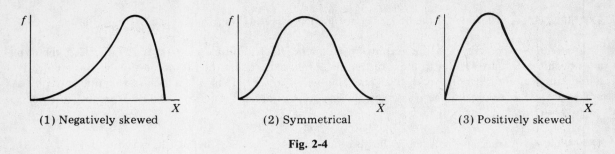

(1) Negatively skewed   (2) Symmetrical   (3) Positively skewed

Fig. 2-4

EXAMPLE 9. Types of frequency curves in terms of kurtosis are shown in Fig. 2-5.

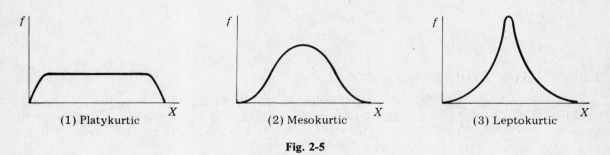

(1) Platykurtic   (2) Mesokurtic   (3) Leptokurtic

Fig. 2-5

2.5 CUMULATIVE FREQUENCY DISTRIBUTIONS

A *cumulative frequency distribution* identifies the cumulative number of observations included below the upper exact limit of each class in the distribution. The cumulative frequency for a class can be determined by adding the observed frequency for that class to the cumulative frequency for the preceding class.

EXAMPLE 10. The calculation of cumulative frequencies is illustrated in Table 2.3.

Table 2.3 Calculation of the Cumulative Frequencies for the Weekly Wage Data of Table 2.2

Weekly wage	Upper exact class limit	Number of workers (f)	Cumulative frequency (cf)
$240–259	$259.50	7	7
260–279	279.50	20	$20 + 7 = 27$
280–299	299.50	33	$33 + 27 = 60$
300–319	319.50	25	$25 + 60 = 85$
320–339	339.50	11	$11 + 85 = 96$
340–359	359.50	4	$4 + 96 = 100$
		Total 100	

The graph of a cumulative frequency distribution is called an *ogive* (pronounced "ō-jive"). For the less-than type of cumulative distribution, this graph indicates the cumulative frequency below each exact class limit of the frequency distribution. When such a line graph is smoothed, it is called an *ogive curve*.

EXAMPLE 11. An ogive curve for the cumulative frequency distribution in Table 2.3 is given in Fig. 2-6.

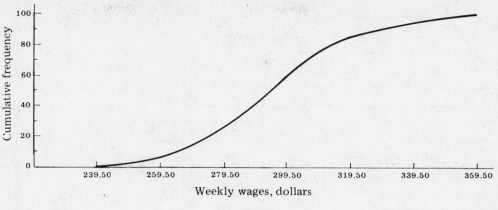

Fig. 2-6

2.6 RELATIVE FREQUENCY DISTRIBUTIONS

A *relative frequency distribution* is one in which the number of observations associated with each class has been converted into a relative frequency by dividing by the total number of observations in the entire distribution. Each relative frequency is thus a proportion, and can be converted into a percentage by multiplying by 100 percent.

One of the advantages associated with constructing a relative frequency distribution is that the cumulative distribution and the ogive for such a distribution indicate the cumulative proportion (or percentage) of observation up to the various possible values of the variable. A *percentile* value is the cumulative percentage of observations up to a designated value of a variable. (See Problems 2.14 and 2.16 to 2.20.)

2.7 THE "AND UNDER" TYPE OF FREQUENCY DISTRIBUTION

Consider the following stated class limits for two adjoining classes:

5 and under 8

8 and under 11

For such "and under" stated class limits, the exact class limits are taken to be identical in value to the respective stated limits. Thus the lower and upper exact limits for the first listed class above are 5.0 and 8.0, respectively. (See Problems 2.21 and 2.22.) The "and under" approach generally is the easier one to implement with computer software, and it sometimes reflects a more "natural" way of data collection when amounts of time are measured. For instance, the ages of people generally are reported as the age at the last birthday, rather than the age at the nearest birthday.

2.8 BAR CHARTS AND LINE GRAPHS

A *bar chart* depicts amounts of frequencies for different categories of data by a series of bars. The difference between a bar chart and a histogram is that a histogram always relates to data in a frequency distribution, whereas a bar chart depicts amounts for any types of categories.

EXAMPLE 12.   The bar chart in Fig. 2-7 depicts factory sales of passenger cars for plants in the United States. The data are categorized by year.

A *component bar chart* portrays subdivisions within the bars on the chart. For example, each bar in Fig. 2-7 could be subdivided into separate parts (and perhaps color-coded) to indicate the relative

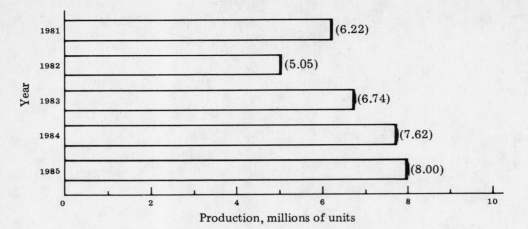

Fig. 2-7   (*Source*: *Survey of Current Business*, U.S. Department of Commerce.)

contribution of each of the automobile manufacturers to the total factory sales (production) for the year. (See Problem 2.24.)

Whenever the categories used represent a time segment, as is true for the data of Fig. 2-7, the data can also be described by means of a *line graph*. A line graph portrays changes in amounts in respect to time by a series of line segments.

EXAMPLE 13.   The data of Fig. 2-7 are presented as a line graph in Fig. 2-8.

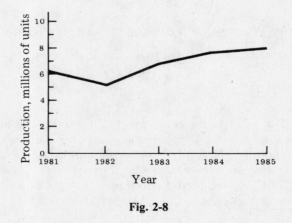

Fig. 2-8

2.9   PIE CHARTS

The *pie chart* is particularly appropriate for portraying the divisions of a total amount, such as the distribution of a company's sales dollar.

A *percentage pie chart* is one in which the values have been converted into percentages in order to make them easier to compare. (See Problem 2.26.)

EXAMPLE 14.   Figure 2-9 is a percentage pie chart depicting the worldwide distribution of the 369,300 employees of the Ford Motor Company in 1985 (data obtained from the *Annual Report*). Conversion of the original numbers to percentages makes comparison much easier in this case.

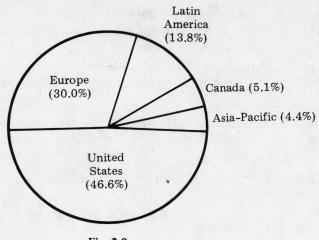

Fig. 2-9

2.10 COMPUTER OUTPUT

Computer software is available to form frequency distributions from the input of ungrouped data and to output the associated histograms. The programs typically will determine a convenient class interval or allow the user to specify the interval size to be used. Because of the convenience in interpreting nonfractional midpoints and exact class limits, the "and under" type of frequency distribution generally is formed. (See Problems 2.27 and 2.28.)

Solved Problems

FREQUENCY DISTRIBUTIONS, CLASS INTERVALS, AND RELATED GRAPHIC METHODS

2.1   With reference to Table 2.4,

   (*a*)   what are the lower and upper stated limits of the first class?

   (*b*)   what are the lower and upper exact limits of the first class?

   (*c*)   the class interval used is the same for all classes of the distribution. What is the interval size?

Table 2.4   Frequency Distribution of Monthly Apartment Rental Rates for 200 Apartments

Rental rate	Number of apartments
$350–379	3
380–409	8
410–439	10
440–469	13
470–499	33
500–529	40
530–559	35
560–589	30
590–619	16
620–649	12
Total	200

(d) what is the midpoint of the first class?

(e) what are the lower and upper exact limits of the class in which the largest number of apartment rental rates was tabulated?

(f) suppose a monthly rental rate of $439.50 was reported. Identify the lower and upper stated limits of the class in which this observation would be tallied.

(a) $350 and $379

(b) $349.50 and $379.50 (*Note*: As in Example 2, two additional digits are expressed in this case instead of the usual one additional digit in exact class limits as compared with stated class limits.)

(c) Focusing on the interval of values in the first class,

$379.50 − $349.50 = $30 (subtraction of the lower exact class limit from the upper exact class limit of the class)

$380 − $350 = $30 (subtraction of the lower stated class limit of the class from the lower stated class limit of the next-higher adjoining class)

(d) $349.50 + $\frac{30}{2}$ = $349.50 + $15.00 = $364.50

(e) $499.50 and $529.50

(f) $440 and $469 (*Note*: $439.50 is first rounded to $440 as the nearest dollar using the "even number" rule described in Section 2.1.)

2.2 Construct a histogram for the data of Table 2.4.

A histogram for the data of Table 2.4 appears in Fig. 2-10.

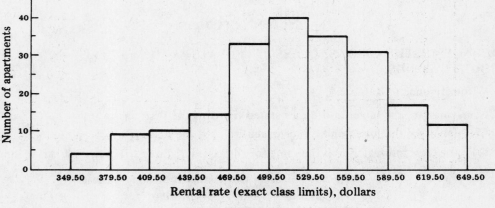

Fig. 2-10

2.3 Construct a frequency polygon and a frequency curve for the data of Table 2.4.

Figure 2-11 is a graphic presentation of the frequency polygon and frequency curve for the data in Table 2.4.

2.4 Describe the frequency curve in Fig. 2-11 from the standpoint of skewness.

The frequency curve appears to be somewhat negatively skewed.

2.5 Construct a cumulative frequency distribution for the data of Table 2.4.

See Table 2.5.

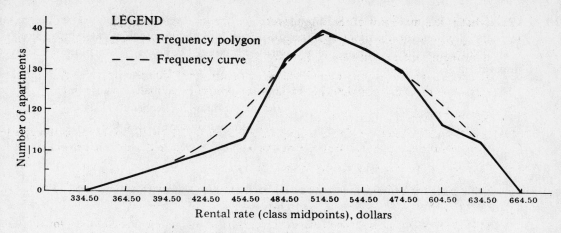

Fig. 2-11

Table 2.5   Cumulative Frequency Distribution of Apartment Rental Rates

Rental rate	Class boundaries	Number of apartments	Cumulative frequency (*cf*)
$350–379	$349.50–379.50	3	3
380–409	379.50–409.50	8	11
410–439	409.50–439.50	10	21
440–469	439.50–469.50	13	34
470–499	469.50–499.50	33	67
500–529	499.50–529.50	40	107
530–559	529.50–559.50	35	142
560–589	559.50–589.50	30	172
590–619	589.50–619.50	16	188
620–649	619.50–649.50	12	200
		Total 200	

2.6   Present the cumulative frequency distribution in Table 2.5 graphically by means of an ogive curve.

The ogive curve for the data of Table 2.5 is shown in Fig. 2-12.

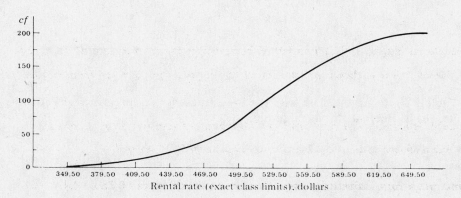

Fig. 2-12

2.7   Listed in Table 2.6 are the required times to complete a sample assembly task for 30 employees who have applied for a promotional transfer to a job requiring precision assembly. Suppose we wish to organize these data into five classes with equal class sizes. Determine the convenient interval size.

$$\text{Approximate interval} = \frac{\left[\begin{array}{c}\text{largest value in}\\ \text{ungrouped data}\end{array}\right] - \left[\begin{array}{c}\text{smallest value in}\\ \text{ungrouped data}\end{array}\right]}{\text{number of classes desired}}$$

$$= \frac{18-9}{5} = 1.80$$

In this case, it is convenient to round the interval to 2.0.

Table 2.6   Assembly Times for 30 Employees, min

10	14	15	13	17
16	12	14	11	13
15	18	9	14	14
9	15	11	13	11
12	10	17	16	12
11	16	12	14	15

2.8   Construct the frequency distribution for the data of Table 2.6 using a class interval of 2.0 for all classes and setting the lower stated limit of the first class at 9 min.

The required construction appears in Table 2.7.

Table 2.7   Frequency Distribution for the Assembly Times

Time, min	Number of employees
9–10	4
11–12	8
13–14	8
15–16	7
17–18	3
Total	30

2.9   In Table 2.7 refer to the class with the lowest number of employees and identify (*a*) its exact limits, (*b*) its interval, (*c*) its midpoint.

(*a*) 16.5–18.5, (*b*) $18.5 - 16.5 = 2.0$, (*c*) $16.5 + 2.0/2 = 17.5$.

2.10   Construct a histogram for the frequency distribution in Table 2.7.

The histogram is presented in Fig. 2-13.

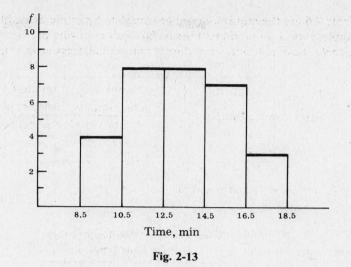

Fig. 2-13

2.11   Construct a frequency polygon and frequency curve for the data in Table 2.7.

The frequency polygon and frequency curve appear in Fig. 2-14.

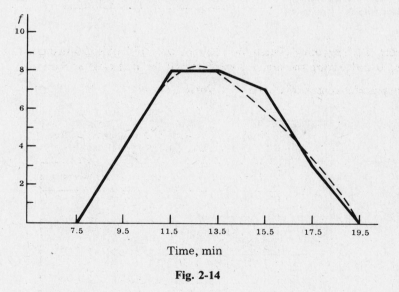

Fig. 2-14

2.12   Describe the frequency curve in Fig. 2-14 in terms of skewness.

The frequency curve is close to being symmetrical, but with slight positive skewness.

2.13   Construct a cumulative frequency distribution for the frequency distribution of assembly times in Table 2.7, using exact limits to identify each class and including cumulative percentages as well as cumulative frequencies in the table.

See Table 2.8 for the cumulative frequency distribution.

**Table 2.8   Cumulative Frequency Distribution for
the Assembly Times**

Time, min	f	cf	Cum. pct.
8.5–10.5	4	4	13.3
10.5–12.5	8	12	40.0
12.5–14.5	8	20	66.7
14.5–16.5	7	27	90.0
16.5–18.5	3	30	100.0

2.14   Refer to the cumulative frequency distribution in Table 2.8.

   (*a*)   Construct the percentage ogive for these data

   (*b*)   At what percentile point is an assembly time of 15 minutes?

   (*c*)   What is the assembly time at the 20th percentile of the distribution?

   (*a*)   The ogive is presented in Fig. 2-15.

   (*b*)   As identified by the dashed lines in the upper portion of the figure, the approximate percentile for
15 minutes of assembly time is 72.

   (*c*)   As identified by the dashed lines in the lower portion of the figure, the approximate time at the 20th
percentile is 11 minutes.

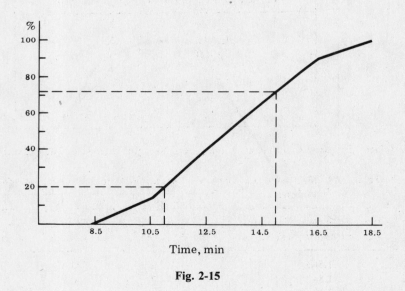

Fig. 2-15

FORMS OF FREQUENCY CURVES

2.15   Assuming that frequency curve (*a*) in Fig. 2-16 is both symmetrical and mesokurtic, describe
curves (*b*), (*c*), (*d*), (*e*), and (*f*) in terms of skewness and kurtosis.

    Curve (*b*) is symmetrical and leptokurtic; curve (*c*), positively skewed and mesokurtic; curve (*d*),
negatively skewed and mesokurtic; curve (*e*), symmetrical and platykurtic; and curve (*f*), positively skewed
and leptokurtic.

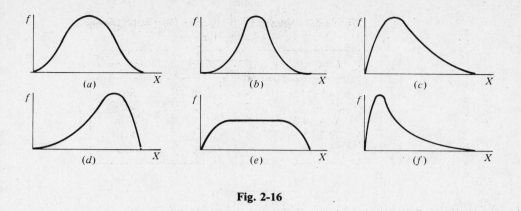

Fig. 2-16

RELATIVE FREQUENCY DISTRIBUTIONS

2.16 Using the instructions in Section 2.6, determine (*a*) the relative frequencies and (*b*) the cumulative proportions for the data in Table 2.9.

Table 2.9   Average Number of Injuries per Thousand Worker-Hours in a Particular Industry

Average number of injuries per thousand worker-hours	Number of firms
1.5–1.7	3
1.8–2.0	12
2.1–2.3	14
2.4–2.6	9
2.7–2.9	7
3.0–3.2	5
	Total 50

The relative frequencies and cumulative proportions for the data in Table 2.9 are given in Table 2.10.

Table 2.10   Relative Frequencies and Cumulative Proportions for Average Number of Injuries

Average number of injuries per thousand worker-hours	Number of firms	(*a*) Relative frequency	(*b*) Cumulative proportion
1.5–1.7	3	0.06	0.06
1.8–2.0	12	0.24	0.30
2.1–2.3	14	0.28	0.58
2.4–2.6	9	0.18	0.76
2.7–2.9	7	0.14	0.90
3.0–3.2	5	0.10	1.00
	Total 50	Total 1.00	

2.17 With reference to Table 2.10, construct (*a*) a histogram for the relative frequency distribution and (*b*) an ogive for the cumulative proportions.

(*a*) See Fig. 2-17.

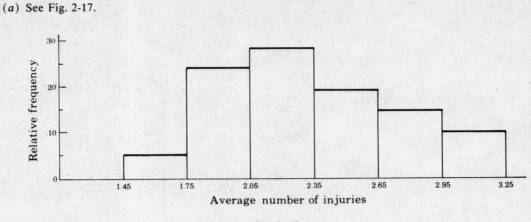

Fig. 2-17

(*b*) See Fig. 2-18.

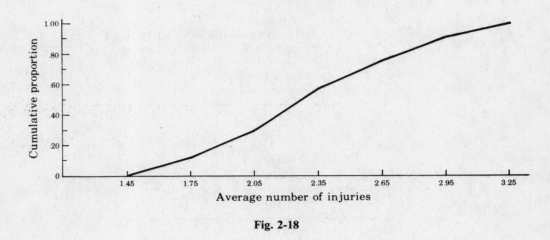

Fig. 2-18

2.18 (*a*) Referring to Table 2.10, what proportion of firms are in the category of having had an average of at least 3.0 injuries per thousand worker-hours? (*b*) What percentage of firms were at or below an average of 2.0 injuries per thousand worker-hours?

(*a*) 0.10, (*b*) 6% + 24% = 30%

2.19 (*a*) Referring to Table 2.10, what is the percentile value associated with an average of 2.95 (approximately 3.0) injuries per thousand worker-hours? (*b*) What is the average number of accidents at the 58th percentile?

(*a*) 90th percentile, (*b*) 2.35

2.20 By graphic interpolation on an ogive curve, we can determine the approximate percentiles for various values of the variable, and vice versa. Referring to Fig. 2-18, (*a*) what is the approximate percentile associated with an average of 2.5 accidents? (*b*) What is the approximate average number of accidents at the 50th percentile point?

(a) 65th percentile (This is the approximate height of the ogive corresponding to 2.50 along the horizontal axis.)

(b) 2.25 (This is the approximate point along the horizontal axis which corresponds to the 0.50 height of the ogive.)

THE "AND UNDER" TYPE OF FREQUENCY DISTRIBUTION

2.21 Identify the exact class limits for the data of Table 2.11.

Table 2.11 Time Required to Process and Prepare Mail Orders

Time, min	Number of orders
5 and under 8	10
8 and under 11	17
11 and under 14	12
14 and under 17	6
17 and under 20	2
Total	47

From Table 2.11,

Table 2.12 Time Required to Process and Prepare Mail Orders (with Exact Class Limits)

Time, min	Number of orders	Exact class limits
5 and under 8	10	5.0–8.0
8 and under 11	17	8.0–11.0
11 and under 14	12	11.0–14.0
14 and under 17	6	14.0–17.0
17 and under 20	2	17.0–20.0
Total	47	

2.22 Construct a frequency polygon for the frequency distribution in Table 2.12.

The frequency polygon appears in Fig. 2-19.

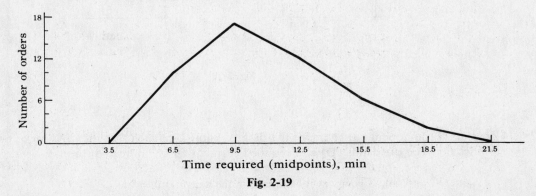

Fig. 2-19

BAR CHARTS

2.23 Table 2.13 reports certain financial results included in the 1985 *Annual Report* of the Tucson Electric Power Company. Construct a vertical bar chart portraying the per-share annual earnings of the company from 1980–1985.

Table 2.13 Per-share Earnings from Continuing Operations, Dividends, and Retained Earnings for the Tucson Electric Power Company

Year	Earnings	Dividends	Retained earnings
1980	$1.61	$1.52	$0.09
1981	2.17	1.72	0.45
1982	2.48	1.92	0.56
1983	3.09	2.20	0.89
1984	4.02	2.60	1.42
1985	4.35	3.00	1.35

The bar chart appears in Fig. 2-20.

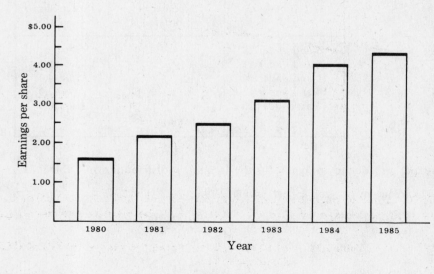

Fig. 2-20

2.24 Construct a component bar chart for the data of Table 2.13 such that the division of per-share earnings between dividends (D) and retained earnings (R) is indicated for each year.

Figure 2-21 presents a component bar chart for the data in Table 2.13.

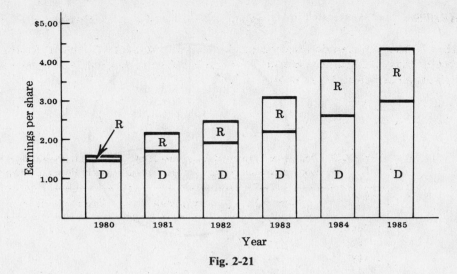

Fig. 2-21

LINE GRAPHS

2.25 Construct a line graph for the per-share earnings reported in Table 2.13.

The line graph is given in Fig. 2-22.

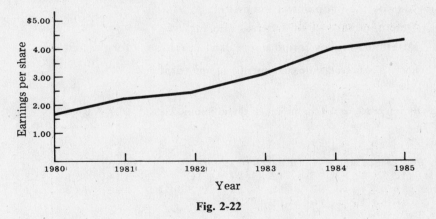

Fig. 2-22

PIE CHARTS

2.26 Table 2.14 is based on values supplied in the *Annual Report* of the Boeing Company for 1985. Construct a percentage pie chart for the order backlog according to category of order.

Table 2.14 Order Backlog in the Boeing Company at the End of 1985, Billions of Dollars

Category	Amount
Commercial transportation products and services	$16.2
Military transportation products and related systems	7.0
Missiles and space	1.2
Other	0.3
Total	$24.7

The pie chart is presented in Fig. 2-23.

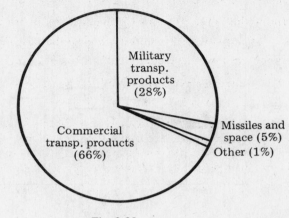

Fig. 2-23

COMPUTER OUTPUT

2.27 Use computer software to form a frequency distribution and to output a histogram for the ungrouped data in Table 2.6, which lists the times it took a sample of 30 employees to complete an assembly task.

(*a*) Identify the midpoint of the first class.

(*b*) Determine the size of the class interval.

(*c*) Determine the lower and upper exact class limits for the first class.

Figure 2-24 presents the computer input and output.

(*a*) 9.0

(*b*) By reference to the midpoints of the adjoining classes: $10.0 - 9.0 = 1.0$.

(*c*) 8.5–9.5

```
MTB > SET ASSEMBLY TIMES INTO C1
DATA>    10    14    15    13    17    16    12    14    11    13    15    18    9
DATA>    14    14     9    15    11    13    11    12    10    17    16    12    11
DATA>    16    12    14    15
DATA> END
MTB > NAME FOR C1 IS 'TIME'
MTB > HISTOGRAM FOR 'TIME'

Histogram of TIME    N = 30

Midpoint    Count
       9        2    **
      10        2    **
      11        4    ****
      12        4    ****
      13        3    ***
      14        5    *****
      15        4    ****
      16        3    ***
      17        2    **
      18        1    *
```

Fig. 2-24 Minitab output for Problem 2.27.

2.28 Refer to Problem 2.27, above. Rerun the analysis with the specification that the midpoint of the first class should be set at 10.0 with a class interval of 2.0.

(*a*) Determine the lower and upper exact class limits for the first class.

(*b*) Identify the type of frequency distribution that has been formed.

(*c*) Determine the lower and upper stated class limits for the first class.

  Figure 2-25 presents the relevant computer input and output using Minitab.

(*a*) 9.0–11.0

(*b*) Although it is not immediately obvious, it is an "and under" type of frequency distribution. By hand-checking the output, for example, one can observe that observed values of "9," at the lower exact limit of the first class, *are* included in that class. However, observed values of "11," at the upper exact limit of the first class, are *not* included in that class.

(*c*) 9 and under 11

```
MTB > HISTOGRAM FOR 'TIME';
SUBC> START AT 10.0;
SUBC> INCREMENT = 2.0.

Histogram of TIME   N = 30

Midpoint   Count
   10.00      4    ****
   12.00      8    ********
   14.00      8    ********
   16.00      7    *******
   18.00      3    ***
```

Fig. 2-25 Minitab output for Problem 2.28.

Supplementary Problems

FREQUENCY DISTRIBUTIONS, CLASS INTERVALS, AND RELATED GRAPHIC METHODS

2.29 Table 2.15 is a frequency distribution for the gasoline mileage obtained for 25 sampled trips for company-owned vehicles. (*a*) What are the lower and upper stated limits of the last class? (*b*) What are the lower and upper exact limits of the last class? (*c*) What class interval is used? (*d*) What is the midpoint of the

Table 2.15 Automobile Mileage for 25 Trips by Company Vehicles

Miles per gallon	Number of trips
24.0–25.9	3
26.0–27.9	5
28.0–29.9	10
30.0–31.9	4
32.0–33.9	2
34.0–35.9	1
Total	25

last class? (e) Suppose the mileage per gallon was found to be 29.9 for a particular trip. Indicate the lower and upper limits of the class in which this result was included.

Ans. (a) 34.0 and 35.9, (b) 33.95 and 35.95, (c) 2.0, (d) 34.95, (e) 28.0 and 29.9

2.30   Construct a histogram for the data of Table 2.15.

2.31   Construct a frequency polygon and a frequency curve for the data of Table 2.15.

2.32   Describe the frequency curve constructed in Problem 2.31 from the standpoint of skewness.

   Ans.   The frequency curve appears to be somewhat positively skewed.

2.33   Construct a cumulative frequency distribution for the data of Table 2.15 and construct an ogive to present this distribution graphically.

2.34   Table 2.16 presents the amounts of 40 personal loans used to finance appliance and furniture purchases. Suppose we wish to arrange the loan amounts in a frequency distribution with a total of seven classes. Assuming equal class intervals, what would be a convenient class interval for this frequency distribution?

   Ans. $400

Table 2.16   The Amounts of 40 Personal Loans

$ 932	$1,000	$ 356	$2,227
515	554	1,190	954
452	973	300	2,112
1,900	660	1,610	445
1,200	720	1,525	784
1,278	1,388	1,000	870
2,540	851	1,890	630
586	329	935	3,000
1,650	1,423	592	334
1,219	727	655	1,590

2.35   Construct the frequency distribution for the data of Table 2.16, beginning the first class at a lower class limit of $300 and using a class interval of $400.

2.36   Prepare a histogram for the frequency distribution constructed in Problem 2.35.

2.37   Construct a frequency polygon and frequency curve for the frequency distribution constructed in Problem 2.35.

2.38   Describe the frequency curve constructed in Problem 2.37 in terms of skewness.

   Ans.   The frequency curve is clearly positively skewed.

2.39   Construct a cumulative frequency distribution for the frequency distribution constructed in Problem 2.35 and construct an ogive curve for these data.

FORMS OF FREQUENCY CURVES

2.40   Describe each of the following curves in terms of skewness or kurtosis, as appropriate: (a) A frequency curve with a "tail" to the right, (b) a frequency curve that is relatively peaked, (c) a frequency curve that is relatively flat, (d) a frequency curve with a "tail" to the left.

   Ans.   (a) Positively skewed, (b) leptokurtic, (c) platykurtic, (d) negatively skewed

RELATIVE FREQUENCY DISTRIBUTIONS

2.41 Construct a relative frequency table for the frequency distribution presented in Table 2.17.

Table 2.17 Lifetime of Cutting Tools in an Industrial Process

Hours before replacement	Number of tools
0.0–24.9	2
25.0–49.9	4
50.0–74.9	12
75.0–99.9	30
100.0–124.9	18
125.0–149.9	4
	Total 70

2.42 Construct a histogram for the relative frequency distribution prepared in Problem 2.41.

2.43 Referring to Table 2.17, (a) what percentage of cutting tools lasted at least 125 hr? (b) What percentage of cutting tools had a lifetime of at least 100 hr?

Ans. (a) 6%, (b) 31%

2.44 Prepare a table of cumulative proportions for the frequency distribution in Table 2.17.

2.45 Referring to the table constructed in Problem 2.44, (a) what is the tool lifetime associated at the 26th percentile of the distribution? (b) What is the percentile associated with a tool lifetime of approximately 100 hr?

Ans. (a) 74.95 hr, $\cong$75 hr (b) 69th percentile

2.46 Construct the ogive for the cumulative proportions determined in Problem 2.44.

2.47 Refer to the ogive prepared in Problem 2.46 and determine the following values approximately by graphic interpolation: (a) The tool lifetime at the 50th percentile of the distribution, (b) the percentile associated with a tool lifetime of 60 hr.

Ans. (a) Approx. 89 hr, (b) approx. 16th percentile

THE "AND UNDER" TYPE OF FREQUENCY DISTRIBUTION

2.48 By reference to the frequency distribution in Table 2.18, determine (a) the lower stated limit of the first

Table 2.18 Ages of a Sample of Applicants for a Training Program

Age	Number of applicants
18 and under 20	5
20 and under 22	18
22 and under 24	10
24 and under 26	6
26 and under 28	5
28 and under 30	4
30 and under 32	2
	Total 50

class, (*b*) the upper stated limit of the first class, (*c*) the lower exact limit of the first class, (*d*) the upper exact limit of the first class, (*e*) the midpoint of the first class.

Ans.   (*a*) 18, (*b*) 20, (*c*) 18.0, (*d*) 20.0, (*e*) 19.0

2.49   Construct a frequency polygon for the frequency distribution in Table 2.18.

BAR CHARTS

2.50   The data in Table 2.19 are taken from the *Survey of Current Business* published by the U.S. Department of Commerce. Construct a vertical bar chart for these data.

Table 2.19   Construction of New One-
Family Structures in the
United States, 1981–1985,
Thousands of Units

Year	Housing starts
1981	705.4
1982	662.6
1983	1,067.6
1984	1,084.2
1985	1,072.4

LINE GRAPHS

2.51   Construct a line graph for the housing starts for single-family structures which are reported in Table 2.19.

PIE CHARTS

2.52   The values in Table 2.20 are taken from the *Annual Report* of First Interstate Bancorp for 1985. Construct a percentage pie chart for the sources of loan fees during that year.

Table 2.20   **First Interstate Bancorp: Sources of Loan Fees in
1985, million dollars**

Commercial, financial, agricultural	$ 45.7
Installment	99.2
Real estate construction	48.9
Real estate mortgage	33.6
Foreign	4.2
Total	$231.6

COMPUTER OUTPUT

2.53   Use computer software to form a frequency distribution and to output a histogram for the ungrouped data in Table 2.16, for the amounts of 40 personal loans.

(*a*)   Identify the midpoint of the first class.

(*b*)   Determine the class interval that is used.

(*c*)   Determine the exact lower and upper class limits for the first class.

Ans.   (*a*) Using Minitab, 400.0, (*b*) 400.0, (*c*) 200.0–600.0

2.54 Refer to Problem 2.53 above. Rerun the analysis with the specification that the midpoint of the first class should be set at 500 with a class interval of 400.

(*a*) Determine the lower and upper exact class limits for the first class.

(*b*) Identify the type of frequency distribution that has been formed.

(*c*) Determine the lower and upper stated class limits for the first class.

Ans. (*a*) 300.0–700.0, (*b*) "and under" type of distribution, (*c*) 300 and under 700

Chapter 3

Describing Business Data: Measures of Location

3.1 MEASURES OF LOCATION IN DATA SETS

A measure of location is a value which is calculated for a group of data and which is used to describe the data in some way. Typically, we wish the value to be representative of all of the values in the group, and thus some kind of *average* is desired. In the statistical sense an "average" is a *measure of central tendency* for a collection of values. This chapter covers the various statistical procedures concerned with measures of location.

3.2 THE ARITHMETIC MEAN

The *arithmetic mean*, or *arithmetic average*, is defined as the sum of the values in the data group divided by the number of values.

In statistics, a descriptive measure of a population, or a *population parameter*, is typically represented by a Greek letter, whereas a descriptive measure of a sample, or a *sample statistic*, is represented by a Roman letter. Thus, the arithmetic mean for a population of values is represented by the symbol μ (read "mew"), while the arithmetic mean for a sample of values is represented by the symbol $\bar{X}$ (read "X bar"). The formulas for the population mean and the sample mean are

$$\mu = \frac{\Sigma X}{N} \qquad (3.1)$$

$$\bar{X} = \frac{\Sigma X}{n} \qquad (3.2)$$

Operationally, the two formulas are identical; in both cases one sums all of the values (ΣX) and then divides by the number of values. However, the distinction in the denominators is that in statistical analysis the uppercase N typically indicates the number of items in the population, while the lowercase n indicates the number of items in the sample.

EXAMPLE 1. During a particular summer month, the eight salesmen in a heating and air-conditioning firm sold the following number of central air-conditioning units: 8, 11, 5, 14, 8, 11, 16, 11. Considering this month as the statistical population of interest, the mean number of units sold is

$$\mu = \frac{\Sigma X}{N} = \frac{84}{8} = 10.5 \text{ units}$$

Note: For reporting purposes, one generally reports the measures of location to one additional digit beyond the original level of measurement.

3.3 THE WEIGHTED MEAN

The *weighted mean* or *weighted average* is an arithmetic mean in which each value is weighted according to its importance in the overall group. The formulas for the population and sample weighted means are identical:

$$\mu_w \text{ or } \bar{X}_w = \frac{\Sigma(wX)}{\Sigma w} \qquad (3.3)$$

Operationally, each value in the group (X) is multiplied by the appropriate weight factor (w), and the products are then summed and divided by the sum of the weights.

EXAMPLE 2. In a multiproduct company, the profit margins for the company's four product lines during the past fiscal year were: line A, 4.2 percent; line B, 5.5 percent; line C, 7.4 percent; and line D, 10.1 percent. The *unweighted* mean profit margin is

$$\mu = \frac{\Sigma X}{N} = \frac{27.2}{4} = 6.8\%$$

However, unless the four products are equal in sales, this unweighted average is incorrect. Assuming the sales totals in Table 3.1, the weighted mean correctly describes the overall average.

Table 3.1   Profit Margin and Sales Volume for Four Product Lines

Product line	Profit margin (X)	Sales (w)	wX
A	4.2%	$30,000,000	$1,260,000
B	5.5	20,000,000	1,100,000
C	7.4	5,000,000	370,000
D	10.1	3,000,000	303,000
		$\Sigma w = \$58,000,000$	$\Sigma(wX) = \$3,033,000$

$$\mu_w = \frac{\Sigma(wX)}{\Sigma w} = \frac{\$3,033,000}{\$58,000,000} = 5.2\%$$

3.4 THE MEDIAN

The *median* of a group of items is the value of the middle item when all the items in the group are arranged in either ascending or descending order, in terms of value. For a group with an even number of items, the median is assumed to be midway between the two values adjacent to the middle. When a large number of values is contained in the group, the following formula to determine the position of the median in the ordered group is useful:

$$\text{Med} = X_{[(n/2)+(1/2)]} \qquad\qquad (3.4)$$

EXAMPLE 3. The eight salesmen described in Example 1 sold the following number of central air-conditioning units, in ascending order: 5, 8, 8, 11, 11, 11, 14, 16. The value of the median is

$$\text{Med} = X_{[(n/2)+(1/2)]} = X_{[(8/2)+(1/2)]} = X_{4.5} = 11.0$$

The value of the median is between the fourth and fifth value in the ordered group. Since both these values equal "11" in this case, the median equals 11.0.

3.5 THE MODE

The *mode* is the value that occurs most frequently in a set of values. Such a distribution is described as being *unimodal*. For a small data set in which no measured values are repeated, there is no mode. When two nonadjoining values are about equal in having maximum frequencies associated with them, the distribution is described as being *bimodal*. Distributions of measurements with several modes are referred to as being *multimodal*.

EXAMPLE 4. The eight salesmen described in Example 1 sold the following number of central air-conditioning units: 8, 11, 5, 14, 8, 11, 16, and 11. The mode for this group of values is the value with the greatest frequency, or mode = 11.0.

3.6 RELATIONSHIP AMONG THE MEAN, MEDIAN, AND MODE

The differences among the values of the mean, median, and mode are indicative of the form of the frequency curve in terms of skewness. For a unimodal distribution which is symmetrical, the mean, median, and mode all coincide in value [see Fig. 3-1(a)]. For a positively skewed distribution, the mean is largest in value and the median is larger than the mode but smaller than the mean [see Fig. 3-1(b)]. For a negatively skewed distribution, the mean is smallest in value and the median is below the mode but above the mean [see Fig. 3-1(c)]. One well-known measure of skewness which utilizes the observed difference between the mean and median of a group of values is Pearson's coefficient of skewness, described in Section 4.9.

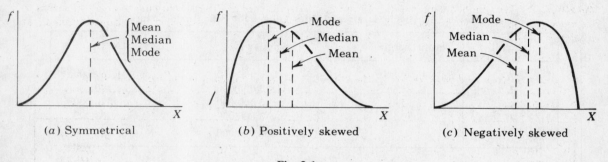

(a) Symmetrical   (b) Positively skewed   (c) Negatively skewed

Fig. 3-1

EXAMPLE 5.   For the sales data considered in Examples 1, 3, and 4, we can observe that the mean is 10.5, the median is 11.0, and the mode is 11.0 units, thus indicating that the distribution is somewhat negatively skewed, or skewed to the left.

3.7 QUARTILES, DECILES, AND PERCENTILES

The quartiles, deciles, and percentiles are very similar to the median in that they also subdivide a distribution of measurements according to the proportion of frequencies observed. Whereas the median divides a distribution into two halves, the quartiles divide it into four quarters, the deciles divide it into 10 tenths, and the percentile points divide it into 100 parts. Formula (3.4) for the median is modified according to the fractional point of interest. For example,

$$Q_1(\text{first quartile}) = X_{[(n/4)+(1/2)]} \qquad (3.5)$$

$$D_3(\text{third decile}) = X_{[(3n/10)+(1/2)]} \qquad (3.6)$$

$$P_{70}(\text{seventieth percentile}) = X_{[(70n/100)+(1/2)]} \qquad (3.7)$$

EXAMPLE 6.   The eight salesmen described in Example 1 sold the following number of central air-conditioning units, in ascending order: 5, 8, 8, 11, 11, 11, 14, 16. Find the position of the third quartile for this distribution.

$$Q_3 = X_{[(3n/4)+(1/2)]} = X_{[(24/4)+(1/2)]} = X_{6.5} = 12.5$$

In this case the value is between the sixth and seventh value in the ordered group.

3.8 THE ARITHMETIC MEAN FOR GROUPED DATA

When data have been grouped in a frequency distribution, the midpoint of each class is used as an approximation of all values contained in the class. The midpoint is represented by the symbol X_c, wherein the subscript c stands for "class," and the symbol f represents the observed frequency of values in each respective class. Thus, the formulas for the population mean and the sample mean computed for grouped data are

$$\mu = \frac{\Sigma(fX_c)}{\Sigma f} \quad \text{or, more simply,} \quad \mu = \frac{\Sigma(fX)}{N} \tag{3.8}$$

$$\bar{X} = \frac{\Sigma(fX_c)}{\Sigma f} \quad \text{or, more simply,} \quad \bar{X} = \frac{\Sigma(fX)}{n} \tag{3.9}$$

Operationally, both formulas indicate that each class midpoint (X_c) is multiplied by the associated class frequency (f), the products are summed (Σ), and then the sum is divided by the total number of observations (Σf) represented in the frequency distribution.

EXAMPLE 7.   The grouped data in Table 3.2 are taken from the frequency distribution presented in Section 2.1. The mean is represented by the symbol $\bar{X}$ because the group of workers is assumed to be a sample from a larger population of workers.

Table 3.2   **Weekly Wages of 100 Unskilled Workers (Rounded to the Nearest Dollar)**

Weekly wage	Class midpoint (X)	Number of workers (f)	fX
$240–259	$249.50	7	$1,746.50
260–279	269.50	20	5,390.00
280–299	289.50	33	9,553.50
300–319	309.50	25	7,737.50
320–339	329.50	11	3,624.50
340–359	349.50	4	1,398.00
		Total 100	$\Sigma(fX) = $29,450.00

$$\bar{X} = \frac{\Sigma(fX)}{n} = \frac{29,450}{100} = \$294.50$$

3.9   THE MEDIAN FOR GROUPED DATA

For grouped data, the *class* which contains the median value has to be determined first, and then the position of the median within the class is determined by interpolation. The class which contains the median is the first class for which the cumulative frequency equals or exceeds one-half the total number of observations. Once this class is identified, the specific value of the median is determined by the formula.

$$\text{Med} = B_L + \left(\frac{\frac{N}{2} - cf_B}{f_c} \right) i \tag{3.10}$$

where B_L = lower boundary, or exact lower limit, of the class containing the median
   N = total number of observations in the frequency distribution (n for a sample)
   cf_B = the cumulative frequency in the class preceding ("before") the class containing the median
   f_c = the number of observations in the class containing the median
   i = the size of the class interval

EXAMPLE 8.   The grouped data in Table 3.3 are repeated from the frequency distribution in Table 3.2. In this case, the class containing the median is the class with the $100/2 = 50$th value. The first class whose cumulative frequency equals or exceeds 50 is the class with the stated limits $280–299; thus, interpolation to determine the specific value of the median is done within this class.

Table 3.3 Weekly Wages of 100 Unskilled Workers

Weekly wage	Number of workers (f)	Cumulative frequency (cf)
$240–259	7	7
260–279	20	27
280–299	33	60
300–319	25	85
320–339	11	96
340–359	4	100
Total	100	

$$\text{Med} = B_L + \left(\frac{\frac{n}{2} - cf_B}{f_c}\right)i = 279.50 + \left(\frac{50-27}{33}\right)20 = \$293.44$$

3.10 THE MODE FOR GROUPED DATA

For data grouped in a frequency distribution with equal class intervals, first the *class* containing the mode is determined by identifying the class with the greatest number of observations. Some statisticians then designate the mode as being at the midpoint of the modal class. However, most statisticians interpolate within the modal class on the basis of the following formula:

$$\text{Mode} = B_L + \left(\frac{d_1}{d_1+d_2}\right)i \qquad (3.11)$$

where B_L = lower boundary, or exact lower limit, of the class containing the mode
d_1 = difference between the frequency in the modal class and the frequency in the preceding class
d_2 = difference between the frequency in the modal class and the frequency in the following class
i = the size of the class interval

EXAMPLE 9. Refer to the grouped data in Table 3.3. The modal class is the class with the stated limits $280–299. Thus,

$$\text{Mode} = B_L + \left(\frac{d_1}{d_1+d_2}\right)i = 279.50 + \left(\frac{13}{13+8}\right)20 = \$291.88$$

3.11 QUARTILES, DECILES, AND PERCENTILES FOR GROUPED DATA

For grouped data, formula (*3.10*) for the median is modified according to the fractional point of interest. In using this modified formula, *first the appropriate class containing the point of interest is determined* by reference to cumulative frequencies, and then the interpolation is done as before. Example formulas in this case are

$$Q_1(\text{first quartile}) = B_L + \left(\frac{\frac{n}{4} - cf_B}{f_c}\right)i \qquad (3.12)$$

$$D_3(\text{third decile}) = B_L + \left(\frac{\frac{3n}{10} - cf_B}{f_c}\right)i \qquad (3.13)$$

$$P_{70}(\text{seventieth percentile}) = B_L + \left(\frac{\frac{70n}{100} - cf_B}{f_c}\right)i \qquad (3.14)$$

EXAMPLE 10. The cumulative frequencies in Table 3.3 can be used to determine the 90th percentile point for these wage rates. In this case, note that the class containing this value is the first class whose cumulative frequency exceeds $\frac{90n}{100}$, or 90. This is the class with the stated limits $320–339, and so the lower boundary used in the formula is $319.50.

$$P_{90} = B_L + \left(\frac{\frac{90n}{100} - cf_B}{f_c}\right) i = 319.50 + \left(\frac{90 - 85}{11}\right) 20 = \$328.59$$

3.12   COMPUTER OUTPUT

The special formulas for grouped data described in the last several sections of this chapter have been important for two different reasons. The first is that data obtained from secondary sources, such as government reports, typically are grouped into frequency distributions, and therefore any calculations utilizing such data would rely on use of the grouped-data formulas. The second use of grouped-data formulas is to make calculations for large data sets easier by first grouping such data into a frequency distribution. Even though such calculated values as the mean and median for grouped data are *approximations* of the respective values for the ungrouped data, because of the grouping, or rounding, error involved, the saving in calculation time has often been considered worth the trade-off. With the widespread availability of computers, the latter reason for using grouped-data formulas has been eliminated.

Computer software is available to determine a variety of measures of average, or central tendency. Problem 3.24 is concerned with determining the mean and the median by the use of a computer.

Solved Problems

THE MEAN, MEDIAN, AND MODE

3.1   For a sample of 15 students at an elementary-school snack bar, the following sales amounts arranged in ascending order of magnitude are observed: $0.10, 0.10, 0.25, 0.25, 0.25, 0.35, 0.40, 0.53, 0.90, 1.25, 1.35, 2.45, 2.71, 3.09, 4.10. Determine the (*a*) mean, (*b*) median, and (*c*) mode for these sales amounts.

(*a*) $\bar{X} = \dfrac{\Sigma X}{n} = \dfrac{18.08}{15} = \$1.21$

(*b*) $\text{Med} = X_{[(n/2)+(1/2)]} = X_{[(15/2)+(1/2)]} = X_8 = \$0.53$

(*c*)   Mode = most frequent value = $0.25

3.2   How would you describe the distribution in Problem 3.1 from the standpoint of skewness?

With the mean being substantially larger than the median and the mode, the distribution of values is clearly positively skewed, or skewed to the right.

3.3   If you were asked for a description of the data in Problem 3.1 by reporting the "typical" amount of purchase per student in the sample, which measure of central tendency, or average, would you report? Why?

For a highly skewed distribution such as this, the arithmetic mean is not descriptive of the typical value in the group. The choice between the median and mode as a better measure depends on the degree to which observations are concentrated at the modal point. In this case, the degree of concentration is not all that great, and so "$0.25" is not really descriptive of the typical purchase amount either. For these reasons, the best choice for these data appears to be the median, and "$0.53" can be considered typical in the sense that approximately half the sales amounts were lower and half were higher.

3.4 A sample of 20 production workers in a small company earned the following wages for a given week, rounded to the nearest dollar and arranged in ascending order: $240, 240, 240, 240, 240, 240, 240, 240, 255, 255, 265, 265, 280, 280, 290, 300, 305, 325, 330, 340. Calculate the (*a*) mean, (*b*) median, and (*c*) mode for this group of wages.

(*a*) $\bar{X} = \dfrac{\Sigma X}{n} = \dfrac{5{,}410}{20} = \$270.50$

(*b*) Median $= X_{[(n/2)+(1/2)]} = X_{[(20/2)+(1/2)]} = X_{10.5} = \$260.00$

(*c*) Mode $=$ most frequent value $= \$240.00$

3.5 For the wage data in Problem 3.4, describe the distribution in terms of skewness.

With the mean being larger than the median and the mode, the distribution can be described as being positively skewed, or skewed to the right.

3.6 Given that you were placed in each of the following positions in turn, indicate which measure of "average" you might be inclined to report for the data in Problem 3.4, and in what sense each value can be considered "typical." (*a*) As vice-president responsible for collective bargaining. (*b*) As the president of the employee bargaining unit.

(*a*) You would be inclined to choose the arithmetic mean to best present the company's position and you could observe that this mean takes every individual's wage into consideration. You might also (incorrectly) be tempted to state that the arithmetic average is the only "real" average.

(*b*) You would be inclined to choose either the mode or the median, in this order, as your description of what is "typical" in this firm. The mode is "typical" in that clearly many more workers are at this wage level than at any other wage level. The median is "typical" in the sense that half of the workers earn less than this amount and half earn more than this amount.

3.7 A work-standards expert observes the amount of time required to prepare a sample of 10 business letters in an office with the following results listed in ascending order to the nearest minute: 5, 5, 5, 7, 9, 14, 15, 15, 16, 18. Determine the (*a*) mean, (*b*) median, and (*c*) mode for this group of values.

(*a*) $\bar{X} = \dfrac{\Sigma X}{n} = \dfrac{109}{10} = 10.9$ min

(*b*) Med $= X_{[(n/2)+(1/2)]} = X_{[(10/2)+(1/2)]} = X_{5.5} = 11.5$ min

(*c*) Mode $=$ most frequent value $= 5.0$ (but see next problem)

3.8 Compare the values of the mean, median, and mode in Problem 3.7 and comment on the form of the distribution.

The results appear unusual in that the mean is not the extreme value (either highest or lowest). The reason for this is that for practical purposes there really is not just one mode. Rather, the values appear to be clustered at two different points in the distribution, forming a bimodal distribution. Because the mean is smaller than the median, the distribution is somewhat negatively skewed.

THE WEIGHTED MEAN

3.9 Suppose that the per-gallon prices of unleaded regular gasoline are listed for 22 metropolitan areas varying considerably in size and in gasoline sales. The (*a*) median and (*b*) arithmetic mean are computed for these data. Describe the meaning that each of these values would have.

(*a*) The median would indicate the average, or typical, price in the sense that half the metropolitan areas would have gasoline prices below this value and half would have prices above this value.

(*b*) The meaning of the arithmetic mean as an unweighted value is questionable at best. Clearly, it would *not* indicate the mean gasoline prices being paid by all the gasoline purchasers living in the 22 metropolitan areas, because the small metropolitan areas would be represented equally with the large areas in such an average. In order to determine a suitable mean for all purchasers in these metropolitan areas, a weighted mean (using gasoline sales as weights) would have to be computed.

3.10 Referring to Table 3.4, determine the overall percentage defective of all items assembled during the sampled week.

Table 3.4 Percentage of Defective Items in an Assembly Department in a Sampled Week

Shift	Percentage defective (X)	Number of items, in thousands (w)	wX
1	1.1	210	231.0
2	1.5	120	180.0
3	2.3	50	115.0
		$\Sigma w = 380$	$\Sigma(wX) = 526.0$

Using formula (*3.3*),

$$\bar{X}_\omega = \frac{\Sigma(wX)}{\Sigma w} = \frac{526.0}{380} = 1.4\% \text{ defective}$$

QUARTILES, DECILES, AND PERCENTILES

3.11 For the data in Problem 3.1, determine the values at the (*a*) second quartile, (*b*) second decile, and (*c*) 40th percentile point for these sales amounts.

(*a*) $Q_2 = X_{[(2n/4)+(1/2)]} = X_{(7.5+0.5)} = X_8 = \$0.53$

(*Note:* By definition, the second quartile is always at the same point as the median.)

(*b*) $D_2 = X_{[(2n/10)+(1/2)]} = X_{(3.0+0.5)} = X_{3.5} = \$0.25$

(*Note:* This is the value midway between the third and fourth sales amounts, arranged in ascending order.)

(*c*) $P_{40} = X_{[(40n/100)+(1/2)]} = X_{(6.0+0.5)} = X_{6.5} = 0.375 \cong \$0.38$

3.12 For the measurements in Problem 3.4, determine the values at the (*a*) third quartile, (*b*) ninth decile, (*c*) 50th percentile point, and (*d*) 84th percentile point for this group of wages.

(*a*) $Q_3 = X_{[(3n/4)+(1/2)]} = X_{15.5} = \$295.00$

(*Note:* This is the value midway between the 15th and 16th wage amounts, arranged in ascending order.)

(*b*) $D_9 = X_{[(9n/10)+(1/2)]} = X_{18.5} = \$327.50$

(*c*) $P_{50} = X_{[(50n/100)+(1/2)]} = X_{10.5} = \$260.00$

(*Note:* By definition the 50th percentile point is always at the same point as the median.)

(*d*) $P_{84} = X_{[(84n/100)+(1/2)]} = X_{(16.8+0.5)} = X_{17.3} = \$311.00$

(*Note:* This is the value which is *three-tenths* of the way between the 17th and 18th wage amounts, arranged in ascending order.)

THE MEAN, MEDIAN, AND MODE FOR GROUPED DATA

3.13 Reproduced in Table 3.5 is the frequency distribution given originally in Problem 2.1. Determine the average rental rate in terms of the (*a*) mean, (*b*) median, and (*c*) mode. Assume that these are *all* the apartments in a given geographic area.

Table 3.5 Frequency Distribution for Monthly Apartment Rental Rates

(1) Rental rate	(2) Class midpoint (X)	(3) Number of apartments (f)	(4) fX	(5) Cumulative frequency (cf)
$350–379	$364.50	3	$1,093.50	3
380–409	394.50	8	3,156.00	11
410–439	424.50	10	4,245.00	21
440–469	454.50	13	5,908.50	34
470–499	484.50	33	15,988.50	67
500–529	514.50	40	20,580.00	107
530–559	544.50	35	19,057.50	142
560–589	574.50	30	17,235.00	172
590–619	604.50	16	9,672.00	188
620–649	634.50	12	7,614.00	200
		Total 200	$\Sigma(fX) = $104,550.00$	

(a) $\quad \mu = \dfrac{\Sigma(fX)}{N} = \dfrac{104{,}550}{200} = \$522.75$

(b) $\quad \text{Med} = B_L + \left(\dfrac{\frac{N}{2} - cf_B}{f_c}\right) i = 499.50 + \left(\dfrac{100 - 67}{40}\right) 30 = \$524.25$

(*Note:* $499.50 is the lower boundary, or exact lower limit, of the class containing the $\frac{N}{2}$, or 100th, measurement.)

(c) $\quad \text{Mode} = B_L + \left(\dfrac{d_1}{d_1 + d_2}\right) i = 499.50 + \left(\dfrac{7}{7+5}\right) 30 = \$517.00$

(*Note:* $499.50 is the lower boundary, or exact lower limit, of the class containing the highest frequency.)

3.14 Comment on the form of the distribution for the apartment rental rates in Problem 3.13.

The frequency polygon for these data as presented in Fig. 2-11 seems to be indicative of some degree of negative skewness. Comparison of the values of the mean and median also supports such a conclusion, since the mean is smaller than the median. However, because the mean is larger than the mode, the amount of skewness is concluded to be minor.

3.15 In conjunction with an annual audit, a public accounting firm makes note of the time required to audit 50 account balances, as indicated in Table 3.6. Compute the (a) mean, (b) median, and (c) mode for the audit time required for this sample of records.

Table 3.6 Time Required to Audit Account Balances

(1) Audit time, nearest minute	(2) Class midpoint (X)	(3) Number of records (f)	(4) fX	(5) Cumulative frequency (cf)
10–19	14.5	3	43.5	3
20–29	24.5	5	122.5	8
30–39	34.5	10	345.0	18
40–49	44.5	12	534.0	30
50–59	54.5	20	1,090.0	50
		Total 50	$\Sigma(fX) = 2{,}135.0$	

(a) $\bar{X} = \dfrac{\Sigma(fX)}{n} = \dfrac{2,135.0}{50} = 42.7$ min

(b)   Med $= B_L + \left(\dfrac{\frac{n}{2} - cf_B}{f_c}\right)i = 39.5 + \left(\dfrac{25-18}{12}\right)10.00 = 45.3$ min

   (*Note:* 39.5 is the lower boundary, or exact lower limit, of the class containing the $\frac{n}{2}$, or 25th, measurement.)

(c)   Mode $= B_L + \left(\dfrac{d_1}{d_1 + d_2}\right)i = 49.5 + \left(\dfrac{8}{8+20}\right)10.0 = 52.36$ min

   (*Note:* 49.5 is the lower boundary, or exact lower limit, of the class with the highest frequency. Also note that $d_2 = 20 - 0 = 20$ in this case.)

3.16   Comment on the form of the distribution for the audit times reported in Problem 3.15.

   The mean is smaller in value than either the median or mode. Thus, the distribution is clearly negatively skewed, or skewed to the left.

3.17   Reproduced in Table 3.7 are the data reported originally in Problem 2.16. Determine the (a) mean, (b) median, and (c) mode for this sample of 50 firms.

Table 3.7   Average Number of Injuries per Thousand Worker-Hours in a Particular Industry

Average number of injuries	Class midpoint (X)	Number of firms (f)	fX	Cumulative frequency (cf)
1.5–1.7	1.6	3	4.8	3
1.8–2.0	1.9	12	22.8	15
2.1–2.3	2.2	14	30.8	29
2.4–2.6	2.5	9	22.5	38
2.7–2.9	2.8	7	19.6	45
3.0–3.2	3.1	5	15.5	50
	Total	50	$\Sigma(fX) = 116.0$	

(a) $\bar{X} = \dfrac{\Sigma(fX)}{n} = \dfrac{116.0}{50} = 2.32$ injuries

(b)   Med $= B_L + \left(\dfrac{\frac{n}{2} - cf_B}{f_c}\right)i = 2.05 + \left(\dfrac{25-15}{14}\right)0.3 = 2.26$ injuries

   (*Note:* 2.05 is the lower boundary, or exact lower limit, of the class containing the $\frac{n}{2}$, or 25th, measurement.)

(c)   Mode $= B_L + \left(\dfrac{d_1}{d_1 + d_2}\right)i = 2.05 + \left(\dfrac{2}{2+5}\right)0.3 = 2.14$ injuries

   (*Note:* 2.05 is the lower boundary, or exact lower limit, of the class containing the $\frac{n}{2}$, or 25th, measurement.)

3.18   Comment on the form of the distribution for the data of Problem 3.17.

   The sample mean is larger than either the median or the mode, and therefore the distribution is positively skewed, or skewed to the right.

3.19   The frequency distribution in Table 3.8 is reproduced from Problem 2.21. Determine the (a) mean, (b) median, and (c) mode for these data. Note the basis for determining class boundaries, or exact limits, and class midpoints for such an "and under" frequency distribution.

Table 3.8   Time Required to Process and Prepare Mail Orders

Time, min	Class boundaries (exact limits)	Class midpoint (X)	Number of orders (f)	fX	Cumulative frequency (cf)
5 and under 8	5.0–8.0	6.5	10	65.0	10
8 and under 11	8.0–11.0	9.5	17	161.5	27
11 and under 14	11.0–14.0	12.5	12	150.0	39
14 and under 17	14.0–17.0	15.5	6	93.0	45
17 and under 20	17.0–20.0	18.5	2	37.0	47
			Total 47	$\Sigma(fX) = 506.5$	

(a) $\bar{X} = \dfrac{\Sigma(fX)}{n} = \dfrac{506.5}{47} = 10.8$ min

(b) $\text{Med} = B_L + \left(\dfrac{\frac{n}{2} - cf_B}{f_c}\right) i = 8.0 + \left(\dfrac{23.5 - 10}{17}\right) 3.0 = 10.4$ min

(*Note:* 8.0 is the lower boundary, or exact lower limit, of the class containing the $\frac{n}{2}$, or 23.5th, measurement.)

(c) $\text{Mode} = B_L + \left(\dfrac{d_1}{d_1 + d_2}\right) i = 8.0 + \left(\dfrac{7}{7+5}\right) 3.0 = 9.75 \cong 9.8$ min

(*Note:* 8.0 is the lower boundary of the class containing the highest frequency.)

3.20   Comment on the form of the distribution for the order processing times analyzed in Problem 3.19.

Since the mean is the largest of the measures of central tendency, the distribution is positively skewed.

QUARTILES, DECILES, AND PERCENTILES FOR GROUPED DATA

3.21   Referring to Table 3.5 (page 38), determine the values at the (a) first quartile, (b) third decile, (c) 80th percentile point, and (d) 17th percentile point for the apartment rental rates.

(a) $Q_1 = B_L + \left(\dfrac{\frac{N}{4} - cf_B}{f_c}\right) i = 469.50 + \left(\dfrac{50 - 34}{33}\right) 30 = \$484.05$

(*Note:* \$469.50 is the lower boundary, or exact lower limit, of the class containing the $\frac{n}{4}$, or 50th, measurement.)

(b) $D_3 = B_L + \left(\dfrac{\frac{3N}{10} - cf_B}{f_c}\right) i = 469.50 + \left(\dfrac{60 - 34}{33}\right) 30 = \$493.14$

(*Note:* \$469.50 is the lower boundary of the class containing the $\frac{3N}{10}$, or 60th, measurement.)

(c) $P_{80} = B_L + \left(\dfrac{\frac{80N}{100} - cf_B}{f_c}\right) i = 559.50 + \left(\dfrac{160 - 142}{30}\right) 30 = \$577.50$

(*Note:* \$559.50 is the lower boundary of the class containing the $\frac{80N}{100}$, or 160th, measurement.)

(d) $P_{17} = B_L + \left(\dfrac{\frac{17N}{100} - cf_B}{f_c}\right) i = 439.50 + \left(\dfrac{34 - 21}{13}\right) 30 = \$469.50$

(*Note 1:* \$439.50 is the lower boundary of the class containing the $\frac{17N}{100}$, or 34th, measurement.)

(*Note 2:* The 17th percentile is at the upper boundary of the class. This is logical, since we need to go through the entire class to get to the 34th item in the distribution.)

3.22   Referring to Table 3.6 (page 38), determine the values at the (a) third quartile, (b) first decile, and (c) 90th percentile point for the time required to audit the account balances.

(a) $Q_3 = B_L + \left(\dfrac{\frac{3n}{4} - cf_B}{f_c}\right)i = 49.5 + \left(\dfrac{37.5 - 30}{20}\right)10 = 53.25 \cong 53.2 \text{ min}$

   (*Note:* 49.5 is the lower boundary, or exact lower limit, of the class containing the $\frac{3n}{4}$, or 37.5th, measurement.)

(b) $D_1 = B_L + \left(\dfrac{\frac{n}{10} - cf_B}{f_c}\right)i = 19.5 + \left(\dfrac{5-3}{5}\right)10 = 23.5 \text{ min}$

   (*Note:* 19.5 is the lower boundary of the class containing the $\frac{n}{10}$, or fifth, measurement.)

(c) $P_{90} = B_L + \left(\dfrac{\frac{90n}{100} - cf_B}{f_c}\right)i = 49.5 + \left(\dfrac{45-30}{20}\right)10 = 57.0 \text{ min}$

   (*Note:* 49.5 is the lower boundary of the class containing the $\frac{90n}{100}$, or 45th, measurement.)

3.23   For the "and under" frequency distribution reported in Table 3.8 (page 40), determine the values of the (a) second quartile, (b) ninth decile, and (c) 75th percentile point for the time required to process and prepare the mail orders.

(a) $Q_2 = B_L + \left(\dfrac{\frac{2n}{4} - cf_B}{f_c}\right)i = 8.0 + \left(\dfrac{23.5 - 10}{17}\right)3.0 = 10.4 \text{ min}$

(b) $D_9 = B_L + \left(\dfrac{\frac{9n}{10} - cf_B}{f_c}\right)i = 14.0 + \left(\dfrac{42.3 - 39}{6}\right)3.0 \cong 15.6 \text{ min}$

(c) $P_{75} = B_L + \left(\dfrac{\frac{75n}{100} - cf_B}{f_c}\right)i = 11.0 + \left(\dfrac{35.25 - 27}{12}\right)3.0 = 13.1 \text{ min}$

COMPUTER OUTPUT

3.24   Use a computer to calculate the mean and median for the data in Table 2.6 (page 16), on the time required to complete an assembly task by a sample of 30 employees. In Problems 2.27 and 2.28 these data were grouped into a frequency distribution by the use of a computer, with the associated output of a histogram.

   Figure 3-2 presents the computer input and output. As can be observed, the mean assembly time is 13.3 min and the median assembly time is 13.5 min.

```
MTB ) SET ASSEMBLY TIMES INTO C1
DATA)    10    14    15    13    17    16    12    14    11    13    15    18     9
DATA)    14    14     9    15    11    13    11    12    10    17    16    12    11
DATA)    16    12    14    15
DATA) END
MTB ) NAME FOR C1 IS 'TIME'
MTB ) MEAN OF 'TIME'
   MEAN      =      13.300
MTB ) MEDIAN OF 'TIME'
   MEDIAN =       13.500
```

Fig. 3-2   Minitab output.

Supplementary Problems

THE MEAN, MEDIAN, AND MODE

3.25   The number of cars sold by each of the 10 salesmen in an automobile dealership during a particular month, arranged in ascending order, is: 2, 4, 7, 10, 10, 10, 12, 12, 14, 15. Determine the population (a) mean, (b) median, and (c) mode for the number of cars sold.

   Ans.   (a) 9.6, (b) 10.0, (c) 10.0

3.26   Which value in Problem 3.25 best describes the "typical" sales volume per salesman?

   Ans.   10.0

3.27   The weights of a sample of outgoing packages in a mailroom, weighed to the nearest ounce, are found to
   be: 21, 18, 30, 12, 14, 17, 28, 10, 16, 25 oz. Determine the (*a*) mean, (*b*) median, and (*c*) mode for these
   weights.

   Ans.   (*a*) 19.1, (*b*) 17.5, (*c*) there is no mode

3.28   How can a mode be obtained for the package weights described in Problem 3.27?

   Ans.   A mode can be obtained by constructing a frequency distribution for the data. In such a case, a
   larger sample size would be required.

3.29   The following examination scores, arranged in ascending order, were achieved by 20 students enrolled in
   a decision analysis course: 39, 46, 57, 65, 70, 72, 72, 75, 77, 79, 81, 81, 84, 84, 84, 87, 93, 94, 97, 97.
   Determine the (*a*) mean, (*b*) median, and (*c*) mode for these scores.

   Ans.   (*a*) 76.7, (*b*) 80.0, (*c*) 84.0

3.30   Describe the distribution of test scores in Problem 3.29 in terms of skewness.

   Ans.   Negatively skewed.

3.31   The number of accidents which occurred during a given month in the 13 manufacturing departments of
   an industrial plant was: 2, 0, 0, 3, 3, 12, 1, 0, 8, 1, 0, 5, 1. Calculate the (*a*) mean, (*b*) median, and (*c*)
   mode for the number of accidents per department.

   Ans.   (*a*) 2.8, (*b*) 1.0, (*c*) 0

3.32   Describe the distribution of accident rates reported in Problem 3.31 in terms of skewness.

   Ans.   Positively skewed.

THE WEIGHTED MEAN

3.33   Suppose the retail prices of the selected items have changed as indicated in Table 3.9. Determine the mean
   percentage change in retail prices *without* reference to the average expenditures included in the table.

   Ans.   4.0%

Table 3.9   Changes in the Retail Prices of Selected Items During a
Particular Year

Item	Percent increase	Average expenditure per month (before increase)
Milk	10%	$20.00
Ground beef	−6	30.00
Apparel	−8	30.00
Gasoline	20	50.00

3.34   Referring to Table 3.9, determine the mean percentage change by weighting the percent increase for each
   item by the average amount per month spent on that item before the increase.

   Ans.   6.0%

3.35   Is the mean percentage price change calculated in Problem 3.33 or 3.34 more appropriate as a measure of the impact of the price changes on this particular consumer? Why?

Ans.   The weighted mean in Problem 3.34 is more appropriate (see Example 2 for an explanation).

QUARTILES, DECILES, AND PERCENTILES

3.36   Determine the values at the (*a*) first quartile, (*b*) second decile, and (*c*) 30th percentile point for the sales amounts in Problem 3.25.

Ans.   (*a*) 7.0, (*b*) 5.5, (*c*) 8.5

3.37   From Problem 3.27, determine the weights at the (*a*) third quartile, (*b*) third decile, and (*c*) 70th percentile point.

Ans.   (*a*) 25.0 oz, (*b*) 15.0 oz, (*c*) 23.0 oz

3.38   Determine the (*a*) second quartile, (*b*) ninth decile, and (*c*) 50th percentile point for the examination scores in Problem 3.29.

Ans.   (*a*) 80.0, (*b*) 95.5, (*c*) 80.0

3.39   In general, which quartile, decile, and percentile point, respectively, are equivalent to the median?

Ans.   Second quartile, fifth decile, and 50th percentile point.

THE MEAN, MEDIAN, AND MODE FOR GROUPED DATA

3.40   With reference to Table 2.15 (page 25), determine the typical mileage which was obtained in terms of the (*a*) mean, (*b*) median, and (*c*) mode.

Ans.   (*a*) 28.95, (*b*) 28.85, (*c*) 28.86

3.41   Comment on the form of the distribution of the mileage figures in Problem 3.40 in terms of skewness.

Ans.   The distribution is relatively symmetrical.

3.42   The frequency distribution in Table 3.10 is based on data supplied originally in Problem 2.35. Compute the (*a*) mean, (*b*) median, and (*c*) mode of the loan amounts, based only on the frequency distribution in Table 3.10.

Ans.   (*a*) $1,109.50, (*b*) $954.05, (*c*) $646.17

Table 3.10   The Amounts of 40 Personal Loans

Loan amount	Number of loans
$ 300–699	13
700–1,099	11
1,100–1,499	6
1,500–1,899	5
1,900–2,299	3
2,300–2,699	1
2,700–3,099	1
Total	40

3.43   Describe the form of the frequency distribution of personal loan amounts in Problem 3.42.

Ans.   Positively skewed.

3.44 From Table 2.17 (page 27), determine the average lifetime of the cutting tools by computing the (*a*) mean, (*b*) median, and (*c*) mode.

Ans. (*a*) 87.45, (*b*) 89.12, (*c*) 89.95

3.45 Describe the frequency distribution of the tool lifetime in Problem 3.44 in terms of skewness.

Ans. There is a slight negative skewness.

3.46 Referring to the "and under" frequency distribution in Table 2.18 (page 27), compute the (*a*) mean, (*b*) median, and (*c*) mode for the applicants' ages.

Ans. (*a*) 23.3, (*b*) 22.4, (*c*) 21.2

3.47 Comment on the form of the distribution in Table 2.18, based on the answers to Problem 3.46.

Ans. The distribution is positively skewed.

QUARTILES, DECILES, AND PERCENTILES FOR GROUPED DATA

3.48 Determine the values at the (*a*) first quartile, (*b*) first decile, and (*c*) 10th percentile point for the automobile mileage data in Table 2.15 (page 25).

Ans. (*a*) 27.25, (*b*) 25.62, (*c*) 25.62

3.49 For the amounts of the personal loans reported in Problem 3.42, determine the values at the (*a*) second quartile, (*b*) second decile, and (*c*) 90th percentile point.

Ans. (*a*) $954.05, (*b*) $545.65, (*c*) $2,032.83

3.50 Determine the values at the (*a*) third quartile, (*b*) seventh decile, and (*c*) 75th percentile for the hours lifetime of cutting tools, as reported in Table 2.17 (page 27).

Ans. (*a*) 106.20, (*b*) 101.34, (*c*) 106.20

3.51 Determine the values at the (*a*) first quartile, (*b*) seventh decile, and (*c*) 80th percentile point for the age data in Table 2.18 (page 27).

Ans. (*a*) 20.8, (*b*) 24.7, (*c*) 26.4

COMPUTER OUTPUT

3.52 Use a computer to calculate the mean and the median for the data in Table 2.16 (page 26), for the amounts of 40 personal loans. In Problems 2.53 and 2.54, these data were grouped into a frequency distribution by the use of a computer, with the associated output of a histogram. Also, approximations of this mean and median based on grouped data were calculated in Problem 3.42.

Ans. The mean loan amount is $1,097.40 while the median is $944.50. These compare with the approximations in Problem 3.42, based on grouped data, of $1,109.50 and $954.05, respectively.

Chapter 4

Describing Business Data: Measures of Variability

4.1 MEASURES OF VARIABILITY IN DATA SETS

The measures of central tendency described in Chapter 3 are useful for identifying the "typical" value in a group of values. In contrast, *measures of variability* are concerned with describing the variability among the values. Several techniques are available for measuring the extent of variability in data sets. The ones which are described in this chapter are the *range, modified ranges, average deviation, variance, standard deviation,* and *coefficient of variation*.

EXAMPLE 1. Suppose that two different packaging machines result in a mean weight of 10.0 oz of cereal being packaged, but that in one case all packages are within 0.10 oz of this weight while in the other case the weights may vary by as much as 1.0 oz in either direction. Measuring the variability, or dispersion, of the amounts being packaged would be every bit as important as measuring the average in this case.

The concept of skewness has been described in Sections 2.4 and 3.6. Pearson's *coefficient of skewness* is described in Section 4.9.

4.2 THE RANGE

The *range*, or *R*, is the difference between the highest and lowest values included in a data set. Thus, when *H* represents the highest value in the group and *L* represents the lowest value, the range for ungrouped data is

$$R = H - L \tag{4.1}$$

EXAMPLE 2. During a particular summer month, the eight salesmen in a heating and air-conditioning firm sold the following numbers of central air-conditioning units: 8, 11, 5, 14, 8, 11, 16, 11. The range of the number of units sold is

$$R = H - L = 16 - 5 = 11.0 \text{ units}$$

Note: For purposes of comparison, we generally report the measures of variability to one additional digit beyond the original level of measurement.

4.3 MODIFIED RANGES

A *modified range* is a range for which a certain percent of the extreme values at each end of the distribution has been eliminated. Typical modified ranges are the *middle 50 percent, middle 80 percent,* and *middle 90 percent.*

The procedure by which a modified range is determined is to first locate the appropriate two percentile points (see Sections 3.7 and 3.11), and then take the difference between the values at these points. For instance, for the middle 80 percent range the appropriate percentile points are the 10th percentile point and the 90th percentile point, because the middle 80 percent of the values are located between these two points.

EXAMPLE 3. The data for the central air-conditioning sales presented in Example 2, in ascending order, are: 5, 8, 8, 11, 11, 11, 14, 16. To compute the middle 50 percent range, we first determine the values at the appropriate percentile points and then subtract the lower from the higher value:

$$P_{75} = X_{[(75n/100)+(1/2)]} = X_{[6+(1/2)]} = X_{6.5} = 12.5$$

$$P_{25} = X_{[(25n/100)+(1/2)]} = X_{[2+(1/2)]} = X_{2.5} = 8.0$$

$$\text{Middle 50\% } R = P_{75} - P_{25} = 12.5 - 8.0 = 4.5 \text{ units}$$

4.4 THE AVERAGE DEVIATION

The *average deviation*, or *AD*, is based on the difference between each value in the data set and the mean of the group. It is the mean of these deviations which is computed (some statisticians use the difference between each value and the median). If the mean of the sum of the plus and minus differences between each value and the arithmetic mean were computed, the answer would in fact always be zero. For this reason, it is the *absolute values* of the differences that are summed.

$$\text{Population } AD = \frac{\Sigma|X - \mu|}{N} \tag{4.2}$$

$$\text{Sample } AD = \frac{\Sigma|X - \bar{X}|}{n} \tag{4.3}$$

EXAMPLE 4. For the air-conditioning sales data given in Example 2, the arithmetic mean is 10.5 units (see Section 3.2). Using the calculations in Table 4.1, the average deviation is determined as follows:

$$AD = \frac{\Sigma|X - \mu|}{N} = \frac{21.0}{8} = 2.625 \cong 2.6 \text{ units}$$

Table 4.1 Worksheet for Calculating the Average Deviation for the Sales Data

X	$X - \mu$	$|X - \mu|$
5	−5.5	5.5
8	−2.5	2.5
8	−2.5	2.5
11	0.5	0.5
11	0.5	0.5
11	0.5	0.5
14	3.5	3.5
16	5.5	5.5
	Total	21.0

Thus, we can say that on the average, a salesman's unit sales of air conditioners differs by 2.6 units from the group mean, in either direction.

4.5 THE VARIANCE AND STANDARD DEVIATION

The *variance* is similar to the average deviation in that it is based on the difference between each value in the data set and the mean of the group. It differs in one very important way: each difference is *squared* before being summed. For a population, the variance is represented by $v(X)$ or, more typically, by the lowercase Greek σ^2 (read "sigma squared"); the formula is

$$v(X) = \sigma^2 = \frac{\Sigma(X - \mu)^2}{N} \tag{4.4}$$

Unlike the situation for other sample statistics we have discussed, the variance for a sample is not computationally exactly equivalent to the variance for a population. Rather, the denominator in the sample variance formula is slightly different. Essentially, a correction factor is included in this formula, so that the sample variance is an unbiased estimator of the population variance (see Section 8.1). The sample variance is represented by s^2; its formula is

$$s^2 = \frac{\Sigma(X - \bar{X})^2}{n - 1} \tag{4.5}$$

In general, it is difficult to interpret the meaning of the value of a variance because the units in which it is expressed are squared values. Partly for this reason, the square root of the variance, represented by Greek σ (or s for a sample) and called the *standard deviation* is more frequently used. The formulas are

Population standard deviation: $$\sigma = \sqrt{\frac{\Sigma(X - \mu)^2}{N}} \tag{4.6}$$

Sample standard deviation: $$s = \sqrt{\frac{\Sigma(X - \bar{X})^2}{n - 1}} \tag{4.7}$$

The standard deviation is particularly useful in conjunction with the so-called normal distribution (see Section 4.7).

EXAMPLE 5.   For the air-conditioning sales data given in Example 2 the arithmetic mean is 10.5 units (see Section 3.2). Considering these monthly sales data to be the statistical population of interest, the standard deviation is determined from the calculations in Table 4.2 as follows:

$$\sigma = \sqrt{\frac{\Sigma(X - \mu)^2}{N}} = \sqrt{\frac{86}{8}} = \sqrt{10.75} = 3.3$$

Table 4.2   Worksheet for Calculating the Population Standard Deviation for the Sales Data

X	$X - \mu$	$(X - \mu)^2$
5	−5.5	30.25
8	−2.5	6.25
8	−2.5	6.25
11	0.5	0.25
11	0.5	0.25
11	0.5	0.25
14	3.5	12.25
16	5.5	30.25
		Total 86.00

4.6   SHORT-CUT CALCULATIONS OF THE VARIANCE AND STANDARD DEVIATION

The formulas in Section 4.5 are called *deviations formulas*, because in each case the specific deviations of individual values from the group mean must be determined. Alternative formulas which are mathematically equivalent but which do not require the determination of each deviation have been developed. Because these formulas are generally easier to use for computations, they are called *computational formulas*.

The computational formulas are

Population variance: $$\sigma^2 = \frac{\Sigma X^2 - N\mu^2}{N} \qquad (4.8)$$

Population standard deviation: $$\sigma = \sqrt{\frac{\Sigma X^2 - N\mu^2}{N}} \qquad (4.9)$$

Sample variance: $$s^2 = \frac{\Sigma X^2 - n\bar{X}^2}{n-1} \qquad (4.10)$$

Sample standard deviation: $$s = \sqrt{\frac{\Sigma X^2 - n\bar{X}^2}{n-1}} \qquad (4.11)$$

EXAMPLE 6. For the air-conditioning sales data presented in Example 2, we calculate the population standard deviation below by the use of the alternative computational formula and Table 4.3 to demonstrate that the answer is the same as the answer obtained with the deviation formula in Example 5. The mean for these data is 10.5 units.

$$\sigma = \sqrt{\frac{\Sigma X^2 - N\mu^2}{N}} = \sqrt{\frac{968 - 8(10.5)^2}{8}} = \sqrt{10.75} = 3.3 \text{ units}$$

Table 4.3   **Worksheet for Calculating the Population Standard Deviation for the Sales Data**

X	X^2
5	25
8	64
8	64
11	121
11	121
11	121
14	196
16	256
Total	968

4.7 USE OF THE STANDARD DEVIATION

The standard deviation is the most important measure of dispersion, in that it is used in conjunction with a number of methods of statistical inference discussed in later chapters of this book. A description of these uses is beyond the scope of the present chapter. However, as one example of the use of the standard deviation, consider a frequency distribution that is both symmetrical and mesokurtic. In statistical analysis, such a frequency curve is called a *normal curve*. For a distribution which is *normally distributed*, it is known that approximately 68 percent of the measurements are located within one standard deviation of the mean and approximately 95 percent of the measurements are located within two standard deviation units of the mean. These observations are presented diagrammatically in Figs. 4-1(a) and (b), respectively.

EXAMPLE 7. The electrical billings in a residential area for the month of June are observed to be normally distributed. If the mean of the billings is calculated to be $84.00 with a standard deviation of $24.00, then it follows that approximately 68 percent of the billed amounts are within $24.00 of the mean, or between $60.00 and $108.00. It also follows that approximately 95 percent of the billed amounts are within $48.00 of the mean, or between $36.00 and $132.00.

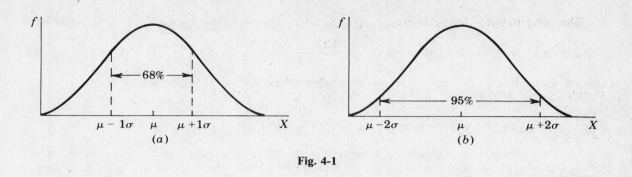

Fig. 4-1

4.8 THE COEFFICIENT OF VARIATION

The *coefficient of variation*, *CV*, indicates the relative magnitude of the standard deviation as compared with the mean of the distribution of measurements. Thus, the formulas are

Population:
$$CV = \frac{\sigma}{\mu} \qquad\qquad (4.12)$$

Sample:
$$CV = \frac{s}{\bar{X}} \qquad\qquad (4.13)$$

The coefficient of variation is useful when we wish to compare the variability of two data sets relative to the general level of values (and thus relative to the mean) in each set.

EXAMPLE 8.   For two common stock issues in the electronics industry, the daily mean closing market price during a one-month period for stock *A* was $150 with a standard deviation of $5. For stock *B*, the mean price was $50 with a standard deviation of $3. On an absolute comparison basis, the variability in the price of stock *A* was greater, because of the larger standard deviation. But relative to the price level, the respective coefficients of variation should be compared:

$$CV(A) = \frac{\sigma}{\mu} = \frac{5}{150} = 0.033 \quad \text{and} \quad CV(B) = \frac{\sigma}{\mu} = \frac{3}{50} = 0.060$$

Therefore, relative to the average price level for each stock issue, we can conclude that stock *B* has been almost twice as variable in price as stock *A*.

4.9 PEARSON'S COEFFICIENT OF SKEWNESS

Pearson's *coefficient of skewness* measures the departure from symmetry by expressing the difference between the mean and the median relative to the standard deviation of the group of measurements. The formulas are

$$\text{Population skewness} = \frac{3(\mu - \text{Med})}{\sigma} \qquad\qquad (4.14)$$

$$\text{Sample skewness} = \frac{3(\bar{X} - \text{Med})}{s} \qquad\qquad (4.15)$$

For a symmetrical distribution the value of the coefficient of skewness will always be zero, because the mean and median are equal to one another in value. For a positively skewed distribution, the mean is always larger than the median; hence, the value of the coefficient is positive. For a negatively skewed distribution, the mean is always smaller than the median; hence, the value of the coefficient is negative.

EXAMPLE 9.   For the air-conditioning sales data presented in Example 2, the mean is 10.5 units, the median is

11.0 units (from Sections 3.2 and 3.4), and the standard deviation is 3.3 units. The coefficient of skewness is

$$\text{Skewness} = \frac{3(\mu - \text{Med})}{\sigma} = \frac{3(10.5 - 11.0)}{3.3} = -0.45$$

Thus the distribution of sales amounts is somewhat negatively skewed, or "skewed to the left."

4.10 THE RANGE AND MODIFIED RANGES FOR GROUPED DATA

For data grouped in a frequency distribution, the range is generally defined as the difference between the upper boundary, or exact limit, of the highest class, $B_U(H)$, and the lower boundary of the lowest-valued class, $B_L(L)$. Thus, the range for grouped data is

$$R = B_U(H) - B_L(L) \qquad (4.16)$$

EXAMPLE 10. The grouped data in Table 4.4 are taken from the frequency distribution of weekly wages for 100 unskilled workers, as presented in Sections 2.1 and 3.8. The range is

$$R = B_U(H) - B_L(L) = 359.50 - 239.50 = \$120.00$$

Table 4.4 Weekly Wages of 100 Unskilled Workers

Weekly wage	Class boundaries	Number of workers (f)	Cumulative frequency (cf)
$240–259	$239.50–259.50	7	7
260–279	259.50–279.50	20	27
280–299	279.50–299.50	33	60
300–319	299.50–319.50	25	85
320–339	319.50–339.50	11	96
340–359	339.50–359.50	4	100
		Total 100	

EXAMPLE 11. The middle 90 percent range for the frequency distribution of wages given in Table 4.4 is

$$\text{Middle } 90\% \ R = P_{95} - P_{05} = 337.68 - 253.79 = \$83.89$$

where

$$P_{95} = B_L + \left(\frac{\frac{95n}{100} - cf_B}{f_c} \right) i = 319.50 + \left(\frac{95 - 85}{11} \right) 20 = \$337.68$$

$$P_{05} = B_L + \left(\frac{\frac{5n}{100} - cf_B}{f_c} \right) i = 239.50 + \left(\frac{5 - 0}{7} \right) 20 = \$253.79$$

4.11 THE AVERAGE DEVIATION FOR GROUPED DATA

For data grouped in a frequency distribution, the midpoint of each class is taken to represent all of the measurements included in that class. This is the same approach which is used in determining the arithmetic mean for grouped data, as described in Section 3.8. Accordingly,

$$\text{Population } AD = \frac{\Sigma(f|X - \mu|)}{N} \qquad (4.17)$$

$$\text{Sample } AD = \frac{\Sigma(f|X - \bar{X}|)}{n} \qquad (4.18)$$

EXAMPLE 12. For the weekly wage data given in Table 4.4, the arithmetic mean is $294.50 (see Section 3.8).

The average deviation is determined as follows, from the calculations in Table 4.5.

$$AD = \frac{\Sigma(f|X - \bar{X}|)}{n} = \frac{1{,}960.00}{100} = \$19.60$$

Table 4.5   Worksheet for Calculating the Average Deviation for Grouped Data

Weekly wage	Class midpoint (X)	Number of workers (f)	$|X - \bar{X}|$	$f|X - \bar{X}|$
$240–259	$249.50	7	$45.00	$ 315.00
260–279	269.50	20	25.00	500.00
280–299	289.50	33	5.00	165.00
300–319	309.50	25	15.00	375.00
320–339	329.50	11	35.00	385.00
340–359	349.50	4	55.00	220.00
		Total 100		Total $1,960.00

4.12   THE VARIANCE AND STANDARD DEVIATION FOR GROUPED DATA

For data grouped in a frequency distribution, the midpoint of each class is taken to represent all of the measurements included in that class. This is the same approach as used for computing the average deviation in Section 4.11. Accordingly, the formulas for grouped population and sample data are

Population variance:
$$\sigma^2 = \frac{\Sigma[f(X - \mu)^2]}{N} \qquad (4.19)$$

Sample variance:
$$s^2 = \frac{\Sigma[f(X - \bar{X})^2]}{n - 1} \qquad (4.20)$$

The formulas for the standard deviation for grouped population and sample data are

Population standard deviation:
$$\sigma = \sqrt{\frac{\Sigma[f(X - \mu)^2]}{N}} \qquad (4.21)$$

Sample standard deviation:
$$s = \sqrt{\frac{\Sigma[f(X - \bar{X})^2]}{n - 1}} \qquad (4.22)$$

EXAMPLE 13.   For weekly wage data presented in Example 10, the sample mean is $294.50, as determined in Section 3.8. From Table 4.6, the sample standard deviation for these grouped values is determined

Table 4.6   Worksheet for Calculating the Sample Standard Deviation for Grouped Data

Weekly wage	Class midpoint (X)	Number of workers (f)	$X - \bar{X}$	$(X - \bar{X})^2$	$f(X - \bar{X})^2$
$240–259	$249.50	7	$−45.00	$2,025	$14,175
260–279	269.50	20	−25.00	625	12,500
280–299	289.50	33	−5.00	25	825
300–319	309.50	25	15.00	225	5,625
320–339	329.50	11	35.00	1,225	13,475
340–359	349.50	4	55.00	3,025	12,100
		Total 100			Total $58,700

as follows:

$$s = \sqrt{\frac{\Sigma[f(X - \bar{X})^2]}{n - 1}} = \sqrt{\frac{58,700}{99}} = \sqrt{592.9293} = \$24.35$$

For grouped data, the short-cut computational formulas are

Population variance:
$$\sigma^2 = \frac{\Sigma(fX^2) - N\mu^2}{N} \qquad (4.23)$$

Population standard deviation:
$$\sigma = \sqrt{\frac{\Sigma(fX^2) - N\mu^2}{N}} \qquad (4.24)$$

Sample variance:
$$s^2 = \frac{\Sigma(fX^2) - n\bar{X}^2}{n - 1} \qquad (4.25)$$

Sample standard deviation:
$$s = \sqrt{\frac{\Sigma(fX^2) - n\bar{X}^2}{n - 1}} \qquad (4.26)$$

EXAMPLE 14.   For the weekly wage data presented in Example 10, we calculate the sample standard deviation below by the use of the alternative computational formula and Table 4.7 to demonstrate that the answer is the same as the answer obtained with the deviation formula in Example 13. The sample mean for these data is $294.50.

$$s = \sqrt{\frac{\Sigma(fX^2) - n\bar{X}^2}{n - 1}} = \sqrt{\frac{8,731,725.00 - 100(294.50)^2}{100 - 1}} = \sqrt{592.9293} = \$24.35$$

Table 4.7   **Worksheet for Calculating the Sample Standard Deviation for Grouped Data**

Weekly wage	Class midpoint (X)	Number of workers (f)	X^2	fX^2
$240–259	$249.50	7	$ 62,250.25	$ 435,751.75
260–279	269.50	20	72,630.25	1,452,605.00
280–299	289.50	33	83,810.25	2,765,738.25
300–319	309.50	25	95,790.25	2,394,756.25
320–339	329.50	11	108,570.25	1,194,272.75
340–359	349.50	4	122,150.25	488,601.00
		Total 100		Total $8,731,725.00

4.13   COMPUTER OUTPUT

As explained in Section 3.12, although grouped-data formulas are useful if the data obtained from secondary sources is available only in the form of frequency distributions, the grouping of data in order to simplify calculations for large data sets is no longer necessary because of the availability of computers.

Computer software is available for a variety of measures of variability. Problem 4.25 is concerned with determining the range and standard deviation by the use of such software.

Solved Problems

THE RANGES, AVERAGE DEVIATION, AND STANDARD DEVIATION

4.1   For a sample of 15 students at an elementary-school snack bar, the following sales amounts, arranged in ascending order of magnitude, are observed: $0.10, 0.10, 0.25, 0.25, 0.25, 0.35, 0.40,

0.53, 0.90, 1.25, 1.35, 2.45, 2.71, 3.09, 4.10. Determine the (*a*) range and (*b*) middle 50 percent range for these sample data.

(*a*) $R = H - L = \$4.10 - 0.10 = \$4.00$

(*b*)   Middle 50% $R = P_{75} - P_{25} = 2.175 - 0.25 = 1.925 \cong \$1.92$

where $P_{75} = X_{[(75n/100)+(1/2)]} = X_{(11.25+0.50)} = X_{11.75} = 1.35 + 0.825 = \$2.175$

(*Note:* This is the interpolated value *three-fourths* of the distance between the 11th and 12th ordered sales amounts.)

$$P_{25} = X_{[(25n/100)+(1/2)]} = X_{(3.75+0.50)} = X_{4.25} = \$0.25$$

4.2   Compute the average deviation for the data in Problem 4.1. The sample mean for this group of values was determined to be $1.21 in Problem 3.1.

Using Table 4.8, the average deviation is calculated as follows:

$$AD = \frac{\Sigma |X - \bar{X}|}{n} = \frac{\$15.45}{15} = \$1.03$$

Table 4.8   Worksheet for Calculating the Average Deviation for the Snack Bar Data

X	$X - \bar{X}$	$|X - \bar{X}|$
$0.10	$-1.11	$ 1.11
0.10	-1.11	1.11
0.25	-0.96	0.96
0.25	-0.96	0.96
0.25	-0.96	0.96
0.35	-0.86	0.86
0.40	-0.81	0.81
0.53	-0.68	0.68
0.90	-0.31	0.31
1.25	0.04	0.04
1.35	0.14	0.14
2.45	1.24	1.24
2.71	1.50	1.50
3.09	1.88	1.88
4.10	2.89	2.89
		Total $15.45

Table 4.9   Worksheet for Calculating the Sample Standard Deviation for the Snack Bar Data

X	$X - \bar{X}$	$(X - \bar{X})^2$	X^2
$0.10	$-1.11	1.2321	0.0100
0.10	-1.11	1.2321	0.0100
0.25	-0.96	0.9216	0.0625
0.25	-0.96	0.9216	0.0625
0.25	-0.96	0.9216	0.0625
0.35	-0.86	0.7396	0.1225
0.40	-0.81	0.6561	0.1600
0.53	-0.68	0.4624	0.2809
0.90	-0.31	0.0961	0.8100
1.25	0.04	0.0016	1.5625
1.35	0.14	0.0196	1.8225
2.45	1.24	1.5376	6.0025
2.71	1.50	2.2500	7.3441
3.09	1.88	3.5344	9.5481
4.10	2.89	8.3521	16.8100
		Total 22.8785	Total 44.6706

4.3   Determine the sample standard deviation for the data in Problems 4.1 and 4.2 by using (*a*) the deviations formula and (*b*) the alternative computational formula, and demonstrate that the answers are equivalent.

From Table 4.9,

(*a*) $s = \sqrt{\dfrac{\Sigma(X - \bar{X})^2}{n-1}} = \sqrt{\dfrac{22.8785}{15-1}} = \sqrt{1.6342} \cong \$1.28$

(*b*) $s = \sqrt{\dfrac{\Sigma X^2 - n\bar{X}^2}{n-1}} = \sqrt{\dfrac{44.6706 - 15(1.21)^2}{15-1}} = \sqrt{1.6221} \cong \$1.27$

The answers are slightly different only because of rounding error, associated with the fact that the sample mean was rounded to two places with respect to the decimal point.

4.4 A sample of 20 production workers in a small company earned the following wages for a given week, rounded to the nearest dollar and arranged in ascending order: $240, 240, 240, 240, 240, 240, 240, 240, 255, 255, 265, 265, 280, 280, 290, 300, 305, 325, 330, 340. Determine the (a) range and (b) middle 80 percent range for this sample.

(a) $R = H - L = \$340 - \$240 = \$100$

(b) Middle 80% $R = P_{90} - P_{10} = \$327.50 - \$240.00 = \$87.50$

where $P_{90} = X_{[(90n/100)+(1/2)]} = X_{[18+(1/2)]} = X_{18.5} = \$325 + \$2.50 = \$327.50$

$P_{10} = X_{[(10n/100)+(1/2)]} = X_{[2+(1/2)]} = X_{2.5} = \$240.00$

4.5 Compute the average deviation for the wages in Problem 4.4. The sample mean for these wages was determined to be $270.50 in Problem 3.4.

From Table 4.10, the average deviation is

$$AD = \frac{\Sigma|X - \bar{X}|}{n} = \frac{\$572.00}{20} = \$28.60$$

Table 4.10 Worksheet for Calculating the Average Deviation for the Wage Data

X	$X - \bar{X}$	$|X - \bar{X}|$
$240	$-30.50	$ 30.50
240	-30.50	30.50
240	-30.50	30.50
240	-30.50	30.50
240	-30.50	30.50
240	-30.50	30.50
240	-30.50	30.50
240	-30.50	30.50
255	-15.50	15.50
255	-15.50	15.50
265	-5.50	5.50
265	-5.50	5.50
280	9.50	9.50
280	9.50	9.50
290	19.50	19.50
300	29.50	29.50
305	34.50	34.50
325	54.50	54.50
330	59.50	59.50
340	69.50	69.50
		Total $572.00

Table 4.11 Worksheet for Calculating the Sample Variance and Standard Deviation for the Wage Data

X	$X - \bar{X}$	$(X - \bar{X})^2$
$240	$-30.50	$ 930.25
240	-30.50	930.25
240	-30.50	930.25
240	-30.50	930.25
240	-30.50	930.25
240	-30.50	930.25
240	-30.50	930.25
240	-30.50	930.25
255	-15.50	240.25
255	-15.50	240.25
265	-5.50	30.25
265	-5.50	30.25
280	9.50	90.25
280	9.50	90.25
290	19.50	380.25
300	29.50	870.25
305	34.50	1,190.25
325	56.50	2,970.25
330	59.50	3,540.25
340	69.50	4,830.25
		Total $21,945.00

4.6 Determine the (a) sample variance and (b) sample standard deviation for the data in Problems 4.4 and 4.5, using the deviations formulas.

With reference to Table 4.11,

(a) $s^2 = \dfrac{\Sigma(X - \bar{X})^2}{n - 1} = \dfrac{21,945.00}{20 - 1} = 1,155.00$

(b) $s = \sqrt{\dfrac{\Sigma(X - \bar{X})^2}{n - 1}} = \sqrt{1,155.00} \cong \$33.99$

4.7   A work-standards expert observes the amount of time required to prepare a sample of 10 business letters in an office with the following results listed in ascending order to the nearest minute: 5, 5, 5, 7, 9, 14, 15, 15, 16, 18. Determine the (*a*) range and (*b*) middle 70 percent range for the sample.

(*a*) $R = H - L = 18 - 5 = 13$ min

(*b*)   Middle 70% $R = P_{85} - P_{15} = 16.0 - 5.0 = 11.0$ min

where $P_{85} = X_{[(85n/100)+(1/2)]} = X_{(8.5+0.5)} = X_9 = 16.0$ min

$P_{15} = X_{[(15n/100)+(1/2)]} = X_{(1.5+0.5)} = X_2 = 5.0$ min

4.8   Compute the average deviation for the preparation time in Problem 4.7. The sample mean was determined to be 10.9 min in Problem 3.7.

From Table 4.12,

$$AD = \frac{\Sigma |X - \bar{X}|}{n} = \frac{47.0}{10} = 4.7 \text{ min}$$

Table 4.12   Worksheet for Calculating the Average Deviation for the Preparation-Time Data

X	$X - \bar{X}$	$|X - \bar{X}|$
5	−5.9	5.9
5	−5.9	5.9
5	−5.9	5.9
7	−3.9	3.9
9	−1.9	1.9
14	3.1	3.1
15	4.1	4.1
15	4.1	4.1
16	5.1	5.1
18	7.1	7.1
		Total 47.0

Table 4.13   Worksheet for Calculating the Sample Variance and Standard Deviation for the Preparation-Time Data

X	X^2
5	25
5	25
5	25
7	49
9	81
14	196
15	225
15	225
16	256
18	324
	Total 1,431

4.9   Determine the (*a*) sample variance and (*b*) sample standard deviation for the preparation-time data in Problems 4.7 and 4.8, using the alternative computational formulas.

With reference to Table 4.13,

(*a*) $s^2 = \dfrac{\Sigma X^2 - n\bar{X}^2}{n-1} = \dfrac{1{,}431 - 10(10.9)^2}{10-1} = 26.99$

(*b*) $s = \sqrt{\dfrac{\Sigma X^2 - n\bar{X}^2}{n-1}} = \sqrt{26.99} \cong 5.2$ min

THE COEFFICIENT OF VARIATION

4.10   Determine the coefficient of variation for the wage data analyzed in Problems 4.4 to 4.7.

Since $\bar{X} = \$270.50$ and $s = \$33.99$,

$$CV = \frac{s}{\bar{X}} = \frac{33.99}{270.50} = 0.126$$

4.11 For the same industrial firm as in Problem 4.10, above, the mean weekly salary for a sample of supervisory employees is $\bar{X} = \$730.75$ with $s = \$45.52$. Determine the coefficient of variation for these salary amounts.

$$CV = \frac{s}{\bar{X}} = \frac{45.52}{730.75} = 0.062$$

4.12 Compare the variability of the production workers' wages in Problem 4.10 with the supervisory salaries in Problem 4.11 (*a*) on an absolute basis, and (*b*) relative to the mean level of weekly income for the two groups of employees.

(*a*) On an absolute basis, there is more variability among the supervisors' salaries ($s = \$45.52$) than there is among the production workers' weekly wages ($s = \$33.99$).

(*b*) On a basis relative to the respective means, the two coefficients of variation are compared. Referring to the solutions to Problems 4.10 and 4.11, we can observe that the coefficient of variation for the hourly wage data (0.126) is twice as large as the respective coefficient for the weekly salaries (0.062), thereby indicating greater *relative* variability for the wage data.

PEARSON'S COEFFICIENT OF SKEWNESS

4.13 Compute the coefficient of skewness for the snack-bar data that were analyzed in Problems 4.1 to 4.3.

Since $\bar{X} = \$1.21$, Med $= \$0.53$ (from Problem 3.1), and $s = \$1.27$,

$$\text{Skewness} = \frac{3(\bar{X} - \text{Med})}{s} = \frac{3(1.21 - 0.53)}{1.27} = 1.61$$

Therefore the distribution of sales amounts is positively skewed, or skewed to the right.

4.14 Compute the coefficient of skewness for the wage data analyzed in Problems 4.4 to 4.7.

Since $\bar{X} = \$270.50$, Med $= \$260.00$ (from Problem 3.4), and $s = \$33.99$,

$$\text{Skewness} = \frac{3(\bar{X} - \text{Med})}{s} = \frac{3(270.50 - 260.00)}{33.99} = 0.93$$

Therefore the distribution of the wage amounts is slightly positively skewed.

THE RANGES, AVERAGE DEVIATION, AND STANDARD DEVIATION FOR GROUPED DATA

4.15 Determine the (*a*) range and (*b*) middle 50 percent range for the rental rates in Table 2.5 (page 15). Assume that these are all of the apartments in a given geographic area.

(*a*) $R = B_U(H) - B_L(L) = 649.50 - 349.50 = \$300.00$

(*b*) Middle 50% $R = P_{75} - P_{25} = 567.50 - 484.05 = \$83.45$

where $P_{75} = B_L + \left(\dfrac{\frac{75N}{100} - cf_B}{f_c}\right) i = 559.50 + \left(\dfrac{150 - 142}{30}\right) 30 = \$567.50$

$P_{25} = B_L + \left(\dfrac{\frac{25N}{100} - cf_B}{f_c}\right) i = 469.50 + \left(\dfrac{50 - 34}{33}\right) 30 = \$484.05$

4.16 Compute the average deviation for the rental rate data in Table 2.5. The population mean was determined to be $\$522.75$ in Problem 3.13.

Using Table 4.14,

$$AD = \frac{\Sigma(f|X-\mu|)}{N} = \frac{\$9,925.50}{200} = \$49.63$$

Table 4.14   Worksheet for Calculating the Average Deviation for Grouped Data

Rental rate	Class midpoint (X)	Number of apartments (f)	$|X-\mu|$	$f|X-\mu|$
$350–379	$364.50	3	$158.25	$ 474.75
380–409	394.50	8	128.25	1,026.00
410–439	424.50	10	98.25	982.50
440–469	454.50	13	68.25	887.25
470–499	484.50	33	38.25	1,262.25
500–529	514.50	40	8.25	330.00
530–559	544.50	35	21.75	761.25
560–589	574.50	30	51.75	1,552.50
590–619	604.50	16	81.75	1,308.00
620–649	634.50	12	111.75	1,341.00
		Total 200		Total $9,925.50

4.17   From Problem 4.11, determine the population standard deviation by using (a) the deviations formula and (b) the alternative computational formula, and demonstrate that the answers are equivalent.

(a)   From Table 4.15,

$$\sigma = \sqrt{\frac{\Sigma[f(X-\mu)^2]}{N}} = \sqrt{\frac{768,487.50}{200}} = \sqrt{3,842.4375} \cong \$61.99$$

Table 4.15   Worksheet for Calculating the Standard Deviation by the Deviation Formula

Rental rate	Class midpoint (X)	Number of apartments (f)	$X-\mu$	$(X-\mu)^2$	$f(X-\mu)^2$
$350–379	$364.50	3	$158.25	$25,043.0625	$ 75,129.1875
380–409	394.50	8	128.25	16,448.0625	131,584.5000
410–439	424.50	10	98.25	9,653.0625	96,530.6250
440–469	454.50	13	68.25	4,658.0625	60,554.8125
470–499	484.50	33	38.25	1,463.0625	48,281.0625
500–529	514.50	40	8.25	68.0625	2,722.5000
530–559	544.50	35	21.75	473.0625	16,557.1875
560–589	574.50	30	51.75	2,678.0625	80,341.8750
590–619	604.50	16	81.75	6,683.0625	106,929.0000
620–649	634.50	12	111.75	12,488.0625	149,856.7500
		Total 200			Total $768,487.5000

(b)   From Table 4.16,

$$\sigma = \sqrt{\frac{\Sigma(fX^2) - N\mu^2}{N}} = \sqrt{\frac{55,422,000.00 - 200(522.75)^2}{200}} = \sqrt{3,842.4375} \cong \$61.99$$

Table 4.16   Worksheet for Calculating the Standard Deviation by the Alternative Computational Formula

Rental rate	Class midpoint (X)	Number of apartments (f)	X^2	fX^2
$350–379	$364.50	3	$132,860.25	$ 398,580.75
380–409	394.50	8	155,630.25	1,245,042.00
410–439	424.50	10	180,200.25	1,802,002.50
440–469	454.50	13	206,570.25	2,685,413.25
470–499	484.50	33	234,740.25	7,776,428.25
500–529	514.50	40	264,710.25	10,588,410.00
530–559	544.50	35	296,480.25	10,376,808.75
560–589	574.50	30	330,050.25	9,901,507.50
590–619	604.50	16	365,420.25	5,846,724.00
620–649	634.50	12	402,590.25	4,831,083.00
		Total 200		Total $55,422,000.00

4.18   In conjunction with an annual audit, a public accounting firm collects the data reported in Table 4.17. Determine the (a) range and (b) middle 80 percent range for this sample of records.

Table 4.17   Time Required to Audit Account Balances

Audit time (in minutes)	Class boundaries	Number of records (f)	Cumulative frequency (cf)
10–19	9.5–19.5	3	3
20–29	19.5–29.5	5	8
30–39	29.5–39.5	10	18
40–49	39.5–49.5	12	30
50–59	49.5–59.5	20	50
		Total 50	

Using Table 4.17,

(a) $R = B_U(H) - B_L(L) = 59.5 - 9.5 = 50.0$ min

(b)   Middle 80% $R = P_{90} - P_{10} = 57.0 - 23.5 = 33.5$ min

where $P_{90} = B_L + \left(\dfrac{\frac{90n}{100} - cf_B}{f_c}\right)i = 49.5 + \left(\dfrac{45 - 30}{20}\right)10 = 57.0$ min

$P_{10} = B_L + \left(\dfrac{\frac{10n}{100} - cf_B}{f_c}\right)i = 19.5 + \left(\dfrac{5 - 3}{5}\right)10 = 23.5$ min

4.19   Compute the average deviation for the audit time data in Table 4.17. The sample mean was determined to be 42.7 min in Problem 3.15.

With reference to Table 4.18,

$$AD = \frac{\Sigma(f|X - \bar{X}|)}{n} = \frac{515.2}{50} \cong 10.3$$

Table 4.18   Worksheet for Calculating the Average Deviation

Audit time (in minutes)	Class midpoint (X)	Number of records (f)	$\lvert X - \bar{X}\rvert$	$f\lvert X - \bar{X}\rvert$
10–19	14.5	3	28.2	84.6
20–29	24.5	5	18.2	91.0
30–39	34.5	10	8.2	82.0
40–49	44.5	12	1.8	21.6
50–59	54.5	20	11.8	236.0
		Total 50		Total 515.2

4.20 Determine the (a) sample variance and (b) sample standard deviation for the data in Table 4.17 by using the deviations formulas.

From Table 4.19,

(a) $s^2 = \dfrac{\Sigma[f(X - \bar{X})^2]}{n-1} = \dfrac{7{,}538}{49} = 153.84$

(b) $s = \sqrt{\dfrac{\Sigma[f(X - \bar{X})^2]}{n-1}} = \sqrt{153.84} \cong 12.4 \text{ min}$

Table 4.19   Worksheet for Computing the Variance and Standard Deviation

Audit time (in minutes)	Class midpoint (X)	Number of records (f)	$X - \bar{X}$	$(X - \bar{X})^2$	$f(X - \bar{X})^2$
10–19	14.5	3	−28.2	795.24	2,385.72
20–29	24.5	5	−18.2	331.24	1,656.20
30–39	34.5	10	−8.2	67.24	672.40
40–49	44.5	12	1.8	3.24	38.88
50–59	54.5	20	11.8	139.24	2,784.80
		Total 50			Total 7,538.00

4.21 Reproduced in Table 4.20 are the average number of injuries per thousand worker-hours, as reported in Problem 2.16. Determine the (a) range and (b) middle 90 percent range for this sample of 50 firms.

Table 4.20   Average Number of Injuries per Thousand Worker-Hours in a Particular Industry

Average number of injuries	Class boundaries	Number of firms (f)	Cumulative frequency (cf)
1.5–1.7	1.45–1.75	3	3
1.8–2.0	1.75–2.05	12	15
2.1–2.3	2.05–2.35	14	29
2.4–2.6	2.35–2.65	9	38
2.7–2.9	2.65–2.95	7	45
3.0–3.2	2.95–3.25	5	50
		Total 50	

(a) $R = B_U(H) - B_L(L) = 3.25 - 1.45 = 1.80$ injuries

(b) Middle 90% $R = P_{95} - P_{05} = 3.10 - 1.70 = 1.40$ injuries

where $P_{95} = B_L - \left(\dfrac{\frac{95n}{100} - cf_B}{f_c} \right) i = 2.95 + \left(\dfrac{47.5 - 45}{5} \right) 0.30 = 3.10$ injuries

$$P_{05} = B_L + \left(\dfrac{\frac{5n}{100} - cf_B}{f_c} \right) 0.30 = 1.45 + \left(\dfrac{2.5 - 0}{3} \right) 0.30 = 1.70 \text{ injuries}$$

4.22 Compute the average deviation for the number of injuries in Table 4.20. The sample mean was determined to be 2.32 injuries per thousand man-hours in Problem 3.17.

With reference to Table 4.21,

$$AD = \frac{\Sigma(f|X - \bar{X}|)}{n} = \frac{17.76}{50} = 0.36 \text{ injury}$$

Table 4.21 Worksheet for Calculating the Average Deviation

Average number of injuries	Class midpoint (X)	Number of firms (f)	$|X - \bar{X}|$	$f|X - \bar{X}|$
1.5–1.7	1.6	3	0.72	2.16
1.8–2.0	1.9	12	0.42	5.04
2.1–2.3	2.2	14	0.12	1.68
2.4–2.6	2.5	9	0.18	1.62
2.7–2.9	2.8	7	0.48	3.36
3.0–3.2	3.1	5	0.78	3.90
		Total 50		Total 17.76

4.23 Determine the (a) sample variance and (b) sample standard deviation for the injuries data in Table 4.20, using the alternative computational formulas.

From Table 4.22,

(a) $s^2 = \dfrac{\Sigma(fX^2) - n\bar{X}^2}{n - 1} = \dfrac{277.94 - 50(2.32)^2}{50 - 1} = \dfrac{8.82}{49} = 0.1800$

(b) $s = \sqrt{\dfrac{\Sigma(fX^2) - n\bar{X}^2}{n - 1}} = \sqrt{0.1800} \cong 0.42$ injury

Table 4.22 Worksheet for Calculating the Variance and Standard Deviation

Average number of injuries	Class midpoint (X)	Number of firms (f)	X^2	fX^2
1.5–1.7	1.6	3	2.56	7.68
1.8–2.0	1.9	12	3.61	43.32
2.1–2.3	2.2	14	4.84	67.76
2.4–2.6	2.5	9	6.25	56.25
2.7–2.9	2.8	7	7.84	54.88
3.0–3.2	3.1	5	9.61	48.05
		Total 50		Total 277.94

4.24 The frequency distribution in Table 4.23 is reproduced from Problem 2.21. Determine the (*a*) range and (*b*) standard deviation, by the use of the alternative computational formula. Note the basis for determining class boundaries and class midpoints for such an "and under" frequency distribution. The sample mean was determined to be 10.8 min in Problem 3.19.

Table 4.23   Time Required to Process and Prepare Mail Orders

Time, min	Class boundaries	Class midpoint (X)	Number of orders (f)	X^2	fX^2
5 and under 8	5.8–8.0	6.5	10	42.25	422.50
8 and under 11	8.0–11.0	9.5	17	90.25	1,534.25
11 and under 14	11.0–14.0	12.5	12	156.25	1,875.00
14 and under 17	14.0–17.0	15.5	6	240.25	1,441.50
17 and under 20	17.0–20.0	18.5	2	342.25	684.50
			Total 47		Total 5,957.75

(*a*) $R = B_U(H) - B_L(L) = 20.0 - 5.0 = 15.0$ min

(*b*) $s = \sqrt{\dfrac{\Sigma(fX^2) - n\bar{X}^2}{n-1}} = \sqrt{\dfrac{5,957.75 - 47(10.8)^2}{47-1}} = \sqrt{10.34} \cong 3.2$ min

COMPUTER OUTPUT

4.25 Use a computer to determine the range and standard deviation for the data in Table 2.6 (page 16), on the time required to complete an assembly task by a sample of 30 employees. The mean and median were determined for these data by the use of a computer in Problem 3.24.

Figure 4-2 presents the computer input and output. As can be observed, $R = 9.0$ and $s = 2.4375 \cong 2.4$ min. Note that in the Minitab software the STANDARD DEVIATION command results in output of the *sample* standard deviation.

```
MTB ) SET ASSEMBLY TIMES INTO C1
DATA)   10    14    15    13    17    16    12    14    11    13    15    18     9
DATA)   14    14     9    15    11    13    11    12    10    17    16    12    11
DATA)   16    12    14    15
DATA) END
MTB ) NAME FOR C1 IS 'TIME'
MTB ) MAXIMUM 'TIME', PUT IN K1
   MAXIMUM =      18.000
MTB ) MINIMUM 'TIME', PUT IN K2
   MINIMUM =       9.0000
MTB ) SUBTRACT K2 FROM K1, PUT RANGE INTO K3
   ANSWER =        9.0000
MTB ) STANDARD DEVIATION OF 'TIME'
   ST.DEV. =       2.4375
```

Fig. 4-2   Minitab output.

Supplementary Problems

THE RANGES, AVERAGE DEVIATION, AND STANDARD DEVIATION FOR UNGROUPED DATA

4.26   The number of cars sold by each of the 10 salesmen in an automobile dealership during a particular month, arranged in ascending order, is: 2, 4, 7, 10, 10, 10, 12, 12, 14, 15. Determine the (*a*) range and (*b*) middle 80 percent range for these data.

Ans.   (*a*) 13, (*b*) 11.5

4.27   Compute the average deviation for the sales data in Problem 4.26. The population mean for these values was determined to be 9.6 in Problem 3.25.

Ans.   3.16 ≅ 3.2

4.28   From Problem 4.26, determine the standard deviation by using the deviations formula and considering the group of values as constituting a statistical population.

Ans.   3.955 ≅ 4.0

4.29   The weights of a sample of outgoing packages in a mailroom, weighted to the nearest ounce, are found to be: 21, 18, 30, 12, 14, 17, 28, 10, 16, 25 oz. Determine the (*a*) range and (*b*) middle 50 percent range for these weights.

Ans.   (*a*) 20.0, (*b*) 11.0

4.30   Compute the average deviation for the sampled packages in Problem 4.29. The sample mean was determined to be 19.1 oz in Problem 3.27.

Ans.   5.52 ≅ 5.5

4.31   Determine the (*a*) sample variance and (*b*) sample standard deviation for the data in Problem 4.29, by use of the computational versions of the respective formulas.

Ans. (*a*) 45.7, (*b*) 6.8

4.32   The following examination scores, arranged in ascending order, were achieved by 20 students enrolled in a decision analysis course: 39, 46, 57, 65, 70, 72, 72, 75, 77, 79, 81, 81, 84, 84, 84, 87, 93, 94, 97, 97. Determine the (*a*) range and (*b*) the middle 90 percent range for these ungrouped data.

Ans.   (*a*) 58.0, (*b*) 54.5

4.33   Compute the average deviation for the examination scores in Problem 4.32. The mean examination score was determined to be 76.7 in Problem 3.29.

Ans.   11.76 ≅ 11.8

4.34   Considering the examination scores in Problem 4.32 to be a statistical population, determine the standard deviation by use of (*a*) the deviations formula and (*b*) the alternative computational formula.

Ans.   (*a*) 15.294 ≅ 15.3, (*b*) 15.294 ≅ 15.3

4.35   The number of accidents which occurred during a given month in the 13 manufacturing departments of an industrial plant was: 2, 0, 0, 3, 3, 12, 1, 0, 8, 1, 0, 5, 1. Determine the (*a*) range and (*b*) middle 50 percent range for the number of accidents.

Ans.   (*a*) 12.0, (*b*) 3.5

4.36   Compute the average deviation for the data in Problem 4.35. The mean number of accidents was determined to be 2.8 in Problem 3.31.

Ans.   2.646 ≅ 2.6

4.37 Considering the accident data in Problem 4.35 to be a statistical population, compute the standard deviation by using the alternative computational formula.

    Ans. $3.465 \cong 3.5$

THE COEFFICIENT OF VARIATION

4.38 Determine the coefficient of variation for the car-sales data analyzed in Problems 4.26 through 4.28.

    Ans. $CV = 0.417$

4.39 Refer to Problem 4.38. In another (larger) dealership, the mean number of cars sold per salesperson during a particular month was $\mu = 17.6$, with a standard deviation of $\sigma = 6.5$. Compare the variability of car sales per salesperson (*a*) on an absolute basis, and (*b*) relative to the mean level of car sales at the two dealerships.

    Ans.   (*a*) The standard deviation in the first dealership (4.0) is smaller than the standard deviation in the second dealership (6.5). (*b*) The coefficient of variation in the first dealership (0.417) is larger than the coefficient of variation in the second dealership (0.369).

PEARSON'S COEFFICIENT OF SKEWNESS

4.40 Compute the coefficient of skewness for the car-sales data analyzed in Problems 4.26 through 4.28. The median for these data was determined to be 10.0 in Problem 3.25.

    Ans.   Skewness $= -0.30$ (Thus the distribution of car sales is slightly negatively skewed, or skewed to the left.)

4.41 Compute the coefficient of skewness for the accident data analyzed in Problems 4.35 through 4.37. The median for these data was determined to be 1.0 in Problem 3.31.

    Ans.   Skewness $= 1.54$ (Thus the distribution of accidents is positively skewed.)

THE RANGES, AVERAGE DEVIATION, AND STANDARD DEVIATION FOR GROUPED DATA

4.42 Determine the (*a*) range and (*b*) middle 80 percent range for the mileage data in Table 2.15 (page 25).

    Ans.   (*a*) 12.00, (*b*) 6.83

4.43 Compute the average deviation for the mileage data in Table 2.15. The mean mileage was determined to be 28.95 in Problem 3.40.

    Ans.   1.76

4.44 Considering the mileage data in Table 2.15 to be for a sample of trips, compute the standard deviation for these data by use of the (*a*) deviations formula and (*b*) alternative computational formula. The mean mileage was determined to be 28.95 in Problem 3.40.

    Ans.   (*a*) $2.517 \cong 2.52$, (*b*) $2.517 \cong 2.52$

4.45 From Table 3.10 (page 43), determine the (*a*) range and (*b*) middle 50 percent range for the amounts of personal loans.

    Ans.   (*a*) \$2,800.00, (*b*) \$892.31

4.46 Compute the average deviation for the personal loan data in Table 3.10. The mean for this sample was determined to be \$1,109.50 in Problem 3.42.

    Ans.   \$512.00

4.47   Determine the sample standard deviation for the data in Table 3.10 by using the alternative computational formula.

   Ans. $627.49

4.48   Determine the (*a*) range and (*b*) middle 90 percent range for the lifetime of the cutting tools in Table 2.17 (page 27).

   Ans.   (*a*) 149.95 hr, (*b*) 93.75 hr

4.49   Compute the average deviation for the lifetime of cutting tools reported in Table 2.17. The mean lifetime was determined to be 87.45 hr in Problem 3.44.

   Ans.   18.57 hr

4.50   Considering the data in Table 2.17 to be a statistical population, compute the (*a*) population variance and (*b*) population standard deviation by using the deviations formulas. The mean lifetime was determined to be 87.45 hr in Problem 3.44.

   Ans.   (*a*) 714.29, (*b*) 26.73 hr

4.51   From Table 2.18 (page 27), determine the (*a*) range and (*b*) average deviation for the age of the applicants. The mean age was determined to be 23.3 in Problem 3.46.

   Ans.   (*a*) 14.0, (*b*) $2.656 \cong 2.7$ yr

4.52   Compute the sample standard deviation for the data in Table 2.18 by using the alternative computational formula. The mean age was determined to be 23.3 in Problem 3.46.

   Ans.   3.4 yr

COMPUTER OUTPUT

4.53   Use a computer to determine the range and standard deviation for the sample of 40 loan amounts in Table 2.16 (page 26). The mean and median for these data were calculated by the use of a computer in Problem 3.52. Also, the approximate range and standard deviation based on the data being grouped were calculated in Problems 4.45 and 4.47, respectively.

   Ans. $R = \$2,700.00$ and $s = \$642.24$. This compares with the approximations of $2,800.00 and $627.49, respectively, based on the grouped loan amounts.

Chapter 5

Probability

5.1 BASIC DEFINITIONS OF PROBABILITY

Historically, three different conceptual approaches have been developed for defining probability and for determining probability values: the classical, relative frequency, and subjective approaches.

By the *classical approach* to probability, if $N(A)$ possible elementary outcomes are favorable to event A, $N(S)$ possible outcomes are included in the sample space, and all the elementary outcomes are equally likely and mutually exclusive, then the probability that event A will occur is

$$P(A) = \frac{N(A)}{N(S)} \tag{5.1}$$

Note that the classical approach to probability is based on the assumption that each outcome is equally likely. Because this approach (when it is applicable) permits determination of probability values before any sample events are observed, it has also been called the *a priori approach*.

EXAMPLE 1. In a well-shuffled deck of cards which contains four aces and 48 other cards, the probability of an ace (A) being obtained on a single draw is

$$P(A) = \frac{N(A)}{N(S)} = \frac{4}{52} = \frac{1}{13}$$

By the *relative frequency approach*, the probability is determined on the basis of the proportion of times that a favorable outcome occurs in a number of observations or experiments. No prior assumption of equal likelihood is involved. Because determination of the probability values is based on observation and collection of data, this approach has also been called the *empirical approach*. The probability that event A will occur by the relative frequency approach is

$$P(A) = \frac{\text{no. of observations of } A}{\text{sample size}} = \frac{n(A)}{n} \tag{5.2}$$

EXAMPLE 2. Before including coverage for certain types of dental problems in health insurance policies for employed adults, an insurance company wishes to determine the probability of occurrence of such problems, so that the insurance rate can be set accordingly. Therefore, the statistician collects data for 10,000 adults in the appropriate age categories and finds that 100 people have experienced the particular dental problem during the past year. The probability of occurrence is thus

$$P(A) = \frac{n(A)}{n} = \frac{100}{10,000} = 0.01, \text{ or } 1\%$$

Both the classical and relative frequency approaches yield *objective* probability values, in the sense that the probability values indicate the relative rate of occurrence of the event in the long run. In contrast, the *subjective approach* to probability is particularly appropriate when there is only one opportunity for the event to occur, and it will either occur or not occur that one time. By the subjective approach, the probability of an event is the *degree of belief* by an individual that the event will occur, based on all evidence available to him. Because the probability value is a personal judgment, the subjective approach has also been called the *personalistic approach*. This approach to probability has been developed relatively recently, and is related to Bayesian decision analysis (see Section 1.3 and Chapters 18 through 20).

EXAMPLE 3. Because of taxes and alternative uses for his funds, an investor has determined that the purchase of land parcels is worthwhile only if there is at least a probability of 0.90 that the land will appreciate in value by 50 percent or more during the next 4 years. In evaluating a certain parcel of land he studies price changes in the area during recent years, considers present price levels, studies the current and likely future status of land development projects, and reviews the statistics concerned with the economic development of the overall geographic area. On the basis of this review he concludes that there is a probability of about 0.75 that the required appreciation in value will in fact occur. Because this probability value is less than the required minimum probability of 0.90, the investment should not be made.

5.2 EXPRESSING PROBABILITY

The symbol P is used to designate the probability of an event. Thus $P(A)$ denotes the probability that event A will occur in a single observation or experiment.

The smallest value that a probability statement can have is 0 (indicating the event is impossible) and the largest value it can have is 1 (indicating the event is certain to occur). Thus, in general:

$$0 \le P(A) \le 1 \qquad (5.3)$$

In a given observation or experiment, an event must either occur or not occur. Therefore, the sum of the probability of occurrence plus the probability of nonoccurrence always equals 1. Thus, where A' indicates the nonoccurrence of event A, we have:

$$P(A) + P(A') = 1 \qquad (5.4)$$

A *Venn diagram* is a diagram related to set theory in mathematics by which the events that can occur in a particular observation or experiment can be portrayed. An enclosed figure represents a sample space, and portions of the area within the space are designated to represent particular elementary or composite events, or event spaces.

EXAMPLE 4. Figure 5.1 represents the probabilities of the two events, A and A' (read "not-A"). Because $P(A) + P(A') = 1$, all of the area within the diagram is accounted for.

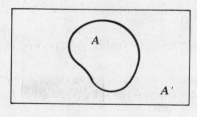

Fig. 5-1

As an alternative to probability values, probabilities can also be expressed in terms of *odds*. The odds ratio favoring the occurrence of an event is the ratio of the relative number of outcomes, designated by a, that are favorable to A, to the relative number of outcomes, designated by b, that are not favorable to A:

$$\text{Odds} = a : b \; (\text{"a to b"}) \qquad (5.5)$$

Odds of $5:2$ (read "five to two") indicate that for every five elementary events constituting success there are two elementary events constituting failure. Note that by the classical approach to probability discussed in Section 5.1 the probability value equivalent to an odds ratio of $5:2$ is

$$P(A) = \frac{N(A)}{N(S)} = \frac{a}{a+b} = \frac{5}{5+2} = \frac{5}{7}$$

EXAMPLE 5. Suppose success is defined as drawing any face card *or* an ace from a well-shuffled deck of 52 cards. Because 16 cards out of 52 are either the jack, queen, king, or ace, the odds associated with success are 16:36, or 4:9. The probability of success is $16/(16+36) = 16/52 = 4/13$.

5.3 MUTUALLY EXCLUSIVE AND NONEXCLUSIVE EVENTS

Two or more events are *mutually exclusive,* or *disjoint,* if they cannot occur together. That is, the occurrence of one event automatically precludes the occurrence of the other event (or events). For instance, suppose we consider the two possible events "ace" and "king" with respect to a card being drawn from a deck of playing cards. These two events are mutually exclusive, because any given card cannot be both an ace and a king.

Two or more events are *nonexclusive* when it is possible for them to occur together. Note that this definition does *not* indicate that such events must necessarily always occur jointly. For instance, suppose we consider the two possible events "ace" and "spade." These events are not mutually exclusive, because a given card can be both an ace and a spade; however, it does not follow that every ace is a spade or every spade is an ace.

EXAMPLE 6. In a study of consumer behavior, an analyst classifies the people who enter a stereo shop according to sex ("male" or "female") and according to age ("under 30" or "30 and over"). The two events, or classifications, "male" and "female" are mutually exclusive, since any given person would be classified in one category or the other. Similarly, the events "under 30" and "30 and over" are also mutually exclusive. However, the events "male" and "under 30" are not mutually exclusive, because a randomly chosen person could have both characteristics.

5.4 THE RULES OF ADDITION

The rules of addition are used when we wish to determine the probability of one event *or* another (or both) occurring in a single observation. Symbolically, we can represent the probability of event A or event B occurring by $P(A$ or $B)$. In the language of set theory this is called the *union* of A and B and the probability is designated by $P(A \cup B)$ (read: "probability of A union B").

There are two variations of the rule of addition, depending on whether the two events are mutually exclusive. The rule of addition for mutually exclusive events is

$$P(A \text{ or } B) = P(A \cup B) = P(A) + P(B) \qquad\qquad (5.6)$$

EXAMPLE 7. When drawing a card from a deck of playing cards, the events "ace" (A) and "king" (K) are mutually exclusive. The probability of drawing either an ace or a king in a single draw is

$$P(A \text{ or } K) = P(A) + P(K) = \frac{4}{52} + \frac{4}{52} = \frac{8}{52} = \frac{2}{13}$$

(*Note:* Problem 5.4 extends the application of this rule to three events.)

For events that are *not* mutually exclusive, the probability of the joint occurrence of the two events is subtracted from the sum of the simple probabilities of the two events. We can represent the probability of joint occurrence by $P(A$ and $B)$. In the language of set theory this is called the *intersection* of A and B and the probability is designated by $P(A \cap B)$ (read: "probability of A intersect B"). Thus, the rule of addition for events that are not mutually exclusive is

$$P(A \text{ or } B) = P(A) + P(B) - P(A \text{ and } B) \qquad\qquad (5.7)$$

Formula (5.7) is also often called the *general rule of addition,* because for events that are mutually exclusive the last term would always be equal to zero. Thus, the formula is then algebraically the same as formula (5.6) for mutually exclusive events.

EXAMPLE 8. When drawing a card from a deck of playing cards, the events "ace" and "spade" are not mutually exclusive. The probability of drawing an ace (A) or space (S) (or both) in a single draw is

$$P(A \text{ or } S) = P(A) + P(S) - P(A \text{ and } S) = \frac{4}{52} + \frac{13}{52} - \frac{1}{52} = \frac{16}{52} = \frac{4}{13}$$

Venn diagrams can be used to portray the rationale underlying the two rules of addition. In Fig. 5-2(a), note that the probability of A or B occurring is conceptually equivalent to adding the proportion of area included in A and B. In Fig. 5-2(b), for events that are not mutually exclusive, some elementary events are included in *both* A and B; thus, there is overlap between these event sets. When the areas included in A and B are added together for such events that are not mutually exclusive, the area of overlap is essentially added in twice. Thus, the rationale of subtracting $P(A \text{ and } B)$ in the rule of addition for nonexclusive events is to correct the sum for the duplicate addition of the intersect area.

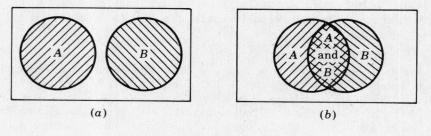

(a)   (b)

Fig. 5-2

5.5 INDEPENDENT EVENTS, DEPENDENT EVENTS, AND CONDITIONAL PROBABILITY

Two events are *independent* when the occurrence or nonoccurrence of one event has no effect on the probability of occurrence of the other event. Two events are *dependent* when the occurrence or nonoccurrence of one event *does* affect the probability of occurrence of the other event.

EXAMPLE 9. The outcomes associated with tossing a fair coin twice in succession are considered to be independent events, because the outcome of the first toss has no effect on the respective probabilities of a head or tail occurring on the second toss. The drawing of two cards *without replacement* from a deck of playing cards are dependent events, because the probabilities associated with the second draw are dependent on the outcome of the first draw. Specifically, if an "ace" occurred on the first draw, then the probability of an "ace" occurring on the second draw is the ratio of the number of aces still remaining in the deck to the total number of cards remaining in the deck, or 3/51.

When two events are dependent, the concept of *conditional probability* is employed to designate the probability of occurrence of the related event. The expression $P(B|A)$ indicates the probability of event B occurring *given* that event A has occurred. Note that "$B|A$" is *not* a fraction.

Conditional probability expressions are not required for independent events because by definition there is no relationship between the occurrence of such events. Therefore, if events A and B are independent, the conditional probability $P(B|A)$ is always equal to the simple (unconditional) probability $P(B)$. Therefore, one approach by which the independence of two events A and B can be tested is by comparing

$$P(B|A) \stackrel{?}{=} P(B) \tag{5.8}$$

or

$$P(A|B) \stackrel{?}{=} P(A) \tag{5.9}$$

(See Problems 5.8 through 5.10.)

If the simple (unconditional) probability of a first event A and the joint probability of two events A and B are known, then the conditional probability $P(B|A)$ can be determined by

$$P(B|A) = \frac{P(A \text{ and } B)}{P(A)} \tag{5.10}$$

(See Problems 5.8 through 5.10.)

There is often some confusion regarding the distinction between mutually exclusive and nonexclusive events on the one hand, and the concepts of independence and dependence on the other hand. Particularly, note the difference between events that are mutually exclusive and events that are independent. Mutual exclusiveness indicates that two events *cannot* both occur, whereas independence indicates that the probability of occurrence of one event is not affected by the occurrence of the other event. Therefore it follows that if two events are mutually exclusive, this is a particular example of highly *dependent* events, because the probability of one event given that the other has occurred would always be equal to zero. See Problem 5.10.

5.6   THE RULES OF MULTIPLICATION

The rules of multiplication are concerned with determining the probability of the joint occurrence of A and B. As explained in Section 5.4, this concerns the intersection of A and B: $P(A \cap B)$. There are two variations of the rule of multiplication, according to whether the two events are independent or dependent. The rule of multiplication for independent events is

$$P(A \text{ and } B) = P(A \cap B) = P(A)P(B) \tag{5.11}$$

EXAMPLE 10.   By formula (5.11), if a fair coin is tossed twice the probability that both outcomes will be "heads" is $(\frac{1}{2}) \times (\frac{1}{2}) = (\frac{1}{4})$.

(*Note:* Problem 5.11 extends the application of this rule to three events.)

The tree diagram is particularly useful as a method of portraying the possible events associated with sequential observations, or sequential trials. Figure 5-3 is an example of such a diagram for the events associated with tossing a coin twice, and identifies the outcomes that are possible and the probability at each point in the sequence.

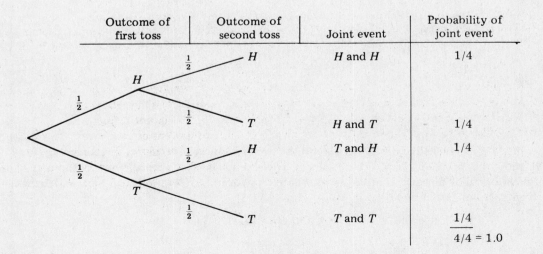

Fig. 5-3

EXAMPLE 11. By reference to Fig. 5-3, we see that there are four types of sequences, or joint events, that are possible: H and H, H and T, T and H, and T and T. By the rule of multiplication for independent events, the probability of joint occurrences for any one of these sequences in this case is $1/4$, or 0.25. Since these are the only sequences which are possible, and since they are mutually exclusive sequences, by the rule of addition the sum of the four joint probabilities should be 1.0, which it is.

For dependent events the probability of the joint occurrence of A and B is the probability of A multiplied by the *conditional* probability of B given A. An equivalent value is obtained if the two events are reversed in position. Thus the rule of multiplication for dependent events is

$$P(A \text{ and } B) = P(A)P(B|A) \tag{5.12}$$

or
$$P(A \text{ and } B) = P(B \text{ and } A) = P(B)P(A|B) \tag{5.13}$$

Formula (5.12) [or (5.13)] is often called the *general rule of multiplication*, because for events that are independent the conditional probability $P(B|A)$ is always equal to the unconditional probability value $P(B)$, resulting in formula (5.12) then being equivalent to formula (5.11) for independent events.

EXAMPLE 12. Suppose that a set of 10 spare parts is known to contain eight good parts (G) and two defective parts (D). Given that two parts are selected randomly without replacement, the sequence of possible outcomes and the probabilities are portrayed by the tree diagram in Fig. 5-4 (subscripts indicate sequential position of outcomes). Based on the multiplication rule for dependent events, the probability that the two parts selected are both good is

$$P(G_1 \text{ and } G_2) = P(G_1)P(G_2|G_1) = \left(\frac{8}{10}\right) \times \left(\frac{7}{9}\right) = \frac{56}{90} = \frac{28}{45}$$

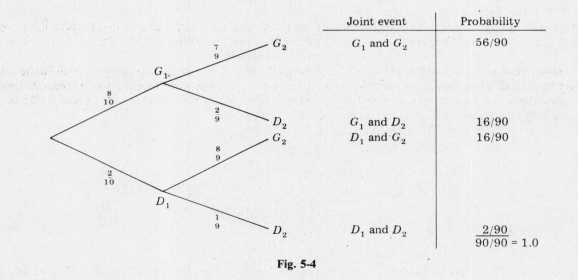

Fig. 5-4

[*Note:* Problems 5.12(b) and 5.13 extend the application of this rule to three events.]

If the probability of joint occurrence of two events is available directly without use of the multiplication rules as such, then as an alternative to formulas (5.8) and (5.9) the independence of two events A and B can be tested by comparing

$$P(A \text{ and } B) \stackrel{?}{=} P(A)P(B) \tag{5.14}$$

EXAMPLE 13. By our knowledge of a playing deck of 52 cards, we know that only one card is both an ace (A) and a spade (S), and thus that $P(A \text{ and } S) = 1/52$. We also know that the probability of drawing any ace is $4/52$

and the probability of drawing any spade is 13/52. We thus can verify that the events "ace" and "spade" are independent events, as follows:

$$P(A \text{ and } S) \overset{?}{=} P(A)P(S)$$

$$\frac{1}{52} \overset{?}{=} \frac{4}{52} \times \frac{13}{52}$$

$$\frac{1}{52} = \frac{1}{52} \text{ (therefore the events } are \text{ independent)}$$

5.7 BAYES' THEOREM

In its simplest algebraic form, Bayes' theorem is concerned with determining the conditional probability of event A given that event B has occurred. The general form of Bayes' theorem is

$$P(A \mid B) = \frac{P(A \text{ and } B)}{P(B)} \qquad (5.15)$$

Formula (5.15) is simply a particular application of the general formula for conditional probability presented in Section 5.5. However, the special importance of Bayes' theorem is that it is applied in the context of sequential events, and further, that the computational version of the formula provides the basis for determining the conditional probability of an event having occurred in the *first* sequential position given that a particular event has been observed in the *second* sequential position. The computational form of Bayes' theorem is

$$P(A \mid B) = \frac{P(A)P(B \mid A)}{P(A)P(B \mid A) + P(A')P(B \mid A')} \qquad (5.16)$$

As illustrated in Problem 5.20(c), the denominator above is the overall (unconditional) probability of the event in the second sequential position; $P(B)$ for the formula above.

EXAMPLE 14.   Suppose there are two urns U_1 and U_2. Urn 1 has eight red balls and two green balls, while urn 2 has four red balls and six green balls. If an urn is selected randomly, and a ball is then selected randomly from that urn, the sequential process and probabilities can be represented by the tree diagram in Fig. 5-5. The tree

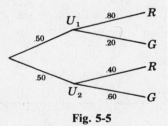

Fig. 5-5

diagram indicates that the probability of choosing either urn is 0.50, and then the conditional probabilities of a red (R) or a green (G) ball being drawn are indicated according to the urn involved. Now, suppose we observe a green ball in step 2 *without* knowing which urn was selected in step 1. What is the probability that urn 1 was selected in step 1? Symbolically, what is $P(U_1 \mid G)$? Substituting U_1 and G for A and B, respectively, in the computational form of Bayes' theorem:

$$P(U_1 \mid G) = \frac{P(U_1)P(G \mid U_1)}{P(U_1)P(G \mid U_1) + P(U_2)P(G \mid U_2)} = \frac{(0.50)(0.20)}{(0.50)(0.20) + (0.50)(0.60)} = \frac{0.10}{0.40} = 0.25$$

In Example 14, note that Bayes' theorem presents the basis for obtaining what might be called a "backward conditional" probability value, since we can determine the probability that a particular urn was selected in step 1 given the observation of a sampled item from that urn in step 2. In Bayesian decision analysis this theorem provides the conceptual basis for revising the probabilities associated with the several events, or states involved in a decision problem. See Section 19.2.

5.8 JOINT PROBABILITY TABLES

A *joint probability table* is a table in which all possible events (or outcomes) for one variable are listed as row headings, all possible events for a second variable are listed as column headings, and the value entered in each cell of the table is the probability of each joint occurrence. Often, the probabilities in such a table are based on observed frequencies of occurrence for the various joint events, rather than being *a priori* in nature. The table of joint-occurrence frequencies which can serve as the basis for constructing a joint probability table is called a *contingency table*.

EXAMPLE 15. Table 5.1(*a*) is a contingency table which describes 200 people who entered a stereo shop according to sex and age, while Table 5.1(*b*) is the associated joint probability table. The frequency reported in each cell of the contingency table is converted into a probability value by dividing by the total number of observations, in this case, 200.

Table 5.1(*a*) Contingency Table for Stereo Shop Customers

	Sex		
Age	Male	Female	Total
Under 30	60	50	110
30 and over	80	10	90
Total	140	60	200

Table 5.1(*b*) Joint Probability Table for Stereo Shop Customers

	Sex		Marginal probability
Age	Male (M)	Female (F)	
Under 30 (U)	0.30	0.25	0.55
30 and over (O)	0.40	0.05	0.45
Marginal probability	0.70	0.30	1.00

In the context of joint probability tables, a *marginal probability* is so named because it is a marginal total of a row or a column. Whereas the probability values in the cells are probabilities of joint occurrence, the marginal probabilities are the unconditional, or simple, probabilities of particular events.

EXAMPLE 16. The probability of 0.30 in row 1 and column 1 of Table 5.1(*b*) indicates that there is a probability of 0.30 that a randomly chosen person from this group of 200 people will be a male and under 30. The marginal probability of 0.70 for column 1 indicates that there is a probability of 0.70 that a randomly chosen person will be a male.

Recognizing that a joint probability table also includes all of the unconditional probability values as marginal totals, we can use formula (*5.10*) for determining any particular conditional probability value.

EXAMPLE 17. Suppose we are interested in the probability that a randomly chosen person in Table 5.1(*b*) is "under 30" (*U*) given that he is a "male" (*M*). The probability is, using formula (*5.10*),

$$P(U|M) = \frac{P(M \text{ and } U)}{P(M)} = \frac{0.30}{0.70} = \frac{3}{7} \cong 0.43$$

5.9 PERMUTATIONS

By the classical approach to determining probabilities presented in Section 5.1, the probability value is based on the ratio of the number of equally likely elementary outcomes that are favorable to the total number of outcomes in the sample space. When the problems are simple, the number of elementary outcomes can be counted directly. However, for more complex problems the methods of permutations and combinations are required to determine the number of possible elementary outcomes.

The number of *permutations* of *n* objects is the number of ways in which the objects can be arranged in terms of order:

$$\text{Permutations of } n \text{ objects} = n! = (n) \times (n-1) \times \cdots \times (2) \times (1) \tag{5.17}$$

The symbol *n*! is read "*n* factorial." In permutations and combinations problems, *n* is always positive. Also, note that by definition $0! = 1$ in mathematics.

EXAMPLE 18. Three members of a social organization have volunteered to serve as officers for the following year, to take positions as President, Treasurer, and Secretary. The number of ways (permutations) in which the three can assume the positions is

$$n! = 3! = (3)(2)(1) = 6 \text{ ways}$$

This result can be portrayed by a sequential diagram. Suppose that the three people are designated as *A*, *B*, and *C*. The number of possible arrangements, or permutations, is presented in Fig. 5-6.

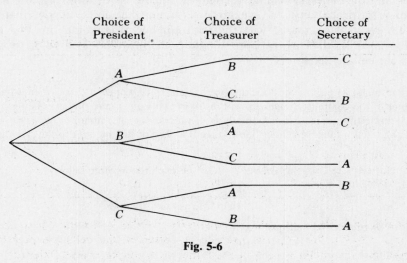

Fig. 5-6

Typically, we are concerned about the number of permutations of some *subgroup* of the *n* objects, rather than all *n* objects as such. That is, we are interested in the number of permutations of *n* objects taken *r* at a time, where *r* is less than *n*:

$$_nP_r = \frac{n!}{(n-r)!} \tag{5.18}$$

EXAMPLE 19. In Example 18, suppose there are 10 members in the social organization and no nominations have yet been presented for the offices of President, Treasurer, and Secretary. The number of different arrangements of three officers elected from the 10 club members is

$$_nP_r = {_{10}}P_3 = \frac{10!}{(10-3)!} = \frac{10!}{7!} = \frac{(10)(9)(8)(7!)}{7!} = (10)(9)(8) = 720$$

5.10 COMBINATIONS

In the case of permutations, the order in which the objects are arranged is important. In the case of *combinations*, we are concerned with the number of different groupings of objects that can occur *without* regard to their order. Therefore, an interest in combinations always concerns the number of different subgroups that can be taken from n objects. The number of combinations of n objects taken r at a time is

$$_nC_r = \frac{n!}{r!(n-r)!} \qquad\qquad (5.19)$$

In many textbooks, the combination of n objects taken r at a time is represented by $\binom{n}{r}$. Note that this is *not* a fraction.

EXAMPLE 20. Suppose that three members from a small social organization containing a total of 10 members are to be chosen to form a committee. The number of different groups of three people which can be chosen, *without regard to the different orders in which each group might be chosen*, is

$$_nC_r = {_{10}}C_3 = \frac{10!}{3!(10-3)!} = \frac{(10)(9)(8)(7!)}{3!(7!)} = \frac{(10)(9)(8)}{(3)(2)} = \frac{720}{6} = 120$$

As indicated in Section 5.9, the methods of permutations and combinations provide a basis for counting the possible outcomes in relatively complex situations. In terms of combinations, we can frequently determine the probability of an event by determining the number of combinations of outcomes which include that event as compared with the total number of combinations that are possible. Of course, this again represents the classical approach to probability and is based on the assumption that all combinations are equally likely.

EXAMPLE 21. Continuing with Example 20, if the group contains six women and four men, what is the probability that a random choice of the committee members will result in two women and one man being selected? The basic approach is to determine the number of combinations of outcomes that contain exactly two women (of the six women) and one man (of the four men) and then to take the ratio of this number to the total number of possible combinations:

Number of committees with $2W$ and $1M = {_6}C_2 \times {_4}C_1$ (see explanatory note below)

$$= \frac{6!}{2!4!} \times \frac{4!}{1!3!} = 15 \times 4 = 60$$

Total number of possible combinations $= {_{10}}C_3$

$$= \frac{10!}{3!7!} = \frac{(10)(9)(8)}{(3)(2)(1)} = \frac{720}{6} = 120$$

$$P(2W \text{ and } 1M) = \frac{{_6}C_2 \times {_4}C_1}{{_{10}}C_3} = \frac{60}{120} = 0.50$$

Note: In Example 21, above, the so-called *method of multiplication* is used. In general, if one event can occur in n_1 ways and a second event can occur in n_2 ways, then

Total number of ways two events can occur in combination $= n_1 \times n_2$ \qquad (5.20)

The value of 60 in the solution in Example 21 is based on multiplying the number of ways that two women can be chosen from the six available women (15) by the number of ways that one man can be chosen from the four available men (4).

Solved Problems

DETERMINING PROBABILITY VALUES

5.1   For each of the following situations, indicate whether the classical, relative frequency, or subjective approach would be most useful for determining the required probability value.

   (*a*)   Probability that there will be a recession next year.

   (*b*)   Probability that a six-sided die will show either a "one" or "six" on a single toss.

   (*c*)   Probability that from a shipment of 20 parts known to contain one defective part, one randomly chosen part will turn out to be defective.

   (*d*)   Probability that a randomly chosen part taken from a large shipment of parts will turn out be defective.

   (*e*)   Probability that a randomly chosen person who enters a large department store will make a purchase in that store.

   (*f*)   Probability that the Dow-Jones Industrial Average will increase by at least 50 points during the next six months.

   (*a*) Subjective, (*b*) classical, (*c*) classical, (*d*) relative frequency (Since there is no information about the overall proportion of the defective parts, the proportion of defective parts in a sample would be used as the basis for estimating the probability value.), (*e*) relative frequency, (*f*) subjective.

5.2   Determine the probability value applicable in each of the following situations.

   (*a*)   Probability of industrial injury in a particular industry on an annual basis. A random sample of 10 firms, employing a total of 8,000 people, reported that 400 industrial injuries occurred during a recent 12-month period.

   (*b*)   Probability of betting on a winning number in the game of roulette. The numbers on the wheel include a "0," "00," and "1" through "36."

   (*c*)   Probability that a fast-foods franchise outlet will be financially successful. The prospective investor obtains data for other units in the franchise system, studies the development of the residential area in which the outlet is to be located, and considers the sales volume required for financial success based on the required capital investment and operational costs. Overall, it is the investor's judgment that there is an 80 percent chance that the outlet will be financially successful and a 20 percent chance that it will not.

   (*a*)   By the relative frequency approach, $P = 400/8,000 = 0.05$. Because this probability value is based on a sample, it is an estimate of the unknown true value. Also, the implicit assumption is made that safety standards have not changed since the 12-month sampled period.

   (*b*)   By the classical approach, $P = 1/38$. This value is based on the assumption that all numbers are equally likely, and therefore a well-balanced wheel is assumed.

   (*c*)   Based on the subjective approach, the value arrived at through the prospective investor's judgment is $P = 0.80$. Note that such a judgment should be based on knowledge of all available information within the scope of the time which is available to collect such information.

5.3   For each of the following reported odds ratios determine the equivalent probability value, and for each of the reported probability values determine the equivalent odds ratio.

(a)   A purchasing agent estimates that the odds are 2:1 that a shipment will arrive on schedule.

(b)   The probability that a new component will not function properly when assembled is assessed as being $P = 1/5$.

(c)   The odds that a new product will succeed are estimated as being 3:1.

(d)   The probability that the home team will win the opening game of the season is assessed as being 1/3.

(a)   The probability that the shipment will arrive on schedule is $P = 2/(2+1) = 2/3 \cong 0.67$.

(b)   The odds that it will not function properly are 1:4.

(c)   The probability that the product will succeed is $P = 3/(3+1) = 3/4 = 0.75$.

(d)   The odds that the team will win are 1:2.

APPLYING THE RULES OF ADDITION

5.4   Determine the probability of obtaining an ace (A), king (K), or a deuce (D) when one card is drawn from a well-shuffled deck of 52 playing cards.

From formula (5.6),

$$P(A \text{ or } K \text{ or } D) = P(A) + P(K) + P(D) = \frac{4}{52} + \frac{4}{52} + \frac{4}{52} = \frac{12}{52} = \frac{3}{13}$$

(*Note:* The events are mutually exclusive.)

5.5   With reference to Table 5.2, what is the probability that a randomly chosen family will have household income (a) between \$18,000 and \$22,999, (b) less than \$23,000, (c) at one of the two extremes of being either less than \$18,000 or at least \$40,000?

Table 5.2   Annual Household Income for 500 Families

Category	Income range	Number of families
1	Less than \$18,000	60
2	\$18,000–\$22,999	100
3	\$23,000–\$29,999	160
4	\$30,000–\$39,999	140
5	\$40,000 and above	40
		Total 500

(a) $P(2) = \dfrac{100}{500} = \dfrac{1}{5} = 0.20$

(b) $P(1 \text{ or } 2) = \dfrac{60}{500} + \dfrac{100}{500} = \dfrac{160}{500} = \dfrac{8}{25} = 0.32$

(c) $P(1 \text{ or } 5) = \dfrac{60}{500} + \dfrac{40}{500} = \dfrac{100}{500} = \dfrac{1}{5} = 0.20$

(*Note:* The events are mutually exclusive.)

5.6   Of 300 business students, 100 are currently enrolled in accounting and 80 are currently enrolled in business statistics. These enrollment figures include 30 students who are in fact enrolled in both courses. What is the probability that a randomly chosen student will be enrolled in either accounting (A) or business statistics (B)?

From formula (5.6),

$$P(A \text{ or } B) = P(A) + P(B) - P(A \text{ and } B) = \frac{100}{300} + \frac{80}{300} - \frac{30}{300} = \frac{150}{300} = \frac{1}{2} = 0.50$$

(*Note:* The events are not mutually exclusive.)

5.7   Of 100 individuals who applied for systems analyst positions with a large firm during the past year, 40 had some prior work experience (W), and 30 had a professional certificate (C). However, 20 of the applicants had both work experience and a certificate, and thus are included in both of the counts.

(*a*)   Construct a Venn diagram to portray these events.

(*b*)   What is the probability that a randomly chosen applicant had either work experience or a certificate (or both)?

(*c*)   What is the probability that a randomly chosen applicant had either work experience or a certificate *but not both*?

(*a*)   See Fig. 5-7.

(*b*) $P(W \text{ or } C) = P(W) + P(C) - P(W \text{ and } C) = 0.40 + 0.30 - 0.20 = 0.50$
(*Note:* The events are not mutually exclusive.)

(*c*) $P(W \text{ or } C, \text{ but not both}) = P(W \text{ or } C) - P(W \text{ and } C) = 0.50 - 0.20 = 0.30$

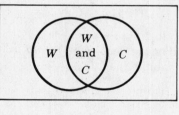

Fig. 5-7

INDEPENDENT EVENTS, DEPENDENT EVENTS, AND CONDITIONAL PROBABILITY

5.8   For Problem 5.7, (*a*) determine the conditional probability that a randomly chosen applicant has a certificate given that he has some previous work experience. (*b*) Apply an appropriate test to determine if work experience and certification are independent events.

(*a*) $P(C \mid W) = \dfrac{P(C \text{ and } W)}{P(W)} = \dfrac{0.20}{0.40} = 0.50$

(*b*) $P(C \mid W) \stackrel{?}{=} P(C)$. Since $0.50 \neq 0.30$, events W and C are dependent. Independence could also be tested by applying the multiplication rule for independent events—see Problem 5.14(*a*).

5.9   Two separate product divisions included in a large firm are Marine Products (M) and Office Equipment (O). The probability that the Marine Products division will have a profit margin of at least 10 percent this fiscal year is estimated to be 0.30, the probability that the Office Equipment division will have a profit margin of at least 10 percent is 0.20, and the probability that both divisions will have a profit margin of at least 10 percent is 0.06.

(a)   Determine the probability that the Office Equipment division will have at least a 10 percent profit margin given that the Marine Products division achieved this profit criterion.

(b)   Apply an appropriate test to determine if achievement of the profit goal in the two divisions is statistically independent.

(a) $P(O|M) = \dfrac{P(O \text{ and } M)}{P(M)} = \dfrac{0.06}{0.30} = 0.20$

(b) $P(O|M) \overset{?}{=} P(O)$. Since $0.20 = 0.20$, the two events are independent. [Independence could also be tested by applying the multiplication rule for independent events—see Problem 5.14(b).]

5.10   Suppose an optimist estimates that the probability of his earning a final grade of "A" in the business statistics course is 0.60 and the probability of a "B" is 0.40. Of course, he cannot earn both grades as final grades, since they are mutually exclusive.

(a)   Determine the conditional probability of his earning a "B" given that he has in fact received the final grade of "A," by use of the appropriate computational formula.

(b)   Apply an appropriate test to demonstrate that such mutually exclusive events are dependent events.

(a) $P(B|A) = \dfrac{P(B \text{ and } A)}{P(A)} = \dfrac{0}{0.60} = 0$

(b) $P(B|A) \overset{?}{=} P(B)$. Since $0 \neq 0.40$, the events are dependent. See Section 5.5.

APPLYING THE RULES OF MULTIPLICATION

5.11   In general, the probability that a prospect will make a purchase when he is contacted by a salesman is $P = 0.40$. If a saleman selects three prospects randomly from a file and makes contact with them, what is the probability that all three prospects will make a purchase?

Since the actions of the prospects are assumed to be independent of one another, the rule of multiplication for independent events is applied.

$P(\text{all 3 are purchasers}) = P(\text{first is a purchaser}) \times P(\text{second is a purchaser}) \times P(\text{third is a purchaser})$
$$= (0.40) \times (0.40) \times (0.40) = 0.064$$

5.12   Of 12 accounts held in a file, four contain a procedural error in posting account balances.

(a)   If an auditor randomly selects two of these accounts (without replacement), what is the probability that neither account will contain a procedural error? Construct a tree diagram to represent this sequential sampling process.

(b)   If the auditor samples three accounts, what is the probability that none of the accounts includes the procedural error?

(a)   In this example the events are dependent, because the outcome on the first sampled account affects the probabilities which apply to the second sampled account. Where E_1' means no error in the first sampled account and E_2' means no error in the second sampled account

$$P(E_1' \text{ and } E_2') = P(E_1')P(E_2'|E_1') = \frac{8}{12} \times \frac{7}{11} = \frac{56}{132} = \frac{14}{33} \cong 0.42$$

In Fig. 5-8, E stands for an account with the procedural error, E' stands for an account with no procedural error, and the subscript indicates the sequential position of the sampled account.

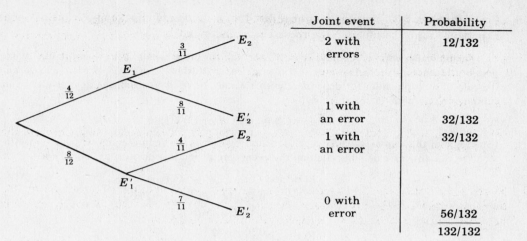

Joint event	Probability
2 with errors	12/132
1 with an error	32/132
1 with an error	32/132
0 with error	56/132
	132/132

Fig. 5-8

(b) $P(E'_1 \text{ and } E'_2 \text{ and } E'_3) = P(E'_1)P(E'_2|E'_1)P(E'_3|E'_1 \text{ and } E'_2)$

$$= \frac{8}{12} \times \frac{7}{11} \times \frac{6}{10} = \frac{336}{1,320} = \frac{42}{165} \cong 0.25$$

5.13 When sampling *without* replacement from a finite population, the probability values associated with various events are dependent on what events (sampled items) have already occurred. On the other hand, when sampling *with* replacement the events are always independent.

(a)   Suppose that three cards are chosen randomly and without replacement from a playing deck of 52 cards. What is the probability that all three cards are aces?

(b)   Suppose that three cards are chosen randomly from a playing deck of 52 cards, but that after each selection the card is replaced and the deck is shuffled before the next selection of a card. What is the probability that all three cards are aces?

(a)   The rule of multiplication for dependent events applies in this case:

$$P(A_1 \text{ and } A_2 \text{ and } A_3) = P(A_1)P(A_2|A_1)P(A_3|A_1 \text{ and } A_2)$$

$$= \frac{4}{52} \times \frac{3}{51} \times \frac{2}{50} = \frac{24}{132,600} = \frac{1}{5,525} \cong 0.0002$$

(b)   The rule of multiplication for independent events applies in this case:

$$P(A_1 \text{ and } A_2 \text{ and } A_3) = P(A)P(A)P(A)$$

$$= \frac{4}{52} \times \frac{4}{52} \times \frac{4}{52} = \frac{64}{140,608} = \frac{1}{2,197} \cong 0.0005$$

5.14 Test the independence (a) of the two events described in Problems 5.7 and 5.8, and (b) for the two events described in Problem 5.9, using the rule of multiplication for independent events.

(a) $P(W \text{ and } C) \stackrel{?}{=} P(W)P(C)$

        $0.20 \stackrel{?}{=} (0.40) \times (0.30)$

        $0.20 \neq 0.12$

    Therefore, events W and C are dependent events. This corresponds with the answer to Problem 5.8(b).

(b) $P(M \text{ and } O) \stackrel{?}{=} P(M)P(O)$

        $0.06 \stackrel{?}{=} (0.30) \times (0.20)$

        $0.06 = 0.06$

    Therefore, events M and O are independent. This corresponds with the answer to Problem 5.9(b).

5.15 From Problem 5.7, what is the probability that a randomly chosen applicant has neither work experience nor a certificate? Are these events independent?

Symbolically, what is required is $P(W'$ and $C')$ for these events that are not mutually exclusive but possibly dependent events. However, in this case neither $P(W'|C')$ nor $P(C'|W')$ is available, and therefore the rule of multiplication for dependent events cannot be used. Instead, the answer can be obtained by subtraction, as follows:

$$P(W' \text{ and } C') = 1 - P(W \text{ or } C) = 1 - 0.50 = 0.50$$

(The logic of this can most easily be understood by drawing a Venn diagram.)

We can now also demonstrate that the events are dependent rather than independent:

$$P(W' \text{ and } C') \overset{?}{=} P(W')P(C')$$
$$0.50 \overset{?}{=} [1 - P(W)][1 - P(C)]$$
$$0.50 \overset{?}{=} (0.60)(0.70)$$
$$0.50 \neq 0.42$$

The conclusion that the events are dependent coincides with the answer to Problem 5.14(*a*), which is directed to the complement of each of these two events.

5.16 Refer to Problem 5.11. (*a*) Construct a tree diagram to portray the sequence of three contacts, using S for sale and S' for no sale. (*b*) What is the probability that the salesman will make *at least* two sales? (*c*) What is the probability that the salesman will make *at least* one sale?

(*a*) See Fig. 5-9.

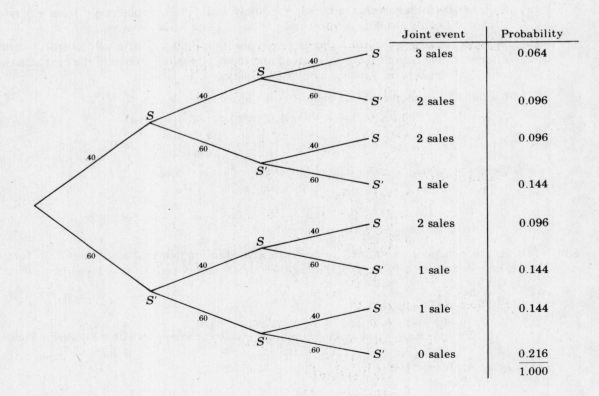

Fig. 5-9

(b) "At least" two sales includes either two *or* three sales. Further, by reference to Fig. 5-9 we note that the two sales can occur by any of three different sequences. Therefore, we use the rule of multiplication for independent events to determine the probability of each sequence and the rule of addition to indicate that any of these sequences constitutes "success":

$$P(\text{at least 2 sales}) = P(S \text{ and } S \text{ and } S) + P(S \text{ and } S \text{ and } S')$$
$$+ P(S \text{ and } S' \text{ and } S) + P(S' \text{ and } S \text{ and } S)$$
$$= (0.064) + (0.096) + (0.096) + (0.096) = 0.352$$

(c) Instead of following the approach in part (b), it is easier to obtain the answer to this question by subtraction:

$$P(\text{at least 1 sale}) = 1 - P(\text{no } S)$$
$$= 1 - P(S' \text{ and } S' \text{ and } S')$$
$$= 1 - 0.216 = 0.784$$

5.17 In Problem 5.12 it was established that 4 of 12 accounts contain a procedural error.

(a) If an auditor samples one account randomly, what is the probability that it will contain the error?

(b) If an auditor samples two accounts randomly, what is the probability that at least one will contain the error?

(c) If an auditor samples three accounts randomly, what is the probability that at least one will contain the error?

(a) $P(E) = \dfrac{\text{No. of accts. with error}}{\text{Total no. of accts.}} = \dfrac{4}{12} = \dfrac{1}{3} \cong 0.33$

(b) $P(\text{at least one } E) = P(E_1 \text{ and } E_2) + P(E_1 \text{ and } E_2') + P(E_1' \text{ and } E_2)$

$$= P(E_1)P(E_2 \mid E_1) + P(E_1)P(E_2' \mid E_1) + P(E_1')P(E_2 \mid E_1')$$

$$= \left(\frac{4}{12}\right)\left(\frac{3}{11}\right) + \left(\frac{4}{12}\right)\left(\frac{8}{11}\right) + \left(\frac{8}{12}\right)\left(\frac{4}{11}\right)$$

$$= \frac{12}{132} + \frac{32}{132} + \frac{32}{132} = \frac{76}{132} = \frac{19}{33} \cong 0.58$$

or

$$P(\text{at least one } E) = 1 - P(\text{no } E)$$
$$= 1 - P(E_1' \text{ and } E_2')$$
$$= 1 - P(E_1')P(E_2' \mid E_1')$$
$$= 1 - \left(\frac{8}{12}\right)\left(\frac{7}{11}\right)$$
$$= 1 - \frac{56}{132} = \frac{76}{132} = \frac{19}{33} \cong 0.58$$

(c) $P(\text{at least one } E) = 1 - P(\text{no } E)$
$$= 1 - P(E_1' \text{ and } E_2' \text{ and } E_3')$$
$$= 1 - P(E_1')P(E_2' \mid E_1')P(E_3' \mid E_1' \text{ and } E_2')$$
$$= 1 - \left(\frac{8}{12}\right)\left(\frac{7}{11}\right)\left(\frac{6}{10}\right)$$
$$= 1 - \frac{336}{1,320} = \frac{984}{1,320} = \frac{123}{165} \cong 0.75$$

BAYES' THEOREM

5.18   Box A is known to contain one penny (P) and one dime (D) while box B contains two dimes. A box is chosen randomly and then a coin is randomly selected from the box. (a) Construct a tree diagram to portray this situation involving sequential events. (b) If box A is selected in the first step, what is the probability that a dime (D) will be selected in the second step? (c) If a dime (D) is selected in the second step, what is the probability that it came from box A? (d) If a penny (P) is selected in the second step, what is the probability that it came from box A?

(a)   See Fig. 5-10.

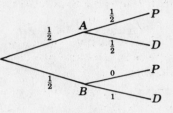

Fig. 5-10

(b) $P(D|A) = \frac{1}{2} = 0.50$

(c) $P(A|D) = \dfrac{P(A \text{ and } D)}{P(D)} = \dfrac{P(A)P(D|A)}{P(A)P(D|A) + P(B)P(D|B)}$

$\qquad = \dfrac{(\frac{1}{2})(\frac{1}{2})}{(\frac{1}{2})(\frac{1}{2}) + (\frac{1}{2})(1)} = \dfrac{\frac{1}{4}}{\frac{1}{4} + \frac{1}{2}} = \dfrac{1}{3} \cong 0.33$

(d) $P(A|P) = \dfrac{P(A \text{ and } P)}{P(P)} = \dfrac{P(A)P(P|A)}{P(A)P(P|A) + P(B)P(P|B)}$

$\qquad = \dfrac{(\frac{1}{2})(\frac{1}{2})}{(\frac{1}{2})(\frac{1}{2}) + (\frac{1}{2})(0)} = \dfrac{\frac{1}{4}}{\frac{1}{4}} = 1$

Thus, if a penny is obtained it must have come from box A.

5.19   An analyst in a photographic concern estimates that the probability is 0.30 that a competing firm plans to begin manufacturing instant photography equipment within the next three years, and 0.70 that the firm does not. If the competing firm has such plans, a new manufacturing facility would definitely be built. If the competing firm does not have such plans, there is still a 60 percent chance that a new manufacturing facility would be built for other reasons.

(a)   Using I for the decision to enter the instant photography field and M for the addition of a new manufacturing facility, portray the possible events by means of a tree diagram.

(b)   Suppose we observe that the competing firm has in fact begun work on a new manufacturing facility. Given this information, what is the probability that the firm has decided to enter the instant photography field?

(a)   See Fig. 5-11.

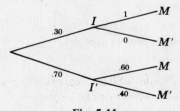

Fig. 5-11

(b) $P(I \mid M) = \dfrac{P(I \text{ and } M)}{P(M)} = \dfrac{P(I)P(M \mid I)}{P(I)P(M \mid I) + P(I')P(M \mid I')}$

$= \dfrac{(0.30)(1)}{(0.30)(1) + (0.70)(0.60)} = \dfrac{0.30}{0.72} \cong 0.42$

5.20 If there is an increase in capital investment next year, the probability that structural steel will increase in price is 0.90. If there is no increase in such investment, the probability of an increase is 0.40. Overall, we estimate that there is a 60 percent chance that capital investment will increase next year.

(a)   Using I and I' for capital investment increasing and not increasing and using R and R' for a rise and nonrise in structural steel prices, construct a tree diagram for this situation involving dependent events.

(b)   What is the probability that structural steel prices will not increase even though there is an increase in capital investment?

(c)   What is the overall (unconditional) probability of an increase in structural steel prices next year?

(d)   Suppose that during the next year structural steel prices in fact increase. What is the probability that there was an increase in capital investment?

(a)   See Fig. 5-12.

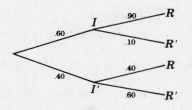

Fig. 5-12

(b) $P(R' \mid I) = 0.10$

(c)   This is the denominator in Bayes' formula:

$$P(R) = P(I \text{ and } R) \text{ or } P(I' \text{ and } R) = P(I)P(R \mid I) + P(I')P(R \mid I')$$
$$= (0.60)(0.90) + (0.40)(0.40) = 0.70$$

(d)   By Bayes' formula:

$$P(I \mid R) = \frac{P(I \text{ and } R)}{P(R)} = \frac{P(I)P(R \mid I)}{P(I)P(R \mid I) + P(I')P(R \mid I')}$$

$$= \frac{(0.60)(0.90)}{(0.60)(0.90) + (0.40)(0.40)} = \frac{0.54}{0.70} \cong 0.77$$

JOINT PROBABILITY TABLES

5.21 Table 5.3 is a contingency table which presents voter reactions to a new property tax plan according to party affiliation. (a) Prepare the joint probability table for these data. (b) Determine the marginal probabilities and indicate what they mean.

Table 5.3 Contingency Table for Voter Reactions to a New Property Tax Plan

Party affiliation	Reaction			Total
	In favor	Neutral	Opposed	
Democratic	120	20	20	160
Republican	50	30	60	140
Independent	50	10	40	100
Total	220	60	120	400

(*a*) See Table 5.4.

Table 5.4 Joint Probability Table for Voter Reactions to a New Property Tax Plan

Party affiliation	Reaction			Marginal probability
	In favor (F)	Neutral (N)	Opposed (O)	
Democratic (D)	0.30	0.05	0.05	0.40
Republican (R)	0.125	0.075	0.15	0.35
Independent (I)	0.125	0.025	0.10	0.25
Marginal probability	0.55	0.15	0.30	1.00

(*b*) Each marginal probability value indicates the unconditional probability of the event identified as the column or row heading. For example, if a person is chosen randomly from this group of 400 voters, the probability that he will be in favor of the tax plan is $P(F) = 0.55$. If a voter is chosen randomly, the probability that he is a Republican is $P(R) = 0.35$.

5.22 Referring to Table 5.4, determine the following probabilities: (*a*) $P(O)$, (*b*) $P(R$ and $O)$, (*c*) $P(I)$, (*d*) $P(I$ and $F)$, (*e*) $P(O|R)$, (*f*) $P(R|O)$, (*g*) $P(R$ or $D)$, (*h*) $P(D$ or $F)$.

(*a*) $P(O) = 0.30$ (the marginal probability)

(*b*) $P(R$ and $O) = 0.15$ (the joint probability in the table)

(*c*) $P(I) = 0.25$ (the marginal probability)

(*d*) $P(I$ and $F) = 0.125$ (the joint probability in the table)

(*e*) $P(O|R) = \dfrac{P(O \text{ and } R)}{P(R)} = \dfrac{0.15}{0.35} = \dfrac{3}{7} \cong 0.43$ (the probability that the voter is opposed to the plan given that he is a Republican)

(*f*) $P(R|O) = \dfrac{P(R \text{ and } O)}{P(O)} = \dfrac{0.15}{0.30} = 0.50$ (the probability that the voter is a Republican given that he is opposed to the plan)

(*g*) $P(R$ or $D) = P(R) + P(D) = 0.35 + 0.40 = 0.75$ (the probability that the voter is either a Democrat or a Republican, which are mutually exclusive events)

(*h*) $P(D$ or $F) = P(D) + P(F) - P(D$ and $F) = 0.40 + 0.55 - 0.30 = 0.65$ (the probability that the voter is either a Democrat or in favor of the proposal, which are not mutually exclusive events)

PERMUTATIONS AND COMBINATIONS

5.23 The five individuals constituting the top management of a small manufacturing firm are to be seated together at a banquet table. (a) Determine the number of different seating arrangements that are possible for the five individuals. (b) Suppose that only three of the five officers will be asked to represent the company at the banquet. How many different arrangements at the banquet table are possible, considering that any three of the five individuals may be chosen?

(a) $_nP_n = n! = (5)(4)(3)(2)(1) = 120$

(b) $_nP_r = \dfrac{n!}{(n-r)!} = \dfrac{5!}{(5-3)!} = \dfrac{(5)(4)(3)(2)(1)}{(2)(1)} = 60$

5.24 For Problem 5.23(b) suppose we are not concerned about the number of different possible seating arrangements, but rather, about the number of different groupings of three officers (out of the five) which might attend the banquet. How many different groupings are there?

Using formula (5.19),

$$_nC_r = \binom{n}{r} = \frac{n!}{r!(n-r)!} = \frac{5!}{3!(5-3)!} = \frac{(5)(4)(3)(2)(1)}{(3)(2)(1)(2)(1)} = 10$$

5.25 A sales representative must visit six cities during a trip.

(a) If there are 10 cities in the geographic area he is to visit, how many different groupings of six cities are there that he might visit?

(b) Suppose that there are 10 cities in the geographic area he is to visit, and further, that the sequence in which he schedules his visits at the six selected cities is also of concern. How many different sequences are there of six cities chosen from the total of 10 cities?

(c) Suppose that the six cities to be visited have been designated, but that the sequence of visiting the six cities has not been designated. How many sequences are possible for the six designated cities?

(a) $_nC_r = \binom{n}{r} = \dfrac{n!}{r!(n-r)!} = \dfrac{10!}{6!(10-6)!} = \dfrac{(10)(9)(8)(7)(6)(5)(4)(3)(2)(1)}{(6)(5)(4)(3)(2)(1)(4)(3)(2)(1)} = 210$

(b) $_nP_r = \dfrac{n!}{(n-r)!} = \dfrac{10!}{(10-6)!} = \dfrac{(10)(9)(8)(7)(6)(5)(4)(3)(2)(1)}{(4)(3)(2)(1)} = 151{,}200$

(c) $_nP_n = n! = 6! = (6)(5)(4)(3)(2)(1) = 720$

5.26 Of the 10 cities described in Problem 5.25, suppose that six are in fact "primary" markets for the product in question while the other four are "secondary" markets. If the salesman chooses the six cities to be visited on a random basis, what is the probability that (a) four of them will be primary market cities and two will be secondary market cities, (b) all six will turn out to be primary market cities?

(a) $P = \dfrac{\text{No. of combinations which include four and two cities, respectively}}{\text{Total number of different combinations of six cities}}$

$$= \frac{_6C_4 \times {_4}C_2}{_{10}C_6} = \frac{\dfrac{6!}{4!2!}\dfrac{4!}{2!2!}}{\dfrac{10!}{6!4!}} = \frac{(15)(6)}{210} = \frac{90}{210} = \frac{3}{7} \cong 0.43$$

(b) $\quad P = \dfrac{{}_6C_6 \times {}_4C_0}{{}_{10}C_6} = \dfrac{\dfrac{6!}{6!0!}\dfrac{4!}{0!4!}}{\dfrac{10!}{6!4!}} = \dfrac{(1)(1)}{210} = \dfrac{1}{210} \cong 0.005$

For this problem, the answer can also be obtained by applying the multiplication rule for dependent events. The probability of selecting a primary market city on the first choice is 6/10. Following this result, the probability on the next choice is 5/9, and so forth. On this basis, the probability that all six will be primary market cities is

$$P = \left(\dfrac{6}{10}\right)\left(\dfrac{5}{9}\right)\left(\dfrac{4}{8}\right)\left(\dfrac{3}{7}\right)\left(\dfrac{2}{6}\right)\left(\dfrac{1}{5}\right) = \left(\dfrac{1}{210}\right) \cong 0.005$$

5.27 In respect to the banquet described in Problem 5.23, determine the probability that the group of three officers chosen from the five will include (a) one particular officer, (b) two particular officers, (c) three particular officers.

(a) $\quad P = \dfrac{\text{No. of combinations which include the particular officer}}{\text{Total number of different combinations of three officers}}$

$\qquad = \dfrac{{}_1C_1 \times {}_4C_2}{{}_5C_3} = \dfrac{\dfrac{1!}{1!0!}\dfrac{4!}{2!2!}}{\dfrac{5!}{3!2!}} = \dfrac{(1)(6)}{10} = \dfrac{6}{10} = 0.60$

In this case, this probability value is equivalent simply to observing that 3/5 of the officers will be chosen, and thus that the probability that any given individual will be chosen is 3/5, or 0.60.

(b) $\quad P = \dfrac{{}_2C_2 \times {}_3C_1}{{}_5C_3} = \dfrac{\dfrac{2!}{2!0!}\dfrac{3!}{1!2!}}{\dfrac{5!}{3!2!}} = \dfrac{(1)(3)}{10} = \dfrac{3}{10} = 0.30$

(c) $\quad P = \dfrac{{}_3C_3 \times {}_2C_0}{{}_5C_3} = \dfrac{\dfrac{3!}{3!0!}\dfrac{2!}{0!2!}}{\dfrac{5!}{3!2!}} = \dfrac{(1)(1)}{10} = \dfrac{1}{10} = 0.10$

Supplementary Problems

DETERMINING PROBABILITY VALUES

5.28 Determine the probability value for each of the following events.

(a) Probability of randomly selecting one account receivable which is delinquent, given that 5 percent of the accounts are delinquent.

(b) Probability that a land investment will be successful. In the given area, only half of such investments are generally successful, but the particular investors' decision methods have resulted in his having a 30 percent better record than the average investor in the area.

(c) Probability that the sum of the dots showing on the face of two dice which are tossed is seven.

Ans. (a) 0.05, (b) 0.65, (c) 1/6

5.29 For each of the following reported odds ratios determine the equivalent probability value, and for each of the reported probability values determine the equivalent odds ratio.

(a) Probability of $P = 2/3$ that a target delivery date will be met.

(b) Probability of $P = 9/10$ that a new product will exceed the breakeven sales level.

(c) Odds of $1:2$ that a competitor will achieve a technological breakthrough.

(d) Odds of $5:1$ that a new product will be profitable.

Ans. (a) $2:1$ (b) $9:1$, (c) $P = 1/3$, (d) $P = 5/6$

APPLYING THE RULES OF ADDITION

5.30 During a given week the probability that a particular common stock issue will increase (I) in price, remain unchanged (U), or decline (D) in price is estimated to be 0.30, 0.20, and 0.50, respectively.

(a) What is the probability that the stock issue will increase in price or remain unchanged?

(b) What is the probability that the price of the issue will change during the week?

Ans. (a) 0.50, (b) 0.80

5.31 Of 500 employees, 200 participate in a company's profit-sharing plan (P), 400 have a major-medical insurance coverage (M), and 200 employees participate in both programs. Construct a Venn diagram to portray the events designated P and M.

5.32 Refer to the Venn diagram prepared in Problem 5.31. What is the probability that a randomly selected employee (a) will be a participant in at least one of the two programs, (b) will not be a participant in either program?

Ans. (a) 0.80, (b) 0.20

5.33 The probability that a new marketing approach will be successful (S) is assessed as being 0.60. The probability that the expenditure for developing the approach can be kept within the original budget (B) is 0.50. The probability that both of these objectives will be achieved is estimated at 0.30. What is the probability that at least one of these objectives will be achieved?

Ans. 0.80

INDEPENDENT EVENTS, DEPENDENT EVENTS, AND CONDITIONAL PROBABILITY

5.34 For the situation described in Problem 5.31, (a) determine the probability that an employee will be a participant in the profit-sharing plan (P) given that he has major-medical insurance coverage (M), and (b) determine if the two events are independent or dependent by reference to the conditional probability value.

Ans. (a) 0.50, (b) dependent

5.35 For Problem 5.33, determine (a) the probability that the new marketing approach will be successful (S) given that the development cost was kept within the original budget (B), and (b) if the two events are independent or dependent by reference to the conditional probability value.

Ans. (a) 0.60, (b) independent

5.36 The probability that automobile sales will increase next month (A) is estimated to be 0.40. The probability that the sale of replacement parts will increase (R) is estimated to be 0.50. The probability that both industries will experience an increase in sales is estimated to be 0.10. What is the probability that (a) automobile sales have increased during the month given that there is information that replacement parts sales have increased, (b) replacement parts sales have increased given information that automobile sales have increased during the month?

Ans. (a) 0.20, (b) 0.25

5.37   For Problem 5.36, determine if the two events are independent or dependent by reference to one of the conditional probability values.

   Ans.   Dependent

APPLYING THE RULES OF MULTIPLICATION

5.38   During a particular period, 80 percent of the common stock issues in an industry which includes just 10 companies have increased in market value. If an investor chose two of these issues randomly, what is the probability that both issues increased in market value during this period?

   Ans. $56/90 \cong 0.62$

5.39   The overall proportion of defective items in a continuous production process is 0.10. What is the probability that (a) two randomly chosen items will both be nondefective (D'), (b) two randomly chosen items will both be defective (D), (c) at least one of two randomly chosen items will be nondefective (D')?

   Ans.   (a) 0.81, (b) 0.01, (c) 0.99

5.40   Test the independence of the two events described in Problem 5.31 by using the rule of multiplication for independent events. Compare your answer with the result of the test in Problem 5.31(b).

   Ans.   Dependent

5.41   Test the independence of the two events described in Problem 5.33 by using the rule of multiplication for independent events. Compare your answer with the result of the test in Problem 5.32(b).

   Ans.   Independent

5.42   From Problem 5.38, suppose an investor chose three of these stock issues randomly. Construct a tree diagram to portray the various possible results for the sequence of three stock issues.

5.43   Referring to the tree diagram prepared in Problem 5.42, determine the probability that (a) only one of the three issues increased in market value, (b) two issues increased in market value, (c) *at least* two issues increased in market value.

   Ans.   (a) $48/720 \cong 0.07$, (b) $336/720 \cong 0.47$, (c) $672/720 \cong 0.93$

5.44   Referring to Problem 5.39, suppose a sample of four items is chosen randomly. Construct a tree diagram to portray the various possible results in terms of individual items being defective (D) or nondefective (D').

5.45   Referring to the tree diagram prepared in Problem 5.44, determine the probability that (a) none of the four items is defective, (b) exactly one item is defective, (c) one or fewer items are defective.

   Ans.   (a) $0.6561 \cong 0.66$, (b) $0.2916 \cong 0.29$, (c) $0.9477 \cong 0.95$

BAYES' THEOREM

5.46   Suppose there are two urns U_1 and U_2. U_1 contains two red balls and one green ball, while U_2 contains one red ball and two green balls.

   (a)   An urn is randomly selected, and then one ball is randomly selected from the urn. The ball is red. What is the probability that the urn selected was U_1?

   (b)   An urn is randomly selected, and then two balls are randomly selected (without replacement) from the urn. The first ball is red and the second ball is green. What is the probability that the urn selected was U_1?

   Ans.   (a) $P(U_1) = 2/3$, (b) $P(U_1) = 1/2$

5.47   Refer to Problem 5.46.

   (*a*)   Suppose an urn is randomly selected, and then two balls are randomly selected (without replacement) from the urn. Both balls are red. What is the probability that the urn selected was U_1?

   (*b*)   Suppose an urn is randomly selected and then two balls are randomly selected, *but with the first selected ball being placed back in the urn before the second ball is drawn*. Both balls are red. What is the probability that the urn selected was U_1?

   Ans.   (*a*) $P(U_1) = 1$, (*b*) $P(U_1) = 4/5$

5.48   Eighty percent of the vinyl material received from Vendor A is of exceptional quality while only 50 percent of the vinyl material received from Vendor B is of exceptional quality. However, the manufacturing capacity of Vendor A is limited, and for this reason only 40 percent of the vinyl material purchased by our firm comes from Vendor A. The other 60 percent comes from Vendor B. An incoming shipment of vinyl material is inspected, and it is found to be of exceptional quality. What is the probability that it came from Vendor A?

   Ans. $P(A) = 0.52$

5.49   Gasoline is being produced at three refineries with daily production levels of 100,000, 200,000, and 300,000 gallons, respectively. The proportion of the output which is below the octane specifications for "name-brand" sale at the three refineries is 0.03, 0.05, and 0.04, respectively. A gasoline tank-truck is found to be carrying gasoline which is below the octane specifications, and therefore the gasoline is to be marketed outside of the name-brand distribution system. Determine the probability that the tank-truck came from each of the three refineries (*a*) without reference to the information that the shipment is below the octane specifications and (*b*) given the additional information that the shipment is below the octane specifications.

   Ans.   (*a*) $P(1) = \frac{1}{6} \cong 0.17$, $P(2) = \frac{2}{6} \cong 0.33$, $P(3) = \frac{3}{6} = 0.50$; (*b*) $P(1) = 0.12$, $P(2) = 0.40$, $P(3) = 0.48$

JOINT PROBABILITY TABLES

5.50   Table 5.5 is a contingency table which presents a classification of 150 sampled companies according to four industry groups and according to whether return on equity is above or below the average return in this sample of 150 firms. Prepare the joint probability table based on these sample data.

Table 5.5   Contingency Table for Return on Equity According to Industry Group

Industry category	Return on equity		Total
	Above average (*A*)	Below average (*B*)	
I	20	40	60
II	10	10	20
III	20	10	30
IV	25	15	40
Total	75	75	150

5.51   Referring to the joint probability table prepared in Problem 5.50, indicate the following probabilities: (*a*) $P(\text{I})$, (*b*) $P(\text{II})$, (*c*) $P(\text{III})$, (*d*) $P(\text{IV})$.

   Ans.   (*a*) 0.40, (*b*) 0.13, (*c*) 0.20, (*d*) 0.27

5.52   Referring to the joint probability table prepared in Problem 5.50, determine the following probabilities:
(a) $P(\text{I and } A)$, (b) $P(\text{II or } B)$, (c) $P(A)$, (d) $P(\text{I or II})$, (e) $P(\text{I and II})$, (f) $P(A \text{ or } B)$, (g) $P(A|\text{I})$,
(h) $P(\text{III}|A)$.

Ans.   (a) 0.13, (b) 0.57, (c) 0.50, (d) 0.53, (e) 0, (f) 1.0, (g) 0.33, (h) 0.27

PERMUTATIONS AND COMBINATIONS

5.53   Suppose there are eight different management trainee positions to be assigned to eight employees in a
company's junior management training program. In how many different ways can the eight individuals be
assigned to the eight different positions?

Ans.   40,320

5.54   Referring to the situation described in Problem 5.53, suppose only six different positions are available for
the eight qualified individuals. In how many different ways can six individuals from the eight be assigned
to the six different positions?

Ans.   20,160

5.55   Referring to the situation described in Problem 5.54, suppose that the six available positions can all be
considered comparable, and not really different for practical purposes. In how may ways can the six
individuals be chosen from the eight qualified people to fill the six positions?

Ans.   28

5.56   A project group of two engineers and three technicians is to be assigned from a departmental group which
includes five engineers and nine technicians. How many different project groups can be assigned from the
fourteen available personnel?

Ans.   840

5.57   For the personnel assignment situation described in Problem 5.56, suppose the five individuals are assigned
randomly from the fourteen personnel in the department, without reference to whether each person is an
engineer or a technician. What is the probability that the project group will include (a) exactly two
engineers, (b) no engineers, (c) no technicians?

Ans.   (a) $P \cong 0.42$, (b) $P \cong 0.06$, (c) $P \cong 0.0005$

Chapter 6

Probability Distributions for Discrete Random Variables:
Binomial, Hypergeometric, and Poisson

6.1 PROBABILITY DISTRIBUTIONS FOR RANDOM VARIABLES

As contrasted to an event, as discussed in Chapter 5, a *random variable* is a *numerical* event whose value is determined by a chance process. When probability values are assigned to all possible numerical values of a random variable X, either by a listing or by a mathematical function, the result is a *probability distribution*. The sum of the probabilities for all the possible numerical outcomes must equal 1.0. Individual probability values may be denoted by the symbol $f(x)$, which recognizes that a mathematical function is involved, by $P(x = X)$, which recognizes that the random variable can have various specific values, or simply by $P(X)$.

For a *discrete random variable* all possible numerical values for the variable can be listed in a table with accompanying probabilities. There are several standard probability distributions that can serve as models for a wide variety of discrete random variables involved in business applications. The standard models described in this chapter are the binomial, hypergeometric, and Poisson probability distributions.

For a *continuous random variable* all possible fractional values of the variable cannot be listed, and therefore the probabilities that are determined by a mathematical function are portrayed graphically by a probability density function, or probability curve. Several standard probability distributions that can serve as models for continuous random variables are described in Chapter 7. See Section 1.4 for an explanation of the difference between discrete and continuous variables.

EXAMPLE 1. The number of vans that have been requested for rental at a car rental agency during a 50-day period is identified in Table 6.1. The observed frequencies have been converted into probabilities for this 50-day period in the last column of the table. Thus, we can observe that the probability of exactly seven vans being requested on a randomly chosen day in this period is 0.20, and the probability of six *or more* being requested is $0.28 + 0.20 + 0.08 = 0.56$.

Table 6.1   Daily Demand for Rental of Vans during a 50-Day Period

Possible demand X	Number of days	Probability $[P(X)]$
3	3	0.06
4	7	0.14
5	12	0.24
6	14	0.28
7	10	0.20
8	4	0.08
	50	1.00

6.2 THE EXPECTED VALUE AND VARIANCE FOR A DISCRETE RANDOM VARIABLE

Just as for collections of sample and population data, it is often useful to describe a random variable in terms of its *mean* (see Section 3.2) and its *variance* (see Section 4.5). The (long-run) mean for a random variable X is called the *expected value* and is denoted by $E(X)$. For a discrete random variable, it is the weighted average of all possible numerical values of the variable with the respective probabilities used as weights. Because the sum of the weights (probabilities) is 1.0, formula (*3.3*) can be simplified, and the expected value for a discrete random variable is

$$E(X) = \sum XP(X) \tag{6.1}$$

EXAMPLE 2. Based on the data in Table 6.1, the calculation of the expected value for the random variable is presented in Table 6.2. The expected value is 5.66 vans. Note that the expected value for a discrete variable can be a fractional value, because it represents the long-run average value, not the specific value for any given observation.

Table 6.2   Expected Value Calculation for the Demand for Vans

Possible demand X	Probability $[P(X)]$	Weighted value $[XP(X)]$
3	0.06	0.18
4	0.14	0.56
5	0.24	1.20
6	0.28	1.68
7	0.20	1.40
8	0.08	0.64
	1.00	$E(X) = 5.66$

The variance of a random variable X is denoted by $V(X)$; it is computed with respect to $E(X)$ as the mean of the probability distribution. The general deviations form of the formula for the variance of a discrete random variable is

$$V(X) = \sum [X - E(X)]^2 P(X) \tag{6.2}$$

The computational form of the formula for the variance of a discrete random variable, which does not require the determination of deviations from the mean, is

$$V(X) = \sum X^2 P(X) - [\sum XP(X)]^2$$
$$= E(X^2) - [E(X)]^2 \tag{6.3}$$

EXAMPLE 3. The worksheet for the calculation of the variance for the demand for van rentals is presented in Table 6.3, using the computational version of the formula. As indicated below, the variance has a value of 1.74.

$$V(X) = E(X^2) - [E(X)]^2 = 33.78 - (5.66)^2 = 33.78 - 32.04 = 1.74$$

Table 6.3   Worksheet for the Calculation of the Variance for the Demand for Vans

Possible demand X	Probability $[P(X)]$	Weighted value $[XP(X)]$	Squared demand (X^2)	Weighted square $[X^2 P(X)]$
3	0.06	0.18	9	0.54
4	0.14	0.56	16	2.24
5	0.24	1.20	25	6.00
6	0.28	1.68	36	10.08
7	0.20	1.40	49	9.80
8	0.08	0.64	64	5.12
		$E(X) = 5.66$		$E(X^2) = 33.78$

6.3 THE BINOMIAL DISTRIBUTION

The binomial distribution is a discrete probability distribution which is applicable as a model for decision-making situations whenever a sampling process can be assumed to conform to a Bernoulli process. A *Bernoulli process* is a sampling process in which:

(1)   Only two mutually exclusive possible outcomes are possible in each trial, or observation. For convenience these are called *success* and *failure*.

(2)   The outcomes in the series of trials, or observations, constitute *independent events*.

(3)   The probability of success, denoted by p, remains constant from trial to trial. That is, the process is stationary.

The binomial distribution can be used to determine the probability of obtaining a designated number of successes in a Bernoulli process. Three values are required: the designated number of successes (X); the number of trials, or observations (n); and the probability of success in each trial (p). Where $q = (1-p)$, the formula for determining the probability of a designated number of successes X for a binomial distribution is

$$P(X|n, p) = {_nC_X} p^X q^{n-X}$$

$$= \frac{n!}{X!(n-X)!} p^X q^{n-X} \qquad (6.4)$$

EXAMPLE 4. The probability that a randomly chosen sales prospect will make a purchase is 0.20. If a salesman calls on six prospects, the probability that he will make exactly four sales is determined as follows:

$$P(X = 4|n = 6, p = 0.20) = {_6C_4}(0.20)^4(0.80)^2 = \frac{6!}{4!2!}(0.20)^4(0.80)^2$$

$$= \frac{6 \times 5 \times 4 \times 3 \times 2}{(4 \times 3 \times 2)(2)}(0.0016)(0.64) = 0.01536 \cong 0.015$$

Often there is an interest in the cumulative probability of "X or more" successes or "X or fewer" successes occurring in n trials. In such a case, the probability of each outcome included within the designated interval must be determined, and then these probabilities are summed.

EXAMPLE 5. For Example 4, the probability that the salesman will make four or more sales is determined as follows:

$$P(X \geq 4|n = 6, p = 0.20) = P(X = 4) + P(X = 5) + P(X = 6)$$

$$= 0.01536 + 0.001536 + 0.000064 = 0.016960 \cong 0.017$$

where $P(X = 4) = 0.01536$ (from Example 4)

$$P(X = 5) = {_6C_5}(0.20)^5(0.80)^1 = \frac{6!}{5!1!}(0.20)^5(0.80) = 6(0.00032)(0.80) = 0.001536$$

$$P(X = 6) = {_6C_6}(0.20)^6(0.80)^0 = \frac{6!}{6!0!}(0.000064)(1) = (1)(0.000064) = 0.000064$$

(*Note:* Recall that any value raised to the zero power is equal to 1.)

Because use of the binomial formula involves considerable arithmetic when the sample is relatively large, tables of binomial probabilities are often used. See Appendix 2.

EXAMPLE 6. If the probability that a randomly chosen sales prospect will make a purchase is 0.20, the probability that a salesman who calls on 15 prospects will make fewer than three sales is

$$P(X < 3 | n = 15, p = 0.20) = P(X \leq 2) = P(X = 0) + P(X = 1) + P(X = 2)$$
$$= 0.0352 + 0.1319 + 0.2309 \text{ (from Appendix 2)}$$
$$= 0.3980 \cong 0.40$$

The values of p referenced in Appendix 2 do not exceed $p = 0.50$. If the value of p in a particular application exceeds 0.50, the problem has to be restated so that the event is defined in terms of the number of "failures" rather than the number of successes (see Problem 6.9).

The expected value (mean) and variance for a given binomial distribution could be determined by listing the probability distribution in a table and applying the formulas presented in Section 6.2. However, the expected number of successes can be computed directly:

$$E(X) = np \qquad (6.5)$$

Where $q = (1 - p)$, the variance of the number of successes can also be computed directly:

$$V(X) = npq \qquad (6.6)$$

EXAMPLE 7.   For Example 6, the expected number of sales (as a long-run average) and the variance associated with making calls on 15 prospects are

$$E(X) = np = 15(0.20) = 3.0 \text{ sales}$$
$$V(X) = np(q) = 15(0.20)(0.80) = 2.4$$

6.4   THE BINOMIAL DISTRIBUTION EXPRESSED BY PROPORTIONS

Instead of expressing the random binomial variable as the number of successes X, we can designate it in terms of the *proportion* of successes $\hat{p}$, which is the ratio of the number of successes to the number of trials:

$$\hat{p} = \frac{X}{n} \qquad (6.7)$$

In such cases, formula (6.4) is modified only with respect to defining the proportion. Thus, the probability of observing exactly $\hat{p}$ proportion of successes in n Bernoulli trials is

$$P\left(\hat{p} = \frac{X}{n} \Big| n, p\right) = {}_nC_X p^X q^{n-X} \qquad (6.8)$$

or

$$P\left(\hat{p} = \frac{X}{n} \Big| n, \pi\right) = {}_nC_X \pi^X (1 - \pi)^{n-X} \qquad (6.9)$$

In formula (6.9), π (Greek "pī") is the equivalent of p except that it specifically indicates that the probability of success in an individual trial is a population parameter.

EXAMPLE 8.   The probability that a randomly selected salaried employee is a participant in a company-sponsored stock investment program is 0.40. If five salaried employees are chosen randomly, the probability that the proportion of participants is exactly 0.60, or 3/5 of the five sampled employees, is

$$P(\hat{p} = 0.60) = P\left(\hat{p} = \frac{3}{5} \Big| n = 5, p = 0.40\right) = {}_5C_3(0.40)^3(0.60)^2 = \frac{5!}{3!2!}(0.064)(0.36) = 0.2304 \cong 0.23$$

When the binomial variable is expressed as a proportion, the distribution is still discrete and not continuous. Only the proportions for which the number of successes X is a whole number can occur. For instance, in Example 8 it is not possible for there to be a proportion of 0.50 participant out of a sample of five. The use of the binomial table with respect to proportions simply requires converting the designated proportion $\hat{p}$ to number of successes X for the given sample size n.

EXAMPLE 9.   The probability that a randomly selected employee is a participant in a company-sponsored stock investment program is 0.40. If 10 employees are chosen randomly, the probability that the proportion of participants is at least 0.70 is

$$P(\hat{p} \geq 0.70) = P(X \geq 7|n = 10, p = 0.10) = P(X = 7) + P(X = 8) + P(X = 9) + P(X = 10)$$

$$= 0.0425 + 0.0106 + 0.0016 + 0.0001 = 0.0548$$

The expected value for a binomial probability distribution expressed by proportions is equal to the population proportion, which may be designated by either p or π:

$$E(\hat{p}) = p \tag{6.10}$$

or

$$E(\hat{p}) = \pi \tag{6.11}$$

The variance of the proportion of successes for a binomial probability distribution, where $q = (1 - p)$, is

$$V(\hat{p}) = \frac{pq}{n} \tag{6.12}$$

or

$$V(\hat{p}) = \frac{\pi(1 - \pi)}{n} \tag{6.13}$$

6.5   THE HYPERGEOMETRIC DISTRIBUTION

When sampling is done *without replacement* of each sampled item taken from a finite population of items, the Bernoulli process does not apply because there is a systematic change in the probability of success as items are removed from the population. When sampling without replacement is used in a situation which would otherwise qualify as a Bernoulli process, the hypergeometric distribution is the appropriate discrete probability distribution.

Given that X is the designated number of successes, N is the total number of items in the population, T is the total number of "successes" included in the population, and n is the number of items in the sample, the formula for determining hypergeometric probabilities is

$$P(X|N, Tn) = \frac{\binom{N-T}{n-X}\binom{T}{X}}{\binom{N}{n}} \tag{6.14}$$

EXAMPLE 10.   Of six employees, three have been with the company five or more years. If four employees are chosen randomly from the group of six, the probability that exactly two will have five or more years seniority is

$$P(X = 2|N = 6, T = 3, n = 4) = \frac{\binom{6-3}{4-2}\binom{3}{2}}{\binom{6}{4}} = \frac{\binom{3}{2}\binom{3}{2}}{\binom{6}{4}} = \frac{\dfrac{3!}{2!1!}\dfrac{3!}{2!1!}}{\dfrac{6!}{4!2!}} = \frac{(3)(3)}{15} = 0.60$$

Note that in Example 10, the required probability value is computed by determining the number of different combinations which would include two high-seniority and two low-seniority employees as a ratio of the total number of combinations of four employees taken from the six. Thus, the hypergeometric formula is a direct application of the rules of combinatorial analysis described in Section 5.10.

When the population is large and the sample is relatively small, the fact that sampling is done without replacement has little effect on the probability of success in each trial. A convenient rule of thumb is that a binomial distribution can be used as an approximation of a hypergeometric probability value when $n < 0.05N$. That is, the sample size should be less than 5 percent of the population size. Different texts may use somewhat different rules for determining when such approximation is appropriate.

6.6 THE POISSON DISTRIBUTION

The *Poisson distribution* can be used to determine the probability of a designated number of events occurring when the events occur in a continuum of time or space. Such a process is called a *Poisson process*; it is similar to the Bernoulli process (see Section 6.3) except that the events occur over a continuum (e.g., during a time interval) rather than occurring on fixed trials or observations. An example of such a process is the arrival of incoming calls at a telephone switchboard. As was the case for the Bernoulli process, it is assumed that the events are independent and that the process is stationary.

Only one value is required to determine the probability of a designated number of events occurring in a Poisson process: the long-run mean number of events for the specific time or space dimension of interest. This mean generally is represented by λ (Greek "lambda"), or possibly by μ. The formula for determining the probability of a designated number of successes X in a Poisson distribution is

$$P(X|\lambda) = \frac{\lambda^X e^{-\lambda}}{X!} \tag{6.15}$$

Here e is the constant 2.7183 that is the base of natural logarithms, and the values of $e^{-\lambda}$ may be obtained from Appendix 3.

EXAMPLE 11. An average of five calls for service per hour are received by a machine repair department. The probability that exactly three calls for service will be received in a randomly selected hour is

$$P(X=3|\lambda=5.0) = \frac{(5)^3 e^{-5}}{3!} = \frac{(125)(0.00674)}{6} = 0.1404$$

Alternatively, a table of Poisson probabilities may be used. Appendix 4 identifies the probability of each designated number of successes for various values of λ.

EXAMPLE 12. We can determine the answer to Example 11 by use of Appendix 4 for Poisson probabilities as follows:

$$P(X=3|\lambda=5.0) = 0.1404$$

When there is an interest in the probability of "X or more" or "X or fewer" successes, the rule of addition for mutually exclusive events is applied.

EXAMPLE 13. If an average of five service calls per hour are received at a machine repair department, the probability that *fewer than* three calls will be received during a randomly chosen hour is determined as follows:

$$P(X<3|\lambda=5.0) = P(X\leq 2) = P(X=0) + P(X=1) + P(X=2)$$

$$= 0.0067 + 0.0337 + 0.0842 = 0.1246$$

where $P(X=0|\lambda=5.0) = 0.0067$ (from Appendix 4)

$P(X=1|\lambda=5.0) = 0.0337$

$P(X=2|\lambda=5.0) = 0.0842$

Because a Poisson process is assumed to be stationary, it follows that the mean of the process is always proportional to the length of the time or space continuum. Therefore, if the mean is available for one length of time, the mean for any other required time period can be determined.

This is important, because the value of λ which is used must apply to the time period of interest.

EXAMPLE 14. On the average, 12 people per hour ask questions of a decorating consultant in a fabric store. The probability that three or more people will approach the consultant with questions during a 10-min period (1/6 of an hour) is determined as follows:

$$\text{Average per hour} = 12$$

$$\lambda = \text{average per 10 min} = \frac{12}{6} = 2.0$$

$$P(X \geq 3 | \lambda = 2.0) = P(X = 3 | \lambda = 2.0) + P(X = 4 | \lambda = 2.0) + P(X = 5 | \lambda = 2.0) + \cdots$$

$$= 0.1804 + 0.0902 + 0.0361 + 0.0120 + 0.0034 + 0.0009 + 0.0002 = 0.3232$$

where $P(X = 3 | \lambda = 2.0) = 0.1804$ (from Appendix 4)

$P(X = 4 | \lambda = 2.0) = 0.0902$

$P(X = 5 | \lambda = 2.0) = 0.0361$

$P(X = 6 | \lambda = 2.0) = 0.0120$

$P(X = 7 | \lambda = 2.0) = 0.0034$

$P(X = 8 | \lambda = 2.0) = 0.0009$

$P(X = 9 | \lambda = 2.0) = 0.0002$

By definition, the expected value (long-run mean) for a Poisson probability distribution is equal to the mean of the distribution:

$$E(X) = \lambda \qquad (6.16)$$

As it happens, the variance of the number of events for a Poisson probability distribution is also equal to the mean of the distribution:

$$V(X) = \lambda \qquad (6.17)$$

6.7 POISSON APPROXIMATION OF BINOMIAL PROBABILITIES

When the number of observations or trials n in a Bernoulli process is large, computations are quite tedious. Further, tabled probabilities for very small values of p are not generally available. Fortunately, the Poisson distribution is suitable as an approximation of binomial probabilities when n is large and p or q is small. A convenient rule is that such approximation can be made when $n \geq 30$, and either $np < 5$ or $nq < 5$. Different texts may use somewhat different rules for determining when such approximation is appropriate.

The mean of the Poisson probability distribution which is used to approximate binomial probabilities is

$$\lambda = np \qquad (6.18)$$

EXAMPLE 15. For a large shipment of transistors from a supplier, 1 percent of the items is known to be defective. If a sample of 30 transistors is randomly selected, the probability that two or more transistors will be defective can be determined by use of the binomial probabilities in Appendix 2:

$$P(X \geq 2 | n = 30, p = 0.01) = P(X = 2) + P(X = 3) + \cdots = 0.0328 + 0.0031 + 0.0002 = 0.0361$$

Where $\lambda = np = 30(0.01) = 0.3$, Poisson approximation of the above probability value is

$$P(X \geq 2 | \lambda = 0.3) = P(X = 2) + P(X = 3) + \cdots = 0.0333 + 0.0033 + 0.0002 = 0.0368$$

Thus, the difference between the Poisson approximation and the actual binomial probability value is just 0.0007.

When n is large but neither np nor nq is less than 5.0, binomial probabilities can be approximated by use of the normal probability distribution (see Section 7.4).

Overall, the availability of computers has made the approximation of probabilities for one model based on another probability model less important.

6.8 COMPUTER APPLICATIONS

Computer software for statistical analysis frequently includes the capability of providing probability tables for the standard discrete probability distributions that are used as models for decision-making situations. Such availability is particularly useful when the particular probabilities are not available in standard tables. Problem 6.21 is concerned with obtaining a table of selected binomial probabilities by the use of a computer.

Solved Problems

DISCRETE RANDOM VARIABLES

6.1   The number of trucks arriving hourly at a warehouse facility has been found to follow the probability distribution in Table 6.4. Calculate (*a*) the expected number of arrivals X per hour and (*b*) the variance of this probability distribution for the discrete random variable.

Table 6.4   Hourly Arrival of Trucks at a Warehouse

Number of trucks (X)	0	1	2	3	4	5	6
Probability [$P(X)$]	0.05	0.10	0.15	0.25	0.30	0.10	0.05

From Table 6.5,

(*a*) $E(X) = 3.15$

(*b*) $V(X) = E(X^2) - [E(X)]^2 = 12.05 - (3.15)^2 = 12.05 - 9.9225 = 2.1275 \cong 2.13$

Table 6.5   Worksheet for the Calculation of the Expected Value and the Variance for Truck Arrivals

Number of trucks (X)	Probability [$P(X)$]	Weighted value [$XP(X)$]	Squared number (X^2)	Weighted square [$X^2P(X)$]
0	0.05	0	0	0
1	0.10	0.10	1	0.10
2	0.15	0.30	4	0.60
3	0.25	0.75	9	2.25
4	0.30	1.20	16	4.80
5	0.10	0.50	25	2.50
6	0.05	0.30	36	1.80
		$E(X) = 3.15$		$E(X^2) = 12.05$

6.2   Table 6.6 identifies the probability that a computer system will be "down" the indicated number of periods per week during the initial installation phase for the system. Calculate (*a*) the expected number of times per week that the computer is inoperative and (b) the variance of this probability distribution.

Table 6.6   Number of Inoperative Periods per Week for a New Computer System

Number of periods (X)	4	5	6	7	8	9
Probability $[P(X)]$	0.01	0.08	0.29	0.42	0.14	0.06

Using Table 6.7,

(a) $E(X) = 6.78$

(b) $V(X) = E(X^2) - [E(X)]^2 = 47.00 - (6.78)^2 = 47.00 - 45.9684 \cong 1.03$

Table 6.7   Worksheet for the Calculation of the Expected Value and the Variance of Computer Malfunction

Number of periods (X)	Probability $[P(X)]$	Weighted value $[XP(X)]$	Squared number (X^2)	Weighted square $[X^2P(X)]$
4	0.01	0.04	16	0.16
5	0.08	0.40	25	2.00
6	0.29	1.74	36	10.44
7	0.42	2.94	49	20.58
8	0.14	1.12	64	8.96
9	0.06	0.54	81	4.86
	1.00	$E(X) = 6.78$		$E(X^2) = 47.00$

6.3   Table 6.8 lists the possible outcomes associated with the toss of 2 six-sided dice and the probability associated with each outcome. These probabilities were determined by use of the rules of addition and multiplication discussed in Sections 5.4 and 5.6. For example, a "3" can be obtained by a combination of a "1" and "2," or a combination of a "2" and "1." Each sequence has a probability of occurrence of $(1/6) \times (1/6) = 1/36$, and since the two sequences are mutually exclusive, $P(X = 3) = 1/36 + 1/36 = 2/36$. Determine (a) the expected number on the throw of two dice and (b) the standard deviation of this distribution.

Table 6.8   Possible Outcomes on the Toss of Two Dice

Number on two dice (X)	2	3	4	5	6	7	8	9	10	11	12
Probability $[P(X)]$	$\frac{1}{36}$	$\frac{2}{36}$	$\frac{3}{36}$	$\frac{4}{36}$	$\frac{5}{36}$	$\frac{6}{36}$	$\frac{5}{36}$	$\frac{4}{36}$	$\frac{3}{36}$	$\frac{2}{36}$	$\frac{1}{36}$

From Table 6.9,

(a) $E(X) = 7$

(b) $V(X) = E(X^2) - [E(X)]^2 = 54.83 - (7)^2 = 54.83 - 49 = 5.83$

$\sigma = \sqrt{V(X)} = \sqrt{5.83} \cong 2.41$

Table 6.9   Worksheet for the Calculation of the Expected Value and the Variance Associated with the Toss of Two Dice

Number (X)	Probability $[P(X)]$	Weighted value $[XP(X)]$	Squared number (X^2)	Weighted square $[X^2 P(X)]$
2	1/36	2/36	4	4/36
3	2/36	6/36	9	18/36
4	3/36	12/36	16	48/36
5	4/36	20/36	25	100/36
6	5/36	30/36	36	180/36
7	6/36	42/36	49	294/36
8	5/36	40/36	64	320/36
9	4/36	36/36	81	324/36
10	3/36	30/36	100	300/36
11	2/36	22/36	121	242/36
12	1/36	12/36	144	144/36
	36/36	$E(X) = 252/36 = 7.0$		$E(X^2) = 1,974/36 \cong 54.83$

THE BINOMIAL DISTRIBUTION

6.4   Because of high interest rates, a firm reports that 30 percent of its accounts receivable from other business firms are overdue. If an accountant takes a random sample of five such accounts, determine the probability of each of the following events by use of the formula for binomial probabilities: (a) none of the accounts is overdue, (b) exactly two accounts are overdue, (c) most of the accounts are overdue, (d) exactly 20 percent of the accounts are overdue.

(a) $P(X = 0 | n = 5, p = 0.30) = {_5}C_0 (0.30)^0 (0.70)^5 = \dfrac{5!}{0!5!} (0.30)^0 (0.70)^5 = (1)(1)(0.16807) = 0.16807$

(b) $P(X = 2 | n = 5, p = 0.30) = {_5}C_2 (0.30)^2 (0.70)^3 = \dfrac{5!}{2!3!} (0.30)^2 (0.70)^3 = (10)(0.09)(0.343) = 0.3087$

(c) $P(X \geq 3 | n = 5, p = 0.30) = P(X = 3) + P(X = 4) + P(X = 5) = 0.1323 + 0.02835 + 0.00243 = 0.16308$

where $P(X = 3) = \dfrac{5!}{3!2!} (0.30)^3 (0.70)^2 = (10)(0.027)(0.49) = 0.1323$

$P(X = 4) = \dfrac{5!}{4!1!} (0.30)^4 (0.70)^1 = (5)(0.0081)(0.70) = 0.02835$

$P(X = 5) = \dfrac{5!}{5!0!} (0.30)^5 (0.70)^0 = (1)(0.00243)(1) = 0.00243$

(d) $P\left(\dfrac{X}{n} = 0.20 | n = 5, p = 0.30\right) = P(X = 1 | n = 5, p = 0.30) = {_5}C_1 (0.30)^1 (0.70)^4 = \dfrac{5!}{1!4!} (0.30)^1 (0.70)^4$

$= (5)(0.30)(0.2401) = 0.36015$

6.5   A mail-order firm has a circular which elicits a 10 percent response rate. Suppose 20 of the circulars are mailed as a market test in a new geographic area. Assuming that the 10 percent response rate is applicable in the new area, determine the probabilities of the following events by use of Appendix 2: (a) no one responds, (b) exactly two people respond, (c) a majority of the people respond, (d) less than 20 percent of the people respond.

(a) $P(X = 0 | n = 20, p = 0.10) = 0.1216$

(b) $P(X = 2 | n = 20, p = 0.10) = 0.2852$

(c) $P(X = 11 | n = 20, p = 0.10) = P(X = 11) + P(X = 12) + \cdots = 0.0000 \cong 0$

(d) $P\left(\dfrac{X}{n} < 0.20 | n = 20, p = 0.10\right) = P(X \leq 3 | n = 20, p = 0.10)$

$$= P(X = 0) + P(X = 1) + P(X = 2) + P(X = 3)$$

$$= 0.1216 + 0.2702 + 0.2852 + 0.1901 = 0.8671$$

6.6   The binomial formula can be viewed as being comprised of two parts: a combinations formula to determine the number of different ways in which the designated event can occur and the rule of multiplication to determine the probability of each sequence. Suppose that three items are selected randomly from a process known to produce 10 percent defectives. Construct a three-step tree diagram portraying the selection of the three items and using D for a defective item being selected and D' for a nondefective item being selected. Also, enter the appropriate probability values in the diagram and use the multiplication rule for independent events to determine the probability of each possible sequence of three events occurring.

See Fig. 6-1.

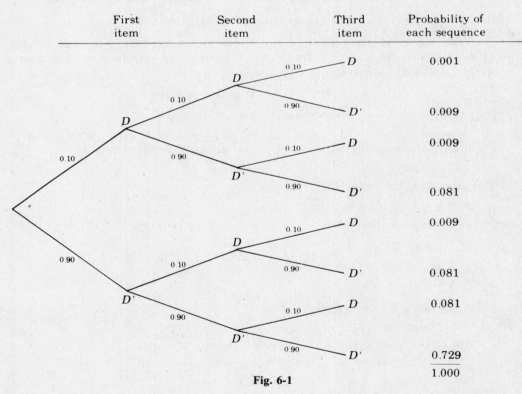

Fig. 6-1

6.7   From Problem 6.6, determine the probability that exactly one of the three sampled items is defective, referring to Fig. 6-1 and using the addition rule for mutually exclusive events.

Beginning from the top of the tree diagram, the fourth, sixth, and seventh sequences include exactly one defective item. Thus,

$$P(X = 1) = (D \text{ and } D' \text{ and } D') + (D' \text{ and } D \text{ and } D') + (D' \text{ and } D' \text{ and } D)$$

$$= 0.081 + 0.081 + 0.081 = 0.243$$

6.8   From Problems 6.6 and 6.7, determine the probability of obtaining exactly one defective item by use of the binomial formula, and note the correspondence between the values in the formula and the values obtained from the tree diagram.

Using formula (6.4),

$$P(X = 1 | n = 3, p = 0.10) = {}_3C_1(0.10)^1(0.90)^2 = \frac{3!}{1!2!}(0.10)(0.81)$$

$$= 3(0.081) = 0.243$$

Thus, the first part of the binomial formula indicates the number of different groups of positions which can include the designated number of successes (in this case there are three ways in which one defective item can be included in the group of three items). The second part of the formula represents the rule of multiplication for the specified independent events.

6.9   During a particular year, 70 percent of the common stocks listed on the New York Stock Exchange increased in market value while 30 percent were unchanged or declined in market value. At the beginning of the year a stock advisory service chose 10 stock issues as being "specially recommended." If the 10 issues represent a random selection, what is the probability that (a) all 10 issues and (b) at least eight issues increase in market value?

(a) $P(X = 10 | n = 10, p = 0.70) = P(X' = 0 | n = 10, q = 0.30) = 0.0282$

(*Note:* When p is greater than 0.50, the problem has to be restated in terms of X' (read "not X") and it follows that $X' = n - X$. Thus, "all 10 increase" is the same event as "none decrease.")

(b) $P(X \geq 8 | n = 10, p = 0.70) = P(X' \leq 2 | n = 10, q = 0.30)$

$$= P(X' = 0) + P(X' = 1) + P(X' = 2)$$

$$= 0.0282 + 0.1211 + 0.2335 = 0.3828$$

(*Note:* When a probability statement is restated in terms of X' instead of X and an inequality is involved, the inequality symbol in the original statement simply is reversed.)

6.10   Using Appendix 2, determine:

(a) $P(X = 5 | n = 9, p = 0.50)$

(b) $P(X = 7 | n = 15, p = 0.60)$

(c) $P(X \leq 3 | n = 20, p = 0.05)$

(d) $P(X \geq 18 | n = 20, p = 0.90)$

(e) $P(X > 8 | n = 10, p = 0.70)$

(a) $P(X = 5 | n = 9, p = 0.50) = 0.2461$

(b) $P(X = 7 | n = 15, p = 0.60) = P(X' = 8 | n = 15, q = 0.40) = 0.1181$

(c) $P(X \leq 3 | n = 20, p = 0.05) = P(X = 0) + P(X = 1) + P(X = 2) + P(X = 3)$

$$= 0.3585 + 0.3774 + 0.1887 + 0.0596 = 0.9842$$

(d) $P(X \geq 18 | n = 20, p = 0.90) = P(X' \leq 2 | n = 20, q = 0.10)$

$$= P(X' = 0) + P(X' = 1) + P(X' = 2)$$

$$= 0.1216 + 0.2702 + 0.2852 = 0.6770$$

(e) $P(X > 8 | n = 10, p = 0.70) = P(X' < 2 | n = 10, q = 0.30)$

$$= P(X' = 0) + P(X' = 1) = 0.0282 + 0.1211 = 0.1493$$

6.11   If a fair coin is tossed five times, the probability distribution with respect to the number of heads observed is based on the binomial distribution with $n = 5$ and $p = 0.50$ (see Table 6.10). Determine

(a) the expected number of heads and (b) the variance of the probability distribution by use of the *general* formulas for discrete random variables.

Table 6.10   Binomial Probability Distribution of the Number of Heads Occurring in Five Tosses of a Fair Coin

Number of heads (X)	0	1	2	3	4	5
Probability [$P(X)$]	0.0312	0.1562	0.3125	0.3125	0.1562	0.0312

Using Table 6.11,

(a) $E(X) = 2.4995 \cong 2.5$

(b) $V(X) = E(X^2) - [E(X)]^2 = 7.4979 - (2.4995)^2 = 7.4979 - 6.2475 = 1.2504 \cong 1.25$

Table 6.11   Worksheet for the Calculation of the Expected Value and the Variance for Problem 6.11

Number of heads (X)	Probability [$P(X)$]	Weighted value [$XP(X)$]	Squared number (X^2)	Weighted square [$X^2P(X)$]
0	0.0312	0	0	0
1	0.1562	0.1562	1	0.1562
2	0.3125	0.6250	4	1.2500
3	0.3125	0.9375	9	2.8125
4	0.1562	0.6248	16	2.4992
5	0.0312	0.1560	25	0.7800
		$E(X) = \overline{2.4995}$		$E(X^2) = \overline{7.4979}$

6.12   Referring to Problem 6.11, determine (a) the expected number of heads and (b) the variance of the probability distribution by use of the *special* formulas applicable for binomial probability distributions. (c) Compare your answers to those in Problem 6.11.

(a) $E(X) = np = 5(0.50) = 2.5$

(b) $V(X) = npq = (5)(0.50)(0.50) = 1.25$

(c)   The answers obtained with the special formulas which are applicable for binomial distributions correspond with the answers obtained by the lengthier general formulas applicable for any discrete random variable.

THE HYPERGEOMETRIC DISTRIBUTION

6.13   A manager randomly selects $n = 3$ individuals from a group of 10 employees in his department for assignment to a wage classification study. Assuming that four of the employees were assigned to a similar project previously, construct a three-step tree diagram portraying the selection of the three individuals in terms of whether each individual chosen has had experience E or has no previous experience E' in such a study. Further, enter the appropriate probability values in the diagram and use the multiplication rule for dependent events to determine the probability of each possible sequence of three events occurring.

See Fig. 6-2.

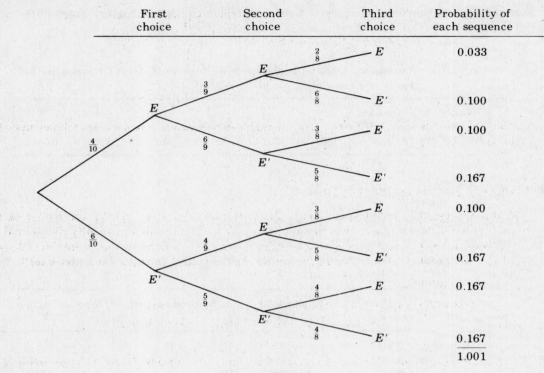

Fig. 6-2

6.14   Referring to Problem 6.13, determine the probability that exactly two of the three employees selected have had previous experience in a wage classification study by reference to Fig. 6-2 and use of the addition rule for mutually exclusive events.

Beginning from the top of the tree diagram, the second, third, and fifth sequences include exactly two employees with experience. Thus, by the rule of addition for these mutually exclusive sequences:

$$P(X = 2) = (E \text{ and } E \text{ and } E') + (E \text{ and } E' \text{ and } E) + (E' \text{ and } E \text{ and } E)$$
$$= 0.100 + 0.100 + 0.100 = 0.30$$

6.15   Referring to Problem 6.13, determine the probability that exactly two of the three employees have had the previous experience, using the formula for determining hypergeometric probabilities.

From formula (*6.14*),

$$P(X|N, T, n) = \frac{\binom{N-T}{n-X}\binom{T}{X}}{\binom{N}{n}}$$

$$P(X = 2|N = 10, T = 4, n = 3) = \frac{\binom{10-4}{3-2}\binom{4}{2}}{\binom{10}{3}} = \frac{\binom{6}{1}\binom{4}{2}}{\binom{10}{3}} = \frac{\left(\frac{6!}{1!5!}\right)\left(\frac{4!}{2!2!}\right)}{\left(\frac{10!}{3!7!}\right)} = \frac{(6)(6)}{120} = 0.30$$

6.16   Section 6.5 states that the hypergeometric formula is a direct application of the rules of combinatorial analysis described in Section 5.9. To demonstrate this, apply the hypergeometric formula to Problem 5.26(*a*).

$$P(X = 4 | N = 10, \, T = 6, \, n = 6) = \frac{\binom{10-6}{6-4}\binom{6}{4}}{\binom{10}{6}} = \frac{\frac{4!}{2!2!} \frac{6!}{4!2!}}{\frac{10!}{6!4!}}$$

$$= \frac{(6)(15)}{210} = \frac{90}{210} = \frac{3}{7} \cong 0.43$$

[*Note:* This result is equivalent to the use of the combinatorial analysis formula in the solution to Problem 5.26(*a*).]

THE POISSON PROBABILITY DISTRIBUTION

6.17   On the average, five people per hour conduct transactions at a "special service" desk in a commercial bank. Assuming that the arrival of such people is independently distributed and equally likely throughout the period of concern, what is the probability that more than 10 people will wish to conduct transactions at the special services desk during a particular hour?

Using Appendix 4,

$$P(X > 10 | \lambda = 5.0) = P(X \geq 11 | \lambda = 5.0) = P(X = 11) + P(X = 12) + \cdots$$
$$= 0.0082 + 0.0034 + 0.0013 + 0.0005 + 0.0002 = 0.0136$$

6.18   On the average, a ship arrives at a certain dock every second day. What is the probability that two or more ships will arrive on a randomly selected day?

Since the average per two days = 1.0, then λ = average per day = $1.0 \times (\frac{1}{2})$ = 0.5. Substituting from Appendix 4,

$$P(X \geq 2 | \lambda = 0.5) = P(X = 2) + P(X = 3) + \cdots$$
$$= 0.0758 + 0.0126 + 0.0016 + 0.0002 = 0.0902$$

6.19   Each 500-ft roll of sheet steel includes two flaws, on the average. A flaw is a scratch or mar which would affect the use of that segment of sheet steel in the finished product. What is the probability that a particular 100-ft segment will include no flaws?

If the average per 500-ft roll = 2.0, then λ = average per 100-ft roll = $2.0 \times (\frac{100}{500})$ = 0.40. Thus, from Appendix 4,

$$P(X = 0 | \lambda = 0.40) = 0.6703$$

6.20   An insurance company is considering the addition of coverage for a relatively rare ailment in the major-medical insurance field. The probability that a randomly selected individual will have the ailment is 0.001, and 3,000 individuals are included in the group which is insured.

(*a*)   What is the expected number of people who will have the ailment in the group?

(*b*)   What is the probability that no one in this group of 3,000 people will have this ailment?

(*a*)   The distribution of the number of people who will have the ailment would follow the binomial probability distribution with $n = 3,000$ and $p = 0.001$.

$$E(X) = np = (3,000)(0.001) = 3.0 \text{ people}$$

(*b*)   Tabled binomial probabilities are not available for $n = 3,000$ and $p = 0.001$. Also, algebraic solution of the binomial formula is not appealing, because of the large numbers which are involved. However, we can use the Poisson distribution to approximate the binomial probability, because $n \geq 30$ and

$np < 5$. Therefore:

$$\lambda = np = (3,000)(0.001) = 3.0$$

$$P_{\text{Binomial}}(X = 0 | n = 3,000, p = 0.001) \cong P_{\text{Poisson}}(X = 0 | \lambda = 3.0) = 0.0498$$

(from Appendix 4).

COMPUTER APPLICATIONS

6.21 The probability that a salesperson will make a sale with a prescreened prospect based on a particular method of product presentation is 0.33.

(a) Using available computer software, obtain the table of binomial probabilities for the number of sales given that $n = 10$ calls are made.

(b) With reference to the table, determine the probability of exactly five sales being made.

(c) With reference to the table, determine the probability that five or more sales are made.

(a) Figure 6-3 for the computer input and output includes the required table of binomial probabilities.

(b) Note that the probabilities reported in Fig. 6-3 are *cumulative*. That is, each probability value is the probability of the designated number of sales *or fewer*. Therefore the probability of exactly five sales occurring is

$$P(X = 5) = P(X \le 5) - P(X \le 4)$$
$$= 0.9268 - 0.7936 = 0.1332$$

(c) $P(X \ge 5) = 1.00 - P(X \le 4)$
$$= 1.00 - 0.7936 = 0.2064$$

```
MTB ) CDF;
SUBC) BINOMIAL FOR N = 10, P = 0.33.

BINOMIAL WITH N =   10   P = 0.330000
    K   P( X LESS OR = K)
    0          0.0182
    1          0.1080
    2          0.3070
    3          0.5684
    4          0.7936
    5          0.9268
    6          0.9815
    7          0.9968
    8          0.9997
    9          1.0000
```

Fig. 6-3 Minitab output.

Supplementary Problems

DISCRETE RANDOM VARIABLES

6.22 The arrival of customers during randomly chosen 10-min intervals at a drive-in facility specializing in photo development and film sales has been found to follow the probability distribution in Table 6.12. Calculate the expected number of arrivals for 10-min intervals and the variance of the arrivals.

Table 6.12   Arrivals of Customers at a Photo-processing Facility during
10-min Intervals

Number of arrivals (X)	0	1	2	3	4	5
Probability [$P(X)$]	0.15	0.25	0.25	0.20	0.10	0.05

Ans. $E(X) = 2.00$, $V(X) = 1.90$

6.23   The newsstand sales of a monthly magazine have been found to follow the probability distribution in Table 6.13. Calculate the expected value and the variance for the magazine sales, in thousands.

Table 6.13   Newsstand Sales of a Monthly Magazine

Number of magazines (X), thousands	15	16	17	18	19	20
Probability [$P(X)$]	0.05	0.10	0.25	0.30	0.20	0.10

Ans. $E(X) = 17.80$, $V(X) = 1.66$

6.24   A salesman has found that the probability of his making various numbers of sales per day, given that he calls on 10 sales prospects, is presented in Table 6.14. Calculate the expected number of sales per day and the variance of the number of sales.

Table 6.14   Sales per Day when 10 Prospects Are Contacted

Number of sales (X)	1	2	3	4	5	6	7	8
Probability [$P(X)$]	0.04	0.15	0.20	0.25	0.19	0.10	0.05	0.02

Ans. $E(X) = 4.00$, $V(X) = 2.52$

6.25   Referring to Problem 6.24, suppose the salesman earns a commission of $25 per sale. Determine his expected daily commission earnings by (*a*) substituting the commission amount for each of the sales numbers in Table 6.14 and calculating the expected commission amount, and (*b*) multiplying the expected sales number calculated in Problem 6.24 by the commission rate.

Ans.   (*a*) $100.00, (*b*) $100.00

THE BINOMIAL DISTRIBUTION

6.26   There is a 90 percent chance that a particular type of component will perform adequately under high temperature conditions. If the device involved has four such components, determine the probability of each of the following events by use of the formula for binomial probabilities.

(*a*)   All of the components perform adequately and therefore the device is operative.

(*b*)   The device is inoperative because exactly one of the four components fails.

(*c*)   The device is inoperative because one *or more* of the components fail.

Ans.   (*a*) 0.6561, (*b*) 0.2916, (*c*) 0.3439

6.27   Verify the answers to Problem 6.26 by constructing a tree diagram and determining the probabilities by the use of the appropriate rules of multiplication and of addition.

6.28   Verify the answers to Problem 6.26 by use of Appendix 2.

6.29   Using the table of binomial probabilities, determine:

    (a) $P(X = 8 | n = 20, p = 0.30)$ (d) $P(X = 5 | n = 10, p = 0.40)$

    (b) $P(X \geq 10 | n = 20, p = 0.30)$ (e) $P(X > 5 | n = 10, p = 0.40)$

    (c) $P(X \leq 5 | n = 20, p = 0.30)$ (f) $P(X < 5 | n = 10, p = 0.40)$

    Ans. (a) 0.1144, (b) 0.0479, (c) 0.4165, (d) 0.2007, (e) 0.1663, (f) 0.6330

6.30   Using the table of binomial probabilities, determine:

    (a) $P(X = 4 | n = 12, p = 0.70)$ (d) $P(X < 3 | n = 8, p = 0.60)$

    (b) $P(X \geq 9 | n = 12, p = 0.70)$ (e) $P(X = 5 | n = 10, p = 0.90)$

    (c) $P(X \leq 3 | n = 8, p = 0.60)$ (f) $P(X > 7 | n = 10, p = 0.90)$

    Ans. (a) 0.0078, (b) 0.4925, (c) 0.1738, (d) 0.0499, (e) 0.0015, (f) 0.9298

6.31   Suppose that 40 percent of the hourly employees in a large firm are in favor of union representation, and a random sample of 10 employees are contacted and asked for an anonymous response. What is the probability that (a) a majority of the respondents, (b) fewer than half of the respondents will be in favor of union representation?

    Ans. (a) 0.1663, (b) 0.6330

6.32   Determine the probabilities in Problem 6.31, if 60 percent in the firm are in favor of union representation.

    Ans. (a) 0.6330, (b) 0.1663

6.33   Refer to the probability distribution in Problem 6.24. Does this probability distribution appear to follow a binomial probability distribution? [*Hint:* Convert the $E(X)$ found in Problem 6.24 into a proportion and use this value as the value of p for comparison with the binomial distribution with $n = 10$.]

    Ans. The two probability distributions correspond quite closely.

THE HYPERGEOMETRIC DISTRIBUTION

6.34   In a class containing 20 students, 15 are dissatisfied with the text used. If a random sample of four students are asked about the text, determine the probability that (a) exactly three, and (b) at least three are dissatisfied with the text.

    Ans. (a) $P \cong 0.47$, (b) $P \cong 0.75$

6.35   Verify the answers to Problem 6.34 by constructing a tree diagram and determining the probabilities by use of the appropriate rules of multiplication and of addition.

6.36   In Section 6.5 it is suggested that the binomial distribution can generally be used to approximate hypergeometric probabilities when $n < 0.05 N$. Demonstrate that the binomial approximation of the probability values requested in Problem 6.34 is quite poor. (*Hint:* Use T/N as the p value for the binomial table, which does *not* conform to the sample size requirement, with $n = 4$ being much *more* than $0.05(20) = 1$.)

6.37   A departmental group includes five engineers and nine technicians. If five individuals are randomly chosen and assigned to a project, what is the probability that the project group will include exactly two engineers? (*Note:* This is a restatement of Problem 5.57(a), for which the answer was determined by combinatorial analysis.)

    Ans. $P \cong 0.42$

THE POISSON PROBABILITY DISTRIBUTION

6.38 On the average, six people per hour use a self-service banking facility during the prime shopping hours in a department store. What is the probability that

(a) exactly six people will use the facility during a randomly selected hour?

(b) fewer than five people will use the facility during a randomly selected hour?

(c) no one will use the facility during a 10-min interval?

(d) no one will use the facility during a 5-min interval?

Ans. (a) 0.1606, (b) 0.2851, (c) 0.3679, (d) 0.6065

6.39 Suppose that the manuscript for a textbook has a total of 50 errors or typos included in the 500 pages of material, and the errors are distributed randomly throughout the text. What is the probability that

(a) a chapter covering 30 pages has two or more errors?

(b) a chapter covering 50 pages has two or more errors?

(c) a randomly selected page has no error?

Ans. (a) 0.8008, (b) 0.9596, (c) 0.9048

6.40 Only one generator per thousand is found to be defective after assembly in a manufacturing plant, and the defective generators are distributed randomly throughout the production run.

(a) What is the probability that a shipment of 500 generators includes no defective generator?

(b) What is the probability that a shipment of 100 generators includes at least one defective generator?

Ans. By the Poisson approximation of binomial probabilities, (a) 0.6065, (b) 0.0952

6.41 Refer to the probability distribution in Problem 6.22. Does this probability distribution of arrivals appear to follow a Poisson probability distribution? [*Hint:* Use the $E(X)$ calculated in Problem 6.22 as the mean (λ) for determining the Poisson distribution with which the probabilities are to be compared.]

Ans. The two probability distributions correspond quite closely.

(*Note:* Problems 7.16–7.22 involve the use of all of the probability distributions covered in Chapters 6 and 7.)

COMPUTER APPLICATIONS

6.42 The probability that a particular electronic component will be defective is 0.005.

(a) Using available computer software, obtain the table of binomial probabilities for the number of defective components if $n = 8$ components are used in a device.

(b) With reference to the table, determine the probability that exactly one component is defective.

(c) With reference to the table, determine the probability that one or more components are defective.

Ans. (b) 0.0386, (c) 0.0393

Chapter 7

Probability Distributions for Continuous
Random Variables:
Normal and Exponential

7.1 CONTINUOUS RANDOM VARIABLES

As contrasted to a discrete random variable, a *continuous random variable* is one that can assume any fractional value within a defined range of values. (See Section 1.4.) Because there is an infinite number of possible fractional measurements, one cannot list every possible value with a corresponding probability. Instead, a *probability density function* is defined. This mathematical expression gives the function of X, represented by the symbol $f(X)$, for any designated value of the random variable X. The plot for such a function is called a *probability curve*, and the area between any two points under the curve indicates the probability of a value between these two points occurring by chance.

EXAMPLE 1. For the continuous probability distribution in Fig. 7-1, the probability that a randomly selected shipment will have a net weight between 6,000 and 8,000 lb is equal to the proportion of the total area under the curve which is included within the shaded area. That is, the total area under the probability density function is defined as being equal to 1, and the proportion of this area which is included between the two designated points can be determined by applying the method of integration (from calculus) in conjunction with the mathematical probability density function for this probability curve.

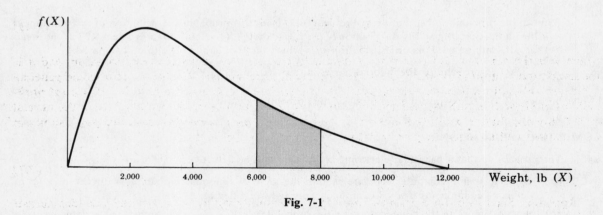

Fig. 7-1

Several standard continuous probability distributions are applicable as models to a wide variety of continuous variables under designated circumstances. Probability tables have been prepared for these standard distributions, making it unnecessary to use the method of integration in order to determine areas under the probability curve for these distributions. The standard continuous probability models described in this chapter are the normal and exponential probability distributions.

7.2 THE NORMAL PROBABILITY DISTRIBUTION

The *normal probability distribution* is a continuous probability distribution that is *both symmetrical* and *mesokurtic* (defined in Section 2.4). The probability curve representing the normal probability distribution is often described as being bell-shaped, as exemplified by the probability curve in Fig. 7-2.

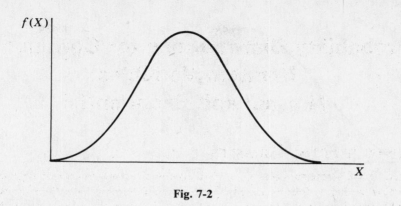

Fig. 7-2

The normal probability distribution is important in statistical inference for three distinct reasons:

(1)   The measurements obtained in many random processes are known to follow this distribution.

(2)   Normal probabilities can often be used to approximate other probability distributions, such as the binomial and Poisson distributions.

(3)   Distributions of such statistics as the sample mean and sample proportion are normally distributed when the sample size is large, regardless of the distribution of the parent population (see Section 8.2).

As for any continuous probability distribution, a probability value can only be determined for an *interval* of values. The height of the density function, or probability curve, for a normally distributed variable is given by

$$f(X) = \frac{1}{\sqrt{2\pi\sigma^2}} \, e^{-[(X-\mu)^2/2\sigma^2]} \tag{7.1}$$

where π is the constant 3.1416, e is the constant 2.7183, μ is the mean of the distribution, and σ is the standard deviation of the distribution. Since every different combination of μ and σ would generate a different normal probability distribution (all symmetrical and mesokurtic), tables of normal probabilities are based on one particular distribution: the *standard normal distribution*. This is the normal probability distribution with $\mu = 0$ and $\sigma = 1$. Any value X from a normally distributed population can be converted into the equivalent standard normal value z by the formula

$$z = \frac{X-\mu}{\sigma} \tag{7.2}$$

Appendix 5 indicates proportions of area for various intervals of values for the standard normal probability distribution, with the lower boundary of the interval always being at the mean. Converting designated values of the variable X into standard normal values makes use of this table possible, and makes use of the method of integration with respect to the equation for the density function unnecessary.

EXAMPLE 2.   The lifetime of an electrical component is known to follow a normal distribution with a mean $\mu = 2,000$ hr and a standard deviation $\sigma = 200$ hr. The probability that a randomly selected component will last between 2,000 and 2,400 hr is determined as follows.

Figure 7-3 portrays the probability curve (density function) for this problem and also indicates the relationship between the hours X scale and the standard normal z scale. Further, the area under the curve corresponding to the interval "2,000 to 2,400" has been shaded.

The lower boundary of the interval is at the mean of the distribution, and therefore is at the value $z = 0$. The upper boundary of the designated interval in terms of a z value is

$$z = \frac{X-\mu}{\sigma} = \frac{2,400 - 2,000}{200} = \frac{400}{200} = +2.0$$

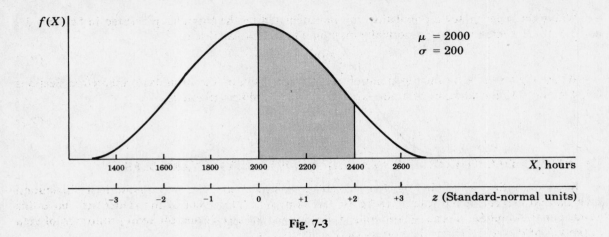

Fig. 7-3

By reference to Appendix 5, we find that

$$P(0 \le z \le +2.0) = 0.4772$$

Therefore, $$P(2{,}000 \le X \le 2{,}400) = 0.4772$$

Of course, not all problems involve an interval with the mean as the lower boundary. However, Appendix 5 can be used to determine the probability value associated with any designated interval, either by appropriate addition or subtraction of areas or by recognizing that the curve is symmetrical. Example 3 and Problems 7.1 through 7.8 include several varieties of such applications.

EXAMPLE 3. With respect to the electrical components described in Example 2, suppose we are interested in the probability that a randomly selected component will last *more* than 2,200 hr.

Note that by definition the total proportion of area to the right of the mean of 2,000 in Fig. 7-4 is 0.5000. Therefore, if we determine the proportion between the mean and 2,200, we can subtract this value from 0.5000 to obtain the probability of the hours X being greater than 2,200, which is shaded in Fig. 7-4.

$$z = \frac{2{,}200 - 2{,}000}{200} = +1.0$$

$$P(0 \le z \le +1.0) = 0.3413 \qquad \text{(from Appendix 5)}$$

$$P(z > +1.0) = 0.5000 - 0.3413 = 0.1587$$

Therefore, $$P(X > 2{,}200) = 0.1587$$

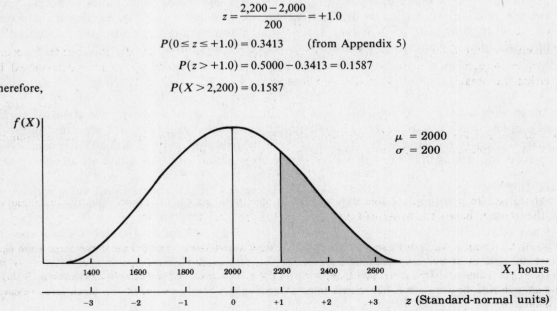

Fig. 7-4

Where the mean μ for a population can be calculated by the formulas presented in Sections 3.2 and 3.8, the expected value of a normally distributed random variable is

$$E(X) = \mu \qquad (7.3)$$

Where the variance σ^2 for a population can be calculated by the formulas presented in Sections 4.5, 4.6, and 4.12 the variance of a normally distributed random variable is

$$V(X) = \sigma^2 \qquad (7.4)$$

7.3 PERCENTILE POINTS FOR NORMALLY DISTRIBUTED VARIABLES

Recall from Section 3.7 that the 90th percentile point, for example, is that point in a distribution such that 90 percent of the values are below this point and 10 percent of the values are above this point. For the standard normal distribution, it is the value of z such that the total proportion of area to the left of this value under the normal curve is 0.90.

EXAMPLE 4.   Figure 7-5 illustrates the position of the 90th percentile point for the standard normal distribution. To determine the required value of z, we use Appendix 5 in the *reverse* of the usual direction, because in this case the area under the curve between the mean and the point of interest is known to be 0.40, and we wish to look up the associated value of z. In Appendix 5 we look in the *body* of the table for the value closest to 0.4000. It is 0.3997. From the row and column headings, the value of z associated with this area is 1.28, and therefore $z_{0.90} = +1.28$.

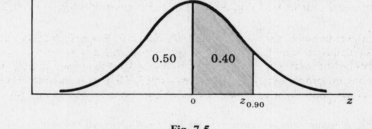

Fig. 7-5

Given the procedure in Example 4 for determining a percentile point for the standard normal distribution, a percentile point for a normally distributed random variable can be determined by converting the obtained value of z to the required value of X by the formula

$$X = \mu + z\sigma \qquad (7.5)$$

EXAMPLE 5.   For the lifetime of the electrical component described in Examples 2 and 3, and utilizing the solution in Example 4, the 90th percentile point for the life of the component is

$$X = \mu + z\sigma = 2{,}000 + (1.28)(200) = 2{,}256 \text{ hr}$$

For percentile points below the 50th percentile, the associated z value will always be negative, since the value is below the mean of 0 for the standard normal distribution.

EXAMPLE 6.   Continuing from Example 5, suppose we wish to determine the lifetime of the component such that only 10 percent of the components will fail before that time (10th percentile point). See Fig. 7-6 for the associated area for the standard normal probability distribution. Just as we did in Example 4, we look in the body of the table in Appendix 5 for the area closest to 0.4000, but in this case the value of z is taken to be negative. The solution is

$$X = \mu + z\sigma = 2{,}000 + (-1.28)(200) = 1{,}744 \text{ hr}$$

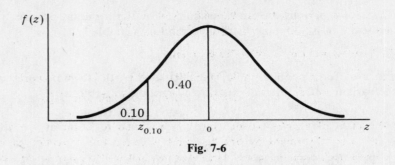

Fig. 7-6

7.4 NORMAL APPROXIMATION OF BINOMIAL PROBABILITIES

When the number of observations or trials n is relatively large, the normal probability distribution can be used to approximate binomial probabilities. A convenient rule is that such approximation is acceptable when $n \geq 30$, and both $np \geq 5$ and $nq \geq 5$. This rule, combined with the one given in Section 6.7 for the Poisson approximation of binomial probabilities, means that whenever $n \geq 30$, binomial probabilities can be approximated by either the normal or the Poisson distribution, depending on the values of np and nq. Different texts may use somewhat different rules for determining when such approximations are appropriate.

When the normal probability distribution is used as the basis for approximating a binomial probability value, the mean and standard deviation are based on the expected value and variance of the number of successes for the binomial distribution, as given in Section 6.3. The mean number of "successes" is

$$\mu = np \tag{7.6}$$

The standard deviation of the number of "successes" is

$$\sigma = \sqrt{npq} \tag{7.7}$$

EXAMPLE 7. For a large group of sales prospects, it has been observed that 20 percent of those contacted personally by a sales representative will make a purchase. If a sales representative contacts 30 prospects, we can determine the probability that 10 or more will make a purchase by reference to the binomial probabilities in Appendix 2:

$$P(X \geq 10 | n = 30, p = 0.20) = P(X = 10) + P(X = 11) + \cdots$$

$$= 0.0355 + 0.0161 + 0.0064 + 0.0022 + 0.0007 + 0.0002$$

$$= 0.0611 \text{ (the binomial probability value)}$$

Now we check to determine if the criteria for normal approximation are satisfied:

Is $n \geq 30$?   Yes, $n = 30$

Is $np \geq 5$?   Yes, $np = 30(0.20) = 6$

Is $nq \geq 5$?   Yes, $nq = 30(0.80) = 24$

The normal approximation of the binomial probability value is

$$\mu = np = (30)(0.20) = 6.0$$

$$\sigma = \sqrt{npq} = \sqrt{(30)(0.20)(0.80)} = \sqrt{4.8} \cong 2.19$$

$$P_{\text{Binomial}}(X \geq 10 | n = 30, p = 0.20) \cong P_{\text{Normal}}(X \geq 9.5 | \mu = 6.0, \sigma = 2.19)$$

(*Note:* This includes the correction for continuity discussed below.)

$$z = \frac{X-\mu}{\sigma} = \frac{9.5-6.0}{2.19} = \frac{3.5}{2.19} \cong +1.60$$

$$P(X \ge 9.5 | \mu = 6.0, \sigma = 2.19) = P(z \ge +1.60)$$

$$= 0.5000 - P(0 \le z \le +1.60) = 0.5000 - 0.4452$$

$$= 0.0548 \quad \text{(the normal approximation)}$$

In Example 7, the class of events "10 or more" is assumed to begin at 9.5 when the normal approximation is used. This adjustment by one-half unit is called the *correction for continuity*. It is required because all the area under the continuous normal distribution has to be assigned to some numeric event, even though no fractional number of successes is possible, for example, between "9 purchasers" and "10 purchasers." Put another way, in the process of normal approximation, a discrete event such as "10 purchasers" has to be considered as a continuous interval of values ranging from an exact lower limit of 9.5 to an exact upper limit of 10.5. Therefore, if Example 7 had asked for the probability of "more than 10 purchasers," the appropriate correction for continuity would involve adding 0.5 to the 10, and determining the area for the interval beginning at 10.5. By the above rationale, the 0.5 is either added or subtracted as a continuity correction according to the form of the probability statement:

(1) Subtract 0.5 from X when $P(X \ge X_i)$ is required.

(2) Subtract 0.5 from X when $P(X < X_i)$ is required.

(3) Add 0.5 to X when $P(X \le X_i)$ is required.

(4) Add 0.5 to X when $P(X > X_i)$ is required.

7.5 NORMAL APPROXIMATION OF POISSON PROBABILITIES

When the mean λ of a Poisson distribution is relatively large, the normal probability distribution can be used to approximate Poisson probabilities. A convenient rule is that such approximation is acceptable when $\lambda \ge 10.0$.

The mean and standard deviation of the normal probability distribution are based on the expected value and the variance of the number of events in a Poisson process, as identified in Section 6.6. This mean is

$$\mu = \lambda \qquad\qquad (7.8)$$

The standard deviation is

$$\sigma = \sqrt{\lambda} \qquad\qquad (7.9)$$

EXAMPLE 8. The average number of calls for service received by a machine repair department per 8-hr shift is 10.0. We can determine the probability that more than 15 calls will be received during a randomly selected 8-hr shift using Appendix 4:

$$P(X > 15 | \lambda = 10.0) = P(X = 16) + P(X = 17) + \cdots$$

$$= 0.0217 + 0.0128 + 0.0071 + 0.0037 + 0.0019 + 0.0009 + 0.0004 + 0.0002 + 0.0001$$

$$= 0.0488 \quad \text{(the Poisson probability)}$$

Because the value of λ is (at least) 10, the normal approximation of the Poisson probability value is acceptable. The normal approximation of the Poisson probability value is

$$\mu = \lambda = 10.0$$

$$\sigma = \sqrt{\lambda} = \sqrt{10.0} \cong 3.16$$

$$P_{\text{Poisson}}(X > 15 | \lambda = 10.0) \cong P_{\text{Normal}}(X \ge 15.5 | \mu = 10.0, \sigma = 3.16)$$

(*Note:* This includes the correction for continuity, discussed below.)

$$z = \frac{X - \mu}{\sigma} = \frac{15.5 - 10.0}{3.16} = \frac{5.5}{3.16} \cong +1.74$$

$$P(z \geq +1.74) = 0.5000 - P(0 \leq z \leq +1.74) = 0.5000 - 0.4591 = 0.0409 \qquad \text{(the normal approximation)}$$

The correction for continuity applied in Example 8 is the same type of correction described for the normal approximation of binomial probabilities. The rules provided in Section 7.4 as to when 0.5 is added to and subtracted from X apply equally to the situation in which the normal probability distribution is used to approximate Poisson probabilities.

7.6 THE EXPONENTIAL PROBABILITY DISTRIBUTION

If events occur in the context of a Poisson process, as described in Section 6.6, then the length of time or space between successive events follows an *exponential probability distribution*. Because the time or space is a continuum, such a measurement is a continuous random variable. As is the case of any continuous random variable, it is not meaningful to ask, "What is the probability that the first request for service will arrive in *exactly* one minute?" Rather, we must designate an *interval* within which the event is to occur, such as by asking, "What is the probability that the first request for service will arrive *within* a minute?"

Since the Poisson process is stationary, with equal likelihood of the event occurring throughout the relevant period of time, the exponential distribution applies whether we are concerned with the time (or space) *until* the very first event, the time *between* two successive events, or the time until the first event occurs after any randomly selected point in time.

Where λ is the mean number of occurrences for the *interval of interest* (see Section 6.6), the exponential probability that the first event will occur *within* the designated interval of time or space is

$$P(T \leq t) = 1 - e^{-\lambda} \qquad (7.10)$$

Similarly, the exponential probability that the first event *will not* occur within the designated interval of time or space is

$$P(T > t) = e^{-\lambda} \qquad (7.11)$$

For both of the above formulas the value of $e^{-\lambda}$ may be obtained from Appendix 3.

EXAMPLE 9.   An average of five calls per hour are received by a machine repair department. Beginning the observation at any point in time, the probability that the first call for service will arrive within a *half hour* is

Average per hour $= 5.0$

$\lambda = $ Average per half hour $= 2.5$

$P = 1 - e^{-\lambda} = 1 - e^{-2.5} = 1 - 0.08208 = 0.91792 \qquad$ (from Appendix 3)

The expected value and the variance of an exponential probability distribution, where the variable is designated as time T, are

$$E(T) = \frac{1}{\lambda} \qquad (7.12)$$

$$V(T) = \frac{1}{\lambda^2} \qquad (7.13)$$

7.7 COMPUTER APPLICATIONS

Computer software for statistical analysis frequently includes the capability of providing probabilities for intervals of values for normally distributed variables. Problems 7.23 and 7.24 are concerned with determining such probabilities and also with determining percentile points for normally distributed random variables by use of a computer.

Solved Problems

THE NORMAL PROBABILITY DISTRIBUTION

7.1   The packaging process in a breakfast cereal company has been adjusted so that an average of $\mu = 13.0$ oz of cereal is placed in each package. Of course, not all packages have precisely 13.0 oz because of random sources of variability. The standard deviation of the actual net weight is $\sigma = 0.1$ oz, and the distribution of weights is known to follow the normal probability distribution. Determine the probability that a randomly chosen package will contain between 13.0 and 13.2 oz of cereal and illustrate the proportion of area under the normal curve which is associated with this probability value.

From Fig. 7-7,

$$z = \frac{X - \mu}{\sigma} = \frac{13.2 - 13.0}{0.1} = +2.0$$

$$P(13.0 \le X \le 13.2) = P(0 \le z \le +2.0) = 0.4772 \qquad \text{(from Appendix 5)}$$

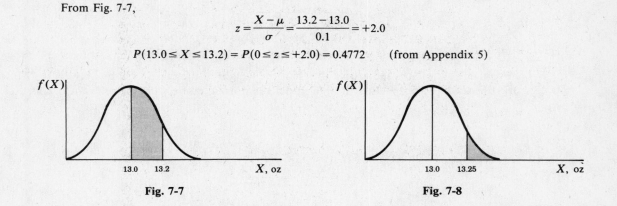

Fig. 7-7   Fig. 7-8

7.2   For the situation described in Problem 7.1, what is the probability that the weight of the cereal will exceed 13.25 oz? Illustrate the proportion of area under the normal curve which is relevant in this case.

With reference to Fig. 7-8,

$$z = \frac{X - \mu}{\sigma} = \frac{13.25 - 13.0}{0.1} = +2.5$$

$$P(X > 13.25) = P(z > +2.5) = 0.5000 - 0.4938 = 0.0062$$

7.3   From Problem 7.1, what is the probability that the weight of the cereal will be between 12.9 and 13.1 oz? Illustrate the proportion of area under the normal curve which is relevant in this case.

Referring to Fig. 7-9,

$$z_1 = \frac{X_1 - \mu}{\sigma} = \frac{12.9 - 13.0}{0.1} = -1.0$$

$$z_2 = \frac{X_2 - \mu}{\sigma} = \frac{13.1 - 13.0}{0.1} = +1.0$$

$$P(12.9 \le X \le 13.1) = P(-1.0 \le z \le +1.0) = 0.3413 + 0.3413 = 0.6826$$

(*Note:* This is the proportion of area from $-1.0z$ to μ plus the proportion from μ to $+1.0z$. Also note that because the normal probability distribution is symmetrical, areas to the left of the mean for negative z values are equivalent to areas to the right of the mean.)

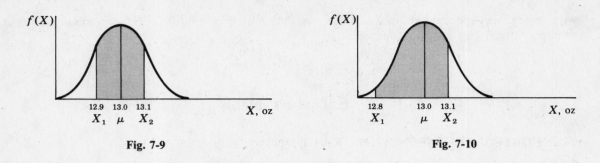

Fig. 7-9   Fig. 7-10

7.4   What is the probability that the weight of the cereal in Problem 7.1 will be between 12.8 and 13.1 oz? Illustrate the proportion of area under the normal curve which is relevant in this case.

With reference to Fig. 7-10,

$$z_1 = \frac{X_1 - \mu}{\sigma} = \frac{12.8 - 13.0}{0.1} = \frac{-0.2}{0.1} = -2.0$$

$$z_2 = \frac{X_2 - \mu}{\sigma} = \frac{13.1 - 13.0}{0.1} = \frac{0.1}{0.1} = +1.0$$

$$P(12.8 \le X \le 13.1) = P(-2.0 \le z \le 1.0) = 0.4772 + 0.3413 = 0.8185$$

7.5   From Problem 7.1, what is the probability that the weight of the cereal will be between 13.1 and 13.2 oz? Illustrate the proportion of area under the normal curve which is relevant in this case.

Referring to Fig. 7-11,

$$z_1 = \frac{X_1 - \mu}{\sigma} = \frac{13.1 - 13.0}{0.1} = +1.0$$

$$z_2 = \frac{X_2 - \mu}{\sigma} = \frac{13.2 - 13.0}{0.1} = +2.0$$

$$P(13.1 \le X \le 13.2) = P(+1.0 \le z \le +2.0) = 0.4772 - 0.3413 = 0.1359$$

(*Note:* The probability is equal to the proportion of area from 13.0 to 13.2 minus the proportion of area from 13.0 to 13.1.)

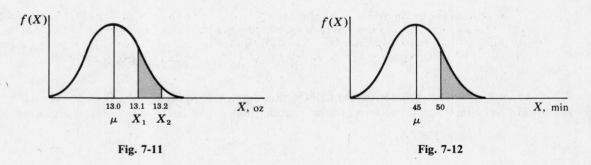

Fig. 7-11   Fig. 7-12

7.6   The amount of time required for a certain type of automobile transmission repair at a service garage is normally distributed with the mean $\mu = 45$ min and the standard deviation $\sigma = 8.0$ min. The service manager plans to have work begin on the transmission of a customer's car 10 min after the car is dropped off, and he tells the customer that the car will be ready within 1 hr total time. What is the probability that he will be wrong? Illustrate the proportion of area under the normal curve which is relevant in this case.

From Fig. 7-12,

$$P(\text{Error}) = P(X > 50 \text{ min}), \text{ since actual work is to begin in 10 min}$$

$$z = \frac{X - \mu}{\sigma} = \frac{50 - 45}{8.0} = \frac{5.0}{8.0} = +0.62$$

$$P(X > 50) = P(z > +0.62) = 0.5000 - 0.2324 = 0.2676$$

PERCENTILE POINTS FOR NORMALLY DISTRIBUTED VARIABLES

7.7   Referring to Problem 7.6, what is the required working time allotment such that there is a 75 percent chance that the transmission repair will be completed within that time? Illustrate the proportion of area that is relevant.

As illustrated in Fig. 7-13, a proportion of area of 0.2500 is included between the mean and the 75th percentile point. Therefore, as the first step in the solution we determine the required z value by finding the area in the body of the table in Appendix 5 that is closest to 0.2500. The closest area is 0.2486, with $z_{0.75} = +0.67$. We then convert this value of z into the required value of X by:

$$X = \mu + z\sigma = 45 + (0.67)(8.0) = 50.36 \text{ min}$$

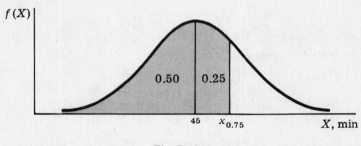

Fig. 7-13

7.8   With reference to Problem 7.6, what is the working time allotment such that there is a probability of just 30 percent that the transmission repair can be completed within that time? Illustrate the proportion of area which is relevant.

Since a proportion of area of 0.30 is to the left of the unknown value of X in Fig. 7-14, it follows that a proportion of 0.20 is between that percentile point and the mean. By reference to Appendix 5, the proportion of area closest to this 0.1985, with $z_{0.30} = -0.52$. The z value is negative because the percentile point is to the left of the mean. Finally, the z value is converted to the required value of X:

$$X = \mu + z\sigma$$

$$X = 45 + (-0.52)(8.0) = 45 - 4.16 = 40.84 \text{ min}$$

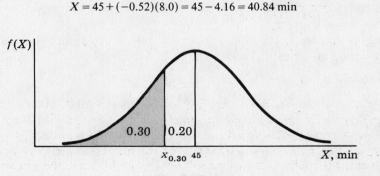

Fig. 7-14

NORMAL APPROXIMATION OF BINOMIAL AND POISSON PROBABILITIES

7.9   Of the people who enter a large shopping mall, it has been found that 70 percent will make at least one purchase. For a sample of $n = 50$ individuals, what is the probability that at least 40 people make one or more purchases each?

The normal approximation of the required binomial probability value can be used, because $n \geq 30$, $np \geq 5$, and $n(q) \geq 5$.

$$\mu = np = (50)(0.70) = 35.0$$

$$\sigma = \sqrt{npq} = \sqrt{(50)(0.70)(0.30)} = \sqrt{10.5} = 3.24$$

$$P_{\text{Binomial}}(X \geq 40 \,|\, n = 50, p = 0.70) \cong P_{\text{Normal}}(X \geq 39.5 \,|\, \mu = 35.0, \sigma = 3.24)$$

(*Note:* The correction for continuity is included, as described in Section 7.4.)

$$z = \frac{X - \mu}{\sigma} = \frac{39.5 - 35.0}{3.24} = \frac{4.5}{3.24} = +1.39$$

$$P(X \geq 39.5) = P(z \geq +1.39) = 0.5000 - 0.4177 = 0.0823$$

7.10   For the situation described in Problem 7.9, what is the probability that fewer than 30 of the 50 sampled individuals make at least one purchase?

Since, from Problem 7.9, $\mu = 35.0$ and $\sigma = 3.24$,

$$P_{\text{Binomial}}(X < 30 \,|\, n = 50, p = 0.70) \cong P_{\text{Normal}}(X \leq 29.5 \,|\, \mu = 35.0, \sigma = 3.24)$$

(*Note:* The correction for continuity is included).

$$z = \frac{X - \mu}{\sigma} = \frac{29.5 - 35.0}{3.24} = \frac{-5.5}{3.24} = -1.70$$

$$P(X \leq 29.5) = P(z \leq -1.70) = 0.5000 - 0.4554 = 0.0446$$

7.11   Calls for service are known to arrive randomly and as a stationary process at an average of five calls per hour. What is the probability that more than 50 calls for service will be received during an 8-hr shift?

Because the mean for the 8-hr period for this Poisson process exceeds $\lambda = 10$, the normal probability distribution can be used to approximate the Poisson probability value. Since $\mu = \lambda = 40.0$ and $\sigma = \sqrt{\lambda} = \sqrt{40.0} = 6.32$,

$$P_{\text{Poisson}}(X > 50 \,|\, \lambda = 40.0) \cong P_{\text{Normal}}(X \geq 50.5 \,|\, \mu = 40.0, \sigma = 6.32)$$

(*Note:* The correction for continuity is included.)

$$z = \frac{X - \mu}{\sigma} = \frac{50.5 - 40.0}{6.32} = \frac{10.5}{6.32} = +1.66$$

$$P(X \geq 50.5) = P(z \geq +1.66) = 0.5000 - 0.4515 = 0.0485$$

7.12   Referring to Problem 7.11, what is the probability that 35 or fewer calls for service will be received during an 8-hr shift?

Since $\mu = 40.0$ and $\sigma = 6.32$,

$$P_{\text{Poisson}}(X \leq 35 \,|\, \lambda = 40.0) \cong P_{\text{Normal}}(X \leq 35.5 \,|\, \mu = 40.0, \sigma = 6.32)$$

(*Note:* The correction for continuity is included.)

$$z = \frac{X - \mu}{\sigma} = \frac{35.5 - 40.0}{6.32} = \frac{-4.5}{6.32} = -0.71$$

$$P(X \leq 35.5) = P(z \leq -0.71) = 0.5000 - 0.2612 = 0.2388$$

THE EXPONENTIAL PROBABILITY DISTRIBUTION

7.13   On the average, a ship arrives at a certain dock every second day. What is the probability that after the departure of a ship 4 days will pass before the arrival of the next ship?

Average per 2 days $= 1.0$

Average per day $= 0.5$

$\lambda =$ average per 4-day period $= 4 \times 0.5 = 2.0$

$P(T > 4) = e^{-\lambda} = e^{-2.0} = 0.13534$ (from Appendix 3)

7.14   Each 500-ft roll of sheet steel includes two flaws, on the average. What is the probability that as the sheet steel is unrolled the first flaw occurs within the first 50-ft segment?

Average per 500-ft roll $= 2.0$

$\lambda =$ average per 50-ft segment $= \dfrac{2.0}{10} = 0.20$

$P(T \leq 50) = 1 - e^{-\lambda} = 1 - e^{-0.20} = 1 - 0.81873 = 0.18127$ (from Appendix 3)

7.15   An application that is concerned with use of the exponential distribution can be transformed into Poisson distribution form, and vice versa. To illustrate such transformation, suppose that an average of four aircraft per 8-hr day arrive for repairs at a repair facility. (*a*) What is the probability that the first arrival does *not* occur during the first hour of work? (*b*) Demonstrate that the equivalent Poisson-oriented problem is the probability that there will be *no* arrivals in the 1-hr period. (*c*) What is the probability that the first arrival occurs within the first hour? (*d*) Demonstrate that the equivalent Poisson-oriented problem is the probability that there will be one or more arrivals during the 1-hr period.

(*a*) $\lambda = 0.5$ (per hour)

$P(T > 1) = e^{-\lambda} = e^{-0.5} = 0.60653$ (from Appendix 3)

(*b*) $P(X = 0 \mid \lambda = 0.5) = 0.6065$ (from Appendix 4)

(*c*) $\lambda = 0.5$ (per hour)

$P(T \leq 1) = 1 - e^{-\lambda} = 1 - e^{-0.5} = 1 - 0.60653 = 0.39347$ (from Appendix 3)

(*d*) $P(X \geq 1 \mid \lambda = 0.5) = 1 - P(X = 0) = 1.0000 - 0.6065 = 0.3935$ (from Appendix 4)

Thus, in the case of both transformations the answers are identical, except for the number of digits which happen to be carried in the two tables.

MISCELLANEOUS PROBLEMS CONCERNING PROBABILITY DISTRIBUTIONS

(*Note:* Problems 7.16 through 7.22 involve the use of all of the probability distributions covered in Chapters 6 and 7.)

7.16   A shipment of 10 engines includes one that is defective. If 7 engines are chosen randomly from this shipment, what is the probability that none of the 7 is defective?

Using the hypergeometric distribution (see Section 6.5),

$$P(X \mid N, T, n) = \frac{\dbinom{N-T}{n-X} \dbinom{T}{X}}{\dbinom{N}{n}}$$

$$P(X = 0 \mid N = 10, T = 1, n = 7) = \frac{\dbinom{10-1}{7-0} \dbinom{1}{0}}{\dbinom{10}{7}} = \frac{\left(\dfrac{9!}{7!2!}\right) \left(\dfrac{1!}{0!1!}\right)}{\left(\dfrac{10!}{7!3!}\right)} = \frac{(36)(1)}{120} = 0.30$$

7.17   Suppose that in Problem 7.16 the overall proportion of engines with some defect is 0.10, but that a very large number are being assembled in an engine-assembly plant. What is the probability that a random sample of 7 engines will include no defective engines?

Using the binomial distribution (see Section 6.3),

$$P(X = 0 \mid n = 7, p = 0.10) = 0.4783 \qquad \text{(from Appendix 2)}$$

7.18   Suppose that the proportion of engines which contain a defect in an assembly operation is 0.10, and a sample of 200 engines is included in a particular shipment. What is the probability that at least 30 of the 200 engines contain a defect?

Use of the normal approximation of the binomial probability distribution described in Section 7.4 is acceptable, because $n \geq 30$, $np \geq 5$, and $n(q) \geq 5$.

$$\mu = np = (200)(0.10) = 20.0$$
$$\sigma = \sqrt{np(q)} = \sqrt{(200)(0.10)(0.90)} = \sqrt{18.00} \cong 4.24$$
$$P_{\text{Binomial}}(X \geq 30 \mid n = 200, p = 0.10) \cong P_{\text{Normal}}(X \geq 29.5 \mid \mu = 20.0, \sigma = 4.24)$$

(*Note:* The correction for continuity is included.)

$$z = \frac{X - \mu}{\sigma} = \frac{29.5 - 20.0}{4.24} = \frac{9.5}{4.24} = +2.24$$

$$P(X \geq 29.5) = P(z \geq +2.24) = 0.5000 - 0.4875 = 0.0125 \qquad \text{(from Appendix 5)}$$

7.19   Suppose that the proportion of engines which contain a defect in an assembly operation is 0.01, and a sample of 200 engines are included in a particular shipment. What is the probability that three or fewer engines contain a defect?

Use of the Poisson approximation of the binomial probability distribution (see Section 6.7) is acceptable in this case because $n \geq 30$ and $np < 5$.

$$\lambda = np = (200)(0.01) = 2.0$$
$$P_{\text{Binomial}}(X \leq 3 \mid n = 200, p = 0.01) \cong P_{\text{Poisson}}(X \leq 3 \mid \lambda = 2.0)$$
$$= 0.1353 + 0.2707 + 0.2707 + 0.1804 = 0.8571$$

$$\text{(from Appendix 4)}$$

7.20   An average of 0.5 customer per minute arrives at a checkout stand. After an attendant opens the stand, what is the probability that he will have to wait at least 3 min before the first customer arrives?

Using the exponential probability distribution described in Section 7.6,

Average per minute $= 0.5$
$\lambda =$ average for 3 min $= 0.5 \times 3 = 1.5$
$P(T > 3) = e^{-\lambda} = e^{-1.5} = 0.22313 \qquad \text{(from Appendix 3)}$

7.21   An average of 0.5 customer per minute arrives at a checkout stand. What is the probability that five or more customers will arrive in a given 5-min interval?

From Section 6.6 on the Poisson probability distribution,

Average per minute $= 0.5$
$\lambda =$ average for 5 min $= 0.5 \times 5 = 2.5$
$P(X \geq 5 \mid \lambda = 2.5) = 0.0668 + 0.0278 + 0.0099 + 0.0031 + 0.0009 + 0.0002 = 0.1087$

$$\text{(from Appendix 4)}$$

7.22 An average of 0.5 customer per minute arrives at a checkout stand. What is the probability that more than 20 customers arrive at the stand during a particular interval of 0.5 hr?

Use of the normal approximation of the Poisson probability distribution described in Section 7.5 is acceptable because $\lambda \geq 10.0$.

$$\text{Average per minute} = 0.5$$
$$\lambda = \text{average for 30 min} = 0.5 \times 30 = 15.0$$
$$\mu = \lambda = 15.0$$
$$\sigma = \sqrt{\lambda} = \sqrt{15.0} \cong 3.87$$
$$P_{\text{Poisson}}(X > 20 \,|\, \lambda = 15.0) \cong P_{\text{Normal}}(X \geq 20.5 \,|\, \mu = 15.0, \sigma = 3.87)$$

(*Note:* The correction for continuity is included.)

$$z = \frac{X - \mu}{\sigma} = \frac{20.5 - 15.0}{3.87} = \frac{5.50}{3.87} = +1.42$$

$$P(X \geq 20.5) = P(z \geq +1.42) = 0.5000 - 0.4222 = 0.0778 \qquad \text{(from Appendix 5)}$$

COMPUTER APPLICATIONS

7.23 From Problem 7.1, the mean weight of cereal per package is $\mu = 13.0$ oz with $\sigma = 0.1$ oz. The weights are normally distributed. Using an available computer program, determine the probability that the weight of a randomly selected package (*a*) exceeds 13.25 oz and (*b*) is between 12.9 and 13.1 oz.

(*a*) Refer to Fig. 7-15. Note that the probability that initially is reported, 0.9938, is the cumulative probability between negative infinity and the designated value, 13.25. Therefore, the logic followed in the remainder of the output is to obtain the required answer by subtraction, as follows:

$$P(X > 13.25) = 1.0000 - P(X \leq 13.25)$$
$$= 1.0000 - 0.9938 = 0.0062$$

(This solution corresponds to the manually derived answer in Problem 7.2.)

```
MTB ) CDF AT 13.25;
SUBC) NORMAL MU = 13.0, SIGMA = 0.1.
    2.50    0.9938
MTB ) SUBTRACT 0.9938 FROM 1.0000, PUT ANSWER IN K1
    ANSWER =        0.0062
```

Fig. 7-15 Minitab output for Problem 7.23(*a*).

(*b*) Refer to Fig. 7-16. Again, subtraction of probabilities is included in the output, as follows:

$$P(12.9 \leq X \leq 13.1) = P(X \leq 13.1) - P(X \leq 12.9)$$
$$= 0.8413 - 0.1587 = 0.6826$$

(This solution corresponds to the manually derived answer in Problem 7.3.)

```
MTB ) CDF AT 12.9;
SUBC) NORMAL MU = 13.0, SIGMA = 0.1.
   -1.00    0.1587
MTB ) CDF AT 13.1;
SUBC) NORMAL MU = 13.0, SIGMA = 0.1.
    1.00    0.8413
MTB ) SUBTRACT 0.1587 FROM 0.8413, PUT ANSWER IN K1
    ANSWER =        0.6826
```

Fig. 7-16 Minitab output for Problem 7.23(*b*).

7.24 From Problem 7.6, the amount of time required for transmission repairs is normally distributed, with $\mu = 45$ min and $\sigma = 8.0$ min. Using an available computer program, determine the time amount at the (*a*) 75th percentile point and (*b*) 30th percentile point.

(*a*) Refer to Fig. 7-17. The answer is 50.3959 min. (This answer corresponds to the manually derived one in Problem 7.7, except for a slight difference due to rounding.)

```
MTB ) INVCDF AT 0.75;
SUBC) NORMAL MU = 45.0, SIGMA = 8.0.
     0.75    50.3959
```

Fig. 7-17   Minitab output for Problem 7.24(*a*).

(*b*) Refer to Fig. 7-18. The answer is 40.8048 min. (This answer corresponds to the manually derived one in Problem 7.8, except for a slight difference due to rounding.)

```
MTB ) INVCDF AT 0.30;
SUBC) NORMAL MU = 45.0, SIGMA = 8.0.
     0.30    40.8048
```

Fig. 7-18   Minitab output for Problem 7.24(*b*).

Supplementary Problems

THE NORMAL PROBABILITY DISTRIBUTION

7.25 The reported scores on a nationally standardized achievement test for high school graduates has a mean of $\mu = 500$ with the standard deviation $\sigma = 100$. The scores are approximately normally distributed. What is the probability that the score of a randomly chosen individual will be (*a*) between 500 and 650? (*b*) Between 450 and 600?

Ans. (*a*) 0.4332, (*b*) 0.5328

7.26 For a nationally standardized achievement test the mean is $\mu = 500$ with $\sigma = 100$. The scores are normally distributed. What is the probability that a randomly chosen individual will have a score (*a*) below 300? (*b*) Above 650?

Ans. (*a*) 0.0228, (*b*) 0.0668

7.27 The useful life of a certain brand of steel-belted radial tires has been found to follow a normal distribution with $\mu = 38,000$ miles and $\sigma = 3,000$ miles. (*a*) What is the probability that a randomly selected tire will have a useful life of at least 35,000 miles? (*b*) What is the probability that it will last more than 45,000 miles?

Ans. (*a*) 0.8413, (*b*) 0.0099

7.28 A dealer orders 500 of the tires specified in Problem 7.27 for resale. Approximately what number of tires will last (*a*) between 40,000 and 45,000 miles? (*b*) 40,000 miles or more?

Ans. (*a*) 121, (*b*) 126

7.29 An individual buys four of the tires described in Problem 7.27. What is the probability that all four tires will last (*a*) at least 38,000 miles? (*b*) At least 35,000 miles? (*Hint:* After obtaining the probability for one tire, use the multiplication rule for independent events from Section 5.6 to determine the probability for all four tires.)

Ans. (*a*) 0.0625, (*b*) 0.5010

7.30   The amount of time required per individual at a bank teller's window has been found to be approximately normally distributed with $\mu = 130$ sec and $\sigma = 45$ sec. What is the probability that a randomly selected individual will (a) require less than 100 sec to complete his transactions? (b) Spend between 2.0 and 3.0 min at the teller's window?

Ans. (a) 0.2514, (b) 0.4536

PERCENTILE POINTS FOR NORMALLY DISTRIBUTED VARIABLES

7.31   For a nationally standardized achievement test the mean is $\mu = 500$ with $\sigma = 100$. The scores are normally distributed. What score is at the (a) 50th percentile point, (b) 30th percentile point, and (c) 90th percentile point?

Ans. (a) 500, (b) 448, (c) 628

7.32   Under the conditions specified in Problem 7.30, (a) within what length of time do the 20 percent of individuals with the simplest transactions complete their business at the window? (b) At least what length of time is required for the individuals in the top 5 percent of required time?

Ans. (a) 92 sec, (b) 204 sec

NORMAL APPROXIMATION OF BINOMIAL AND POISSON PROBABILITIES

7.33   For the several thousand items stocked by a mail order firm, there is an overall probability of 0.08 that a particular item (including specific size and color, etc.) is out of stock. If a shipment covers orders for 120 different items, what is the probability that 15 or more items are out of stock?

Ans. 0.0495

7.34   For the shipment described in Problem 7.33, what is the probability that there are between 10 and 15 items out of stock?

Ans. 0.4887

7.35   During the 4 P.M. to 6 P.M. peak period in an automobile service station, one car enters the station every 3 min, on the average. What is the probability that at least 25 cars enter the station for service between 4 P.M. and 5 P.M.?

Ans. 0.1562

7.36   For the service-station arrivals in Problem 7.35, what is the probability that fewer than 30 cars enter the station between 4 P.M. and 6 P.M. of a randomly selected day?

Ans. 0.0485

THE EXPONENTIAL PROBABILITY DISTRIBUTION

7.37   On the average, six people per hour use a self-service banking facility during the prime shopping hours in a department store.

   (a)   What is the probability that at least 10 min will pass between the arrival of two customers?

   (b)   What is the probability that after a customer leaves, another customer does not arrive for at least 20 min?

   (c)   What is the probability that a second customer arrives within 1 min after a first customer begins his banking transaction?

   Ans. (a) 0.36788, (b) 0.13534, (c) 0.09516

7.38   Suppose that the manuscript for a textbook has a total of 50 errors or typos included in the 500 pages of material, and the errors are distributed randomly throughout the text. As the technical proofreader begins reading a particular chapter, what is the probability that the first error in that chapter (a) is included within the first five pages? (b) Occurs beyond the first 15 pages?

Ans.   (a) 0.39347, (b) 0.22313

MISCELLANEOUS PROBLEMS

(*Note:* Problems 7.39 through 7.46 involve the use of all of the probability distributions covered in Chapters 6 and 7.)

7.39   The frequency distribution of the length of stay in a community hospital has been found to be approximately symmetrical and mesokurtic, with $\mu = 8.4$ days and $\sigma = 2.6$ days (with fractions of days measured). What is the probability that a randomly chosen individual will be in the hospital for (a) less than 5.0 days? (b) More than 8.0 days?

Ans.   (a) 0.0951, (b) 0.5596

7.40   A firm which manufactures and markets a wide variety of low-priced innovative toys (such as a ball which will bounce in unexpected directions) has found that in the long run 40 percent of the toys which it develops have at least moderate market success. If six new toys have been developed for market introduction next summer, what is the probability that at least three of them will have moderate market success?

Ans.   0.4557

7.41   The firm in Problem 7.40 has 60 toy ideas in the process of development for introduction during the next few years. If all 60 of these are eventually marketed, what is the probability that at least 30 of them will have moderate market success?

Ans.   0.0735

7.42   From Problems 7.40 and 7.41, above, suppose 5 percent of the toys which are marketed turn out to be outstanding sales successes. If 60 new toys are introduced during the next few years, what is the probability that none of them will turn out to be an outstanding sales success?

Ans.   0.0498

7.43   Customers arrive at a refreshment stand located in a sports arena at an average rate of two per minute. What is the probability that five or more people will approach the stand for refreshments during a randomly chosen minute?

Ans.   0.0526

7.44   From Problem 7.43, what is the probability that after the refreshment stand opens two full minutes pass before the first customer arrives?

Ans.   0.01832

7.45   For the situation described in Problem 7.43, what is the probability that more than 50 people will come to the stand during a half-hour period?

Ans.   0.8907

7.46   Of the eight hotels located in a resort area, three can be described as being mediocre in terms of customer services. A travel agent chooses two of the hotels randomly for two clients planning vacations in the resort area. What is the probability that at least one of the clients will end up in one of the mediocre hotels?

Ans.   0.6429

COMPUTER APPLICATIONS

7.47   From Problem 7.26, the scores on a nationally standardized achievement test are normally distributed, with $\mu = 500$ and $\sigma = 100$. Using an available computer program, determine the probability that a randomly chosen individual will have a score (a) below 300 and (b) above 650.

   Ans.   (a) 0.0228, (b) 0.0668 (which correspond to the manual solutions in Problem 7.26)

7.48   From Problem 7.31, the scores on a nationally standardized achievement test are normally distributed, with $\mu = 500$ and $\sigma = 100$. Using an available computer program, determine the value at the (a) 30th percentile point and (b) 90th percentile point.

   Ans.   (a) 448, (b) 628 (which correspond to the manual solutions in Problem 7.31)

Chapter 8

Sampling Distributions and Confidence Intervals for the Mean

8.1 POINT ESTIMATION

Because of such factors as time and cost, the parameters of a population frequently are estimated on the basis of sample statistics. As defined in Section 1.2, a *population parameter* is a summary measure of a population, whereas a summary measure of a sample is called a *sample statistic*. In order for the sample data to be used for statistical inference, including estimation, the sample that was taken must be a *random sample*, as described in Section 1.6.

EXAMPLE 1. The mean μ and standard deviation σ of a population of measurements are population parameters. The mean $\bar{X}$ and standard deviation s of a sample of measurements are sample statistics.

A *point estimator* is a single numeric value based on random sample data that is used to estimate the value of a population parameter. One of the most important characteristics of the sample statistic that is used as an estimator is that it be unbiased. An *unbiased estimator* is a sample statistic whose expected value is equal to the parameter being estimated. As explained in Section 6.2, an *expected value* is the long-run mean average of the sample statistic.

Table 8.1 presents some frequently used point estimators of population parameters. In every case, the appropriate estimator of a population parameter simply is the corresponding sample statistic. However, note that in Section 4.5, formula (*4.5*) for the sample variance includes a "correction factor." Without this correction, the sample variance would be a biased estimator of the population variance.

Table 8.1 Frequently Used Point Estimators

Population parameter	Estimator
Mean, μ	$\bar{X}$
Difference between the means of two populations, $\mu_1 - \mu_2$	$\bar{X}_1 - \bar{X}_2$
Proportion, π	$\hat{p}$
Difference between the proportions in two populations, $\pi_1 - \pi_2$	$\hat{p}_1 - \hat{p}_2$
Variance, σ^2	s^2
Standard deviation, σ	s

8.2 SAMPLING DISTRIBUTION OF THE MEAN

A population distribution represents the distribution of a population of values, and a sample distribution represents the distribution of a sample of values collected from a population. In contrast to such distributions of individual measurements, a *sampling distribution* is a probability distribution which applies to the possible values of a sample statistic. Thus, the *sampling distribution of the mean* is the probability distribution for the possible values of the sample mean $\bar{X}$ based on a particular sample size.

For any given sample size n taken from a population with mean μ, the value of the sample mean $\bar{X}$ will vary somewhat from sample to sample. This variability serves as the basis for the sampling distribution. The sampling distribution of the mean is described by determining the expected value $E(\bar{X})$, or mean, of the distribution and the standard deviation of the distribution of means $\sigma_{\bar{x}}$. Because this standard deviation indicates the accuracy of the sample mean as a point estimator, $\sigma_{\bar{x}}$ is usually called the *standard error of the mean*. In general, the expected value of the mean and the standard error of the mean are defined as

$$E(\bar{X}) = \mu \qquad (8.1)$$

$$\sigma_{\bar{x}} = \frac{\sigma}{\sqrt{n}} \qquad (8.2)$$

EXAMPLE 2. Suppose the mean of a very large population is $\mu = 50.0$ and the standard deviation of the measurements is $\sigma = 12.0$. We determine the sampling distribution of the sample means for a sample size of $n = 36$, in terms of the expected value and the standard error of the distribution, as follows:

$$E(\bar{X}) = \mu = 50.0$$

$$\sigma_{\bar{x}} = \frac{\sigma}{\sqrt{n}} = \frac{12.0}{\sqrt{36}} = \frac{12.0}{6} = 2.0$$

When sampling from a population which is finite, you should include a *finite correction factor* in the formula for the standard error of the mean. As a general rule, the correction is negligible and may be omitted when $n < 0.05N$, that is, when the sample size is less than 5 percent of the population size. Many texts and computer programs do not include this correction because the assumption is made that the population always is very large, or possibly even infinite, in size. The formula for the standard error of the mean with the finite correction factor included is

$$\sigma_{\bar{x}} = \frac{\sigma}{\sqrt{n}} \sqrt{\frac{N-n}{N-1}} \qquad (8.3)$$

If the standard deviation of the population is not known, the standard error of the mean can be estimated by using the sample standard deviation as an estimator of the population standard deviation. To differentiate this standard error from one based on a known σ, it is designated by the symbol $s_{\bar{x}}$ (or by $\hat{\sigma}_{\bar{x}}$ in some texts). The formula for the estimated standard error of the mean is

$$s_{\bar{x}} = \frac{s}{\sqrt{n}} \qquad (8.4)$$

The formula for the estimated standard error of the mean with the finite correction factor included is

$$s_{\bar{x}} = \frac{s}{\sqrt{n}} \sqrt{\frac{N-n}{N-1}} \qquad (8.5)$$

EXAMPLE 3. An auditor takes a random sample of size $n = 16$ from a set of $N = 100$ accounts receivable. The standard deviation of the amounts of the receivables for the entire group of 100 accounts is not known. However, the sample standard deviation is $s = \$57.00$. We determine the value of the standard error for the sampling distribution of the mean as follows:

$$s_{\bar{x}} = \frac{s}{\sqrt{n}} \sqrt{\frac{N-n}{N-1}} = \frac{57.00}{\sqrt{16}} \sqrt{\frac{100-16}{100-1}} = \frac{57}{4} \sqrt{\frac{84}{99}}$$

$$= 14.25\sqrt{0.8484} = 14.25(0.9211) = 13.126 \cong \$13.13$$

The standard error of the mean is estimated on the basis of the sample standard deviation in this example, and use of the finite correction factor is required because it is *not* true that $n < 0.05N$, that is, $16 > 0.05(100)$.

The standard error of the mean provides the principal basis for statistical inference concerning an unknown population mean, as presented in this and the following chapters. A theorem in statistics that leads to the usefulness of the standard error of the mean is

Central limit theorem: As the sample size is increased, the sampling distribution of the mean approaches the normal distribution in form *regardless of the form of the population distribution of*

the individual measurements. For practical purposes, the sampling distribution of the mean can be assumed to be approximately normal whenever the sample size is $n \geq 30$.

Thus, given a "large" sample of $n \geq 30$, we can always use the normal probability distribution in conjunction with the standard error of the mean. Further, if the population is normally distributed and σ is known, the normal distribution can be used in statistical inference with small samples as well. The requirement that σ be known is explained in Section 8.5.

EXAMPLE 4. An auditor takes a random sample of size $n = 36$ from a population of 1,000 accounts receivable. The mean value of the accounts receivable for the population is $\mu = \$260.00$ with the population standard deviation $\sigma = \$45.00$. What is the probability that the sample mean will be less than \$250.00?

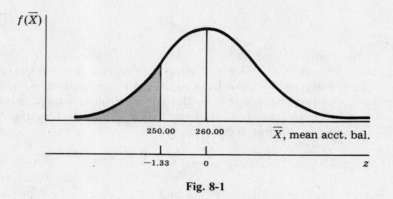

Fig. 8-1

Figure 8-1 portrays the probability curve. The sampling distribution is described by the mean and standard error:

$$E(\bar{X}) = \mu = 260.00 \quad \text{(as given)}$$

$$\sigma_{\bar{x}} = \frac{\sigma}{\sqrt{n}} = \frac{45.00}{\sqrt{36}} = \frac{45.00}{6} = 7.50$$

[*Note:* The finite correction factor is not required because $n < 0.05 N$, that is, 36 is less than 0.05(1,000).]

$$z = \frac{\bar{X} - \mu}{\sigma_{\bar{x}}} = \frac{250.00 - 260.00}{7.50} = \frac{-10.00}{7.50} = -1.33$$

Therefore,

$$P(\bar{X} < 250.00 \mid \mu = 260.00, \sigma_{\bar{x}} = 7.50) = P(z < -1.33)$$

$$P(z < -1.33) = 0.5000 - P(-1.33 \leq z \leq 0)$$

$$= 0.5000 - 0.4082 = 0.0918$$

EXAMPLE 5. With reference to Example 4, what is the probability that the sample mean will be within \$15.00 of the population mean?

Figure 8-2 portrays the probability curve for the sampling distribution.

$$P(245.00 \leq \bar{X} \leq 275.00 \mid \mu = 260.00, \sigma_{\bar{x}} = 7.50)$$

where

$$z_1 = \frac{245.00 - 260.00}{7.50} = -2.00$$

$$z_2 = \frac{275.00 - 260.00}{7.50} = +2.00$$

$$P(-2.00 \leq z \leq +2.00) = 0.4772 + 0.4772 = 0.9544 \cong 95\%$$

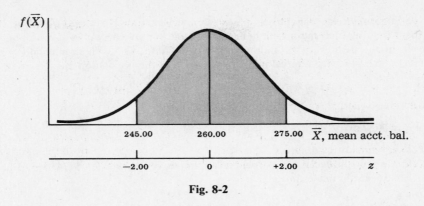

Fig. 8-2

8.3 CONFIDENCE INTERVALS FOR THE MEAN USING THE NORMAL DISTRIBUTION

Examples 4 and 5 above are concerned with determining the probability that the sample mean will have various values given that the population mean and standard deviation are known. What is involved is *deductive reasoning* (reasoning from the general truth to a specific instance) with respect to the sample result based on known population parameters. We now concern ourselves with *inductive reasoning* (reasoning from a particular instance to conclude the general truth) by using sample data to make statements about the value of the population mean.

The methods of interval estimation in this section are based on the assumption that the normal probability distribution can be used. As discussed in Sections 8.2 and 8.5, this assumption is warranted (1) whenever $n \geq 30$, because of the central limit theorem, or (2) when $n < 30$ but the population is normally distributed and σ is known.

Although the sample mean is useful as an unbiased estimator of the population mean, there is no way of expressing the degree of accuracy of a point estimator. In fact, mathematically speaking, the probability that the sample mean is *exactly* correct as an estimator is $P = 0$. A *confidence interval* for the mean is an estimate interval constructed with respect to the sample mean by which the likelihood that the interval includes the value of the population mean can be specified. The *degree of confidence* associated with a confidence interval indicates the long-run percentage of such intervals which would include the parameter being estimated.

Confidence intervals for the mean typically are constructed with the unbiased estimator $\bar{X}$ at the midpoint of the interval. However, Problems 8.14 and 8.15 demonstrate the construction of a so-called "one-sided" confidence interval, for which the sample mean is not at the midpoint of the interval. When use of the normal probability distribution is warranted, the confidence interval for the mean is determined by

$$\bar{X} \pm z\sigma_{\bar{x}} \qquad (8.6)$$

or

$$\bar{X} \pm zs_{\bar{x}} \qquad (8.7)$$

The most frequently used confidence intervals are the 90 percent, 95 percent, and 99 percent confidence intervals. The values of z required in conjunction with such intervals are given in Table 8.2.

Table 8.2   Selected Proportions of Area under the Normal Curve

z (the number of standard deviation units from the mean)	Proportion of area in the interval $\mu \pm z\sigma$
1.645	0.90
1.96	0.95
2.58	0.99

EXAMPLE 6. For a given week, a random sample of 30 hourly employees selected from a very large number of employees in a manufacturing firm has a sample mean wage of $\bar{X} = \$180.00$ with a sample standard deviation of $s = \$14.00$. We estimate the mean wage for all hourly employees in the firm with an interval estimate such that we can be 95 percent confident that the interval includes the value of the population mean, as follows:

$$\bar{X} \pm 1.96 s_{\bar{x}} = 180.000 \pm 1.96(2.56) = \$174.98 \text{ to } \$185.02$$

where $\bar{X} = \$180.00$ (as given)

$$s_{\bar{x}} = s/\sqrt{n} = 14.00/\sqrt{30} = 2.56$$

(*Note:* s is used as an estimator of σ.)

Thus, we can state that the mean wage level for all employees is between $174.98 and $185.02, with a 95 percent degree of confidence in this estimate.

In addition to estimating the value of the population mean as such, there is sometimes an interest in estimating the total quantity or amount in the population. See Problem 8.11(*b*).

8.4 DETERMINING THE REQUIRED SAMPLE SIZE FOR ESTIMATING THE MEAN

Suppose the desired size of a confidence interval and the degree of confidence to be associated with it are specified. If σ is known or can be estimated, such as from the results of similar studies, the required sample size based on use of the normal distribution is

$$n = \left(\frac{z\sigma}{E}\right)^2 \tag{8.8}$$

In formula (*8.8*), z is the value used for the specified degree of confidence, σ is the standard deviation of the population (or estimate thereof), and E is the "plus and minus" error factor allowed in the interval (always one-half the total confidence interval). (*Note:* When solving for sample size, any fractional result is always rounded up. Further, unless σ is known *and* the population is normally distributed, any computed sample size below 30 should be increased to 30 because formula (*8.8*) is based on use of the normal distribution.)

EXAMPLE 7. A personnel department analyst wishes to estimate the mean number of training hours annually for foremen in a division of the company within 3.0 hr (plus or minus) and with 90 percent confidence. Based on data from other divisions, he estimates the standard deviation of training hours to be $\sigma = 20.0$ hr. The minimum required sample size is

$$n = \left(\frac{z\sigma}{E}\right)^2 = \left[\frac{(1.645)(20.0)}{3.0}\right]^2 = \left(\frac{32.9}{3.0}\right)^2 = 120.27 \cong 121$$

8.5 STUDENT'S *t* DISTRIBUTION AND CONFIDENCE INTERVALS FOR THE MEAN

In Section 8.3 we indicated that use of the normal distribution in estimating a population mean is warranted for any large sample ($n \geq 30$), and for a small sample ($n < 30$) only if the population is normally distributed *and* σ is known. In this section we handle the situation in which the sample is small and the population is normally distributed, but σ is not known.

If a population is normally distributed, the sampling distribution of the mean for any sample size will also be normally distributed; this is true whether σ is known or not. However, in the process of inference each value of the mean is converted to a standard normal value, *and herein lies the problem.* If σ is unknown, the conversion formula $(\bar{X} - \mu)/s_{\bar{x}}$ includes a variable in the denominator, because s, and therefore $s_{\bar{x}}$, will be somewhat different from sample to sample. The result is that including the variable $s_{\bar{x}}$ rather than the constant $\sigma_{\bar{x}}$ in the denominator results in converted values that are *not* distributed as z values. Instead, the values are distributed according to Student's t distribution, which is platykurtic (flat) as compared with the normal distribution. Appendix 6 indicates proportions of area under the t distribution, with the specific distribution being based on the degrees of freedom (df) involved. For the case of a single sample, $df = n - 1$.

The distribution is appropriate for inferences concerning the mean whenever σ is not known and the population is normally distributed, regardless of sample size. However, as the sample size (and df) is increased, the t distribution approaches the normal distribution in form. A general rule is that a t distribution can be approximated by the normal distribution when $n \geq 30$ (or $df \geq 29$) for a single sample. This substitution is a different matter from that covered by the central limit theorem, and the fact that a sample of $n \geq 30$ is required in both cases is convenient.

Note that the values of t reported in Appendix 6 indicate the proportion in the upper "tail" of the distribution, rather than the proportion between the mean and a given point, as in Appendix 5 for the normal distribution. Where $df = n - 1$, the confidence interval for estimating the population mean when σ is not known, $n < 30$, and the population is normally distributed is

$$\bar{X} \pm t_{df} s_{\bar{x}} \qquad (8.9)$$

EXAMPLE 8.   The mean operating life for a random sample of $n = 10$ light bulbs is $\bar{X} = 4,000$ hr with the sample standard deviation $s = 200$ hr. The operating life of bulbs in general is assumed to be approximately normally distributed. We estimate the mean operating life for the population of bulbs from which this sample was taken, using a 95 percent confidence interval, as follows:

$$95\% \text{ Int.} = \bar{X} \pm t_{df} s_{\bar{x}}$$
$$= 4,000 \pm (2.262)(63.3)$$
$$= 3,856.8 \text{ to } 4,143.2 \cong 3,857 \text{ to } 4,143 \text{ hr}$$

where $\bar{X} = 4,000$ (as given)

$t_{df} = t_{n-1} = t_9 = 2.262$

$s_{\bar{x}} = \dfrac{s}{\sqrt{n}} = \dfrac{200}{\sqrt{10}} = \dfrac{200}{3.16} = 63.3$

8.6   SUMMARY TABLE FOR INTERVAL ESTIMATION OF THE POPULATION MEAN

Table 8.3   Interval Estimation of the Population Mean

Population	Sample size	σ known	σ unknown
Normally distributed	Large ($n \geq 30$)	$\bar{X} \pm z\sigma_{\bar{x}}$	$\bar{X} \pm zs_{\bar{x}}$**
	Small ($n < 30$)	$\bar{X} \pm z\sigma_{\bar{x}}$	$\bar{X} \pm ts_{\bar{x}}$
Not normally distributed	Large ($n \geq 30$)	$\bar{X} \pm z\sigma_{\bar{x}}$*	$\bar{X} \pm zs_{\bar{x}}$†
	Small ($n < 30$)	Nonparametric procedures directed toward the median generally would be used. (See Chapter 21.)	

*Central limit theorem is invoked.

**z is used as an approximation of t.

†Central limit theorem is invoked, and z is used as an approximation of t.

8.7   COMPUTER OUTPUT

Computer software used for statistical analysis generally allows the user to specify the percent confidence interval for the mean that is desired based on random sample data. Problems 8.18 and 8.19 are concerned with use of the computer for determining confidence intervals for the population mean.

Solved Problems

SAMPLING DISTRIBUTION OF THE MEAN

8.1   For a particular brand of TV picture tube, it is known that the mean operating life of the tubes is $\mu = 9,000$ hr with a standard deviation of $\sigma = 500$ hr. Determine the expected value and standard error of the sampling distribution of the mean given a sample size of $n = 25$. Interpret the meaning of the computed values.

$$E(\bar{X}) = \mu = 9,000$$

$$\sigma_{\bar{x}} = \frac{\sigma}{\sqrt{n}} = \frac{500}{\sqrt{25}} = \frac{500}{5} = 100$$

These calculations indicate that in the long run the mean of a large group of sample means, each based on a sample size of $n = 25$, will be equal to 9,000 hr. Further, the variability of these sample means in respect to the expected value of 9,000 hr is expressed by a standard deviation of 100 hr.

8.2   For a large population of normally distributed account balances, the mean balance is $\mu = \$150.00$ with standard deviation $\sigma = \$35.00$. What is the probability that *one* randomly sampled account has a balance that exceeds $160.00?

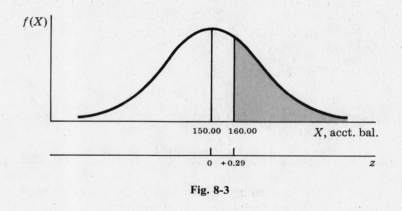

Fig. 8-3

Figure 8-3 portrays the probability curve for the variable.

$$z = \frac{X - \mu}{\sigma} = \frac{160.00 - 150.00}{35.00} = +0.29$$

$$P(X > 160.00 \mid \mu = 150.00, \ \sigma = 35.00) = P(z > +0.29)$$

$$= 0.5000 - P(0 \le z \le +0.29) = 0.5000 - 0.1141 = 0.3859$$

8.3   With reference to Problem 8.2 above, what is the probability that the *mean* for a random sample of $n = 40$ accounts will exceed $160.00?

Figure 8-4 portrays the probability curve for the sampling distribution $E(\bar{X}) = \mu = \$150.00$.

$$\sigma_{\bar{x}} = \frac{\sigma}{\sqrt{n}} = \frac{35.00}{\sqrt{40}} = \$5.53$$

$$z = \frac{\bar{X} - \mu}{\sigma_{\bar{x}}} = \frac{160.00 - 150.00}{5.53} = \frac{10.00}{5.53} = +1.81$$

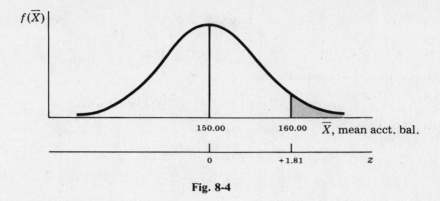

Fig. 8-4

Therefore,

$$P(\bar{X} > 160.00 \mid \mu = 150.00, \sigma_{\bar{x}} = 5.53) = P(z > +1.81)$$
$$= 0.5000 - P(0 \le z \le +1.81)$$
$$= 0.5000 - 0.4649 = 0.0351$$

8.4   This problem and the following two problems serve to illustrate the meaning of the sampling distribution of the mean by reference to a highly simplified population. Suppose a population consists of just the four values 3, 5, 7, and 8. Compute (a) the population mean μ and (b) the population standard deviation σ.

With reference to Table 8.4,

(a) $\mu = \dfrac{\Sigma X}{N} = \dfrac{23}{4} = 5.75$

(b) $\sigma = \sqrt{\dfrac{\Sigma X^2}{N} - \left(\dfrac{\Sigma X}{N}\right)^2} = \sqrt{\dfrac{147}{4} - \left(\dfrac{23}{4}\right)^2}$
$= \sqrt{36.75 - (5.75)^2} = \sqrt{36.75 - 33.0625} = 1.92$

Table 8.4   Worksheet for Problem 8.4

X	X^2
3	9
5	25
7	49
8	64
$\Sigma X = 23$	$\Sigma X^2 = 147$

8.5   For the population described in Problem 8.4, suppose that simple random samples of size $n = 2$ each are taken from this population. For each sample the first sampled item is *not* replaced in the population before the second item is sampled.

(a)   List all possible pairs of values which can constitute a sample.

(b)   For each of the pairs identified in (a), compute the sample mean $\bar{X}$ and demonstrate that the mean of all possible sample means $\mu_{\bar{x}}$ is equal to the mean of the population from which the samples were selected.

(*a*) and (*b*)   From Table 8.5,

$$\mu_{\bar{x}} = \frac{\Sigma \bar{X}}{N_{\text{samples}}} = \frac{34.5}{6} = 5.75$$

[which equals μ as computed in Problem 8.4(*a*)].

Table 8.5   Possible Samples and Sample Means for Problem 8.5

Possible samples	$\bar{X}$
3, 5	4.0
3, 7	5.0
3, 8	5.5
5, 7	6.0
5, 8	6.5
7, 8	7.5
	$\Sigma \bar{X} = 34.5$

8.6   For the sampling situation described in Problems 8.4 and 8.5, compute the standard error of the mean by determining the standard deviation of the six possible sample means identified in Problem 8.5 in respect to the population mean μ. Then compute the standard error of the mean based on σ being known and sampling from a finite population, using the appropriate formula from this chapter. Verify that the two standard error values are the same.

With reference to Table 8.5,

Table 8.6   Worksheet for Problem 8.6

$\bar{X}$	$\bar{X}^2$
4.0	16.00
5.0	25.00
5.5	30.25
6.0	36.00
6.5	42.25
7.5	56.25
$\Sigma \bar{X} = 34.5$	$\Sigma \bar{X}^2 = 205.75$

First method:

$$\sigma_{\bar{x}} = \sqrt{\frac{\Sigma \bar{X}^2}{N_S} - \left(\frac{\Sigma \bar{X}}{N_S}\right)^2} = \sqrt{\frac{205.75}{6} - \left(\frac{34.5}{6}\right)^2}$$
$$= \sqrt{34.2917 - (5.75)^2} = \sqrt{34.2917 - 33.0625} = 1.11$$

Second method:

$$\sigma_{\bar{x}} = \frac{\sigma}{\sqrt{n}} \sqrt{\frac{N-n}{N-1}} = \frac{1.92}{\sqrt{2}} \sqrt{\frac{4-2}{4-1}}$$
$$= \frac{1.92}{1.414} \sqrt{0.6666} = 1.358(0.816) = 1.11$$

Of course, the second method is the one always used for determining the standard error of the mean in actual data situations. But conceptually, the first method illustrates more directly the meaning of the standard error of the mean.

CONFIDENCE INTERVALS FOR THE MEAN USING THE NORMAL DISTRIBUTION

8.7   Suppose that the standard deviation of the tube life for a particular brand of TV picture tube is known to be $\sigma = 500$, but that the mean operating life is not known. Overall, the operating life of the tubes is assumed to be approximately normally distributed. For a sample of $n = 15$, the mean operating life is $\bar{X} = 8,900$ hr. Construct (a) the 95 percent and (b) the 90 percent confidence intervals for estimating the population mean.

The normal probability distribution can be used in this case because the population is normally distributed and σ is known.

(a) $\bar{X} \pm z\sigma_{\bar{x}} = 8,900 \pm 1.96 \dfrac{\sigma}{\sqrt{n}}$

$$= 8,900 \pm 1.96 \frac{500}{\sqrt{15}} = 8,900 \pm 1.96 \left(\frac{500}{3.87}\right)$$

$$= 8,900 \pm 1.96(129.20) = 8,647 \text{ to } 9,153$$

(b) $\bar{X} \pm z\sigma_{\bar{x}} = 8,900 \pm 1.645(129.20) = 8,687 \text{ to } 9,113 \text{ hr}$

8.8   With respect to Problem 8.7, suppose that the population of tube life cannot be assumed to be normally distributed. However, the sample mean of $\bar{X} = 8,900$ is based on a sample of $n = 35$. Construct the 95 percent confidence interval for estimating the population mean.

The normal probability distribution can be used in this case by invoking the central limit theorem, which indicates that for $n \geq 30$ the sampling distribution can be assumed to be normally distributed even though the population is not normally distributed. Thus,

$$\bar{X} \pm z\sigma_{\bar{x}} = 8,900 \pm 1.96 \frac{\sigma}{\sqrt{n}} = 8,900 \pm 1.96 \frac{500}{\sqrt{35}}$$

$$= 8,900 \pm 1.96 \left(\frac{500}{5.92}\right) = 8,900 \pm 1.96(84.46) = 8,734 \text{ to } 9,066 \text{ hr}$$

8.9   With respect to Problem 8.8, suppose that the population can be assumed to be normally distributed, but that the population standard deviation is not known. Rather, the sample standard deviation $s = 500$ and $\bar{X} = 8,900$. Estimate the population mean using a 90 percent confidence interval.

Because $n \geq 30$ the normal distribution can be used as an approximation of the t distribution. However, because the population is normally distributed, the central limit theorem need not be invoked. Therefore,

$$\bar{X} \pm zs_{\bar{x}} = 8,900 \pm 1.645 \left(\frac{500}{\sqrt{35}}\right) = 8,900 \pm 1.645(84.46) = 8,761 \text{ to } 9,039 \text{ hr}$$

8.10   With respect to Problems 8.8 and 8.9, suppose that the population *cannot* be assumed to be normally distributed and, further, that the population σ is not known. As before, $n = 35$, $s = 500$, and $\bar{X} = 8,900$. Estimate the population mean using a 99 percent confidence interval.

In this case the central limit theorem is invoked, as in Problem 8.8, and z is used as an approximation of t, as in Problem 8.9.

$$\bar{X} \pm zs_{\bar{x}} = 8,900 \pm 2.58 \left(\frac{500}{\sqrt{35}}\right) = 8,900 \pm 2.58(84.46) = 8,682 \text{ to } 9,118 \text{ hr}$$

8.11   A marketing research analyst collects data for a random sample of 100 customers out of the 400 who purchased a particular "coupon special." The 100 people spent an average of $\bar{X} = \$24.57$

in the store with a standard deviation of $s = \$6.60$. Using a 95 percent confidence interval, estimate (*a*) the mean purchase amount for all 400 customers and (*b*) the total dollar amount of purchases by the 400 customers.

(*a*) $s_{\bar{x}} = \dfrac{s}{\sqrt{n}} \sqrt{\dfrac{N-n}{N-1}} = \dfrac{6.60}{\sqrt{100}} \sqrt{\dfrac{400-100}{400-1}}$

$= \dfrac{6.60}{10} \sqrt{0.7519} = 0.660(0.867) = 0.57$

$\bar{X} \pm z s_{\bar{x}} = 24.57 \pm 1.96(0.57) = \$23.45 \text{ to } \$25.69$

(*b*) $N(\bar{X} \pm z s_{\bar{x}}) = 400(\$23.45 \text{ to } \$25.69) = \$9,380 \text{ to } \$10,276$

or

$N\bar{X} \pm N(z s_{\bar{x}}) = 400(24.57) \pm 400(1.12)$

$= 9,828 \pm 448 = \$9,380 \text{ to } \$10,276$

(*Note:* The confidence interval for the dollar amount of purchases is simply the total number of customers *in the population* multiplied by the confidence limits for the mean purchase amount per customer. Such a population value is referred to as the *total quantity* in some textbooks.)

DETERMINING THE REQUIRED SAMPLE SIZE FOR ESTIMATING THE MEAN

8.12   A prospective purchaser wishes to estimate the mean dollar amount of sales per customer at a toy store located at an airlines terminal. Based on data from other similar airports, the standard deviation of such sales amounts is estimated to be about $\sigma = \$3.20$. What size of random sample should she collect, as a minimum, if she wants to estimate the mean sales amount within $1.00 and with 99 percent confidence?

$$n = \left(\frac{z\sigma}{E}\right)^2 = \left[\frac{(2.58)(3.20)}{1.00}\right]^2 = (8.256)^2 = 68.16 \cong 69$$

8.13   Referring to Problem 8.12, what is the minimum required sample size if the distribution of sales amounts is not assumed to be normal and the purchaser wishes to estimate the mean sales amount within $2.00 with 99 percent confidence?

$$n = \left(\frac{z\sigma}{E}\right)^2 = \left[\frac{(2.58)(3.20)}{2.00}\right]^2 = (4.128)^2 = 17.04 \cong 18$$

However, because the population is not assumed to be normally distributed, the minimum sample size is $n = 30$, so that the central limit theorem can be invoked as the basis for using the normal probability distribution for constructing the confidence interval.

ONE-SIDED CONFIDENCE INTERVALS FOR THE POPULATION MEAN

8.14   Occasionally, a *one-sided confidence interval* may be of greater interest than the usual two-sided interval. Such would be the case if we are interested only in the highest (or only in the lowest) value of the mean at the indicated degree of confidence. An "upper 95 percent interval" extends from a computed lower limit to positive infinity, with a proportion of 0.05 of the area under the normal curve being to the left of the lower limit. Similarly, a "lower 95 percent confidence interval" extends from negative infinity to a computed upper limit, with a proportion of 0.05 of the area under the normal curve being to the right of the upper limit.

Suppose a prospective purchaser of a toy store at an airlines terminal observes a random sample of $n = 64$ sales and finds that the sample mean is $\bar{X} = \$14.63$ with the sample standard deviation $s = \$2.40$. Determine the upper 95 percent confidence interval so that the *minimum* value of the population mean is identified with a 95 percent degree of confidence.

$$s_{\bar{x}} = \frac{s}{\sqrt{n}} = \frac{2.40}{\sqrt{64}} = \frac{2.40}{8} = 0.30$$

Upper 95% int. $= \bar{X} - zs_{\bar{x}} = 14.63 - 1.645(0.30) = \$14.14$ or higher

Thus, with a 95 percent degree of confidence we can state that the mean sales amount for the population of all customers is equal to or greater than $14.14.

8.15 With 99 percent confidence, what is the estimate of the maximum value of the mean sales amount in Problem 8.14?

Since $\bar{X} = \$14.63$ and $s_{\bar{x}} = 0.30$,

Lower 99% int. $= \bar{X} + zs_{\bar{x}} = 14.63 + 2.33(0.30) = \$15.33$ or less

Thus, with a 99 percent degree of confidence we can state that the mean sales amount is no larger than $15.33.

CONFIDENCE INTERVALS FOR THE MEAN BASED ON USING THE t DISTRIBUTION

8.16 In Problem 8.7 we constructed confidence intervals for estimating the mean operating life of a particular brand of TV picture tube based on the assumption that the operating life of all tubes is approximately normally distributed and $\sigma = 500$, and given a sample of $n = 15$ with $\bar{X} = 8,900$ hr. Suppose that σ is not known, but rather, that the sample standard deviation is $s = 500$.

(*a*) Construct the 95 percent confidence interval for estimating the population mean and compare this interval with the answer to Problem 8.7(*a*).

(*b*) Construct the 90 percent confidence interval for estimating the population mean and compare this interval with the answer to Problem 8.7(*b*).

(*Note:* Use of a t distribution is appropriate in this case because the population is assumed to be normally distributed, σ is not known, and the sample is small $(n < 30)$.)

(*a*) $\bar{X} \pm t_{df} s_{\bar{x}} = 8,900 \pm 2.145 \dfrac{s}{\sqrt{n}} = 8,900 \pm 2.145 \dfrac{500}{\sqrt{15}}$

$= 8,900 \pm 2.145 \left(\dfrac{500}{3.87} \right) = 8,900 \pm 2.145(129.199) = 8,623$ to $9,177$ hr

The confidence interval is wider than the one in Problem 8.7(*a*), reflecting the difference between the t distribution with $df = 15 - 1 = 14$, and the normal probability distribution.

(*b*) $\bar{X} \pm t_{df} s_{\bar{x}} = 8,900 \pm 1.761(129.199) = 8,672$ to $9,128$ hr

Again, the confidence interval is wider than the one in Problem 8.7(*b*).

8.17 As a commercial buyer for a private supermarket brand, suppose you take a random sample of 12 No. 303 cans of string beans at a canning plant. The net weight of the drained beans in each can is reported in Table 8.7. Determine (*a*) the mean net weight of string beans being packed

Table 8.7 Net Weight of Beans Packed in 12 No. 303 Cans

Ounces per can	15.7	15.8	15.9	16.0	16.1	16.2
No. of cans	1	2	2	3	3	1

in each can for this sample and (*b*) the sample standard deviation. (*c*) Assuming that the net weights per can are normally distributed, estimate the mean weight per can of beans being packed using a 95 percent confidence interval.

Table 8.8   Worksheet for Problem 8.17

X per can	No. of cans	Total X	X^2 per can	Total X^2
15.7	1	15.7	246.49	246.49
15.8	2	31.6	249.64	499.28
15.9	2	31.8	252.81	505.62
16.0	3	48.0	256.00	768.00
16.1	3	48.3	259.21	777.63
16.2	1	16.2	262.44	262.44
	$n = 12$	$\Sigma X = 191.6$		$\Sigma X^2 = 3{,}059.46$

Referring to Table 8.8,

(*a*) $\bar{X} = \dfrac{\Sigma X}{n} = \dfrac{191.6}{12} = 15.97 \text{ oz}$

(*b*) $s = \sqrt{\dfrac{n \, \Sigma X^2 - (\Sigma X)^2}{n(n-1)}} = \sqrt{\dfrac{12(3{,}059.46) - (191.6)^2}{12(11)}} = \sqrt{0.0224} = 0.15$

(*c*) $\bar{X} \pm t_{df} s_{\bar{x}} = 15.97 \pm t_{11} \dfrac{s}{\sqrt{n}} = 15.97 \pm 2.201 \left(\dfrac{0.15}{\sqrt{12}} \right)$

$\qquad = 15.97 \pm 2.201 \left(\dfrac{0.15}{3.46} \right) = 15.97 \pm 2.201(0.043) = 15.88 \text{ to } 16.06 \text{ oz}$

COMPUTER OUTPUT

8.18   Refer to the data in Table 2.6 (page 16) on the time required to complete an assembly task. Assuming that the data are based on a random sample, use an available computer program to obtain the 90 percent confidence interval for the population mean.

Figure 8-5 presents the input and output. The output includes the sample size, sample mean, sample standard deviation, and standard error of the mean, as well as the designated confidence interval. As indicated, the 90 percent confidence interval for the population mean assembly time is 12.544 to 14.056 min.

```
MTB ) SET ASSEMBLY TIMES INTO C1
DATA)    10    14    15    13    17    16    12    14    11    13    15    18    9
DATA)    14    14     9    15    11    13    11    12    10    17    16    12    11
DATA)    16    12    14    15
DATA) END
MTB ) NAME FOR C1 IS 'TIME'
MTB ) TINTERVAL WITH 90 PERCENT CONFIDENCE FOR 'TIME'

              N      MEAN     STDEV   SE MEAN     90.0 PERCENT C. I.
TIME         30    13.300     2.437     0.445   ( 12.544,   14.056)
```

Fig. 8-5   Minitab output for Problem 8.18.

8.19   Refer to the data reported in Table 8.7, concerning the net weight of beans packed in a random sample of $n = 12$ cans. Using available computer software, (*a*) determine the 95 percent confidence interval for the population mean and (*b*) compare the answers included in the output with those obtained by the hand calculations in Problem 8.17.

With reference to Figure 8-6,

(*a*)   95 percent conf. int. = 15.8715 to 16.0618 oz

(*b*)   Compared with the solutions in Problem 8.17, the values of the sample mean, standard deviation, standard error of the mean, and confidence limits differ just slightly because of the greater rounding of values included in the hand calculations.

```
MTB > SET WEIGHTS INTO C1
DATA) 15.7  15.8  15.8  15.9  15.9  16.0  16.0  16.0  16.1  16.1  16.1
DATA) 16.2
DATA) END
MTB > NAME FOR C1 IS 'WEIGHT'
MTB > TINTERVAL WITH 95 PERCENT CONFIDENCE FOR 'WEIGHT'

            N      MEAN    STDEV   SE MEAN    95.0 PERCENT C.I.
WEIGHT     12    15.9667   0.1497   0.0432  ( 15.8715, 16.0618)
```

Fig. 8-6   Minitab output for Problem 8.19.

Supplementary Problems

SAMPLING DISTRIBUTION OF THE MEAN

8.20   The mean dollar value of the sales amounts for a particular consumer product last year is known to be normally distributed with $\mu = \$3,400$ per retail outlet with a standard deviation of $\sigma = \$200$. If a large number of outlets handle the product, determine the standard error of the mean for a sample of size $n = 25$.

Ans. $40.00

8.21   Refer to Problem 8.20. What is the probability that the sales amount for *one* randomly sampled retail outlet will be (*a*) greater than $3,500? (*b*) Between $3,350 and $3,450?

Ans.   (*a*) 0.3085, (*b*) 0.1974

8.22   Refer to Problem 8.20. What is the probability that the *mean* will be (*a*) greater than $3,500? (*b*) Between $3,350 and $3,450? Compare your answers with the answers to Problem 8.21.

Ans.   (*a*) 0.0062, (*b*) 0.7888

8.23   Referring to Problem 8.20, suppose only 100 retail outlets handle the product. Determine the standard error of the mean for the sample of $n = 25$ in this case, and compare your answer with the answer to Problem 8.21.

Ans. $34.80

CONFIDENCE INTERVALS FOR THE MEAN

8.24   Suppose that you wish to estimate the mean sales amount per retail outlet for a particular consumer product during the past year. The number of retail outlets is large. Determine the 95 percent confidence interval given that the sales amounts are assumed to be normally distributed, $\bar{X} = \$3,425$, $\sigma = \$200$, and $n = 25$.

Ans. $3,346.60 to $3,503.40

8.25   Referring to Problem 8.24, determine the 95 percent confidence interval given that the population is assumed to be normally distributed, $\bar{X} = \$3,425$, $s = \$200$, and $n = 25$.

   Ans.   \$3,342.44 to \$3,507.56

8.26   For Problem 8.24, determine the 95 percent confidence interval given that the population is *not* assumed to be normally distributed, $\bar{X} = \$3,425$, $s = \$200$, and $n = 50$.

   Ans.   \$3,369.55 to \$3,480.45

8.27   For a sample of 50 firms taken from a particular industry the mean number of employees per firm is 420.4 with a sample standard deviation of 55.7. There is a total of 380 firms in this industry. Determine the standard error of the mean to be used in conjunction with estimating the population mean by a confidence interval.

   Ans.   7.33

8.28   For Problem 8.27, determine the 90 percent confidence interval for estimating the average number of workers per firm in the industry.

   Ans.   408.3 to 432.5

8.29   For the situations described in Problems 8.27 and 8.28, determine the 90 percent confidence interval for estimating the total number of workers employed in the industry.

   Ans.   155,154 to 164,350

8.30   An analyst in a personnel department randomly selects the records of 16 hourly employees and finds that the mean wage rate per hour is \$9.50. The wage rates in the firm are assumed to be normally distributed. If the standard deviation of the wage rates is known to be \$1.00, estimate the mean wage rate in the firm using an 80 percent confidence interval.

   Ans.   \$9.18 to \$9.82

8.31   Referring to Problem 8.30, suppose that the standard deviation of the population is not known, but that the standard deviation of the sample is \$1.00. Estimate the mean wage rate in the firm using an 80 percent confidence interval.

   Ans.   \$9.16 to \$9.84

8.32   The mean diameter of a sample of $n = 12$ cylindrical rods included in a shipment is 2.350 mm with a standard deviation of 0.050 mm. The distribution of the diameters of all of the rods included in the shipment is assumed to be approximately normal. Determine the 99 percent confidence interval for estimating the mean diameter of all of the rods included in the shipment.

   Ans.   2.307 to 2.393 mm

8.33   The mean diameter of a sample of $n = 100$ rods included in a shipment is 2.350 mm with a standard deviation of 0.050 mm. Estimate the mean diameter of all rods included in the shipment if the shipment contains 500 rods, using a 99 percent confidence interval.

   Ans.   2.338 to 2.362 mm

8.34   The mean weight per rod for the sample of 100 rods in Problem 8.33 is 8.45 g with a standard deviation of 0.25 g. Estimate the total weight of the entire shipment (exclusive of packing materials), using a 99 percent confidence interval.

   Ans.   4,195 to 4,255 g

DETERMINING THE REQUIRED SAMPLE SIZE FOR ESTIMATING THE MEAN

8.35   From historical records, the standard deviation of the sales level per retail outlet for a consumer product is known to be $\sigma = \$200$, and the population of sales amounts per outlet is assumed to be normally distributed. What is the minimum sample size required to estimate the mean sales per outlet within \$100 and with 95 percent confidence?

   Ans. $15.37 \cong 16$

8.36   An analyst wishes to estimate the mean hourly wage of workers in a particular company within 25¢ and 90 percent confidence. The standard deviation of the wage rates is estimated as being no larger than \$1.00. What is the number of personnel records that should be sampled, as a minimum, to satisfy this research objective?

   Ans. $43.56 \cong 44$

ONE-SIDED CONFIDENCE INTERVALS FOR THE POPULATION MEAN

8.37   Instead of the two-sided confidence interval constructed in Problem 8.24, suppose we wish to estimate the *minimum* value of the mean level of sales per retail outlet for a particular consumer product during the past year. As before, the distribution of sales amounts per store is assumed to be approximately normal. Determine the minimum value of the mean using a 95 percent confidence interval given that $\bar{X} = \$3,425$, $\sigma = \$200$, and $n = 25$. Compare your confidence interval with the one constructed in Problem 8.24.

   Ans.   Est. $\mu \geq \$3,359$

8.38   Using the data in Problem 8.32, determine the 99 percent lower confidence interval for estimating the mean diameter of all the rods included in the shipment. Compare the interval with the one constructed in Problem 8.32.

   Ans.   Est. $\mu \leq 2.388$ mm

COMPUTER OUTPUT

8.39   Refer to Table 2.16 (page 26) for the amounts of 40 personal loans. Assuming that these are random sample data, use available computer software to determine the 99 percent confidence interval for the mean loan amount in the population.

   Ans.   \$822 to \$1,372

Chapter 9

Other Confidence Intervals

9.1 CONFIDENCE INTERVALS FOR THE DIFFERENCE BETWEEN TWO POPULATION MEANS USING THE NORMAL DISTRIBUTION

There is often a need to estimate the difference between two population means, such as the difference between the wage levels in two firms. As indicated in Section 8.1, the unbiased point estimate of $(\mu_1 - \mu_2)$ is $(\bar{X}_1 - \bar{X}_2)$. The confidence interval is constructed in a manner similar to that used for estimating the mean, except that the relevant standard error for the sampling distribution is the standard error of the *difference* between means. Use of the normal distribution is based on the same conditions as for the sampling distribution of the mean (see Section 8.2), except that two samples are involved. The formula used for estimating the difference between two population means with confidence intervals is

$$(\bar{X}_1 - \bar{X}_2) \pm z\sigma_{\bar{x}_1 - \bar{x}_2} \qquad (9.1)$$

or

$$(\bar{X}_1 - \bar{X}_2) \pm zs_{\bar{x}_1 - \bar{x}_2} \qquad (9.2)$$

When the standard deviations of the two populations are known, the standard error of the difference between means is

$$\sigma_{\bar{x}_1 - \bar{x}_2} = \sqrt{\sigma_{\bar{x}_1}^2 + \sigma_{\bar{x}_2}^2} \qquad (9.3)$$

When the standard deviations of the populations are not known, the estimated standard error of the difference between means given that use of the normal distribution is appropriate is

$$s_{\bar{x}_1 - \bar{x}_2} = \sqrt{s_{\bar{x}_1}^2 + s_{\bar{x}_2}^2} \qquad (9.4)$$

The values of the standard errors of the respective means included in these formulas are calculated by the formulas given in Section 8.2, including the possibility of using finite correction factors when appropriate.

EXAMPLE 1. The mean weekly wage for a sample of $n = 30$ employees in a large manufacturing firm is $\bar{X} = \$280.00$ with a sample standard deviation of $s = \$14.00$. In another large firm a random sample of $n = 40$ hourly employees has a mean weekly wage of \$270.00 with a sample standard deviation of $s = \$10.00$. The 99 percent confidence interval for estimating the difference between the mean weekly wage levels in the two firms is

$$99\% \text{ Int.} = (\bar{X}_1 - \bar{X}_2) \pm zs_{\bar{x}_1 - \bar{x}_2}$$
$$= \$10.00 \pm 2.58(3.01)$$
$$= \$10.00 \pm 7.77$$
$$= \$2.23 \text{ to } \$17.77$$

where $\bar{X}_1 - \bar{X}_2 = \$280.00 - \$270.00 = \$10.00$

$z = 2.58$

$$s_{\bar{x}_1} = \frac{s_1}{\sqrt{n_1}} = \frac{14.00}{\sqrt{30}} = \frac{14.00}{5.477} = 2.56$$

$$s_{\bar{x}_2} = \frac{s_2}{\sqrt{n_2}} = \frac{10.00}{\sqrt{40}} = \frac{10.00}{6.325} = 1.58$$

$$s_{\bar{x}_1 - \bar{x}_2} = \sqrt{s_{\bar{x}_1}^2 + s_{\bar{x}_2}^2} = \sqrt{(2.56)^2 + (1.58)^2} = \sqrt{6.5536 + 2.4964} \cong 3.01$$

Thus, we can state the average weekly wage in the first firm is greater than the average in the second firm by an amount somewhere between \$2.23 and \$17.77, with 99 percent confidence in this interval estimate.

In addition to the two-sided confidence interval, a one-sided confidence interval for the difference between means can also be constructed. (See Problem 9.4.)

9.2 STUDENT'S t DISTRIBUTION AND CONFIDENCE INTERVALS FOR THE DIFFERENCE BETWEEN TWO POPULATION MEANS

As explained in Section 8.5, use of Student's t distribution is necessary when

(1) Population standard deviations σ are not known.

(2) Samples are small ($n < 30$). If samples are large, then t values can be approximated by the standard normal z.

(3) Populations are assumed to be approximately normally distributed (note that the central limit theorem cannot be invoked for small samples).

However, when the t distribution is used to define confidence intervals for the difference between two means, rather than for one population mean, an additional required assumption is

(4) The two (unknown) population variances are equal, $\sigma_1^2 = \sigma_2^2$.

Because of the above equality assumption, the first step in determining the standard error of the difference between means when the t distribution is to be used is to pool the two sample variances:

$$\hat{\sigma}^2 = \frac{(n_1 - 1)s_1^2 + (n_2 - 1)s_2^2}{n_1 + n_2 - 2} \qquad (9.5)$$

The standard error of the difference between means based on using the pooled variance estimate $\hat{\sigma}^2$ is

$$\hat{\sigma}_{\bar{x}_1 - \bar{x}_2} = \sqrt{\frac{\hat{s}^2}{n_1} + \frac{\hat{s}^2}{n_2}} \qquad (9.6)$$

Where $df = n_1 + n_2 - 2$, the confidence interval is

$$(\bar{X}_1 - \bar{X}_2) \pm t_{df}\hat{\sigma}_{\bar{x}_1 - \bar{x}_2} \qquad (9.7)$$

EXAMPLE 2. For a random sample of $n_1 = 10$ bulbs, the mean bulb life is $\bar{X}_1 = 4,600$ hr with $s_1 = 250$ hr. For another brand of bulbs the mean bulb life and standard deviation for a sample of $n_2 = 8$ bulbs are $\bar{X}_2 = 4,000$ hr and $s_2 = 200$ hr. The bulb life for both brands is assumed to be normally distributed. The 90 percent confidence interval for estimating the difference between the mean operating life of the two brands of bulbs is

$$90\% \text{ Int.} = (\bar{X}_1 - \bar{X}_2) \pm t_{16}\hat{\sigma}_{\bar{x}_1 - \bar{x}_2}$$

$$= 600 \pm 1.746(108.847) = 600 \pm 190 = 410 \text{ to } 790 \text{ hr}$$

where $\bar{X}_1 - \bar{X}_2 = 4,600 - 4,000 = 600$

$$t_{df} = t_{n_1 + n_2 - 2} = t_{16} = 1.746$$

$$\hat{\sigma}^2 = \frac{(n_1 - 1)s_1^2 + (n_2 - 1)s_2^2}{n_1 + n_2 - 2} = \frac{9(250)^2 + 7(200)^2}{10 + 8 - 2} = 52,656.25$$

$$\hat{\sigma}_{\bar{x}_1 - \bar{x}_2} = \sqrt{\frac{\hat{\sigma}^2}{n_1} + \frac{\hat{\sigma}^2}{n_2}} = \sqrt{\frac{52,656.25}{10} + \frac{52,656.25}{8}} = 108.847$$

Thus, we can state with 90 percent confidence that the first brand of bulbs has a mean life that is greater than that of the second brand by an amount between 410 and 790 hr.

Note that in the two-sample case it is possible for each sample to be small ($n < 30$), and yet the normal distribution could be used to approximate the t because $df \geq 29$. However, in such use the two populations must be assumed to be approximately normally distributed, because the central limit theorem cannot be invoked with respect to the small samples.

9.3   CONFIDENCE INTERVALS FOR THE PROPORTION USING THE NORMAL DISTRIBUTION

As explained in Section 6.4, the probability distribution which is applicable to proportions is the binomial probability distribution. However, the mathematics associated with constructing a confidence interval for an unknown population proportion on the basis of the Bernoulli process described in Section 6.3 are rather involved. Therefore, most textbooks utilize the normal distribution as an approximation of the binomial for constructing confidence intervals for proportions. As explained in Section 7.4, such approximation is appropriate when $n \geq 30$ and both $np \geq 5$ and $nq \geq 5$ (where $q = 1 - p$). *However, when the population proportion p (or π) is not known, most statisticians suggest that a sample of $n \geq 100$ should be taken.* Note that in the context of statistical estimation π is not known, but is estimated by $\hat{p}$.

The variance of the distribution of proportions (see Section 6.4) serves as the basis for the standard error. Given an observed sample proportion, $\hat{p}$, the estimated standard error of the proportion is

$$s_{\hat{p}} = \sqrt{\frac{\hat{p}(1-\hat{p})}{n}} \qquad (9.8)$$

In the context of statistical estimation, the population p (or π) would not be known because that is the value being estimated. If the population is finite then use of the finite correction factor is appropriate (see Section 8.2). As was the case for the standard error of the mean, use of this correction is generally not considered necessary if $n < 0.05 N$. The formula for the standard error of the proportion which includes the finite correction factor is

$$s_{\hat{p}} = \sqrt{\frac{\hat{p}(1-\hat{p})}{n}} \sqrt{\frac{N-n}{N-1}} \qquad (9.9)$$

Finally, the approximate confidence interval for a population proportion is

$$\hat{p} \pm z s_{\hat{p}} \qquad (9.10)$$

In addition to the two-sided confidence interval, a one-sided confidence interval for the population proportion can also be constructed. (See Problem 9.12.)

EXAMPLE 3.   A marketing research firm contacts a random sample of 100 men in a large community and finds that a sample proportion of 0.40 prefer the razor blades manufactured by the client firm to all other brands. The 95 percent confidence interval for the proportion of all men in the community who prefer the client firm's razor blades is determined as follows:

$$s_{\hat{p}} = \sqrt{\frac{\hat{p}(1-\hat{p})}{n}} = \sqrt{\frac{(0.40)(0.60)}{100}} = \sqrt{\frac{0.24}{100}} = \sqrt{0.0024} \cong 0.05$$

$$\hat{p} \pm z s_{\hat{p}} = 0.40 \pm 1.96(0.05)$$
$$= 0.40 \pm 0.098 \cong 0.40 \pm 0.10 = 0.30 \text{ to } 0.50$$

Therefore, with 95 percent confidence we estimate the proportion of all men in the community who prefer the client firm's blades to be somewhere between 0.30 and 0.50.

9.4   DETERMINING THE REQUIRED SAMPLE SIZE FOR ESTIMATING THE PROPORTION

Before a sample is actually collected, the minimum required sample size can be determined by specifying the degree of confidence required, the error which is acceptable, and by making an initial estimate of π, the unknown population proportion:

$$n = \frac{z^2 \pi (1 - \pi)}{E^2} \qquad (9.11)$$

In (9.11), z is the value used for the specified confidence interval, π is the initial estimate of the population proportion, and E is the "plus and minus" error factor allowed in the interval (always one-half the total confidence interval).

If an initial estimate of π is not possible, then it should be estimated as being 0.50. Such an estimate is "conservative" in that it is the value for which the largest sample size would be required. Under such an assumption, the general formula for sample size is simplified as follows:

$$n = \left(\frac{z}{2E}\right)^2 \tag{9.12}$$

(*Note:* When solving for sample size, any fractional result is always rounded up. Further, any computed sample size below 100 should be increased to 100 because formulas (9.11) and (9.12) are based on use of the normal distribution.)

EXAMPLE 4.   For the study in Example 3, suppose that before data were collected it was specified that the 95 percent interval estimate should be within ±0.05 and no prior judgment was made about the likely value of π. The minimum sample size which should be collected is

$$n = \left(\frac{z}{2E}\right)^2 = \left(\frac{1.96}{2(0.05)}\right)^2 = \left(\frac{1.96}{0.10}\right)^2 = (19.6)^2 = 384.16 = 385$$

In addition to estimating the population proportion, the *total number* in a category of the population can also be estimated. [See Problem 9.7(b).]

9.5   CONFIDENCE INTERVALS FOR THE DIFFERENCE BETWEEN TWO POPULATION PROPORTIONS

In order to estimate the difference between the proportions in two populations, the unbiased point estimate of $(\pi_1 - \pi_2)$ is $(\hat{p}_1 - \hat{p}_2)$. See Section 8.1. The confidence interval involves use of the standard error of the *difference* between proportions. Use of the normal distribution is based on the same conditions as for the sampling distribution of the proportion in Section 9.3, except that two samples are involved and the requirements apply to each of the two samples. The confidence interval for estimating the difference between two population proportions is

$$(\hat{p}_1 - \hat{p}_2) \pm z s_{\hat{p}_1 - \hat{p}_2} \tag{9.13}$$

The standard error of the difference between proportions is determined by (9.14), wherein the value of each respective standard error of the proportion is calculated as described in Section 9.3:

$$s_{\hat{p}_1 - \hat{p}_2} = \sqrt{s_{\hat{p}_1}^2 + s_{\hat{p}_2}^2} \tag{9.14}$$

EXAMPLE 5.   In Example 3 it was reported that a proportion of 0.40 men out of a random sample of 100 in a large community preferred the client firm's razor blades to all others. In another large community, 60 men out of a random sample of 200 men prefer the client firm's blades. The 90 percent confidence interval for the difference in the proportion of men in the two communities preferring the client firm's blades is

$$
\begin{aligned}
90\% \text{ Int.} &= (\hat{p}_1 - \hat{p}_2) \pm z s_{\hat{p}_1 - \hat{p}_2} \\
&= 0.100 \pm 1.645(0.059) \\
&= 0.100 \pm 0.097 \\
&= 0.003 \text{ to } 0.197
\end{aligned}
$$

where $\bar{p}_1 - \bar{p}_2 = 0.40 - 0.30 = 0.10$

$z = 1.645$

$$s_{\hat{p}_1}^2 = \frac{\hat{p}_1(1 - \hat{p}_1)}{n_1} = \frac{(0.40)(0.60)}{100} = 0.0024$$

$$s_{\hat{p}_2}^2 = \frac{\hat{p}_2(1-\hat{p}_2)}{n_2} = \frac{(0.30)(0.70)}{200} = 0.00105$$

$$s_{\hat{p}_1-\hat{p}_2} = \sqrt{s_{\hat{p}_1}^2 + s_{\hat{p}_2}^2} = \sqrt{0.0024 + 0.00105} = \sqrt{0.00345} \cong 0.059$$

9.6 THE χ^2 (CHI-SQUARE) DISTRIBUTION AND CONFIDENCE INTERVALS FOR THE VARIANCE AND STANDARD DEVIATION

Given a normally distributed population of values, the χ^2 (chi-square) distributions can be shown to be the appropriate probability distributions for the ratio $(n-1)s^2/\sigma^2$. There is a different chi-square distribution according to the value of $n-1$, which represents the degrees of freedom (df). Thus,

$$\chi_{df}^2 = \frac{(n-1)s^2}{\sigma^2} \qquad (9.15)$$

Because the sample variance is the unbiased estimator of the population variance, the long-run expected value of the above ratio is equal to the degrees of freedom, or $n-1$. However, in any given sample the sample variance generally is *not* identical in value to the population variance. Since the ratio above is known to follow a chi-square distribution, this probability distribution can be used for statistical inference concerning an unknown variance or standard deviation.

Chi-square distributions are not symmetrical. Therefore, a two-sided confidence interval for a variance or standard deviation involves the use of two different χ^2 values, rather than the "plus-and-minus" approach used with the confidence intervals based on the normal distribution. The formula for constructing a confidence interval for the population variance is

$$\frac{(n-1)s^2}{\chi_{df,\text{upper}}^2} \le \sigma^2 \le \frac{(n-1)s^2}{\chi_{df,\text{lower}}^2} \qquad (9.16)$$

The confidence interval for the population standard deviation is

$$\sqrt{\frac{(n-1)s^2}{\chi_{df,\text{upper}}^2}} \le \sigma \le \sqrt{\frac{(n-1)s^2}{\chi_{df,\text{lower}}^2}} \qquad (9.17)$$

Appendix 7 indicates the proportions of area under the chi-square distributions according to various degrees of freedom, or df. In the general formula above, the subscripts "upper" and "lower" identify the percentile points on the particular χ^2 distribution to be used for constructing the confidence interval. For example, for a 90 percent confidence interval the "upper" is $\chi_{0.95}^2$ and the "lower" is $\chi_{0.05}^2$. By excluding the highest 5 percent and lowest 5 percent of the chi-square distribution, what remains is the "middle" 90 percent.

EXAMPLE 6.   The mean weekly wage for a sample of 30 hourly employees in a large firm is $\bar{X} = \$280.00$ with a sample standard deviation of $s = \$14.00$. The weekly wage amounts in the firm are assumed to be approximately normally distributed. The 95 percent confidence interval for estimating the standard deviation of weekly wages is

$$\sqrt{\frac{(n-1)s^2}{\chi_{df,\text{upper}}^2}} \le \sigma \le \sqrt{\frac{(n-1)s^2}{\chi_{df,\text{lower}}^2}}$$

$$\sqrt{\frac{(29)(196.00)}{\chi_{29,0.975}^2}} \le \sigma \le \sqrt{\frac{(29)(196.00)}{\chi_{29,0.025}^2}}$$

$$\sqrt{\frac{5,684.00}{45.72}} \le \sigma \le \sqrt{\frac{5,684.00}{16.05}}$$

$$\sqrt{124.3220} \le \sigma \le \sqrt{354.1433}$$

$$11.15 \le \sigma \le 18.82$$

In the above example, note that because the column headings in **Appendix 7** are right-tail probabilities, rather than percentile values, the column headings which are used in the table are the complementary values of the required "upper" and "lower" percentile values.

As an alternative to a two-sided confidence interval, a one-sided confidence interval for the variance or standard deviation can also be constructed. (See Problem 9.14.)

9.7 COMPUTER OUTPUT

Computer software for statistical analysis generally includes the capability to determine a number of types of confidence intervals, in addition to confidence intervals for the population mean. Problem 9.15 illustrates the use of the computer to obtain a confidence interval for the difference between two population means.

Solved Problems

CONFIDENCE INTERVALS FOR THE DIFFERENCE BETWEEN TWO POPULATION MEANS USING THE NORMAL DISTRIBUTION

9.1 For the study reported in Problem 8.11, suppose that there were 900 customers who did not purchase the "coupon special" but who did make other purchases in the store during the period of the study. For a sample of 200 of these customers, the mean purchase amount was $\bar{X} = \$19.60$ with a sample standard deviation of $s = \$8.40$.

(a) Estimate the mean purchase amount for the noncoupon customers, using a 95 percent confidence interval.

(b) Estimate the difference between the mean purchase amount of coupon and noncoupon customers, using a 90 percent confidence interval.

(a) $s_{\bar{x}} = \dfrac{s}{\sqrt{n}} \sqrt{\dfrac{N-n}{N-1}} = \dfrac{8.40}{\sqrt{200}} \sqrt{\dfrac{900-200}{900-1}} = \dfrac{8.40}{14.142} \sqrt{0.7786} = (0.594)(0.882) = 0.52$

$\bar{X} \pm z s_{\bar{x}} = 19.60 \pm 1.96(0.52) = \$19.60 \pm 1.02 = \$18.58 \text{ to } \$20.62$

(b) $s_{\bar{x}_1-\bar{x}_2} = \sqrt{s_{\bar{x}_1}^2 + s_{\bar{x}_2}^2} = \sqrt{(0.57)^2 + (0.52)^2} = \sqrt{0.3249 + 0.2704} = \sqrt{0.5953} = 0.772$

$(\bar{X}_1 - \bar{X}_2) \pm z s_{\bar{x}_1-\bar{x}_2} = (24.57 - 19.60) \pm 1.645(0.772) = \$4.97 \pm 1.27 = \$3.70 \text{ to } \$6.24$

Thus, we can state with 90 percent confidence that the mean level of sales for coupon customers exceeds that for noncoupon customers by an amount somewhere between $3.70 and $6.24.

9.2 A random sample of 50 households in community A has a mean household income of $\bar{X} = \$34,600$ with a standard deviation $s = \$2,200$. A random sample of 50 households in community B has a mean of $\bar{X} = \$33,800$ with a standard deviation of $s = \$2,800$. Estimate the difference in the average household income in the two communities using a 95 percent confidence interval.

$$s_{\bar{x}_1} = \frac{s_1}{\sqrt{n_1}} = \frac{2,200}{\sqrt{50}} = \frac{2,200}{7.07} = \$311.17$$

$$s_{\bar{x}_2} = \frac{s_2}{\sqrt{n_2}} = \frac{2,800}{\sqrt{50}} = \frac{2,800}{7.07} = \$396.04$$

$$s_{\bar{x}_1-\bar{x}_2} = \sqrt{s_{\bar{x}_1}^2 + s_{\bar{x}_2}^2} = \sqrt{(311.17)^2 + (396.04)^2} = \sqrt{96,826.77 + 156,847.68} = \$503.66$$

$$(\bar{X}_1 - \bar{X}_2) \pm z s_{\bar{x}_1-\bar{x}_2} = (34,600 - 33,800) \pm 1.96(503.66)$$

$$= 800 \pm 987.17 = -\$187.17 \text{ to } \$1,787.17$$

With a 95 percent degree of confidence, the limits of the confidence interval indicate that the mean in the first community might be less than the mean in the second community by \$187.17, while at the other limit the mean of the first community might exceed that in the second by as much as \$1,787.17. Note that the possibility that there is no actual difference between the two population means is included within this 95 percent confidence interval.

CONFIDENCE INTERVALS FOR THE DIFFERENCE BETWEEN TWO MEANS USING THE t DISTRIBUTION

9.3   In one canning plant the average net weight of string beans being packed in No. 303 cans for a sample of $n = 12$ cans is $\bar{X}_1 = 15.97$ oz, with $s_1 = 0.15$ oz. At another canning plant the average net weight of string beans being packed in No. 303 cans for a sample of $n_2 = 15$ cans is $\bar{X}_2 = 16.14$ oz with a standard deviation of $s_2 = 0.09$ oz. The distributions of the amounts packed are assumed to be approximately normal. Estimate the difference between the average net weight of beans being packed in No. 303 cans at the two plants, using a 90 percent confidence interval.

$$\hat{\sigma}^2 = \frac{(n_1-1)s_1^2 + (n_2-1)s_2^2}{n_1+n_2-2} = \frac{11(0.15)^2 + 14(0.09)^2}{12+15-2} = 0.014436$$

$$\hat{\sigma}_{\bar{x}_1-\bar{x}_2} = \sqrt{\frac{\hat{\sigma}^2}{n_1} + \frac{\hat{\sigma}^2}{n_2}} = \sqrt{\frac{0.014436}{12} + \frac{0.014436}{15}} = 0.047$$

$$(\bar{x}_1 - \bar{x}_2) \pm t_{df}\hat{\sigma}_{\bar{x}_1-\bar{x}_2} = (15.97 - 16.14) \pm t_{25}(0.047)$$

$$= (-0.17) \pm 1.708(0.047) = (-0.17) \pm 0.08 = -0.25 \text{ to } -0.09$$

In other words, with 90 percent confidence we can state the average net weight being packed at the *second* plant is somewhere between 0.09 and 0.25 oz more than at the first plant.

ONE-SIDED CONFIDENCE INTERVALS FOR THE DIFFERENCE BETWEEN TWO MEANS

9.4   Just as for the mean, as explained in Problem 8.14, a difference between means can be estimated by the use of a one-sided confidence interval. Referring to the data in Problem 9.1(*b*), estimate the minimum difference between the mean purchase amounts of "coupon" and "noncoupon" customers by constructing a 90 percent upper confidence interval.

Since, from Problem 9.1(*b*), $\bar{X}_1 - \bar{X}_2 = \$4.97$ and $s_{\bar{x}_1-\bar{x}_2} = 0.772$,

$$\text{Est. } (\mu_1 - \mu_2) \geq (\bar{X}_1 - \bar{X}_2) - zs_{\bar{x}_1-\bar{x}_2}$$

$$\geq \$4.97 - (1.28)(0.772)$$

$$\geq \$3.98$$

9.5   For the income data reported in Problem 9.2, estimate the maximum difference between the mean income levels in the first and second community by constructing a 95 percent lower confidence interval.

Since, from Problem 9.2, $\bar{X}_1 - \bar{X}_2 = \$800$ and $s_{\bar{x}_1-\bar{x}_2} = \$503.66$,

$$\text{Est. } (\mu_1 - \mu_2) \leq (\bar{X}_1 - \bar{X}_2) + zs_{\bar{x}_1-\bar{x}_2}$$

$$\leq \$800 + 1.645(503.66)$$

$$\leq \$1,628.52$$

CONFIDENCE INTERVALS FOR ESTIMATING THE POPULATION PROPORTION

9.6   A college administrator collects data on a nationwide random sample of 230 students enrolled in graduate programs in business administration, and finds that 54 of these students have undergraduate degrees in business. Estimate the proportion of such students nationwide who have undergraduate degrees in business administration, using a 90 percent confidence interval.

$$\hat{p} = \frac{54}{230} = 0.235$$

$$s_{\hat{p}} = \sqrt{\frac{\hat{p}(1-\hat{p})}{n}} = \sqrt{\frac{(0.235)(0.765)}{230}} = \sqrt{\frac{0.179775}{230}} = \sqrt{0.0007816} = 0.028$$

$$\hat{p} \pm z s_{\hat{p}} = 0.235 \pm 1.645(0.028)$$
$$= 0.235 \pm 0.046 \cong 0.19 \text{ to } 0.28$$

It is assumed here that the number of such students nationwide is large enough that the finite correction factor is not required.

9.7   In a large metropolitan area in which a total of 800 gasoline service stations are located, for a random sample of $n = 36$ stations, 20 of them carry a particular nationally advertised brand of oil. Using a 95 percent confidence interval estimate (a) the proportion of all stations in the area which carry the oil and (b) the total number of service stations in the area which carry the oil.

(a) $\hat{p} = \dfrac{20}{36} = 0.5555 \cong 0.56$

$$s_{\hat{p}} = \sqrt{\frac{\hat{p}(1-\hat{p})}{n}} = \sqrt{\frac{(0.56)(0.44)}{36}} = \sqrt{\frac{0.2464}{36}} = \sqrt{0.006844} = 0.083$$

$\hat{p} \pm z s_{\hat{p}} = 0.56 \pm 1.96(0.083) = 0.40 \text{ to } 0.72$

(b)   [*Note:* As was the case in the solution to Problem 8.11(b) for the mean and the total quantity, the *total number* in a category of the population is determined by multiplying the confidence limits for the proportion by the total number of all elements in the population.]

$$N(\hat{p} \pm z s_{\hat{p}}) = 800(0.40 \text{ to } 0.72) = 320 \text{ to } 576 \text{ stations}$$

or $$N(\hat{p}) \pm N(z s_{\hat{p}}) = 800(0.56) \pm 800(0.16) = 320 \text{ to } 576 \text{ stations}$$

DETERMINING THE REQUIRED SAMPLE SIZE FOR ESTIMATING THE PROPORTION

9.8   A university administrator wishes to estimate the proportion of students enrolled in graduate programs in business administration who also have undergraduate degrees in business administration within ±0.05 and with 90 percent confidence. What sample size should he collect, as a minimum, if there is no basis for estimating the approximate value of the proportion before the sample is taken?

Using formula (*9.12*),

$$n = \left(\frac{z}{2E}\right)^2 = \left(\frac{1.645}{2(0.05)}\right)^2 = (16.45)^2 = 270.60 \cong 271$$

9.9   With respect to Problem 9.8, what is the minimum sample size required if prior data and information indicate that the proportion will be no larger than 0.30?

From formula (*9.11*),

$$n = \frac{z^2 \pi (1-\pi)}{E^2} = \frac{(1.645)^2(0.30)(0.70)}{(0.05)^2} = 227.31 \cong 228$$

CONFIDENCE INTERVALS FOR THE DIFFERENCE BETWEEN TWO POPULATION PROPORTIONS

9.10 In attempting to gauge voter sentiment regarding a school bonding proposal, a superintendent of schools collects random samples of $n = 100$ in each of the two major residential areas included within the school district. In the first area 70 of the 100 sampled voters indicate that they intend to vote for the proposal, while in the second area 50 of 100 sampled voters indicate this intention. Estimate the difference between the actual proportions of voters in the two areas who intend to vote for the proposal, using 95 percent confidence limits.

$$s_{\hat{p}_1}^2 = \frac{\hat{p}_1(1-\hat{p}_1)}{n_1} = \frac{(0.70)(0.30)}{100} = \frac{0.21}{100} = 0.0021$$

$$s_{\hat{p}_2}^2 = \frac{\hat{p}_2(1-\hat{p}_2)}{n_2} = \frac{(0.50)(0.50)}{100} = \frac{0.25}{100} = 0.0025$$

Therefore,

$$s_{\hat{p}_1-\hat{p}_2} = \sqrt{s_{\hat{p}_1}^2 + s_{\hat{p}_2}^2} = \sqrt{0.0021 + 0.0025} = \sqrt{0.0046} = 0.068$$

$$(\hat{p}_1 - \hat{p}_2) \pm z s_{\hat{p}_1-\hat{p}_2} = (0.70 - 0.50) \pm 1.96(0.068) = 0.20 \pm 0.13 = 0.07 \text{ to } 0.33$$

Thus, the difference in the population proportions is somewhere between 0.07 and 0.33 (or 7 to 33 percent). In this solution it is assumed that the number of voters in each area is large enough that the finite correction factor is not required.

ONE-SIDED CONFIDENCE INTERVALS FOR PROPORTIONS

9.11 Just as for the mean and difference between means (see Problems 8.14, 9.4, and 9.5), a proportion or difference between proportions can be estimated by the use of a one-sided confidence interval. For the data of Problem 9.6, find the *minimum* proportion of the graduate students who have an undergraduate degree in business administration, using a 90 percent confidence interval.

Since, from Problem 9.6, $\hat{p} = 0.235$ and $s_{\hat{p}} = 0.028$,

$$\hat{p} - z s_{\hat{p}} = 0.235 - 1.28(0.028) = 0.199 \text{ or higher}$$

9.12 For the data of Problem 9.10, what is the upper 95 percent confidence interval for the difference in the proportions of people in the first and second neighborhoods who intend to vote for the bonding proposal?

$$(\hat{p}_1 - \hat{p}_2) - z s_{\hat{p}_1-\hat{p}_2} = (0.70 - 0.50) - 1.645(0.068) = 0.088 \text{ or higher}$$

Thus, with a 95 percent degree of confidence we can state that the *minimum* difference between the proportions of voters in the two neighborhoods who intend to vote in favor of the school bonding proposal is 0.088, or 8.8 percent.

CONFIDENCE INTERVALS FOR THE VARIANCE AND STANDARD DEVIATION

9.13 For the random sample of $n = 12$ cans of string beans in Problem 8.17, the mean was $\bar{X} = 15.97$ oz, the variance was $s^2 = 0.0224$, and the standard deviation was $s = 0.15$. Estimate the (*a*) variance and (*b*) standard deviation for all No. 303 cans of beans being packed in the plant, using 90 percent confidence intervals.

(*a*) $\quad \dfrac{(n-1)s^2}{\chi_{df,\text{upper}}^2} \le \sigma^2 \le \dfrac{(n-1)s^2}{\chi_{df,\text{lower}}^2}$

$\quad \dfrac{(11)(0.0224)}{\chi_{11,0.95}^2} \le \sigma^2 \le \dfrac{(11)(0.0224)}{\chi_{11,0.05}^2}$

$$\frac{0.2464}{19.68} \leq \sigma^2 \leq \frac{0.2464}{4.57}$$

$$0.0125 \leq \sigma^2 \leq 0.0539$$

(b) $\sqrt{0.0125} \leq \sigma \leq \sqrt{0.0539}$

$$0.11 \leq \sigma \leq 0.23$$

9.14   Just as for other confidence intervals (see Problems 8.14, 9.4, 9.5, and 9.11), a population variance or standard deviation can be estimated by the use of a one-sided confidence interval. The usual concern is with respect to the upper limit of the variance or standard deviation, and thus the lower confidence interval is the most frequent type of one-sided interval. For the data of Problem 9.13, what is the lower 90 percent confidence interval for estimating the population standard deviation?

$$\text{Est. } \sigma \leq \sqrt{\frac{(n-1)s^2}{\chi^2_{\text{df,lower}}}} \leq \sqrt{\frac{(11)(0.0224)}{\chi^2_{11,0.90}}}$$

$$\leq \sqrt{\frac{0.2464}{5.58}} \leq \sqrt{0.04416} \leq 0.21$$

Thus, with a 90 percent degree of confidence we can state that the standard deviation of the population is no larger than 0.21.

COMPUTER OUTPUT

9.15   Table 9.1 presents the amounts of 10 randomly selected automobile damage claims for each of two geographic areas, as obtained from a large number of such claims from the records of an insurance company. The amounts of the claims in each area are assumed to be approximately normally distributed, and the variance of the claims is assumed to be about the same in the two areas. Using available computer software, obtain the 95 percent confidence interval for the difference between the mean dollar amount of claims in the two areas.

Table 9.1   Automobile Damage Claims in Two Geographic areas

Area 1		Area 2	
$1,033	$1,069	$1,177	$1,146
1,274	1,121	258	1,096
1,114	1,269	715	742
924	1,150	1,027	796
1,421	921	871	905

As can be observed in the relevant portion of the output in Fig. 9-1, with 95 percent confidence we can conclude that the mean claim amount in Area 1 is greater than that in Area 2 by an amount between $47 and $466. The wide confidence interval is associated with the large amount of variability in the data, as well as the small sample sizes. Note that in computer software such as that illustrated in Fig. 9-1, when the population variances are not known, then the t distribution would be used even for large samples, because there is no need to use normal approximation when the exact t values for large degrees of freedom can be determined by the computer. (The part of the output in Fig. 9-1 that follows the confidence interval output will be discussed in Problem 11.20, which is concerned with hypothesis testing.)

```
MTB ) SET CLAIMS FOR AREA-ONE INTO C1
DATA) 1033  1274  1114   924  1421  1069  1121  1269  1150   921
DATA) END
MTB ) SET CLAIMS FOR AREA-TWO INTO C2
DATA) 1177   258   715  1027   871  1146  1096   742   796   905
DATA) END
MTB ) TWOSAMPLE TTEST 95% CONFIDENCE INTERVAL,FOR C1 AND C2;
SUBC) POOLED.

TWOSAMPLE T FOR C1 VS C2
        N       MEAN     STDEV    SE MEAN
C1  10         1130       158        50
C2  10          873       272        86

95 PCT CI FOR MU C1 - MU C2: (47, 466)
TTEST MU C1 = MU C2 (VS NE): T=2.57 P=0.019 DF=18.0
```

Fig. 9-1 Minitab output.

Supplementary Problems

CONFIDENCE INTERVALS FOR THE DIFFERENCE BETWEEN TWO POPULATION MEANS

9.16 For a particular consumer product, the mean dollar sales per retail outlet last year in a sample of $n_1 = 10$ stores was $\bar{X}_1 = \$3,425$ with $s_1 = \$200$. For a second product the mean dollar sales per outlet in a sample of $n_2 = 12$ stores was $\bar{X}_2 = \$3,250$ with $s_2 = \$175$. The sales amounts per outlet are assumed to be normally distributed for both products. Estimate the difference between the mean level of sales per outlet last year using a 95 percent confidence interval.

Ans. $8.28 to $341.72

9.17 From the data in Problem 9.16, suppose the two sample sizes were $n_1 = 20$ and $n_2 = 24$. Determine the 95 percent confidence interval for the difference between the two means based on the assumption that the two population variances are not equal.

Ans. $62.83 to $287.17

9.18 Using the data in Problem 9.16, suppose that we are interested in only the minimum difference between the sales levels of the first and second product. Determine the lower limit of such an estimation interval at the 95 percent degree of confidence.

Ans. $37.13 or more

9.19 For a sample of 50 firms taken from a particular industry, the mean number of employees per firm is $\bar{X}_1 = 420.4$ with $s_1 = 55.7$. There is a total of 380 firms in this industry. In a second industry which includes 200 firms, the mean number of employees in a sample of 50 firms is $\bar{X}_2 = 392.5$ employees with $s_2 = 87.9$. Estimate the difference in the mean number of employees per firm in the two industries, using a 95 percent confidence interval.

Ans. 2.3 to 53.5 employees

9.20 Construct the 99 percent confidence interval for the difference between means for Problem 9.19.

Ans. −5.8 to 61.6 employees

9.21   For a sample of 30 employees in one large firm, the mean hourly wage is $\bar{X}_1 = \$7.50$ with $s_1 = \$1.00$. In a second large firm, the mean hourly wage for a sample of 40 employees is $\bar{X}_2 = \$9.05$ with $s_2 = \$1.20$. Estimate the difference between the mean hourly wage at the two firms, using a 90 percent confidence interval.

  Ans. $0.02 to $0.88 per hour

9.22   For the data in Problem 9.21, suppose we are concerned with determining the maximum difference between the mean wage rates, using a 90 percent confidence interval. Construct such a lower confidence interval.

  Ans.   Est. $(\mu_1 - \mu_2) \leq \$0.78$ per hour

CONFIDENCE INTERVALS FOR ESTIMATING THE POPULATION PROPORTION

9.23   For a random sample of 100 households in a large metropolitan area, the number of households in which at least one adult is currently unemployed and seeking a full-time job is 12. Estimate the percentage of households in the area in which at least one adult is unemployed, using a 95 percent confidence interval. (*Note:* Percentage limits can be obtained by first determining the confidence interval for the proportion, and then multiplying these limits by 100.)

  Ans.   5.7% to 18.3%

9.24   Suppose the confidence interval obtained in Problem 9.23 is considered to be too wide for practical purposes (i.e., it is lacking in precision). Instead, we desire that the 95 percent confidence interval be within two percentage points of the true percentage of households with at least one adult unemployed. What is the minimum sample size required to satisfy this specification (*a*) if we make no assumption about the true percentage before collecting a larger sample and (*b*) if, on the basis of the sample collected in Problem 9.22, we assume that the true percentage is no larger than 18 percent?

  Ans.   (*a*) 2,401, (*b*) 1,418

9.25   A small manufacturer has purchased a batch of 200 small electronic parts from the "excess inventory" of a larger firm. For a random sample of 50 of the parts, five are found to be defective. Estimate the proportion of all of the parts in the shipment which are defective, using a 95 percent confidence interval.

  Ans.   0.03 to 0.17

9.26   For Problem 9.25, estimate the total number of parts in the shipment which are defective, using a 90 percent confidence interval.

  Ans.   8 to 32

9.27   For the situation in Problem 9.25, suppose the price of the electronics parts was such that the small manufacturer would be satisfied with his purchase as long as the true proportion defective does not exceed 0.20. Construct a one-sided 95 percent confidence interval and observe whether the upper limit of this interval exceeds the proportion 0.20.

  Ans.   Est. $\pi \leq 0.16$

CONFIDENCE INTERVALS FOR THE DIFFERENCE BETWEEN TWO POPULATION PROPORTIONS

9.28   In contrast to the data in Problem 9.23, in a second metropolitan area a random sample of 100 households yields only six households in which at least one adult is unemployed and seeking a full-time job. Estimate the difference in the percentage of households in the two areas which include an unemployed adult, using a 90 percent confidence interval.

  Ans.   −0.6% to +12.6%

9.29   Referring to Problem 9.28, what is the maximum percentage by which the household unemployment in the first metropolitan area exceeds the percentage unemployment in the second area, using a 90 percent one-sided confidence interval?

   Ans.   Est. Dif. $\leq 11.1\%$

CONFIDENCE INTERVALS FOR THE VARIANCE AND STANDARD DEVIATION

9.30   For a particular consumer product, the mean dollar sales per retail outlet last year in a sample of $n = 10$ stores was $\bar{X} = \$3,425$ with $s = \$200$. The sales amounts per outlet are assumed to be normally distributed. Estimate the (*a*) variance and (*b*) standard deviation of dollar sales of this product in all stores last year, using a 90 percent confidence interval.

   Ans.   (*a*) $21,278 \leq \sigma^2 \leq 108,271$, (*b*) $145.9 \leq \sigma \leq 329.0$

9.31   With reference to Problem 9.30, there is particular concern about how *large* the standard deviation of dollar sales might be. Construct the 90 percent one-sided confidence interval which identifies this value.

   Ans. $\sigma \leq 293.9$

COMPUTER OUTPUT

9.32   A business firm that processes many of its orders by telephone has two types of customers: general and commercial. Table 9.2 reports the required per-item telephone order times for a random sample of 12 general-customer calls and 10 commercial-customer calls. The amounts of time required for each type of call are assumed to be approximately normally distributed. Using an available computer program, obtain the 95 percent confidence interval for the difference in the mean amount of per-item time required for each type of call.

   Ans. -23 to 52 sec

Table 9.2   Time (in Seconds) Required per Item Order

General customers	Commercial customers
48	81
66	137
106	107
84	110
146	107
139	40
154	154
150	142
177	34
156	165
122	
121	

Chapter 10

Testing Hypotheses Concerning the Value of the Population Mean

10.1 BASIC STEPS IN HYPOTHESIS TESTING

In hypothesis testing we begin with an assumed (hypothesized) value of a population parameter. After a random sample is collected, we compare the sample statistic, such as the sample mean ($\bar{X}$), with the hypothesized parameter, such as the hypothesized population mean (μ). Then, we either *accept* or *reject* the hypothesized value as being correct. The hypothesized value is rejected only if the sample result clearly is unlikely to occur when the hypothesis is true.

Step 1: Formulate the null hypothesis and the alternative hypothesis. The *null hypothesis* (H_0) is the hypothesized parameter value which is compared with the sample result. It is rejected *only if* the sample result is unlikely to have occurred given the correctness of the hypothesis. The *alternative hypothesis* (H_1) is accepted only if the null hypothesis is rejected.

EXAMPLE 1. An auditor wishes to test the assumption that the mean value of all accounts receivable in a given firm is \$260.00 by taking a sample of $n = 36$ and computing the sample mean. He wishes to reject the assumed value of \$260.00 only if it is clearly contradicted by the sample mean, and thus the hypothesized value should be given the "benefit of the doubt" in the testing procedure. The null and alternative hypotheses for this test are $H_0: \mu = \$260.00$ and $H_1: \mu \neq \$260.00$.

Step 2: Specify the level of significance to be used. The level of significance is the statistical standard which is specified for rejecting the null hypothesis. If a 5 percent level of significance is specified, then the null hypothesis is rejected only if the sample result is so different from the hypothesized value that a difference of that amount or larger would occur by chance with a probability of 0.05 or less.

Note that if the 5 percent level of significance is used, there is a probability of 0.05 of rejecting the null hypothesis when it is in fact true. This is called *Type I error*. The probability of Type I error is always equal to the level of significance that is used as the standard for rejecting the null hypothesis; it is designated by the lowercase Greek α ("alpha"), and thus α also designates the level of significance. The most frequently used levels of significance in hypothesis testing are the 5 percent and 1 percent levels.

A *Type II error* occurs if the null hypothesis is accepted when it is in fact false. Determining the probability of Type II error is explained in Section 10.3. Table 10.1 summarizes the types of decisions and the possible consequences of the decisions which are made in hypothesis testing.

Table 10.1 Consequences of Decisions in Hypothesis Testing

Possible decisions	Possible states	
	Null hypothesis true	Null hypothesis false
Accept null hypothesis	Correctly accepted	Type II error
Reject null hypothesis	Type I error	Correctly rejected

Step 3: *Select the test statistic.* The test statistic will either be the sample statistic (the unbiased estimator of the parameter being tested), or a transformed version of the sample statistic. For example, in order to test a hypothesized value of the population mean, the mean of a random sample taken from that population could serve as the test statistic. However, if the sampling distribution of the mean is normally distributed, then the value of the sample mean typically is transformed into a z value, which then serves as the test statistic.

Step 4: *Establish the critical value or values of the test statistic.* Having specified the null hypothesis, the level of significance, and the test statistic to be used, we now establish the critical value(s) of the test statistic. There may be one or two such values, depending on whether a so-called one-tail or two-tail test is involved (see Section 10.2). In either case, a *critical value* identifies the value of the test statistic that is required to reject the null hypothesis.

Step 5: *Determine the actual value of the test statistic.* For example, in testing a hypothesized value of the population mean a random sample is collected and the value of the sample mean is determined. If the critical value was established as a z value, then the sample mean is converted into a z value.

Step 6: *Make the decision.* The observed value of the sample statistic is compared with the critical value (or values) of the test statistic. The null hypothesis is then either accepted or rejected. If the null hypothesis is rejected, the alternative hypothesis is accepted. In turn, this decision will have relevance to other decisions to be made by operating managers, such as whether a standard of performance is being maintained or which of two marketing strategies should be used.

10.2 TESTING A HYPOTHESIZED VALUE OF THE MEAN USING THE NORMAL DISTRIBUTION

The normal probability distribution can be used for testing a hypothesized value of the population mean (1) whenever $n \geq 30$, because of the central limit theorem, or (2) when $n < 30$ but the population is normally distributed and σ is known. (See Section 8.2.)

A *two-tail test* is used when we are concerned about a possible deviation in *either* direction from the hypothesized value of the mean. The formula used to establish the critical values of the sample mean is similar to the formula for determining confidence limits for estimating the population mean (see Section 8.3), except that the hypothesized value of the population mean μ_0 is the reference point rather than the sample mean. The critical values of the sample mean for a two-tailed test, according to whether σ is known, are

$$\bar{X}_{CR} = \mu_0 \pm z\sigma_{\bar{x}} \tag{10.1}$$

or

$$\bar{X}_{CR} = \mu_0 \pm zs_{\bar{x}} \tag{10.2}$$

EXAMPLE 2. For the null hypothesis formulated in Example 1, determine the critical values of the sample mean for testing the hypothesis at the 5 percent level of significance. Given that the standard deviation of the accounts receivable amounts is known to be $\sigma = \$43.00$, the critical values are

Hypotheses: $H_0: \mu = \$260.00$; $H_1: \mu \neq \$260.00$

Level of significance: $\alpha = 0.05$

Test statistic: $\bar{X}$ based on a sample of $n = 36$ and with $\sigma = 43.00$

$\bar{X}_{CR}$ = critical values of the sample mean

$$\bar{X}_{CR} = \mu_0 \pm z\sigma_{\bar{x}} = 260.00 \pm 1.96 \frac{\sigma}{\sqrt{n}} = 260 \pm 1.96 \frac{43.00}{\sqrt{36}}$$

$$= 260.00 \pm 1.96(7.17) = 260.00 \pm 14.05 = \$245.95 \text{ and } \$274.05$$

Therefore, in order to reject the null hypothesis the sample mean must have a value which is less than $245.95 *or* greater than $274.05. Thus, there are two regions of rejection in the case of a two-tail test (see Fig. 10-1). The

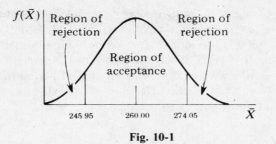

Fig. 10-1

z values of ±1.96 are used to establish the critical limits because for the standard normal distribution a proportion of 0.05 of the area remains in the two tails, which corresponds to the specified $\alpha = 0.05$.

Instead of establishing critical values in terms of the sample mean as such, the critical values in hypothesis testing typically are specified in terms of z values. For the 5 percent level of significance the critical values z for a two-tail test are -1.96 and $+1.96$, for example. When the value of the sample mean is determined, it is transformed to a z value so that it can be compared with the critical values of z. The transformation formula, according to whether σ is known, is

$$z = \frac{\bar{X} - \mu_0}{\sigma_{\bar{x}}} \qquad (10.3)$$

or

$$z = \frac{\bar{X} - \mu_0}{s_{\bar{x}}} \qquad (10.4)$$

EXAMPLE 3.   For the hypothesis-testing problem in Examples 1 and 2, suppose the sample mean is $\bar{X} = \$240.00$. We determine whether the null hypothesis should be accepted or rejected by transforming this mean to a z value and comparing it to the critical values of ±1.96 as follows:

$$\sigma_{\bar{x}} = 7.17 \qquad \text{(from Example 2)}$$

$$z = \frac{\bar{X} - \mu_0}{\sigma_{\bar{x}}} = \frac{240.00 - 260.00}{7.17} = \frac{-20.00}{7.17} = -2.79$$

This value of z is in the left-tail region of rejection of the hypothesis-testing model portrayed in Fig. 10-2. Thus, the null hypothesis is rejected and the alternative, that $\mu \neq \$260.00$, is accepted. Note that the same conclusion is reached by comparing the sample mean of $\bar{X} = \$240.00$ with the critical limits for the mean identified in Fig. 10-1.

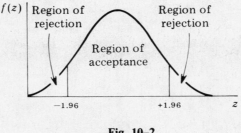

Fig. 10–2

A *one-tail test* is appropriate when we are concerned about possible deviations in only one direction from the hypothesized value of the mean. The auditor in Example 1 may not be concerned that the true average of all accounts receivable exceeds $\$260.00$, but only that it might be less than $\$260.00$. Thus, if he gives the benefit of the doubt to the stated claim that the true mean is *at least* $\$260.00$, the null and alternative hypotheses are

$$H_0: \mu \geq \$260.00 \qquad \text{and} \qquad H_1: \mu < \$260.00$$

There is only one region of rejection for a one-tail test, and for the above example the test is a lower-tail test. The region of rejection for a one-tail test is always in the tail which represents support of the *alternative* hypothesis. As for a two-tail test, the critical value can be determined for the mean as such or in terms of a z value. However, critical values for one-tail tests differ from those for two-tail tests because the given proportion of area is all in one tail of the distribution. Table 10.2 presents the values of z needed for one-tail and two-tail tests. The general formula to establish the critical value of the sample mean for a one-tail test, according to whether σ is known, is

$$\bar{X}_{CR} = \mu_0 + z\sigma_{\bar{x}} \tag{10.5}$$

or

$$\bar{X}_{CR} = \mu_0 + zs_{\bar{x}} \tag{10.6}$$

In formulas (*10.5*) and (*10.6*) above, note that z can be negative, resulting in a subtraction of the second term in each formula.

Table 10.2   Critical Values of z in Hypothesis Testing

Level of significance	Type of test	
	One-tail	Two-tail
5%	+1.645 (or −1.645)	±1.96
1%	+2.33 (or −2.33)	±2.58

EXAMPLE 4.   Assume that the auditor in Examples 1 through 3 began with the null hypothesis that the mean value of all accounts receivable is at least $260.00. Given that the sample mean is $240.00, we test this hypothesis at the 5 percent level of significance by the following two separate procedures.

(1) Determining the critical value of the sample mean, where $H_0: \mu \geq \$260.00$ and $H_1: \mu < \$260.00$.

$$\bar{X}_{CR} = \mu_0 + z\sigma_{\bar{x}} = 260 + (-1.645)(7.17) = \$248.21$$

Since $\bar{X} = \$240.00$, it is in the region of rejection. The null hypothesis is therefore rejected and the alternative hypothesis, that $\mu < \$260.00$, is accepted.

(2) Specifying the critical value in terms of z, where critical z ($\alpha = 0.05$) $= -1.645$:

$$z = \frac{\bar{X} - \mu_0}{\sigma_{\bar{x}}} = \frac{240.00 - 260.00}{7.17} = -2.79$$

Since $z = -2.79$ is in the region of rejection (to the left of the critical value of -1.645), the null hypothesis is rejected. Figure 10-3 portrays the critical value for this one-tail test in terms of $\bar{X}$ and z.

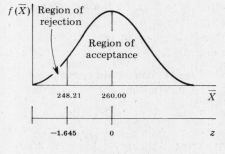

Fig. 10-3

10.3 TYPE I AND TYPE II ERRORS IN HYPOTHESIS TESTING

In this section Type I and Type II errors (defined in Section 10.1) are discussed entirely with respect to one-tail tests of a hypothesized mean. However, the basic concepts illustrated here apply to other hypothesis-testing models as well.

The probability of Type I error is always equal to the level of significance used in testing the null hypothesis. This is so because by definition the proportion of area in the region of rejection is equal to the proportion of sample results that would occur in that region given that the null hypothesis is true.

The probability of Type II error is generally designated by the Greek β ("beta"). The only way it can be determined is with respect to a *specific* value included within the range of the *alternative* hypothesis.

EXAMPLE 5. As in Example 4, the null hypothesis to be tested is that the mean of all accounts receivable is at least $260.00, and this test is to be carried out at the 5 percent level of significance. Further, the auditor indicates that he would consider an actual mean of $240.00 (or less) to be an important and material difference from the hypothesized value of the mean. As before, $\sigma = \$43.00$ and the sample size is $n = 36$ accounts. The determination of the probability of Type II error requires that we

  (1) formulate the null and alternative hypotheses for this testing situation,

  (2) determine the critical value of the sample mean to be used in testing the null hypothesis at the 5 percent level of significance,

  (3) identify the probability of Type I error associated with using the critical value computed above as the basis for the decision rule,

  (4) identify the probability of Type II error associated with the decision rule given the specific alternative mean value of $240.00.

The complete solution is

  (1) $$H_0: \mu \geq \$260.00 \qquad H_1: \mu < \$260.00$$

  (2) $$\bar{X}_{CR} = \mu_0 + z\sigma_{\bar{x}} = 260.00 + (-1.645)(7.17) = \$248.21$$

where $\sigma_{\bar{x}} = \dfrac{\sigma}{\sqrt{n}} = \dfrac{43.00}{\sqrt{36}} = \dfrac{43.00}{6} = 7.17$

  (3) The probability of Type I error equals 0.05 (the level of significance used in testing the null hypothesis).

  (4) The probability of Type II error is the probability that the mean of the random sample will equal or exceed $248.21 given that the mean of all accounts is actually at $240.00.

$$z = \frac{\bar{X}_{CR} - \mu_1}{\sigma_{\bar{x}}} = \frac{248.21 - 240.00}{7.17} = \frac{8.21}{7.17} = +1.15$$

$$P(\text{Type II error}) = P(z \geq +1.15) = 0.5000 - 0.3749 = 0.1251 \cong 0.13$$

Figure 10-4 illustrates the approach followed in Example 5. In general, the critical value of the mean determined with respect to the null hypothesis is "brought down" and used as the critical value with respect to the specific alternative hypothesis. Problem 10.13 illustrates the determination of the probability of Type II error for a two-tail test.

With the level of significance and sample size held constant, the probability of Type II error decreases as the specific alternative value of the mean is set farther from the value in the null hypothesis. It increases as the alternative value is set closer to the value in the null hypothesis. *An operating characteristic (OC) curve* portrays the probability of accepting the null hypothesis given various alternative values of the true mean. Figure 10-5 is the *OC* curve applicable to any *lower-tail test* of a hypothesized mean carried out at the 5 percent level of significance and based on the use of the normal probability distribution. Note that it is applicable to *any* such test because the values on the horizontal axis are stated in units of the standard error of the mean. For any values to the left of μ_0, the probability of acceptance indicates the probability of Type II error. To the right of μ_0, the probabilities indicate

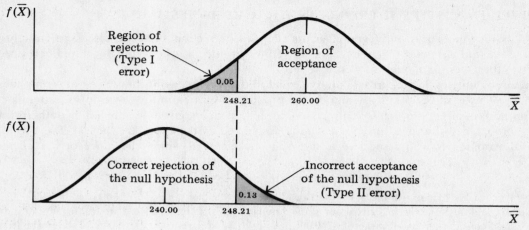

Fig. 10-4

correct acceptance of the null hypothesis. As indicated by the dashed lines, when in fact $\mu = \mu_0$, then the probability of accepting the null hypothesis is $1 - \alpha$, or in this case, $1 - 0.05 = 0.95$.

EXAMPLE 6.   We can verify the probability of Type II error determined in Example 5 by reference to Fig. 10-5, as follows:

As identified in Example 5, $\mu_0 = \$260.00$, $\mu_1 = \$240.00$, and $\sigma_{\bar{x}} = 7.17$. Therefore, the difference between the two designated values of the mean *in units of the standard error* is

$$z = \frac{\mu_1 - \mu_0}{\sigma_{\bar{x}}} = \frac{240 - 260}{7.17} = -2.8$$

By reference to Fig. 10-5, the height of the curve at a horizontal-axis value of $\mu_0 - 2.8\sigma_{\bar{x}}$ can be seen to be just above 0.10, as indicated by the dotted lines. The actual computed value in Example 5 is 0.13.

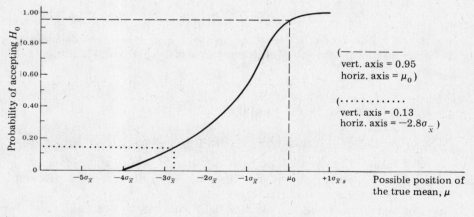

Fig. 10-5

In hypothesis testing, the concept of *power* refers to the probability of rejecting a null hypothesis given a specific alternative value of the parameter (in our examples, the population mean). Where the probability of type II error is designated β, it follows that the power of the test is always $1 - \beta$. Referring to Fig. 10-5, note that the power for alternative values of the mean is the difference between the value indicated by the *OC* curve and 1.0, and thus a *power curve* can be obtained "by subtraction," by using the *OC* curve.

EXAMPLE 7.   Referring to Example 5, we can determine the power of the test given the specific alternative value of the mean of $240.00 as follows:

Since $\beta = P(\text{Type II error}) = 0.13$ (from Example 5),

$$\text{Power} = 1 - \beta = 1.00 - 0.13 = 0.87$$

(*Note:* This is the probability of correctly rejecting the null hypothesis when $\mu = \$240.00$.)

10.4 DETERMINING THE REQUIRED SAMPLE SIZE FOR TESTING THE MEAN

Before a sample is actually collected, the required sample size can be determined by specifying (1) the hypothesized value of the mean, (2) a specific alternative value of the mean such that the difference from the null-hypothesized value is considered important, (3) the level of significance to be used in the test, (4) the probability of Type II error which is to be permitted, and (5) the value of the population standard deviation σ. The formula for determining the minimum sample size required in conjunction with testing a hypothesized value of the mean, based on use of the normal distribution, is

$$n = \frac{(z_0 - z_1)^2 \sigma^2}{(\mu_1 - \mu_0)^2} \qquad (10.7)$$

In (10.7), z_0 is the critical value of z used in conjunction with the specified level of significance (α level) while z_1 is the value of z with respect to the designated probability of Type II error (β level). The value of σ must either be known or be estimated on some basis. Formula (10.7) can be used for either one-tail or two-tail tests. The only value that differs for the two types of tests is the value of z_0 which is used (see Examples 8 and 9).

[*Note*: When solving for minimum sample size, any fractional result is always rounded up. Further, unless σ is known *and* the population is normally distributed, any computed sample size below 30 should be increased to 30 because (10.7) is based on the use of the normal distribution.]

EXAMPLE 8.   An auditor wishes to test the null hypothesis that the mean value of all accounts receivable is at least $260.00. He considers that the difference would be material and important if the true mean is at the specific alternative of $240.00 (or less). The acceptable levels of Type I error (α) and Type II error (β) are 0.05 and 0.10 respectively. The standard deviation of the accounts receivable amounts is known to be $\sigma = \$43.00$. The size of the sample which should be collected, as a minimum, to carry out this test is

$$n = \frac{(z_0 - z_1)^2 \sigma^2}{(\mu_1 - \mu_0)^2} = \frac{(-1.645 - 1.28)^2 (43.00)^2}{(240.00 - 260.00)^2} = \frac{(8.5556)(1,849)}{400} = 39.55 \cong 40$$

(*Note:* Because z_0 and z_1 would always have opposite algebraic signs, the result is that the two z values are always accumulated in the numerator above. If the accumulated value is a negative value, the process of squaring results in a positive value.)

EXAMPLE 9.   Suppose the auditor in Example 8 is concerned about a discrepancy in *either* direction from the null-hypothesized value of $260.00, and that a discrepancy of $20 in either direction would be considered important. Given the other information and specifications in Example 8, the minimum size of the sample which should be collected is

$$n = \frac{(z_0 - z_2)^2 \sigma^2}{(\mu_1 - \mu_0)^2} = \frac{(-1.96 - 1.28)^2 \sigma^2}{(240.00 - 260.00)^2} \quad \text{or} \quad \frac{[1.96 - (-1.28)]^2 \sigma^2}{(280.00 - 260.00)^2}$$

$$= \frac{(-3.24)^2 (43.00)^2}{(-20)^2} \quad \text{or} \quad \frac{(3.24)^2 (43.00)^2}{(20)^2}$$

$$= \frac{(10.4976)(1,849)}{400} = 48.53 \cong 49$$

(*Note:* Because any deviation from the hypothesized value can only be in one direction or the other, we use either the $+1.96$ or the -1.96 as the value of z_0 in conjunction with the then relevant value of z_1. As in Example 8, the two z values will in effect always be accumulated before being squared.)

10.5 TESTING A HYPOTHESIZED VALUE OF THE MEAN USING STUDENT'S t DISTRIBUTION

The t distribution (see Section 8.5) is appropriate for use as the test statistic when the sample is small ($n < 30$), the population is normally distributed, and σ is not known. The procedure used for testing an assumed value of a population mean is identical to that described in Section 10.2, except for the use of t as the test statistic. The test statistic used is

$$t = \frac{\bar{X} - \mu_0}{s_{\bar{x}}} \qquad (10.8)$$

EXAMPLE 10. The null hypothesis has been formulated that the mean operating life of light bulbs of a particular brand is at least 4,200 hr. The mean operating life for a random sample of $n = 10$ light bulbs is $\bar{X} = 4,000$ hr with a sample standard deviation of $s = 200$ hr. The operating life of bulbs in general is assumed to be normally distributed. We test the null hypothesis at the 5 percent level of significance as follows:

$$H_0: \mu \geq 4,200 \qquad H_1: \mu < 4,200$$

$$\text{Critical } t \ (df = 9, \ \alpha = 0.05) = -1.833$$

$$s_{\bar{x}} = \frac{s}{\sqrt{n}} = \frac{200}{\sqrt{10}} = \frac{200}{3.16} = 63.3 \text{ hr}$$

$$t = \frac{\bar{X} - \mu_0}{s_{\bar{x}}} = \frac{4,000 - 4,200}{63.3} = \frac{-200}{63.3} = -3.16$$

Because -3.16 is in the left-tail region of rejection (to the left of the critical value -1.833), the null hypothesis is rejected and the alternative hypothesis, that the true mean operating life is less than 4,200 hours, is accepted.

10.6 THE P-VALUE APPROACH TO TESTING NULL HYPOTHESES CONCERNING THE POPULATION MEAN

By the P-value approach, instead of comparing the observed value of a test statistic with a critical value, the probability of the occurrence of the test statistic, given that the null hypothesis is true, is determined and compared to the level of significance α. The null hypothesis is rejected if the P value is *less* than the designated α.

EXAMPLE 11. Refer to Example 4, in which $H_0: \mu \geq \$260.00$, $H_1: \mu < \$260.00$, $\alpha = 0.05$, $\bar{X} = \$240.00$, and the calculated test statistic is $z = -2.79$. Because the z statistic *is* in the direction of the region of rejection, we determine the probability of a z value this small (or smaller) occurring by chance:

$$P(z \leq -2.79) = 0.500 - 0.4974 = 0.0026$$

Because the P value of 0.0026 is less than the designated level of significance of $\alpha = 0.05$, the null hypothesis is rejected.

For two-tail tests, the calculated P value associated with one tail of the distribution is *doubled*, and then compared with the level of significance α. (See Problem 10.19.)

When the t statistic is used as the test statistic and a standard table for the t distribution is used, an exact P value cannot be determined. Rather, a range of probability values is identified. (See Problem 10.21.)

10.7 THE CONFIDENCE INTERVAL APPROACH TO TESTING NULL HYPOTHESES CONCERNING THE POPULATION MEAN

By this approach, a confidence interval for the population mean is constructed based on the sample results, and then we observe whether the hypothesized value of the population mean is included within the confidence interval. If the hypothesized value is included within the interval, then the null hypothesis cannot be rejected. If the hypothesized value is not included in the interval, then the null hypothesis is rejected. Where α is the level of significance to be used for the test, the $1 - \alpha$ confidence interval is constructed.

EXAMPLE 12.   Refer to Example 3, in which $H_0: \mu = \$260.00$, $H_1: \mu \neq \$260.00$, $\alpha = 0.05$, $\bar{X} = \$240.00$, and $\sigma_{\bar{x}} = 7.17$. We can test the null hypothesis at the 5 percent level of significance by constructing the 95 percent confidence interval:

$$\bar{X} \pm z\sigma_{\bar{x}} = 240.00 \pm 1.96(7.17) = 240 \pm 14.05$$
$$= \$225.95 \text{ to } \$254.05$$

Because the hypothesized value of $\$260.00$ is not included within the 95 percent confidence interval, the null hypothesis is rejected at the 5 percent level of significance.

For a one-tail test, a one-sided confidence interval is constructed. (See Problem 10.23.)

10.8   SUMMARY TABLE FOR TESTING A HYPOTHESIZED VALUE OF THE MEAN

Table 10.3   Testing a Hypothesized Value of the Mean

Population	Sample size	σ known	σ unknown
Normally distributed	Large ($n \geq 30$)	$z = \dfrac{\bar{X} - \mu_0}{\sigma_{\bar{x}}}$	$z = \dfrac{\bar{X} - \mu_0}{s_{\bar{x}}}$**
	Small ($n < 30$)	$z = \dfrac{\bar{X} - \mu_0}{\sigma_{\bar{x}}}$	$t = \dfrac{\bar{X} - \mu_0}{s_{\bar{x}}}$
Not normally distributed	Large ($n \geq 30$)	$z = \dfrac{\bar{X} - \mu_0}{\sigma_{\bar{x}}}$*	$z = \dfrac{\bar{X} - \mu_0}{s_{\bar{x}}}$†
	Small ($n < 30$)	Nonparametric tests directed toward the median generally would be used. (See Chapter 21.)	

*Central limit theorem is invoked.
**z is used as an approximation of t.
†Central limit theorem is invoked and z is used as an approximation of t.

10.9   COMPUTER OUTPUT

Computer software for testing hypotheses concerning the mean allow the user to specify the hypothesized value and whether a one-tail or two-tail test is desired. The P-value approach to hypothesis testing generally is used, as described in Section 10.6. Problems 10.24 and 10.25 illustrate the use of such software for a two-tail test and a one-tail test, respectively.

Solved Problems

TESTING A HYPOTHESIZED VALUE OF THE MEAN USING THE NORMAL DISTRIBUTION

10.1   A representative of a community group informs the prospective developer of a shopping center that the average income per household in the area is $\$25,000$. Suppose that for the type of area involved household income can be assumed to be approximately normally distributed, and that the standard deviation can be accepted as being equal to $\sigma = \$2{,}000$, based on an earlier study. For a random sample of $n = 15$ households, the mean household income is found to be $\bar{X} = \$24{,}000$. Test the null hypothesis that $\mu = \$25{,}000$ by establishing critical limits of the sample mean in terms of dollars, using the 5 percent level of significance.

(*Note:* The normal probability distribution can be used even though the sample is small, because the population is assumed to be normally distributed *and* σ is known.)

Since $H_0: \mu = \$25{,}000$ and $H_1: \mu \neq \$25{,}000$, the critical limits of $\bar{X}$ ($\alpha = 0.05$) are

$$\bar{X}_{CR} = \mu_0 \pm z\sigma_{\bar{x}} = \mu_0 \pm z\left(\frac{\sigma}{\sqrt{n}}\right) = 25{,}000 \pm 1.96\left(\frac{2{,}000}{\sqrt{15}}\right)$$

$$= 25{,}000 \pm 1.96\left(\frac{2{,}000}{3.87}\right) = 25{,}000 \pm 1.96(516.80) = \$23{,}987 \text{ and } \$26{,}013$$

Since the sample mean of $\bar{X} = \$24{,}000$ is between the two critical limits and in the region of acceptance of the null hypothesis, the community representative's claim cannot be rejected at the 5 percent level of significance.

10.2   Test the hypothesis in Problem 10.1, by using the standard normal variable z as the test statistic.

$$H_0: \mu = \$25{,}000 \qquad H_1: \mu \neq \$25{,}000$$

$$\text{Critical } z \ (\alpha = 0.05) = \pm 1.96$$

Thus,
$$\sigma_{\bar{x}} = \frac{\sigma}{\sqrt{n}} = \frac{2{,}000}{\sqrt{15}} = \frac{2{,}000}{3.87} = \$516.80$$

$$z = \frac{\bar{X} - \mu_0}{\sigma_{\bar{x}}} = \frac{24{,}000 - 25{,}000}{516.80} = \frac{-1{,}000}{516.80} = -1.93$$

Since the computed z of -1.93 is in the region of acceptance of the null hypothesis, the community representative's claim cannot be rejected at the 5 percent level of significance.

10.3   With reference to Problems 10.1 and 10.2, the prospective developer is not really concerned about the possibility that the average household income is higher than the claimed $25,000, but only that it might be lower. Accordingly, reformulate the null and alternate hypotheses and carry out the appropriate statistical test, still giving the benefit of the doubt to the community representative's claim.

$$H_0: \mu \geq \$25{,}000 \qquad H_1: \mu < \$25{,}000$$

$$\text{Critical } z \ (\alpha = 0.05) = -1.645$$

$$z = -1.93 \qquad \text{(from Problem 10.2)}$$

The computed z of -1.93 is less than the critical z of -1.645 for this lower-tail test. Therefore the null hypothesis is rejected at the 5 percent level of significance, and the alternative hypothesis, that the mean household income is less than $25,000, is accepted. The reason for the change in the decision from Problem 10.2 is that the 5 percent region of rejection is located entirely in one tail of the distribution.

10.4   For Problem 10.3, suppose that the population standard deviation is not known, which typically would be the case, and the population of income figures is not assumed to be normally distributed. For a sample of $n = 30$ households, the sample standard deviation is $s = \$2{,}000$ and the sample mean remains $\bar{X} = \$24{,}000$. Test the null hypothesis that the mean household income in the population is at least $25,000, using the 5 percent level of significance.

(*Note:* The normal probability distribution can be used both because of the central limit theorem and because z can be used as an approximation of t when $n \geq 30$.)

$$H_0: \mu \geq \$25,000 \qquad H_1: \mu < \$25,000$$
$$\text{Critical } z \ (\alpha = 0.05) = -1.645$$

$$s_{\bar{x}} = \frac{s}{\sqrt{n}} = \frac{2,000}{\sqrt{30}} = \frac{2,000}{5.48} = \$364.96$$

$$z = \frac{\bar{X} - \mu_0}{s_{\bar{x}}} = \frac{24,000 - 25,000}{364.96} = \frac{-1,000}{364.96} = -2.74$$

The computed value of z of -2.74 is less than the critical z of -1.645 for this lower-tail test. Therefore the null hypothesis is rejected at the 5 percent level of significance. Notice that the computed value of z in this case is arithmetically smaller in value and more clearly in the region of rejection as compared with Problem 10.3. This is due entirely to the increase in sample size from $n = 15$ to $n = 30$, which results in a smaller value for the standard error of the mean.

10.5 A manufacturer contemplating the purchase of new tool-making equipment has specified that on the average the equipment should not require more than 10 min of set-up time per hour of operation. The purchasing agent visits a company where the equipment being considered is installed; from records there he notes that 40 randomly selected hours of operation included a total of 7 hr and 30 min of set-up time, and the standard deviation of set-up time per hour was 3.0 min. Based on this sample result, can the assumption that the equipment meets set-up time specifications be rejected at the 1 percent level of significance?

$$H_0: \mu \leq 10.0 \text{ min (per hour)} \qquad H_1: \mu > 10.0 \text{ min (per hour)}$$
$$\text{Critical } z \ (\alpha = 0.01) = +2.33$$

$$\bar{X} = \frac{\Sigma X}{n} = \frac{450 \text{ min}}{40} = 11.25 \text{ min}$$

$$s_{\bar{x}} = \frac{s}{\sqrt{n}} = \frac{3.0}{\sqrt{40}} = \frac{3.0}{6.32} = 0.47 \text{ min}$$

$$z = \frac{\bar{X} - \mu_0}{s_{\bar{x}}} = \frac{11.25 - 10.0}{0.47} = \frac{1.25}{0.47} = +2.66$$

The calculated z of $+2.66$ is greater than the critical value of $+2.33$ for this upper-tail test. Therefore, the null hypothesis is rejected at the 1 percent level of significance, and the alternative hypothesis, that the average set-up time for this equipment is greater than 10 min per hour of operation, is accepted.

10.6 The standard deviation of the tube life for a particular brand of ultraviolet tube is known to be $\sigma = 500$ hr, and the operating life of the tubes is normally distributed. The manufacturer claims that average tube life is at least 9,000 hr. Test this claim at the 5 percent level of significance by designating it as the null hypothesis and given that for a sample of $n = 15$ tubes the mean operating life was $\bar{X} = 8,800$ hr.

$$H_0: \mu \geq 9,000 \qquad H_1: \mu < 9,000$$
$$\text{Critical } z \ (\alpha = 0.05) = -1.645$$

$$\sigma_{\bar{x}} = \frac{\sigma}{\sqrt{n}} = \frac{500}{\sqrt{15}} = \frac{500}{3.87} = 129.20$$

$$z = \frac{\bar{X} - \mu_0}{\sigma_{\bar{x}}} = \frac{8,800 - 9,000}{129.20} = \frac{-200}{129.20} = -1.55$$

The calculated z of -1.55 is *not* less than the critical z of -1.645 for this lower-tail test. Therefore the null hypothesis cannot be rejected at the 5 percent level of significance.

10.7 With respect to Problem 10.6, suppose the sample data were obtained for a sample of $n = 35$ sets. Test the claim at the 5 percent level of significance.

$$H_0: \mu \geq 9,000 \qquad H_1: \mu < 9,000$$

$$\text{Critical } z \ (\alpha = 0.05) = -1.645$$

$$\sigma_{\bar{x}} = \frac{\sigma}{\sqrt{n}} = \frac{500}{\sqrt{35}} = \frac{500}{5.92} = 84.46$$

$$z = \frac{\bar{X} - \mu_0}{\sigma_{\bar{x}}} = \frac{8,800 - 9,000}{84.46} = \frac{-200}{84.46} = -2.37$$

The calculated z of -2.37 is less than the critical z of -1.645 for this lower-tail test. Therefore the null hypothesis cannot be rejected at the 5 percent level of significance.

10.8 A marketing research analyst collects data for a random sample of 100 customers out of the 400 who purchased a particular "coupon special." The 100 people spent an average of $\bar{X} = \$24.57$ in the store with a standard deviation of $s = \$6.60$. Before seeing these sample results, the marketing manager had claimed that the average purchase by those responding to the coupon offer would be at least \$25.00. Can his claim be rejected, using the 5 percent level of significance?

$$H_0: \mu \geq \$25.00 \qquad H_1: \mu < \$25.00$$

$$\text{Critical } z \ (\alpha = 0.05) = -1.645$$

$$s_{\bar{x}} = \frac{s}{\sqrt{n}} \sqrt{\frac{N-n}{N-1}} = \frac{6.60}{\sqrt{100}} \sqrt{\frac{400-100}{400-1}} = \frac{6.60}{10} \sqrt{0.7519} = 0.660(0.867) = 0.57$$

(The finite correction factor is required because $n > 0.05 N$.)

$$z = \frac{\bar{X} - \mu_0}{s_{\bar{x}}} = \frac{24.57 - 25.00}{0.57} = \frac{-0.43}{0.57} = -0.75$$

The calculated z of -0.75 is *not* less than the critical z of -1.645 for this lower-tail test. Therefore, the claim cannot be rejected at the 5 percent level of significance.

TYPE I AND TYPE II ERRORS IN HYPOTHESIS TESTING

10.9 For Problem 10.3, suppose the prospective developer would consider it an important discrepancy if the average household income is at or below \$23,500, as contrasted to the \$25,000 claimed income level. Identify (a) the probability of Type I error, (b) the probability of Type II error, (c) the power associated with this lower-tail test.

(a) $P(\text{Type I error}) = 0.05$ (the α-level, or level of significance)

(b) $P(\text{Type II error}) = P(\text{critical limit of } \bar{X} \text{ will be exceeded given } \mu = \$23,500)$

Lower critical limit of $\bar{X} = \mu_0 + z\sigma_{\bar{x}} = 25,000 + (-1.645)(516.80) = \$24,149.86$

where $\mu_0 = \$25,000$

$$z = -1.645$$

$$\sigma_{\bar{x}} = \frac{\sigma}{\sqrt{n}} = \frac{2,000}{\sqrt{15}} = \frac{2,000}{3.87} = \$516.80$$

$$P(\text{Type II error}) = P(\bar{X} \geq 24,149.86 \,|\, \mu_1 = 23,500, \ \sigma_{\bar{x}} = 516.80)$$

$$z_1 = \frac{\bar{X}_{CR} - \mu_1}{\sigma_{\bar{x}}} = \frac{24,149.86 - 23,500}{516.80} = \frac{649.86}{516.80} = 1.26$$

$$P(\text{Type II error}) = P(z_1 \geq +1.26) = 0.5000 - 0.3962 = 0.1038 \cong 0.10$$

(c) Power $= 1 - P(\text{Type II error}) = 1.00 - 0.10 = 0.90$

10.10 With respect to Problems 10.3 and 10.9, the specific null-hypothesized value of the mean is $\mu_0 = \$25,000$ and the specific alternative value is $\mu_1 = \$23,500$. Based on the sample of $n = 15$, $\sigma_{\bar{x}} = 516.80$ and the critical value of the sample mean is $\bar{X}_{CR} = \$24,149.86$. ($a$) Explain what is meant by calling the last value above a "critical" value. (b) If the true value of the population mean is $\mu = \$23,000$, what is the power of the statistical test?

(a)   The $24,149.86 value is the "critical" value of the sample mean in that if the sample mean equals or exceeds this value, the null hypothesis ($H_0: \mu \geq \$25,000$) will be accepted, whereas if the sample mean is less than this value, the null hypothesis will be rejected and the alternative hypothesis ($H_1: \mu < \$25,000$) will be accepted.

(b)   Power $= P$(Rejecting a false null hypothesis given a specific value of μ). In this case, Power $= P(\bar{X} < 24,149.86 \,|\, \mu_1 = 23,000, \sigma_{\bar{x}} = 516.80)$

$$z = \frac{\bar{X}_{CR} - \mu_1}{\sigma_{\bar{x}}} = \frac{24,149.86 - 23,000}{516.80} = \frac{1,149.86}{516.80} = +2.22$$

Power $= P(z < +2.22) = 0.5000 + 0.4868 = 0.9868 \cong 0.99$

In other words, given a population mean of $\mu = \$23,000$, there is about a 99 percent chance that the null hypothesis will be (correctly) rejected, and about a 1 percent chance that it will be (incorrectly) accepted (Type II error).

10.11 Referring to Problem 10.10,

(a)   What is the power of the statistical test if the population mean is $\mu = \$24,000$? Compare this answer with the value obtained for $\mu = \$23,000$ in the answer to Problem 10.10(b).

(b)   What is the power of the statistical test if the population mean is $\mu = \$24,500$? Compare this answer with the values obtained for $\mu = \$23,000$ in Problem 10.10(b) and for $\mu = \$24,000$ in (a) above.

(a)   Power $= P(\bar{X} < 24,149.86 \,|\, \mu = 24,000, \sigma_{\bar{x}} = 516.80) = P(z < 0.29)$

$= 0.5000 + 0.1141 = 0.6141 \cong 0.61$

where $z = \dfrac{\bar{X}_{CR} - \mu_1}{\sigma_{\bar{x}}} = \dfrac{24,149.86 - 24,000}{516.80} = \dfrac{149.86}{516.80} = +0.29$

The power of the test in this case is substantially lower than the power of 0.99 in Problem 10.10(b). This would be expected, since the present mean of $\mu = \$24,000$ is closer in value to the null-hypothesized $\mu \geq \$25,000$.

(b)   Power $= P(\bar{X} < 24,149.86 \,|\, \mu = 24,500, \sigma_{\bar{x}} = 516.80) = P(z < -0.68)$

$= 0.5000 - 0.2518 = 0.2482 \cong 0.25$

where $z = \dfrac{\bar{X}_{CR} - \mu_1}{\sigma_{\bar{x}}} = \dfrac{24,149.86 - 24,500}{516.80} = \dfrac{-350.14}{516.80} = -0.68$

The power of the test is substantially lower than the previous values of 0.99 and 0.61.

10.12 In Problems 10.9, 10.10, and 10.11, the power of the one-tail statistical test associated with alternative values of the population mean of $24,500, $24,000, $23,500, and $23,000 was determined to be 0.25, 0.61, 0.89, and 0.99, respectively. Determine the OC values associated with these alternative values of the population mean, and also for the case where $\mu = \$25,000$ (with $H_0: \mu \geq \$25,000$).

Since OC value $= P$ (accepting the null hypothesis given a specific value of μ), OC value $= 1 - $ power (for each specific value of μ), as reported in Table 10.4.

Table 10.4 Power and *OC* Values

Value of μ	Power	*OC* Value
$25,000	0.05	0.95
24,500	0.25	0.75
24,000	0.61	0.39
23,500	0.90	0.10
23,000	0.99	0.01

In Table 10.4, since $25,000 is the point at which the range of the null-hypothesized values begin, the "power" is in fact the probability of incorrect rejection of the null hypothesis and is also therefore equal to the level of significance and the probability of Type I error. Similarly, when μ is less than $25,000 in the table, the *OC* value is the probability of incorrectly accepting the null hypothesis and is thus the probability of Type II error.

10.13 For Problems 10.1 and 10.2, suppose that the prospective developer would consider it important if the average household income per year differs from the claimed $25,000 by $1,500 or more in *either* direction. Given that the null hypothesis is being tested at the 5 percent level of significance, identify the probability of Type I and Type II error.

(*Note:* This solution is an extension of the explanation given in Section 10.3 in that a two-tail test rather than a one-tail test is involved.)

$$P(\text{Type I error}) = 0.05 \quad \text{(the } \alpha\text{-level, or level of significance)}$$

$$P(\text{Type II error}) = P(\text{lower critical limit of } \bar{X} \text{ will be exceeded given } \mu = \$23,500) \quad \text{or}$$

$$= P(\bar{X} \text{ will be below the upper critical limit given } \mu = \$26,500)$$

(*Note: Either* calculation will yield the solution for the probability of Type II error. These two values are *not* accumulated because the specific alternative value can be at only one point in a given situation. The calculation below is by the first of the two alternative approaches.)

Since, from Problem 10.1, lower critical limit of $\bar{X} = \$23,987$ and $\sigma_{\bar{x}} = \$516.80$,

$$P(\text{Type II error}) = P(\bar{X} \geq 23,987 \mid \mu_1 = 23,500, \sigma_{\bar{x}} = 516.80) = P(z_1 \geq +0.94)$$

$$= 0.5000 - 0.3264 = 0.1736 \cong 0.17$$

$$\text{where } z_1 = \frac{\bar{X}_{CR} - \mu_1}{\sigma_{\bar{x}}} = \frac{23,987 - 23,500}{516.80} = \frac{487}{516.80} = +0.94$$

DETERMINING THE REQUIRED SAMPLE SIZE FOR TESTING THE MEAN

10.14 Suppose the prospective developer of the shopping center in Problem 10.3 wishes to test the null hypothesis $H_0: \mu \geq \$25,000$ at the 5 percent level of significance and he considers it an important difference if the average household income level is at (or below) $23,500. Because he is particularly concerned about the error of developing a shopping center in an area where it cannot be supported, he sets the desired level of Type II error at $\beta = 0.01$. If the standard deviation for such income data is assumed to be $\sigma = \$2,000$, determine the size of the sample required to achieve the developer's objectives in regard to Type I and Type II error if no assumption is made about the normality of the population.

$$n = \frac{(z_0 - z_1)^2 \sigma^2}{(\mu_1 - \mu_0)^2} = \frac{[-1.645 - (+2.33)]^2 (2,000)^2}{(23,500 - 25,000)^2} = \frac{(-3.975)^2 (4,000,000)}{(-1,500)^2}$$

$$= \frac{63,202,500}{2,250,000} = 28.09 = 29 \text{ households}$$

However, because the population is not assumed to be normally distributed, the required sample size is increased to $n = 30$, so that the central limit theorem can be invoked as the basis for using the normal probability distribution.

10.15 The manufacturer contemplating the purchase of tool-making equipment in Problem 10.5 has specified in the null hypothesis that the average set-up time for new equipment being considered is equal to or less than 10 min per hour ($H_0: \mu \leq 10.0$). Suppose he would be particularly concerned if the average set-up time per hour is 12.0 min or more. He estimates that the standard deviation is about $\sigma = 3.0$ min, and assumes that a variable such as set-up time is likely to be normally distributed. Because of the long-run cost implications of this decision, he designates that the probability of both Type I error and Type II error should be held to 0.01. How many randomly selected hours of equipment operation should be sampled, as a minimum, to satisfy his testing objectives?

$$n = \frac{(z_0 - z_1)^2 \sigma^2}{(\mu_1 - \mu_0)^2} = \frac{[2.33 - (-2.33)]^2 (3.0)^2}{(12.0 - 10.0)^2}$$

$$= \frac{(4.66)^2 (9)}{(2.0)^2} = \frac{21.7156(9)}{4.0} = \frac{195.4404}{4.0} = 48.86 = 49 \text{ hr}$$

TESTING A HYPOTHESIZED VALUE OF THE MEAN USING STUDENT'S t DISTRIBUTION

10.16 As a modification of Problem 10.3, a representative of a community group informs the prospective developer of a shopping center that the average household income in the community is *at least* $\mu = \$25,000$, and this claim is given the benefit of doubt in the hypothesis-testing procedure. As before, the population of income figures in the community is assumed to be normally distributed. For a random sample of $n = 15$ households taken in the community the sample mean is $\bar{X} = \$24,000$ and the sample standard deviation is $s = \$2,000$. Test the null hypothesis at the 5 percent level of significance.

(*Note:* The t distribution is appropriate because the sample is small ($n < 30$), σ is not known, and the population is assumed to be normally distributed.)

$$H_0: \mu \geq \$25,000 \qquad H_1: \mu < \$25,000$$
$$\text{Critical } t \ (df = 14, \ \alpha = 0.05) = -1.761$$

$$s_{\bar{x}} = \frac{s}{\sqrt{n}} = \frac{2,000}{\sqrt{15}} = \frac{2,000}{3.87} = 516.80$$

$$t = \frac{\bar{X} - \mu_0}{s_{\bar{x}}} = \frac{24,000 - 25,000}{516.80} = \frac{-1,000}{516.80} = -1.935$$

The calculated t of -1.935 is less than the critical t of -1.761 for this lower-tail test. Therefore the null hypothesis is rejected at the 5 percent level of significance. The decision is the same as in Problem 10.3 in which σ was known, except that in the former case z was the test statistic, and the critical value was $z = -1.645$.

10.17 Of 100 juniors and seniors majoring in accounting in a college of business administration, a random sample of $n = 12$ students has a mean grade-point average of 2.70 (where A = 4.00) with a sample standard deviation of $s = 0.40$. Grade-point averages for juniors and seniors are assumed to be normally distributed. Test the hypothesis that the overall grade-point average for all students majoring in accounting is at least 3.00, using the 1 percent level of significance.

$$H_0: \mu \geq 3.00 \qquad H_1: \mu < 3.00$$

$$\text{Critical } t \ (df = 11, \ \alpha = 0.01) = -2.718$$

$$s_{\bar{x}} = \frac{s}{\sqrt{n}} \sqrt{\frac{N-n}{N-1}} = \frac{0.40}{\sqrt{12}} \sqrt{\frac{100-12}{100-1}}$$

$$= \frac{0.40}{3.46} \sqrt{\frac{88}{99}} = 0.116(0.943) = 0.109$$

(The finite correction factor is required because $n > 0.05 N$.)

$$t = \frac{\bar{X} - \mu_0}{s_{\bar{x}}} = \frac{2.70 - 3.00}{0.109} = \frac{-0.30}{0.109} = -2.752$$

The calculated t of -2.752 is just less than the critical value of -2.718 for this lower-tail test. Therefore the null hypothesis is rejected at the 1 percent level of significance. If we had neglected to use the finite correction factor the computed t value would have been -2.586, and we would have inappropriately accepted the null hypothesis.

10.18 As a commercial buyer for a private supermarket brand, suppose you take a random sample of 12 No. 303 cans of string beans at a canning plant. The average weight of the drained beans in each can is found to be $\bar{X} = 15.97$ oz, with $s = 0.15$. The claimed minimum average net weight of the drained beans per can is 16.0 oz. Can this claim be rejected at the 10 percent level of significance?

$$H_0: \mu \geq 16.0 \qquad H_1: \mu < 16.0$$

$$\text{Critical } t \ (df = 11, \ \alpha = 0.10) = -1.363$$

$$s_{\bar{x}} = \frac{s}{\sqrt{n}} = \frac{0.15}{\sqrt{12}} = \frac{0.15}{3.46} = 0.043 \text{ oz}$$

$$t = \frac{\bar{X} - \mu_0}{s_{\bar{x}}} = \frac{15.97 - 16.00}{0.043} = \frac{-0.03}{0.043} = -0.698$$

The calculated t of -0.698 is *not* less than the critical value of -1.363 for this lower-tail test. Therefore the claim cannot be rejected at the 10 percent level of significance.

THE *P*-VALUE APPROACH TO TESTING NULL HYPOTHESES CONCERNING THE POPULATION MEAN

10.19 Using the *P*-value approach, test the null hypothesis in Problem 10.2 at the 5 percent level of significance.

$$H_0: \mu = \$25{,}000 \qquad H_1: \mu \neq \$25{,}000$$

$$z = -1.93 \qquad \text{(from Problem 10.2)}$$

$$P(z \leq -1.93) = 0.5000 - 0.4732 = 0.0268$$

Finally, because this is a two-tail test:

$$P = 2(0.0268) = 0.0536$$

Because the P value of 0.0536 is larger than the level of significance of 0.05, the null hypothesis *cannot* be rejected at this level. Therefore we cannot reject the claim that the mean household income in the population is $25,000.

10.20 Using the *P*-value approach, test the null hypothesis in Problem 10.3 at the 5 percent level of significance.

$$H_0: \mu \geq \$25{,}000 \qquad H_1: \mu < \$25{,}000$$

$$z = -1.93 \qquad \text{(from Problem 10.3)}$$

(Note that the z statistic *is* in the direction of the region of rejection for this left-tail test.)

$$P(z \leq -1.93) = 0.5000 - 0.4732 = 0.0268$$

Because the P value of 0.0268 is smaller than the level of significance of 0.05, we reject the null hypothesis. Therefore the alternative hypothesis, that the mean household income in the population is less than $25,000, is accepted.

10.21 Using the P-value approach, test the null hypothesis in Problem 10.16 at the 5 percent level of significance.

$$H_0: \mu \geq \$25,000 \qquad H_1: \mu < \$25,000$$

$$t = -1.93 \qquad \text{(from Problem 10.16)}$$

For this left-tail test, and with $df = 14$:

$$P(t \leq -1.93) = (0.025 < P < 0.05)$$

The probability of the t statistic of -1.93 occurring by chance when the null hypothesis is true is somewhere between 0.025 and 0.05, because a t value of -2.145 is required for a left-tail probability of 0.025 while a t value of -1.761 is required for a left-tail probability of 0.05. Therefore, the null hypothesis is rejected at the 5 percent level of significance, and we conclude that the mean household income in the population is less than $25,000.

THE CONFIDENCE INTERVAL APPROACH TO TESTING NULL HYPOTHESES CONCERNING THE POPULATION MEAN

10.22 Apply the confidence interval approach to testing the null hypothesis in Problem 10.2, using the 5 percent level of significance.

$$H_0: \mu = \$25,000 \qquad H_1: \mu \neq \$25,000$$

$$\bar{X} = \$24,000$$

$$\sigma_{\bar{x}} = \$516.80 \qquad \text{(from Problem 10.2)}$$

$$95\% \text{ Conf. Int.} = \bar{X} \pm z\sigma_{\bar{x}} = 24,000 \pm 1.96(516.80)$$

$$= 24,000 \pm 1,012.93 = \$22,987.07 \text{ to } \$25,012.93$$

Because the 95 percent confidence interval includes the hypothesized value of $25,000, the null hypothesis cannot be rejected at the 5 percent level of significance.

10.23 Apply the confidence interval approach to test the null hypothesis in Problem 10.3, using the 5 percent level of significance.

$$H_0: \mu \geq \$25,000 \qquad H_1: \mu < \$25,000$$

$$\bar{X} = \$24,000$$

$$\sigma_{\bar{x}} = \$516.80 \qquad \text{(from Problem 10.3)}$$

$$\text{Upper Conf. Limit} = \bar{X} + z\sigma_{\bar{x}} = 24,000 + 1.645(516.80)$$

$$= 24,000 + 850.14 = \$24,850.14$$

With 95 percent confidence, we conclude that the population mean can be as large as $24,850.14. Because this one-sided confidence interval does not include the hypothesized value of $25,000 (or greater), the null hypothesis is rejected at the 5 percent level of significance.

COMPUTER OUTPUT

10.24 Table 10.5 presents the dollar amounts of automobile damage claims filed by a random sample of 10 insurers involved in automobile collisions in a particular geographic area. Use available computer software to test the null hypothesis that the mean amount of the claims in the sampled population is $1,000.

**Table 10.5   Automobile
Damage Claims**

$1,033	$1,069
1,274	1,121
1,114	1,269
924	1,150
1,421	921

$$H_0: \mu = \$1,000 \qquad H_1: \mu \neq \$1,000 \qquad \alpha = 0.05$$

Refer to Fig. 10-6. The P-value is reported as being 0.029. Because this probability is less than the designated level of significance of 0.05, the null hypothesis is rejected, and we conclude that the mean claim amount in the population is different from $1,000 based on this two-tail test.

```
MTB ) SET DAMAGE CLAIMS INTO C1
DATA) 1033 1274 1114  924 1421 1069 1121 1269 1150   921
DATA) END
MTB ) NAME FOR C1 IS 'CLAIM'
MTB ) TTEST OF MU = 1000 FOR 'CLAIM'

TEST OF MU = 1000.0 VS MU N.E. 1000.0

           N      MEAN    STDEV   SE MEAN       T    P VALUE
CLAIM     10    1129.6    158.0      50.0    2.59     0.029
```

Fig. 10-6   Minitab output for Problem 10.24.

10.25 Refer to the data in Table 10.5. Use available computer software to test the null hypothesis that the mean amount of claims in the target population is *no greater* than $1,000, using the 5 percent level of significance.

$$H_0: \mu \leq \$1,000 \qquad H_1: \mu > \$1,000 \qquad \alpha = 0.05$$

Refer to Fig. 10-7. The P value is reported as being 0.015. Because this probability is less than the designated level of significance of 0.05, the null hypothesis is rejected, and we conclude that the mean claim amount in the population is greater than $1,000 based on this upper-tail test. (*Note*: In Minitab an upper-tail test is specified by the subcommand "ALTERNATIVE = 1", while a lower-tail test is specified by "ALTERNATIVE = −1".)

```
MTB ) SET DAMAGE CLAIMS INTO C1
DATA) 1033 1274 1114  924 1421 1069 1121 1269 1150  921
DATA) END
MTB ) NAME FOR C1 IS 'CLAIM'
MTB ) TTEST OF MU = 1000 FOR 'CLAIM';
SUBC) ALTERNATIVE = 1.

TEST OF MU = 1000.0 VS MU G.T. 1000.0

           N      MEAN    STDEV   SE MEAN       T    P VALUE
CLAIM     10    1129.6    158.0      50.0    2.59     0.015
```

Fig. 10-7   Minitab output for Problem 10.25.

Supplementary Problems

TESTING A HYPOTHESIZED VALUE OF THE MEAN

10.26 A fast-foods chain will build a new outlet in a proposed location if at least 200 cars per hour pass the location during certain hours. For 20 randomly sampled hours during the designated hours, the average number of cars passing the location is $\bar{X} = 208.5$ with $s = 30.0$. The statistical population is assumed to be approximately normal. The management of the chain conservatively adopted the null hypothesis that the traffic volume does *not* satisfy their requirement, that is $H_0: \mu \leq 200.0$. Can this hypothesis be rejected at the 5 percent level of significance?

Ans. No.

10.27 Suppose the sample results in Problem 10.26 are based on a sample of $n = 50$ hr. Can the null hypothesis be rejected at the 5 percent level of significance?

Ans. Yes.

10.28 The mean sales amount per retail outlet for a particular consumer product during the past year is found to be $\bar{X} = \$3,425$ in a sample of $n = 25$ outlets. Based on sales data for other similar products, the distribution of sales is assumed to be normal and the standard deviation of the population is assumed to be $\sigma = \$200$. Suppose it was claimed that the true sales amount per outlet is at least $3,500. Test this claim at the (*a*) 5 percent and (*b*) 1 percent level of significance.

Ans. (*a*) Reject H_0, (*b*) accept H_0.

10.29 In Problem 10.28, suppose that no assumption was made about the population standard deviation, but that $s = \$200$. Test the claim at the (*a*) 5 percent and (*b*) 1 percent level of significance.

Ans. (*a*) Reject H_0, (*b*) accept H_0.

10.30 For a sample of 50 firms taken from a particular industry the mean number of employees per firm is 420.4 with a sample standard deviation of 55.7. There are a total of 380 firms in this industry. Before the data were collected, it was hypothesized that the mean number of employees per firm in this industry does not exceed 408 employees. Test this hypothesis at the 5 percent level of significance.

Ans. Reject H_0.

10.31 Suppose the analyst in Problem 10.30 neglected to use the finite correction factor in determining the value of the standard error of the mean. What would be the result of the test, still using the 5 percent level of significance?

Ans. Accept H_0.

10.32 The manufacturer of a new compact car claims that the car will average at least 35 miles per gallon in general highway driving. For 40 test runs, the car averaged 34.5 miles per gallon with a standard deviation of 2.3 miles per gallon. Can the manufacturer's claim be rejected at the 5 percent level of significance?

Ans. No.

10.33 Referring to Problem 10.32, before the highway tests were carried out, a consumer advocate claimed that the compact car would *not* exceed 35 miles per gallon in general highway driving. Can this claim be rejected at the 5 percent level of significance? Consider the implications of your answer to this question and in Problem 10.32 regarding which hypothesis is designated as the null hypothesis.

Ans. No.

10.34 An analyst in a personnel department randomly selects the records of 16 hourly employees and finds that the mean wage rate is $\bar{X} = \$7.50$ with a standard deviation of $s = \$1.00$. The wage rates in the firm are assumed to be normally distributed. Test the null hypothesis $H_0: \mu = \$8.00$, using the 10 percent level of significance.

Ans. Reject H_0.

10.35 A random sample of 30 employees at the Secretary II level in a large organization take a standardized typing test. The sample results are $\bar{X} = 63.0$ wpm (words per minute) with $s = 5.0$ wpm. Test the null hypothesis that the secretaries in general do *not* exceed a typing speed of 60 wpm, using the 1 percent level of significance.

Ans. Reject H_0.

10.36 An automatic soft ice cream dispenser has been set to dispense 4.00 oz per serving. For a sample of $n = 10$ servings, the average amount of ice cream is $\bar{X} = 4.05$ oz with $s = 0.10$ oz. The amounts being dispensed are assumed to be normally distributed. Basing the null hypothesis on the assumption that the process is "in control," should the dispenser be reset as a result of a test at the 5 percent level of significance?

Ans. No.

10.37 A shipment of 100 defective machines has been received in a machine-repair department. For a random sample of 10 machines, the average repair time required is $\bar{X} = 85.0$ min with $s = 15.0$ min. Test the null hypothesis $H_0: \mu = 100.0$ min, using the 10 percent level of significance and based on the assumption that the distribution of repair time is approximately normal.

Ans. Reject H_0.

TYPE I AND TYPE II ERRORS IN HYPOTHESIS TESTING

10.38 With reference to Problem 10.35, suppose it is considered an important difference from the hypothesized value of the mean if the average typing speed is at least 64.0 wpm. Determine the probability of (*a*) Type I error and (*b*) Type II error.

Ans. (*a*) $\alpha = 0.01$, (*b*) $\beta = 0.0192 \cong 0.02$.

10.39 For Problem 10.38, determine the probability of Type II error (*a*) if the level of significance is changed to the 5 percent level and (*b*) if the level of significance is kept at the 1 percent level, but the sample size was $n = 60$ instead of $n = 30$.

Ans. (*a*) $\beta = 0.003$, (*b*) $\beta < 0.001$.

10.40 For the testing procedure described in Problem 10.32, suppose it is considered an important discrepancy from the claim if the average mileage is 34.0 miles per gallon or less. Given this additional information, determine (*a*) the minimum OC value if the null hypothesis is true, and (*b*) the minimum power associated with the statistical test if the discrepancy from the claim is an important one.

Ans. (*a*) $OC = 0.95$, (*b*) power $= 0.8729 \cong 0.87$.

10.41 Determine the minimum power of the test in Problem 10.40, given that it is considered an important discrepancy if the true mileage is (*a*) 0.5 mile per gallon and (*b*) 0.1 mile per gallon less than the claim. Comparing the several power values, consider the implication of the standard used to define an "important discrepancy."

Ans. (*a*) Power $= 0.4013 \cong 0.40$, (*b*) power $= 0.0869 \cong 0.09$.

DETERMINING THE REQUIRED SAMPLE SIZE FOR TESTING THE MEAN

10.42 Before collecting any sample data, the prospective investor in Problem 10.26 stipulates that the level of Type I error should be no larger than $\alpha = 0.01$, and that if the number of cars passing the site is at or above $\mu = 210$ per hour, then the level of Type II error should also not exceed $\beta = 0.01$. He estimates that the population standard deviation is no larger than $\sigma = 40$. What sample size is required to achieve his objectives?

Ans. $n = 347.45 = 348$.

10.43 For the testing situation described in Problem 10.28 it is considered an important discrepancy if the mean sales amount per outlet is $100 less than the claimed amount. What sample size is required if the test is to be carried out at the 1 percent level of significance, allowing a maximum probability of Type II error of $\beta = 0.05$?

   Ans. $n = 63.20 = 64$.

THE *P*-VALUE APPROACH TO TESTING NULL HYPOTHESES CONCERNING THE POPULATION MEAN

10.44 Using the *P*-value approach, test the null hypothesis in Problem 10.27 at the 5 percent level of significance.

   Ans.   Reject H_0 ($P = 0.0228$).

10.45 Using the *P*-value approach, test the null hypothesis in Problem 10.28 at the 5 percent level of significance.

   Ans.   Reject H_0 ($P = 0.0307$).

THE CONFIDENCE INTERVAL APPROACH TO TESTING NULL HYPOTHESES CONCERNING THE MEAN

10.46 Apply the confidence interval approach to testing the null hypothesis in Problem 10.27, using the 5 percent level of significance.

   Ans.   Reject H_0.

10.47 Apply the confidence interval approach to testing the null hypothesis in Problem 10.34, using the 10 percent level of significance.

   Ans.   Reject H_0.

COMPUTER OUTPUT

10.48 Refer to Table 2.16 (page 26) for the amounts of 40 personal loans. Assuming that these are random sample data, use available computer software to test the null hypothesis that the mean loan amount in the population is no greater than $1,000.00, using the 1 percent level of significance.

   Ans.   Accept H_0 ($P = 0.17$).

Chapter 11

Testing Other Hypotheses

11.1 TESTING THE DIFFERENCE BETWEEN TWO MEANS USING THE NORMAL DISTRIBUTION

The procedure associated with testing the difference between two means is similar to that for testing a hypothesized value of the mean (see Sections 10.1 and 10.2), except that the standard error of the *difference* between means is used as the basis for determining the z value associated with the sample result. Use of the normal distribution is based on the same conditions as in the one-sample case, except that two independent samples are involved. The general formula for determining the z value for testing the difference between two means, according to whether the σ values for the two populations are known, is

$$z = \frac{(\bar{X}_1 - \bar{X}_2) - (\mu_1 - \mu_2)_0}{\sigma_{\bar{x}_1 - \bar{x}_2}} \tag{11.1}$$

or

$$z = \frac{(\bar{X}_1 - \bar{X}_2) - (\mu_1 - \mu_2)_0}{s_{\bar{x}_1 - \bar{x}_2}} \tag{11.2}$$

As implied by (*11.1*) and (*11.2*), we may begin with any assumed difference, $(\mu_1 - \mu_2)_0$, which is to be tested. However, the usual null hypothesis tested is that the two samples have been obtained from populations with means that are equal. In this case $(\mu_1 - \mu_2)_0 = 0$, and the above formulas are simplified as follows:

$$z = \frac{\bar{X}_1 - \bar{X}_2}{\sigma_{\bar{x}_1 - \bar{x}_2}} \tag{11.3}$$

or

$$z = \frac{\bar{X}_1 - \bar{X}_2}{s_{\bar{x}_1 - \bar{x}_2}} \tag{11.4}$$

In general, the standard error of the difference between means is computed as described in Section 9.1 [see formulas (*9.3*) and (*9.4*)]. However, in testing the difference between two means, the null hypothesis of interest is generally not only that the sample means were obtained from populations with equal means, but that the two samples were in fact obtained from the *same* population of values. This means that $\sigma_1 = \sigma_2$, which we can simply designate σ. Thus, the assumed common variance is often estimated by pooling the two sample variances, and the estimated value of σ^2 is then used as the basis for the standard error of the difference. The pooled estimate of the population variance is

$$\hat{\sigma}^2 = \frac{(n_1 - 1)s_1^2 + (n_2 - 1)s_2^2}{n_1 + n_2 - 2} \tag{11.5}$$

The estimated standard error of the difference based on the assumption that the population standard deviations are equal is

$$\hat{\sigma}_{\bar{x}_1 - \bar{x}_2} = \sqrt{\frac{\hat{\sigma}^2}{n_1} + \frac{\hat{\sigma}^2}{n_2}} \tag{11.6}$$

The assumption that the two sample variances were obtained from populations with equal variances can itself be tested as the null hypothesis (see Section 11.9).

Tests concerned with the difference between means can be either two-tail or one-tail, as illustrated in the following examples.

EXAMPLE 1. The mean weekly wage for a sample of $n_1 = 30$ employees in a large manufacturing firm is $\bar{X}_1 = \$280.00$ with a sample standard deviation of $s_1 = \$14.00$. In another large firm a random sample of $n_2 = 40$

178

employees has a mean wage of $\bar{X}_2 = \$270.00$ with a sample standard deviation of $s_2 = \$10.00$. The standard deviations of the two populations of wage amounts are not assumed to be equal. We test the hypothesis that there is no difference between the mean weekly wage amounts in the two firms, using the 5 percent level of significance, as follows:

$$H_0: (\mu_1 - \mu_2) = 0 \qquad \text{or, equivalently,} \qquad \mu_1 = \mu_2 \qquad \bar{X}_1 = \$280.00 \qquad \bar{X}_2 = \$270.00$$

$$H_1: (\mu_1 - \mu_2) \neq 0 \qquad \text{or, equivalently,} \qquad \mu_1 \neq \mu_2 \qquad s_1 = \$14.00 \qquad s_2 = \$10.00$$

$$n_1 = 30 \qquad n_2 = 40$$

$$\text{Critical } z \ (\alpha = 0.05) = \pm 1.96$$

$$z = \frac{\bar{X}_1 - \bar{X}_2}{s_{\bar{x}_1 - \bar{x}_2}}$$

$$= \frac{280 - 270}{3.01} = \frac{10.0}{3.01} = +3.32$$

where $\quad s_{\bar{x}_1} = \dfrac{s_1}{\sqrt{n_1}} = \dfrac{14.00}{\sqrt{30}} = \dfrac{14.00}{5.477} = 2.56$

$$s_{\bar{x}_2} = \frac{s_2}{\sqrt{n_2}} = \frac{10.00}{\sqrt{40}} = \frac{10.00}{6.325} = 1.58$$

$$s_{\bar{x}_1 - \bar{x}_2} = \sqrt{s_{\bar{x}_1}^2 + s_{\bar{x}_2}^2} = \sqrt{(2.56)^2 + (1.58)^2} = \sqrt{6.5536 + 2.4964} = 3.01$$

The computed z of $+3.32$ is in the region of rejection of the hypothesis testing model portrayed in Fig. 11-1. Therefore the null hypothesis is rejected, and the alternative hypothesis, that the average weekly wage in the two firms is different, is accepted.

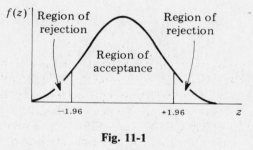

Fig. 11-1

EXAMPLE 2. Before seeing the sample results in Example 1, a wage analyst believed that the average wage in the first firm was greater than the average wage in the second firm. In order to subject his belief to a critical test, he gives the benefit of the doubt to the opposite possibility, and formulates the null hypothesis that the average wage in the first firm is equal to or less than the average in the second firm. We test this hypothesis at the 1 percent level of significance, again without assuming that the standard deviations of the two populations are equal, as follows:

$$H_0: (\mu_1 - \mu_2) \leq 0 \qquad \text{or, equivalently,} \qquad \mu_1 \leq \mu_2$$

$$H_1: (\mu_1 - \mu_2) > 0 \qquad \textbf{or, equivalently,} \qquad \mu_1 > \mu_2$$

$$\textbf{Critical } z \ (\alpha = 0.01) = +2.33$$

$$\text{Computed } z = +3.32 \quad \textbf{(from Example 1)}$$

The computed z of $+3.32$ is **greater than the critical** value of $+2.33$ for this upper-tail test, as portrayed in Fig. 11-2. Therefore the null hypothesis is rejected and the alternative hypothesis, that the average wage in the first firm is *greater than* the average wage in the second firm, is accepted.

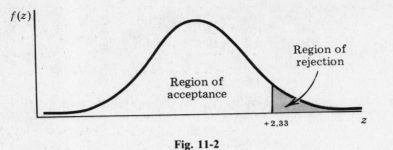

Fig. 11-2

11.2 TESTING THE DIFFERENCE BETWEEN TWO MEANS USING STUDENT'S t DISTRIBUTION

When the difference between two means is tested by the use of the t distribution, a necessary assumption is that the variances of the two populations are equal. Therefore, in such a test, the estimated standard error of the mean is calculated from (11.5) and (11.6). The several requirements associated with the appropriate use of the t distribution are described in Sections 8.5 and 10.5.

EXAMPLE 3. For a random sample of $n_1 = 10$ light bulbs, the mean bulb life is $\bar{X}_1 = 4,000$ hr with $s_1 = 200$. For another brand of bulbs whose useful life is also assumed to be normally distributed, a random sample of $n_2 = 8$ has a sample mean of $\bar{X}_2 = 4,300$ hr and a sample standard deviation of $s = 250$. We test the hypothesis that there is no difference between the mean operating life of the two brands of bulbs, using the 1 percent level of significance, as follows:

$$H_0: (\mu_1 - \mu_2) = 0 \qquad \bar{X}_1 = 4,000 \text{ hr} \qquad \bar{X}_2 = 4,300 \text{ hr}$$

$$H_1: (\mu_1 - \mu_2) \neq 0 \qquad s_1 = 200 \text{ hr} \qquad s_2 = 250 \text{ hr}$$

$$n_1 = 10 \qquad n_2 = 8$$

$$df = n_1 + n_2 - 2 = 10 + 8 - 2 = 16$$

Critical t ($df = 16$, $\alpha = 0.01$) $= \pm 2.921$

$$\hat{\sigma}^2 = \frac{(n_1-1)s_1^2 + (n_2-1)s_2^2}{n_1 + n_2 - 2} = \frac{(9)(200)^2 + (7)(250)^2}{10 + 8 - 2} = \frac{360,000 + 437,500}{16} \doteq 49,843.75$$

$$\hat{\sigma}_{\bar{x}_1 - \bar{x}_2} = \sqrt{\frac{\hat{\sigma}^2}{n_1} + \frac{\hat{\sigma}^2}{n_2}} = \sqrt{\frac{49,843.75}{10} + \frac{49,843.75}{8}} = \sqrt{11,214.843} = 105.9$$

$$t = \frac{\bar{X}_1 - \bar{X}_2}{\hat{\sigma}_{\bar{x}_1 - \bar{x}_2}} = \frac{4,000 - 4,300}{105.9} = \frac{-300}{105.9} = -2.833$$

The computed t of -2.833 is in the region of acceptance of the null hypothesis. Therefore the null hypothesis cannot be rejected at the 1 percent level of significance.

11.3 TESTING THE DIFFERENCE BETWEEN TWO MEANS BASED ON PAIRED OBSERVATIONS

The procedures in Sections 11.1 and 11.2 are based on the assumption that the two samples were collected as two independent random samples. However, in many situations the samples are collected as pairs of values, such as when determining the productivity level of each worker before and after a training program. These are referred to as *paired observations*, or *matched pairs*. Also, as contrasted to independent samples, two samples that contain paired observations often are called *dependent samples*.

For paired observations, the appropriate approach for testing the difference between the means of the two samples is to first determine the difference d between each pair of values, and then test the null hypotheses that the mean population *difference* is zero. Thus, from the computational standpoint the test is applied to the *one* sample of d values.

The mean and standard deviation of the sample d values are obtained by use of the basic formulas in Chapters 3 and 4, except that d is substituted for X. The mean difference for a set of paired observations is

$$\bar{d} = \frac{\Sigma d}{n} \tag{11.7}$$

The deviations formula and the computational formula for the standard deviation of the differences between paired observations are, respectively,

$$s_d = \sqrt{\frac{\Sigma(d - \bar{d})^2}{n - 1}} \tag{11.8}$$

$$s_d = \sqrt{\frac{\Sigma d^2 - n\bar{d}^2}{n - 1}} \tag{11.9}$$

The standard error of the mean difference between paired observations is obtained by formula (8.4) for the standard error of the mean, except that d is again substituted for X:

$$s_{\bar{d}} = \frac{s_d}{\sqrt{n}} \tag{11.10}$$

Because the standard error of the mean difference is computed on the basis of the differences observed in the paired samples (that is, the population value σ_d is unknown) and because values of d generally can be assumed to be normally distributed, the t distribution is appropriate for testing the null hypothesis that $\mu_d = 0$.

The degrees of freedom is the number of *pairs* of observed values minus one, or $n - 1$. As discussed in Section 8.5, the standard normal z distribution can be used in place of the t distributions when $n \geq 30$. Example 4 illustrates a two-tail test, whereas Problem 11.5 illustrates a one-tail test. Thus, the test statistic used to test the hypothesis that there is no difference between the means of a set of paired observations is

$$t = \frac{\bar{d}}{s_d} \tag{11.11}$$

EXAMPLE 4. An automobile manufacturer collects mileage data for a sample of $n = 10$ cars in various weight categories by use of a standard grade of gasoline with and without a particular additive. Of course, the engines were tuned to the same specifications before each run, and the same drivers were used for the two gasoline conditions (with the driver in fact being unaware of which gasoline was being used on a particular run). Given the mileage data in Table 11.1, we test the hypothesis that there is no difference between the mean mileage obtained

Table 11.1   Automobile Mileage Data and Worksheet for Computing the Mean
Difference and the Standard Deviation of the Difference

Automobile	Mileage with additive	Mileage without additive	d	d^2
1	36.7	36.2	0.5	0.25
2	35.8	35.7	0.1	0.01
3	31.9	32.3	−0.4	0.16
4	29.3	29.6	−0.3	0.09
5	28.4	28.1	0.3	0.09
6	25.7	25.8	−0.1	0.01
7	24.2	23.9	0.3	0.09
8	22.6	22.0	0.6	0.36
9	21.9	21.5	0.4	0.16
10	20.3	20.0	0.3	0.09
Total	276.8	275.1	+1.7	1.31

with and without the additive, using the 5 percent level of significance, as follows:

$$\text{Average with additive} = \frac{276.8}{10} = 27.68 \text{ mpg}$$

$$\text{Average without additive} = \frac{275.1}{10} = 27.51 \text{ mpg}$$

$$H_0: \quad \mu_d = 0$$

$$H_1: \quad \mu_d \neq 0$$

$$\text{Critical } t \ (df = 9, \alpha = 0.05) = \pm 2.262$$

$$\bar{d} = \frac{\Sigma d}{n} = \frac{1.7}{10} = 0.17$$

$$s_d = \sqrt{\frac{\Sigma d^2 - n\bar{d}^2}{n-1}} = \sqrt{\frac{1.31 - 10(0.17)^2}{10-1}} = \sqrt{\frac{1.31 - 10(0.0289)}{9}} = \sqrt{0.1134} = 0.337$$

$$s_{\bar{d}} = \frac{s_d}{\sqrt{n}} = \frac{0.337}{\sqrt{10}} = \frac{0.337}{3.16} = 0.107$$

$$t = \frac{\bar{d}}{s_{\bar{d}}} = \frac{0.17}{0.107} = +1.59$$

The computed t of $+1.59$ is in the region of acceptance of the null hypothesis. Therefore the null hypothesis that there is no difference in the miles per gallon obtained with as compared to without the additive is accepted.

11.4 TESTING A HYPOTHESIZED VALUE OF THE POPULATION PROPORTION USING THE BINOMIAL DISTRIBUTION

When a sampling process can be assumed to conform to a Bernoulli process (see Section 6.3), the binomial distribution can be used in conjunction with testing hypotheses regarding the population proportion.

Typically, **tests** of proportions based on using the binomial distribution are one-tail tests. Given the hypothesized value of the population proportion, the "region" of rejection is the set of sample observations which deviates from the hypothesized value and for which the probability of occurrence by chance does not exceed the specified level of significance. That is, essentially the P-value approach to hypothesis testing is used (see Section 11.10). The test procedure for a one-tail test is illustrated by Example 5. See Problem 11.8 for a two-tail test using a binomial distribution.

EXAMPLE 5. The director of a college placement office claims that by March 1 at least 50 percent of the graduating seniors will have obtained full-time jobs. A random sample of 10 graduating seniors are polled on March 1, and only two indicate that they have concluded their job arrangements. Can the director's claim be rejected, using the 5 percent level of significance? $H_0: \pi \geq 0.50$ and $H_1: \pi < 0.50$.

Based on the binomial distribution, the probability values associated with fewer than five students having obtained jobs, given a population proportion of 0.50, are given in Table 11.2 (from Appendix 2, with $n = 10$ and $p = 0.50$).

Critical values of test statistic: The test statistic is the number of students in the sample of $n = 10$ who have already obtained employment. In order to reject the null hypothesis at the 5 percent level of significance, only "0" or "1" student would have to be observed to have jobs. This is because the probabilities are accumulated in the "lower tail" of this binomial distribution to determine the region of rejection. To attempt to include "2" students in the region of rejection results in a cumulative probability (for "0, 1, or 2") of $0.0010 + 0.0098 + 0.0439 = 0.0547$, which just exceeds the designated 0.05 test level.

Result of test: Based on the critical values identified above, observing that only two students out of a sample of 10 have obtained jobs is not low enough to reject the director's claim at the 5 percent level of significance.

Table 11.2 Probability Values Associated with Fewer than Five out of Ten Students Having Obtained Jobs

Number of students	Probability
0	0.0010
1	0.0098
2	0.0439
3	0.1172
4	0.2051

(*Note:* With a larger sample, the same relative difference from the hypothesized value might indeed lead to the rejection of the null hypothesis. See Problem 11.7.)

11.5 TESTING A HYPOTHESIZED VALUE OF THE POPULATION PROPORTION USING THE NORMAL DISTRIBUTION

As explained in Section 7.4, the normal distribution can be used as an approximation of a binomial distribution when $n \geq 30$ and both $np \geq 5$ and $n(q) \geq 5$, where $q = 1 - p$. This is the basis upon which confidence intervals for the proportion are constructed in Section 9.3, where the standard error of the proportion is also discussed. However, in the case of confidence intervals, a sample size of at least $n = 100$ is required, as explained in Section 9.3.

In hypothesis testing, the value of the standard error of the proportion used in conjunction with hypothesis testing is based on using the hypothesized value π_0:

$$\sigma_{\hat{p}} = \sqrt{\frac{\pi_0(1 - \pi_0)}{n}} \tag{11.12}$$

The formula for the standard error of the proportion which includes the finite correction factor is

$$\sigma_{\hat{p}} = \sqrt{\frac{\pi_0(1 - \pi_0)}{n}} \sqrt{\frac{N - n}{N - 1}} \tag{11.13}$$

The procedure associated with testing a hypothesized value of the population proportion is identical to that described in Section 10.2, except that a value of the proportion rather than of the mean is being tested. Thus, the formula for the z statistic for testing a hypothesized value of the proportion is

$$z = \frac{\hat{p} - \pi_0}{\sigma_{\hat{p}}} \tag{11.14}$$

EXAMPLE 6. In Example 5, the placement director claimed that at least 50 percent of the graduating seniors had finalized job arrangements by March 1. Suppose a random sample of $n = 30$ seniors are polled, rather than the 10 in Example 5, and only 10 of the students indicate that they have concluded their job arrangements by March 1. Can the placement director's claim be rejected at the 5 percent level of significance? We utilize z as the test statistic, as follows:

$$H_0: \ \pi \geq 0.50 \qquad H_1: \ \pi < 0.50$$

$$\text{Critical } z \ (\alpha = 0.05) = -1.645$$

[Use of the normal distribution is warranted because $n \geq 30$, $n\pi_0 \geq 5$, and $n(1 - \pi_0) \geq 5$.]

$$\sigma_{\hat{p}} = \sqrt{\frac{\pi_0(1 - \pi_0)}{n}} = \sqrt{\frac{(0.50)(0.50)}{30}} = \sqrt{\frac{0.25}{30}} = \sqrt{0.0083} = 0.09$$

(It is assumed that the sample is less than 5 percent of the population size, and so the finite correction factor is not used.)

$$z = \frac{\hat{p} - \pi_0}{\sigma_{\hat{p}}} = \frac{0.33 - 0.50}{0.09} = \frac{-0.17}{0.09} = -1.88$$

The computed z of -1.88 is less than the critical value of -1.645 for this lower-tail test. Therefore the director's claim is rejected at the 5 percent level of significance.

11.6 DETERMINING REQUIRED SAMPLE SIZE FOR TESTING THE PROPORTION

Before a sample is actually collected, the required sample size for testing a hypothesized value of the proportion can be determined by specifying (1) the hypothesized value of the proportion, (2) a specific alternative value of the proportion such that the difference from the null-hypothesized value is considered important, (3) the level of significance to be used in the test, and (4) the probability of Type II error which is to be permitted. The formula for determining the minimum sample size required for testing a hypothesized value of the proportion is

$$n = \left[\frac{z_0 \sqrt{\pi_0 (1 - \pi_0)} - z_1 \sqrt{\pi_1 (1 - \pi_1)}}{\pi_1 - \pi_0} \right]^2 \qquad (11.15)$$

In (11.15), z_0 is the critical value of z used in conjunction with the specified level of significance (α level) while z_1 is the value of z with respect to the designated probability of Type II error (β level). As was true in Section 10.4, on determining sample size for testing the mean, z_0 and z_1 always have opposite algebraic signs. The result is that the two products in the numerator will always be accumulated. Also, formula (11.15) can be used in conjunction with either one-tail or two-tail tests and any fractional sample size is rounded up. In addition, the sample size should be large enough to warrant use of the normal probability distribution in conjunction with π_0 and π_1, as reviewed in Section 11.5.

EXAMPLE 7.   A Congressman wishes to test the hypothesis that at least 60 percent of his constituents are in favor of labor legislation being introduced in Congress, using the 5 percent level of significance. He considers the discrepancy from this hypothesis to be important if only 50 percent (or fewer) favor the legislation, and is willing to accept a risk of Type II error of $\beta = 0.05$. The sample size which he should arrange to collect, as a minimum, to satisfy these decision-making specifications is

$$n = \left[\frac{z_0 \sqrt{\pi_0 (1 - \pi_0)} - z_1 \sqrt{\pi_1 (1 - \pi_1)}}{\pi_1 - \pi_0} \right]^2 = \left[\frac{-1.645 \sqrt{(0.60)(0.40)} - (+1.645) \sqrt{(0.50)(0.50)}}{0.50 - 0.60} \right]^2$$

$$= \left[\frac{-1.645(0.49) - 1.645(0.50)}{-0.10} \right]^2 = \left(\frac{-0.806 - 0.822}{-0.10} \right)^2 = (16.28)^2 = 265.04 = 266$$

11.7 TESTING THE DIFFERENCE BETWEEN TWO POPULATION PROPORTIONS

When we wish to test the hypothesis that the proportions in two populations are not different, the two sample proportions are pooled as a basis for determining the standard error of the difference between proportions. Note that this differs from the procedure used in Section 9.5 on statistical estimation, in which the assumption of no difference was *not* made. Further, the present procedure is conceptually similar to that presented in Section 11.1, in which the two sample variances are pooled as the basis for computing the standard error of the difference between means. The pooled estimate of the population proportion, based on the proportions obtained in two independent samples, is

$$\hat{\pi} = \frac{n_1 \hat{p}_1 + n_2 \hat{p}_2}{n_1 - n_2} \qquad (11.16)$$

The standard error of the difference between proportions used in conjunction with testing the assumption of no difference is

$$\hat{\sigma}_{\hat{p}_1 - \hat{p}_2} = \sqrt{\frac{\hat{\pi}(1 - \hat{\pi})}{n_1} + \frac{\hat{\pi}(1 - \hat{\pi})}{n_2}} \qquad (11.17)$$

The formula for the z statistic for testing the difference between two proportions is

$$z = \frac{\hat{p}_1 - \hat{p}_2}{\hat{\sigma}_{\hat{p}_1 - \hat{p}_2}} \qquad (11.18)$$

A test of the difference between proportions can be carried out as either a one-tail test (see Problem 11.11) or a two-tail test (see Example 8).

EXAMPLE 8.   A sample of 50 households in one community shows that 10 of them are watching a TV special on the national economy. In a second community, 15 of a random sample of 50 households are watching the TV special. We test the hypothesis that the overall proportion of viewers in the two communities does not differ, using the 1 percent level of significance, as follows:

$$H_0: (\pi_1 - \pi_2) = 0 \qquad \text{or, equivalently,} \qquad \pi_1 = \pi_2$$
$$H_1: (\pi_1 - \pi_2) \neq 0 \qquad \text{or, equivalently,} \qquad \pi_1 \neq \pi_2$$

$$\text{Critical } z \ (\alpha = 0.01) = \pm 2.58$$

$$\hat{\pi} = \frac{n_1 \hat{p}_1 + n_2 \hat{p}_2}{n_1 + n_2} = \frac{50(0.20) + 50(0.30)}{50 + 50} = \frac{10 + 15}{100} = 0.25$$

$$\hat{\sigma}_{\hat{p}_1 - \hat{p}_2} = \sqrt{\frac{\hat{\pi}(1 - \hat{\pi})}{n_1} + \frac{\hat{\pi}(1 - \hat{\pi})}{n_2}} = \sqrt{\frac{(0.25)(0.75)}{50} + \frac{(0.25)(0.75)}{50}} = \sqrt{0.00375 + 0.00375} = 0.087$$

$$z = \frac{\hat{p}_1 - \hat{p}_2}{\hat{\sigma}_{\hat{p}_1 - \hat{p}_2}} = \frac{0.20 - 0.30}{0.087} = \frac{-0.10}{0.087} = -1.15$$

The computed z of -1.15 is in the region of acceptance of the null hypothesis. Therefore the hypothesis that there is no difference in the proportion of viewers in the two areas cannot be rejected.

11.8 TESTING A HYPOTHESIZED VALUE OF THE VARIANCE USING THE CHI-SQUARE DISTRIBUTION

As explained in Section 9.6, for a normally distributed population the ratio $(n-1)s^2/\sigma^2$ follows a χ^2 probability distribution, with there being a different chi-square distribution according to degrees of freedom $(n-1)$. Therefore, the statistic which is used to test a hypothesized value of the population variance is

$$\chi^2 = \frac{(n-1)s^2}{\sigma_0^2} \qquad (11.19)$$

The test based on (11.19) can be either a one-tail test or a two-tail test, although most often hypotheses about a population variance relate to one-tail tests. Appendix 7 can be used to determine the critical value(s) of the chi-square statistic for various levels of significance.

EXAMPLE 9.   The mean operating life for a random sample of $n = 10$ light bulbs is $\bar{X} = 4,000$ hr with a standard deviation of $s = 200$ hr. The operating life of bulbs in general is assumed to be normally distributed. Suppose that before the sample was collected, it was hypothesized that the population standard deviation is no larger than $\sigma = 150$. Based on the sample results, this hypothesis is tested at the 1 percent level of significance as follows:

$$H_0: \sigma^2 \leq 22{,}500 \text{ [because } \sigma_0^2 = (150)^2 = 22{,}500] \qquad H_1: \sigma^2 > 22{,}500$$

$$\text{Critical } \chi^2 \, (df = 9, \, \alpha = 0.01) = 21.67$$

$$\chi^2 = \frac{(n-1)s^2}{\sigma_0^2} = \frac{(9)(40{,}000)}{22{,}500} = \frac{360{,}000}{22{,}500} = 16.0$$

Because the calculated test statistic of 16.0 does *not* exceed the critical value of 21.67 for this upper-tail test, the null hypothesis that $\sigma \leq 150$ cannot be rejected at the 1 percent level of significance.

11.9 THE F DISTRIBUTION AND TESTING THE DIFFERENCE BETWEEN TWO VARIANCES

The F distribution can be shown to be the appropriate probability model for the ratio of the variances of two samples taken independently from the same normally distributed population, with there being a different F distribution for every combination of the degrees of freedom df associated with each sample. For each sample, $df = n - 1$. Thus, the statistic which is used to test the null hypothesis that there is no difference between two variances is

$$F_{df_1, df_2} = \frac{s_1^2}{s_2^2} \tag{11.20}$$

Since each sample variance is an unbiased estimator of the population variance, the long-run expected value of the above ratio is about 1.0. (*Note:* The expected value is not exactly 1.0, but rather is $df_2/(df_2 - 2)$, for mathematical reasons that are outside of the scope of this outline.) However, for any given pair of samples the sample variances are not likely to be identical in value, even though the null hypothesis is true. Since this ratio is known to follow an F distribution, this probability distribution can be used in conjunction with testing the difference between two variances. Although a necessary mathematical assumption is that the two populations are normally distributed, the F test has been demonstrated to be relatively insensitive to departures from normality when each population is at least unimodal and the sample sizes are about equal.

Appendix 8 indicates the values of F exceeded by proportions of 0.05 and 0.01 of the distribution of F values. The degrees of freedom df associated with the numerator of the calculated F ratio are the column headings of this table and the degrees of freedom for the denominator are the row headings. The table does not identify any critical values of F for the lower tail of the distribution, partly because the F distribution is typically used in conjunction with one-tail tests. This is particularly true for the use of the F distribution in the analysis of variance (see Chapter 13). Another reason for providing only upper-tail F values is that lower-tail values of F can be calculated by the so-called *reciprocal property* of the F distribution, as follows:

$$F_{df_1, df_2 \text{lower}} = \frac{1}{F_{df_2, df_1, \text{upper}}} \tag{11.21}$$

In applying formula (*11.21*), an F value at the lower 5 percent point is determined by entering an upper-tail value at the 5 percent point in the denominator. Note, however, that the two df values in the denominator are the *reverse* of the order in the required F value.

EXAMPLE 10. For the data in Example 3, bulb life is assumed to be normally distributed. We test the null hypothesis that the samples were obtained from populations with equal variances, using the 10 percent level of significance for the test, by use of the F distribution:

$$H_0: \sigma_1^2 = \sigma_2^2 \qquad s_1 = 200 \text{ hr} \qquad s_2 = 250 \text{ hr}$$
$$H_1: \sigma_1^2 \neq \sigma_2^2 \qquad s_1^2 = 40{,}000 \qquad s_2^2 = 62{,}500$$
$$n_1 = 10 \qquad\qquad n_2 = 8$$

For the test at the 10 percent level of significance, the upper 5 percent point for F and the lower 5 percent point for F are the critical values.

$$\text{Critical } F_{9,7} \text{ (upper 5\%)} = 3.68$$

$$\text{Critical } F_{9,7} \text{ (lower 5\%)} = \frac{1}{F_{7,9}(\text{upper 5\%})} = \frac{1}{3.29} = 0.304$$

$$F_{df_1, df_2} = \frac{s_1^2}{s_2^2}$$

$$F_{9,7} = \frac{40,000}{62,500} = 0.64$$

Since the computed F ratio is neither smaller than 0.304 nor larger than 3.68, it is in the region of acceptance of the null hypothesis. Thus, the assumption that the variances of the two populations are equal cannot be rejected at the 10 percent level of significance.

11.10   ALTERNATIVE APPROACHES TO TESTING NULL HYPOTHESES

As described in Sections 10.6 and 10.7, the P-value approach and the confidence interval approach are alternatives to the critical value approach to hypothesis testing used in the preceding sections of this chapter.

By the P-value approach, instead of comparing the observed value of a test statistic with a critical value, the probability of the occurrence of the test statistic, given that the null hypothesis is true, is determined and compared to the level of significance α. The null hypothesis is rejected if the P value is *less* than the designated α. Problems 11.16 and 11.17 illustrate the application of this approach to two-tail and one-tail tests, respectively, concerning the difference between means.

By the confidence interval approach, the $1 - \alpha$ confidence interval is constructed for the parameter value of concern. If the hypothesized value of the parameter is not included in the interval, then the null hypothesis is rejected. Problems 11.18 and 11.19 illustrate the application of this approach to two-tail and one-tail tests, respectively, concerning the difference between means.

11.11   COMPUTER OUTPUT

Computer software is available for most of the hypothesis-testing situations described in this chapter. The P-value approach to hypothesis testing generally is used, as described in Section 11.10. The use of computer software for testing the difference between the means of two independent samples is illustrated in Problem 11.20, while the use of such software with the paired observations design for testing the difference between means is illustrated in Problem 11.21.

Solved Problems

TESTING THE DIFFERENCE BETWEEN TWO MEANS USING THE NORMAL DISTRIBUTION

11.1   A developer is considering two alternative sites for a regional shopping center. Since household income in the community is one important consideration in such site selection, he wishes to test the null hypothesis that there is no difference between the mean household income amounts in the two communities. Consistent with this hypothesis, he assumes that the standard deviation of household income is also the same in the two communities. For a sample of $n_1 = 30$ households in the first community, the average annual income is $\bar{X}_1 = \$35,500$ with the sample standard deviation $s_1 = \$1,800$. For a sample of $n_2 = 40$ households in the second community, $\bar{X}_2 = \$34,600$ and $s_2 = \$2,400$. Test the null hypothesis at the 5 percent level of significance.

$$H_0: (\mu_1 - \mu_2) = 0 \qquad H_1: (\mu_1 - \mu_2) \neq 0$$

$$\bar{X}_1 = \$35,500 \qquad \bar{X}_2 = \$34,600$$

$$s_1 = \$1,800 \qquad s_2 = \$2,400$$

$$n_1 = 30 \qquad n_2 = 40$$

Critical z $(\alpha = 0.05) = \pm 1.96$

$$\hat{\sigma}^2 = \frac{(n_1-1)s_1^2 + (n_2-1)s_2^2}{n_1 + n_2 - 2} = \frac{29(1,800)^2 + 39(2,400)^2}{30+40-2} = \frac{318,600,000}{68} = \$4,685,294$$

(The variances are pooled because of the assumption that the standard deviation values in the two populations are equal.)

$$\hat{\sigma}_{\bar{x}_1-\bar{x}_2} = \sqrt{\frac{\hat{\sigma}^2}{n_1} + \frac{\hat{\sigma}^2}{n_2}} = \sqrt{\frac{4,685,294}{30} + \frac{4,685,294}{40}} = \sqrt{156,176.46 + 117,132.35} = \$522.79$$

$$z = \frac{\bar{X}_1 - \bar{X}_2}{\hat{\sigma}_{\bar{x}_1-\bar{x}_2}} = \frac{35,500 - 34,600}{522.79} = \frac{900}{522.79} = +1.72$$

The computed z value of $+1.72$ is in the region of acceptance of the null hypothesis. Therefore the null hypothesis cannot be rejected at the 5 percent level of significance, and the hypothesis that the average income per household in the two communities does not differ is accepted.

11.2   With reference to Problem 11.1, before any data were collected it was the developer's judgment that income in the first community might be higher. In order to subject this judgment to a critical test, he gave the benefit of doubt to the other possibility and formulated the null hypothesis H_0: $(\mu_1 - \mu_2) \leq 0$. Test this hypothesis at the 5 percent level of significance with the further assumption that the standard deviation values for the two populations are *not* necessarily equal.

$$H_0: (\mu_1 - \mu_2) \leq 0 \qquad H_1: (\mu_1 - \mu_2) > 0$$

Critical z $(\alpha = 0.05) = +1.645$

$$s_{\bar{x}_1} = \frac{s_1}{\sqrt{n_1}} = \frac{1,800}{\sqrt{30}} = \frac{1,800}{5.48} = \$328.47$$

$$s_{\bar{x}_2} = \frac{s_2}{\sqrt{n_2}} = \frac{2,400}{\sqrt{40}} = \frac{2,400}{6.32} = \$379.75$$

$$s_{\bar{x}_1-\bar{x}_2} = \sqrt{s_{\bar{x}_1}^2 + s_{\bar{x}_2}^2} = \sqrt{(328.47)^2 + (379.75)^2} = \sqrt{252,102.60} = \$502.10$$

$$z = \frac{\bar{X}_1 - \bar{X}_2}{s_{\bar{x}_1-\bar{x}_2}} = \frac{35,500 - 34,600}{502.10} = \frac{900}{502.10} = +1.79$$

The computed z value of $+1.79$ is larger than the critical value of $+1.645$ for this upper-tail test. Therefore the null hypothesis is rejected at the 5 percent level of significance, and the alternative hypothesis, that average household income is larger in the first community as compared with the second, is accepted.

11.3   With respect to Problems 11.1 and 11.2, before any data were collected it was the developer's judgment that the average income in the first community exceeds the average in the second community by at least $1,500 per year. In this case, give this judgment the benefit of the doubt and test the assumption as the null hypothesis using the 5 percent level of significance. The population standard deviations are not assumed to be equal.

$$H_0: (\mu_1 - \mu_2) \geq 1,500 \qquad H_1: (\mu_1 - \mu_2) < 1,500$$
$$\text{Critical } z \ (\alpha = 0.05) = -1.645$$

$$s_{\bar{x}_1 - \bar{x}_2} = \$502.10 \quad \text{(from Problem 11.2)}$$

$$z = \frac{(\bar{X}_1 - \bar{X}_2) - (\mu_1 - \mu_2)_0}{s_{\bar{x}_1 - \bar{x}_2}} = \frac{(35,500 - 34,600) - 1,500}{502.10} = \frac{-600}{502.10} = -1.19$$

The calculated z value of -1.19 is *not* less than the critical value of -1.645 for this lower-tail test. Therefore the null hypothesis cannot be rejected at the 5 percent level of significance. Although the sample difference of $900 does not meet the $1,500 difference hypothesized by the developer, it is not different enough from his assumption when the assumption is given the benefit of the doubt by being designated as the null hypothesis.

TESTING THE DIFFERENCE BETWEEN TWO MEANS USING THE t DISTRIBUTION

11.4   Of 100 juniors and seniors majoring in accounting in a college of business administration, a random sample of $n_1 = 12$ students has a mean grade-point average of 2.70 (where A = 4.00) with a sample standard deviation of 0.40. For the 50 juniors and seniors majoring in computer information systems, a random sample of $n_2 = 10$ students has a mean grade-point average of $\bar{X}_2 = 2.90$ with a standard deviation of 0.30. The grade-point values are assumed to be normally distributed. Test the null hypothesis that the mean grade-point average for the two categories of students is not different, using the 5 percent level of significance.

$$H_0: (\mu_1 - \mu_2) = 0 \qquad H_1: (\mu_1 - \mu_2) \neq 0$$

$$\bar{X}_1 = 2.70 \qquad\qquad \bar{X}_2 = 2.90$$

$$s_1 = 0.40 \qquad\qquad s_2 = 0.30$$

$$n_1 = 12 \qquad\qquad n_2 = 10$$

$$N_1 = 100 \qquad\qquad N_2 = 50$$

$$\text{Critical } t \ (df = 20, \ \alpha = 0.05) = \pm 2.086$$

(*Note:* Section 11.2 specifies that a necessary assumption when using a t distribution for testing a difference between means is that the variances are equal.) Therefore the two sample variances are pooled:

$$\hat{\sigma}^2 = \frac{(n_1 - 1)s_1^2 + (n_2 - 1)s_2^2}{n_1 + n_2 - 2} = \frac{(11)(0.40)^2 + (9)(0.30)^2}{12 + 10 - 2} = \frac{1.76 + 0.81}{20} = \frac{2.57}{20} = 0.128$$

$$\hat{\sigma}_{\bar{x}_1 - \bar{x}_2} = \sqrt{\frac{\hat{\sigma}^2}{n_1}\left(\frac{N_1 - n_1}{N_1 - 1}\right) + \frac{\hat{\sigma}^2}{n_2}\left(\frac{N_2 - n_2}{N_2 - 1}\right)} = \sqrt{\frac{0.128}{12}\left(\frac{100 - 12}{100 - 1}\right) + \frac{0.128}{10}\left(\frac{50 - 10}{50 - 1}\right)}$$

$$= \sqrt{0.011\left(\frac{88}{99}\right) + 0.013\left(\frac{40}{49}\right)} = \sqrt{0.011(0.889) + 0.013(0.816)} = \sqrt{0.020387} = 0.143$$

(For each sample, $n < 0.05N$, and thus use of the finite correction factor is required.)

$$t = \frac{\bar{X}_1 - \bar{X}_2}{\hat{\sigma}_{\bar{x}_1 - \bar{x}_2}} = \frac{2.70 - 2.90}{0.143} = \frac{-0.20}{0.143} = -1.399$$

The calculated value of t of -1.399 is in the region of acceptance of the null hypothesis. Therefore the null hypothesis that there is no difference between the grade-point averages for the two populations of students cannot be rejected at the 5 percent level of significance.

TESTING THE DIFFERENCE BETWEEN MEANS BASED ON PAIRED OBSERVATIONS

11.5   A company training director wishes to compare a new approach to technical training, involving a combination of tutorial computer disks and laboratory problem-solving, with the traditional lecture-discussion approach. She matches 12 pairs of trainees according to prior background and academic performance, and assigns one member of each pair to the traditional class and the other to the new approach. At the end of the course, the level of learning is determined by an examination covering basic information as well as the ability to apply the information. Because the training director wishes to give the benefit of the doubt to the established instructional system, she formulates the null hypothesis that the mean performance for the established system is equal to or greater than the mean level of performance for the new system. Test this hypothesis at the 5 percent level of significance. The sample performance data are presented in the first three columns of Table 11.3.

Table 11.3   Training Program Data and Worksheet for Computing the Mean Difference and the Standard Deviation of the Difference

Trainee pair	Traditional method (X_1)	New approach (X_2)	d $(X_1 - X_2)$	d^2
1	89	94	−5	25
2	87	91	−4	16
3	70	68	2	4
4	83	88	−5	25
5	67	75	−8	64
6	71	66	5	25
7	92	94	−2	4
8	81	88	−7	49
9	97	96	1	1
10	78	88	−10	100
11	94	95	−1	1
12	79	87	−8	64
Total	988	1,030	−42	378

$$\text{Mean performance (traditional method)} = \frac{988}{12} = 82.33$$

$$\text{Mean performance (new approach)} = \frac{1,030}{12} = 85.83$$

$$H_0: \mu_d \geq 0 \qquad H_1: \mu_d < 0$$
$$\text{Critical } t \ (df = 11, \ \alpha = 0.05) = -1.796$$

$$\bar{d} = \frac{\Sigma d}{n} = \frac{-42}{12} = -3.5$$

$$s_d = \sqrt{\frac{\Sigma d^2 - n\bar{d}^2}{n-1}} = \sqrt{\frac{378 - 12(-3.5)^2}{11}} = \sqrt{\frac{378 - 147}{11}} = \sqrt{\frac{231}{11}} = \sqrt{21} = 4.58$$

$$s_{\bar{d}} = \frac{s_d}{\sqrt{n}} = \frac{4.58}{\sqrt{12}} = \frac{4.58}{3.46} = 1.32$$

$$t = \frac{\bar{d}}{s_{\bar{d}}} = \frac{-3.5}{1.32} = -2.652$$

The computed value of t of -2.652 is less than the critical value of -1.796 for this lower-tail test. Therefore the null hypothesis is rejected at the 5 percent level of significance, and we conclude that the mean level of performance for those trained by the new approach is superior to those trained by the traditional method.

TESTING A HYPOTHESIZED VALUE OF THE PROPORTION USING THE BINOMIAL DISTRIBUTION

11.6   When a production process is in control, no more than 1 percent of the components are defective and have to be removed in the inspection process. For a random sample of $n = 10$ components, one is found to be defective. On the basis of this sample result, can the null hypothesis that the process is in control be rejected at the 5 percent level of significance?

For the hypotheses H_0: $\pi \leq 0.01$ and H_1: $\pi > 0.01$, based on the binomial distribution, the probability of obtaining one or more defectives by chance given that $\pi = 0.01$ is 1.0 minus the probability of obtaining zero defectives (from Appendix 1, with $n = 10$, $p = 0.01$):

$$P(X \geq 1 \mid n = 10, p = 0.01) = 1.000 - 0.9044 = 0.0956$$

Since this probability is greater than 0.05, the null hypothesis cannot be rejected. For this problem, two or more items would have to be found defective to reject the null hypothesis, because the probability associated with this "tail" of the distribution is less than 0.05. Further, the probability of two or more items being defective is also less than 0.01:

$$P(X \geq 2 \mid n = 10, p = 0.01) = 0.0042 + 0.0001 + 0.0000 + \cdots = 0.0043$$

11.7   With reference to Example 5 (page 182), suppose that a sample of $n = 20$ students are polled and only four indicate having jobs by March 1 (the same sample proportion as in Example 5). Can the director's claim be rejected in this case, using the 5 percent level of significance?

For the hypotheses, H_0: $\pi \geq 0.05$ and H_1: $\pi < 0.50$, based on the binomial distribution, the probabilities of sample results divergent from the claim, and not exceeding a cumulative probability of 0.05, are as follows (from Appendix 2, with $n = 20$ and $p = 0.50$):

Number obtaining jobs	Probability	Cumulative probability
0	0.0000 ⎫	
1	0.0000 ⎪	
2	0.0002 ⎪	
3	0.0011 ⎬	0.0207
4	0.0046 ⎪	
5	0.0148 ⎭	
6	0.0370	

Therefore, for a one-tail test at the 5 percent level of significance (in fact, the 2.07 percent level) the critical number for rejection is five or fewer. To include the category "6" would result in a probability greater than 0.05.

Given the sample result that only *four* students reported having jobs, the null hypothesis is rejected. Note that even though the sample proportion is the same as in Example 5, the larger sample size is associated with a lower sampling error and leads to a test that is more sensitive to detecting a difference.

11.8   It is hypothesized that 40 percent of the voters in a primary election will vote for the incumbent, and that the other 60 percent of the votes will be distributed among three other candidates. Of a random sample of 20 registered voters who plan to vote in the primary election, 12 indicate that they will vote for the incumbent. Test the hypothesis that the overall proportion of voters who will cast their ballots for the incumbent is $\pi = 0.40$, using the 5 percent level of significance.

Where H_0: $\pi = 0.40$ and H_1: $\pi \neq 0.40$, based on the binomial distribution, the probabilities of extreme observations in either "tail" of the distribution, and not exceeding a cumulative probability of 0.025 in each tail, are as follows (from Appendix 1, with $n = 20$, $p = 0.40$):

Number for the candidate	Probability	Cumulative probability
0	0.0000 ⎫	
1	0.0005	
2	0.0031 ⎬	0.0159
3	0.0123 ⎭	
4	0.0350	
⋮	⋮	
12	0.0355	
13	0.0146 ⎫	
14	0.0049	
15	0.0013	
16	0.0003 ⎬	0.0211
17	0.0000	
18	0.0000	
19	0.0000	
20	0.0000 ⎭	

Therefore, for the two-tail test the overall level of significance without exceeding 0.025 in each tail is in fact at $\alpha = 0.037$. The critical number in the sample for rejection is "three or fewer" or "13 or more." Because 12 voters indicated they intend to vote for the incumbent, the null hypothesis cannot be rejected.

TESTING A HYPOTHESIZED VALUE OF THE PROPORTION USING THE NORMAL DISTRIBUTION

11.9 It is hypothesized that no more than 5 percent of the parts being produced in a manufacturing process are defective. For a random sample of $n = 100$ parts, 10 are found to be defective. Test the null hypothesis at the 5 percent level of significance.

$$H_0: \pi \le 0.05 \qquad H_1: \pi > 0.05$$
$$\text{Critical } z \ (\alpha = 0.05) = +1.645$$

(Use of the normal distribution is warranted because $n \ge 30$, $n\pi_0 \ge 5$, and $n(1 - \pi_0) \ge 5$.)

$$\sigma_{\hat{p}} = \sqrt{\frac{\pi_0(1 - \pi_0)}{n}} = \sqrt{\frac{(0.05)(0.95)}{100}} = \sqrt{\frac{0.0475}{100}} = \sqrt{0.000475} = 0.022$$

$$z = \frac{\hat{p} - \pi_0}{\sigma_{\hat{p}}} = \frac{0.10 - 0.05}{0.022} = \frac{0.05}{0.022} = +2.27$$

The calculated value of z of $+2.27$ is greater than the critical value of $+1.645$ for this upper-tail test. Therefore with 10 parts out of 100 found to be defective, the hypothesis that the proportion defective in the population is at or below 0.05 is rejected, using the 5 percent level of significance in the test.

11.10 For Problem 11.9, the manager stipulates that the probability of stopping the process for adjustment when it is in fact not necessary should be at only a 1 percent level, while the probability of *not* stopping the process when the true proportion defective is at $\pi = 0.10$ can be set at 5 percent. What sample size should be obtained, as a minimum, to satisfy these test objectives?

$$n = \left[\frac{z_0\sqrt{\pi_0(1-\pi_0)} - z_1\sqrt{\pi_1(1-\pi_1)}}{\pi_1 - \pi_0} \right]^2 = \left[\frac{2.33\sqrt{(0.05)(0.95)} - (-1.645)\sqrt{(0.10)(0.90)}}{0.10 - 0.05} \right]^2$$

$$= \left[\frac{2.33(0.218) + 1.645(0.300)}{0.05} \right]^2 = \left(\frac{1.0014}{0.05} \right)^2 = (20.03)^2 = 401.2 = 402 \text{ parts}$$

For industrial sampling purposes this is a rather large sample, so the manager might well want to reconsider the test objectives in regard to the designated P (Type I error) of 0.01 and P (Type II error) of 0.05.

TESTING THE DIFFERENCE BETWEEN TWO POPULATION PROPORTIONS

11.11 A manufacturer is evaluating two types of equipment for the fabrication of a component. A random sample of $n_1 = 50$ is collected for the first brand of equipment, and five items are found to be defective. A random sample of $n_2 = 80$ is collected for the second brand and six items are found to be defective. The fabrication rate is the same for the two brands. However, because the first brand costs substantially less, the manufacturer gives this brand the benefit of the doubt and formulates the hypothesis $H_0: \pi_1 \le \pi_2$. Test this hypothesis at the 5 percent level of significance.

$$H_0: (\mu_1 - \mu_2) \le 0 \qquad H_1: (\mu_1 - \mu_2) > 0$$
$$\text{Critical } z \ (\alpha = 0.05) = +1.645$$

$$\hat{\pi} = \frac{n_1\hat{p}_1 + n_2\hat{p}_2}{n_1 + n_2} = \frac{50(0.10) + 80(0.075)}{50 + 80} = \frac{5 + 6}{130} = 0.085$$

$$\hat{\sigma}_{\hat{p}_1 - \hat{p}_2} = \sqrt{\frac{\hat{\pi}(1-\hat{\pi})}{n_1} + \frac{\hat{\pi}(1-\hat{\pi})}{n_2}} = \sqrt{\frac{(0.085)(0.915)}{50} + \frac{(0.085)(0.915)}{80}}$$

$$= \sqrt{\frac{0.0778}{50} + \frac{0.0778}{80}} = \sqrt{0.0016 + 0.0010} = 0.051$$

$$z = \frac{\hat{p}_1 - \hat{p}_2}{\hat{\sigma}_{\hat{p}_1 - \hat{p}_2}} = \frac{0.10 - 0.075}{0.051} = \frac{0.025}{0.051} = +0.49$$

The calculated value of z of $+0.49$ is not greater than $+1.645$ for this upper-tail test. Therefore the null hypothesis cannot be rejected at the 5 percent level of significance.

TESTING A HYPOTHESIZED VALUE OF THE VARIANCE USING THE CHI-SQUARE DISTRIBUTION

11.12 Suppose it is hypothesized that the standard deviation of household income per year in a particular community is $3,000. For a sample of $n = 15$ randomly selected households, the standard deviation is $s = \$2,000$. The household income figures for the population are assumed to be normally distributed. On the basis of this sample result, can the null hypothesis be rejected using the 5 percent level of significance?

$$H_0: \sigma^2 = (\$3,000)^2 = \$9,000,000 \qquad H_1: \sigma^2 \ne \$9,000,000$$
$$\text{Critical } \chi^2 \ (df = 14, \ \alpha = 0.05) = 5.63 \text{ and } 26.12 \text{ (respectively, for the two-tail test)}$$

$$\chi^2 = \frac{(n-1)s^2}{\sigma_0^2} = \frac{(14)(2,000)^2}{(3,000)^2} = \frac{14(4,000,000)}{9,000,000} = \frac{56,000,000}{9,000,000} = 6.22$$

The computed value of 6.22 is neither less than the lower critical value of 5.63 nor larger than the upper critical value of 26.12 for this two-tail test. Therefore the null hypothesis that $\sigma = \$3,000$ is not rejected at the 5 percent level of significance.

11.13 For Problem 11.12, suppose the null hypothesis was that the population standard deviation is *at least* \$3,000. Test this hypothesis at the 5 percent level of significance.

$$H_0: \sigma^2 \geq \$9,000,0009 \qquad H_1: \sigma^2 < \$9,000,000$$
$$\text{Critical } \chi^2 \,(df = 14, \alpha = 0.05) = 6.57 \text{ (lower-tail critical value)}$$
$$\chi^2 = 6.22 \qquad \text{(from Problem 11.12)}$$

The calculated test statistic of 6.22 is just less than the critical value of 6.57 for this lower-tail test. Therefore the null hypothesis is rejected at the 5 percent level of significance, and the alternative hypothesis, that $\sigma^2 < \$9,000,000$ (that $\sigma < \$3,000$), is accepted.

TESTING THE DIFFERENCE BETWEEN TWO VARIANCES

11.14 In Problem 11.4, concerned with testing the difference between two sample means by using a t distribution, a necessary assumption was that the two population variances are equal. The two sample variances were $s_1^2 = 0.40^2 = 0.16$ and $s_2^2 = 0.30^2 = 0.09$, with $n_1 = 12$ and $n_2 = 10$, respectively. Test the null hypothesis that the two population variances are equal, using a 10 percent level of significance.

$$H_0: \sigma_1^2 = \sigma_2^2 \qquad s_1^2 = 0.16 \qquad s_2^2 = 0.09$$
$$H_1: \sigma_1^2 \neq \sigma_2^2 \qquad n_1 = 12 \qquad n_2 = 10$$
$$\text{Critical } F_{11,9} \,(\text{upper } 5\%) = 3.10 \qquad \text{(from Appendix 8)}$$

$$\text{Critical } F_{11,9} \,(\text{lower } 5\%) = \frac{1}{F_{9,11}(\text{upper } 5\%)} = \frac{1}{2.90} = 0.345$$

$$F_{df_1, df_2} = \frac{s_1^2}{s_2^2} = \frac{0.16}{0.09} = 1.78$$

The calculated F statistic of 1.78 is neither less than the lower critical value of 0.345 nor greater than the upper critical value of 3.10. Therefore the hypothesis of no difference between the variances cannot be rejected.

11.15 In Problem 11.1, the assumption was made that the variance of household income was not different in the two communities. Test the null hypothesis that the two variances are equal, using the 10 percent level of significance.

$$H_0: \sigma_1^2 = \sigma_2^2 \qquad s_1^2 = (1,800)^2 = 3,240,000 \qquad s_2^2 = (2,400)^2 = 5,760,000$$
$$H_1: \sigma_1^2 \neq \sigma_2^2 \qquad n_1 = 30 \qquad n_2 = 40$$

(*Note:* Because of limitations in Appendix 8, the specific F values for 29 and 39 degrees of freedom cannot be determined. Therefore the approximate F values are determined by using the nearest degrees of freedom of 30 and 40, respectively.)

$$\text{Critical } F_{30,40} \,(\text{upper } 5\%) = 1.74$$

$$\text{Critical } F_{30,40} \,(\text{lower } 5\%) = \frac{1}{F_{40,30} \,(\text{upper } 5\%)} = \frac{1}{1.79} = 0.559$$

$$F_{df_1, df_2} = \frac{s_1^2}{s_2^2} = \frac{3,240,000}{5,760,000} = 0.562$$

The computed F statistic is neither less than the lower critical value of 0.559 nor greater than the upper critical value of 1.74. Therefore the F statistic is just within the region of acceptance of the null hypothesis at the 10 percent level of significance.

ALTERNATIVE APPROACHES TO TESTING NULL HYPOTHESES

11.16 Using the *P*-value approach, test the null hypothesis in Problem 11.1 at the 5 percent level of significance.

$$H_0: (\mu_1 - \mu_2) = 0 \qquad H_1: (\mu_1 - \mu_2) \neq 0$$

$$z = +1.72 \qquad \text{(from Problem 11.1)}$$

$$P(z \geq +1.72) = 0.5000 - 0.4573 = 0.0427$$

Finally, because this is a two-tail test,

$$P = 2(0.0427) = 0.0854$$

Because the *P* value of 0.0854 is larger than the level of significance of 0.05, the null hypothesis *cannot* be rejected at this level. Therefore we conclude that there is no difference between the mean household income amounts in the two communities.

11.17 Using the *P*-value approach, test the null hypothesis in Problem 11.2 at the 5 percent level of significance.

$$H_0: (\mu_1 - \mu_2) \leq 0 \qquad H_1: (\mu_1 - \mu_2) > 0$$

$$z = +1.79 \qquad \text{(from Problem 11.2)}$$

(Note that the *z* statistic *is* in the direction of the region of rejection for this upper-tail test.)

$$P(z \geq +1.79) = 0.5000 - 0.4633 = 0.0367$$

Because the *P* value of 0.0367 is smaller than the level of significance of 0.05, we reject the null hypothesis and conclude that the mean level of income in the first community is greater than the mean income in the second community.

11.18 Apply the confidence interval approach to testing the null hypothesis in Problem 11.1, using the 5 percent level of significance.

$$H_0: (\mu_1 - \mu_2) = 0 \qquad H_1: (\mu_1 - \mu_2) \neq 0$$

$$\bar{X}_1 = \$35,500 \qquad \bar{X}_2 = \$34,600$$

$$\hat{\sigma}_{\bar{x}_1 - \bar{x}_2} = \$522.79 \qquad \text{(from Problem 11.1)}$$

$$95\% \text{ Conf. Int.} = (\bar{X}_1 - \bar{X}_2) \pm z\hat{\sigma}_{\bar{x}_1 - \bar{x}_2}$$
$$= (35,500 - 34,600) \pm 1.96(522.79)$$
$$= 900 \pm 1,024.67 = -\$124.67 \text{ to } \$1,924.67$$

Because the 95 percent confidence interval includes the hypothesized difference of $0, the null hypothesis cannot be rejected at the 5 percent level of significance.

11.19 Apply the confidence interval approach to testing the null hypothesis in Problem 11.2.

$$H_0: (\mu_1 - \mu_2) \leq 0 \qquad H_1: (\mu_1 - \mu_2) > 0$$

$$\bar{X}_1 = \$35,500 \qquad \bar{X}_2 = \$34,600$$

$$s_{\bar{x}_1 - \bar{x}_2} = \$502.10 \qquad \text{(from Problem 11.2)}$$

$$\text{Lower Conf. Limit} = (\bar{X}_1 - \bar{X}_2) - zs_{\bar{x}_1 - \bar{x}_2}$$
$$= (35,500 - 34,600) - 1.645(502.10)$$
$$= 900 - 825.95 = \$74.05$$

With 95 percent confidence, we conclude that the difference between the population means can be as small as $74.05. Because this one-sided confidence interval does not include the hypothesized value of $0 (or less), the null hypothesis is rejected at the 5 percent level of significance, and we conclude that the average household income in the first community is greater than the average household income in the second community.

COMPUTER OUTPUT

11.20 Refer to the computer input and output associated with Problem 9.15 (page 153). The data are for two random samples of automobile damage claims, each sample taken from a different geographic area. The first part of the output, discussed in Problem 9.15, presents the 95 percent confidence interval for the difference between the two population means. Refer to the last portion of the computer output and test the null hypothesis that there is no difference between the means of the two populations, using the 5 percent level of significance for the test.

$$H_0: (\mu_1 - \mu_2) = 0 \qquad H_1: (\mu_1 - \mu_2) \neq 0 \qquad \alpha = 0.05$$

Referring to the last portion of the output in Fig. 9-1, the reported P value for the two-tail test is 0.019. Because this probability is less than the designated level of significance of 0.05, the null hypothesis is rejected. We conclude that there *is* a difference in the mean level of dollar claims in the two areas. Note that in this output we can also use the confidence interval approach to testing the null hypothesis. Because the 95 percent confidence interval does *not* include the difference of 0, the null hypothesis that the difference between the population means is 0 is rejected at the 5 percent level of significance.

11.21 Refer to the data in Table 11.3, that were associated with testing the difference between means by the paired observations design in Problem 11.5. Test the null hypothesis presented in that problem at the 5 percent level of significance by using available computer software.

$$H_0: \mu_d \geq 0 \qquad H_1: \mu_d < 0 \qquad \alpha = 0.05$$

Refer to Fig. 11-3. The P value is reported as being 0.011. Because this probability is less than the designated level of significance of 0.05, the null hypothesis is rejected at that level of significance, and we conclude that the mean level of performance for those trained by the new approach is superior to those trained by the traditional method. This result coincides with the hand solution using the critical value approach in Problem 11.5. (*Note*: The subcommand "ALTERNATIVE = −1" specifies a lower-tail test in Minitab.)

```
MTB )  READ OLD METHOD AND NEW METHOD INTO C1 AND C2
DATA)   89    94
DATA)   87    91
DATA)   70    68
DATA)   83    88
DATA)   67    75
DATA)   71    66
DATA)   92    94
DATA)   81    88
DATA)   97    96
DATA)   78    88
DATA)   94    95
DATA)   79    87
DATA)  END
MTB )  NAME C1 = 'OLD', C2 = 'NEW', C3 = 'CHANGE'
MTB )  LET 'CHANGE' = 'OLD' - 'NEW'
MTB )  TTEST OF MU = 0 FOR 'CHANGE';
SUBC)  ALTERNATIVE = -1.

TEST OF MU = 0.00 VS MU L.T. 0.00

             N     MEAN    STDEV   SE MEAN        T    P VALUE
CHANGE      12    -3.50     4.58      1.32    -2.65     0.011
```

Fig. 11-3 Minitab output

Supplementary Problems

TESTING THE DIFFERENCE BETWEEN TWO MEANS

11.22 As reported in Problem 9.16, for one consumer product the mean dollar sales per retail outlet last year in a sample of $n_1 = 10$ stores was $\bar{X}_1 = \$3,425$ with $s_1 = \$200$. For a second product the mean dollar sales per outlet in a sample of $n_2 = 12$ stores was $\bar{X}_2 = \$3,250$ with $s_2 = \$175$. The sales amounts per outlet are assumed to be normally distributed for both products. Test the null hypothesis that there is no difference between the mean dollar sales for the two products using the 1 percent level of significance.

Ans. Accept H_0.

11.23 For the data in Problem 11.22, suppose the two sample sizes were $n_1 = 20$ and $n_2 = 24$. Test the difference between the two means at the 1 percent level of significance.

Ans. Reject H_0.

11.24 For a sample of 30 employees in one large firm, the mean hourly wage is $\bar{X}_1 = \$9.50$ with $s_1 = \$1.00$. In a second large firm, the mean hourly wage for a sample of 40 employees is $\bar{X}_2 = \$9.05$ with $s_2 = \$1.20$. Test the hypothesis that there is no difference between the average wage rate being earned in the two firms, using the 5 percent level of significance and assuming that the variances of the two populations are not necessarily equal.

Ans. Accept H_0.

11.25 In Problem 11.24, suppose the null hypothesis to be tested is that the average wage in the second firm was equal to or greater than the average wage rate in the first firm. Can this hypothesis be rejected at the 5 percent level of significance?

Ans. Yes.

11.26 A random sample of $n_1 = 10$ salesmen are placed on one incentive system while a random sample $n_2 = 10$ other salesmen are placed under a second incentive system. During the comparison period, the salesmen under the first system have average weekly sales of $\bar{X}_1 = \$5,000$ with a standard deviation of $s_1 = \$1,200$ while the salesmen under the second system have sales of $\bar{X}_2 = \$4,600$ with a standard deviation of $\$1,000$. Test the null hypothesis that there is no difference between the mean sales per week for the two incentive systems, using the 5 percent level of significance.

Ans. Accept H_0.

11.27 In order to compare two computer software packages, a manager has 10 individuals use each software package to perform a standard set of tasks typical of those encountered in the office. Of course, in carrying out the comparison the manager was careful to use individuals who did not have an established preference or skill with either type of software, and five individuals were randomly selected to use Software A first while the other five used Software B first. The time required to perform the standard set of tasks, to the nearest minute, is reported in Table 11.4. Test the null hypothesis that there is no difference between the mean time required to perform the standard tasks by the two software packages, using the 5 percent level of significance.

Ans. Reject H_0.

Table 11.4   Time Required to Perform a Standard Set of Tasks Using Two Software Packages (Nearest Minute)

Individual	1	2	3	4	5	6	7	8	9	10
Software A	12	16	15	13	16	10	15	17	14	12
Software B	10	17	18	16	19	12	17	15	17	14

TESTING A HYPOTHESIZED PROPORTION USING THE BINOMIAL DISTRIBUTION

11.28 Suppose we hypothesize that a coin is fair in the absence of having the opportunity to examine it directly. The coin is tossed, and the result is that it comes up "heads" all five times. Test the null hypothesis at the (a) 5 percent and (b) 10 percent level of significance.

Ans. (a) Accept H_0, (b) reject H_0.

11.29 A salesman claims that on the average he obtains orders from at least 30 percent of his prospects. For a random sample of 10 prospects he is able to obtain just one order. Can his claim be rejected on the basis of the sample result, at the 5 percent level of significance?

Ans. No.

11.30 The sponsor of a television "special" expected that at least 40 percent of the viewing audience would watch the show in a particular metropolitan area. For a random sample of 20 households with television sets turned on, only four households are watching the program. Based on this limited sample size, test the null hypothesis that at least 40 percent of the viewing audience are watching the program, using the 10 percent level of significance.

Ans. Reject H_0.

TESTING PROPORTIONS USING THE NORMAL DISTRIBUTION

11.31 With reference to Problem 11.29, suppose the salesman is able to obtain orders from 20 out of 100 randomly selected prospects. Can his claim be rejected at the (a) 5 percent and (b) 1 percent level of significance?

Ans. (a) Yes, (b) no.

11.32 With reference to Problem 11.30, the sample is expanded so that 100 households with sets turned on are contacted. Of the 100 households, 30 households are viewing the special. Can the sponsor's assumption that at least 40 percent of the households would watch the program be rejected at the (a) 10 percent and (b) 5 percent level of significance?

Ans. (a) Yes, (b) yes.

11.33 For Problems 11.30 and 11.32, suppose the sponsor specifies that as the result of the study the probability of rejecting a true claim should be no larger than $P = 0.02$, and the probability of accepting the claim given that the percentage viewing the program is actually 30 percent or less should be no larger than $P = 0.05$. What sample size is required in the study, as a minimum, to satisfy this requirement?

Ans. 311 households

11.34 For Problems 11.30 and 11.32, it was suggested that the program might appeal differently to urban versus suburban residents, but there was a difference of opinion among the production staff regarding the direction of the difference. For a random sample of 50 urban households, 20 reported watching the program. For a random sample of 50 suburban households, 30 reported watching the program. Can the difference be considered significant at the (a) 10 percent and (b) 5 percent level?

Ans. (a) Yes, (b) yes.

TESTING A HYPOTHESIZED VALUE OF THE VARIANCE AND THE DIFFERENCE BETWEEN TWO VARIANCES

11.35 Based on the specifications provided by the process designer, the standard deviation of casting diameters is hypothesized to be no larger than 3.0 mm. For a sample of $n = 12$ castings, the sample standard deviation is $s = 4.2$ mm. The distribution of the diameters is assumed to be approximately normal. Can the null hypothesis that the true standard deviation is no larger than 3.0 mm be rejected at the (a) 5 percent and (b) 1 percent level of significance?

Ans. (a) Yes, (b) no.

11.36 In Problem 11.22, under the necessary assumption that the variances of the two populations were equal, the null hypothesis that the means are equal could not be rejected using the t test at the 1 percent level of significance. At the 10 percent level of significance, was the assumption that the two variances are not different warranted?

Ans. Yes.

11.37 A new molding process is designed to reduce the variability of casting diameters. To test the new process, we conservatively formulate the hypothesis that the variance of casting diameters under the new process is equal to or greater than the variance for the old process. Rejection of this null hypothesis would then permit us to accept the alternative hypothesis that the variance associated with the new process is smaller than that associated with the old process. For a sample of $n_1 = 8$ castings produced by the new process $s_1 = 4.2$ mm. For a sample of $n_2 = 10$ castings produced by the old process $s_2 = 5.8$ mm. Can the null hypothesis be rejected at the 5 percent level of significance?

Ans. No.

ALTERNATIVE APPROACHES TO TESTING NULL HYPOTHESES

11.38 Using the P-value approach, test the null hypothesis in Problem 11.24 at the 5 percent level of significance.

Ans. Accept H_0 ($P = 0.0836$).

11.39 Using the P-value approach, test the null hypothesis in Problem 11.25 at the 5 percent level of significance.

Ans. Reject H_0 ($P = 0.0418$).

11.40 Apply the confidence interval approach to testing the null hypothesis in Problem 11.24, using the 5 percent level of significance.

Ans. Accept H_0.

11.41 Apply the confidence interval approach to testing the null hypothesis in Problem 11.25, using the 5 percent level of significance.

Ans. Reject H_0.

COMPUTER OUTPUT

11.42 Refer to the data in Table 9.2, which was associated with using computer software to construct the 95 percent confidence interval for the difference between means in Problem 9.32. Using available computer software, test the null hypothesis that the mean amount of per-item time does not differ for the two types of calls, using the 5 percent level of significance.

Ans. Accept H_0 ($P = 0.42$).

11.43 Refer to the data in Table 11.4, concerned with the time required to perform a standard set of tasks using two different software packages. The null hypothesis that there is no difference between the average time required for the two packages was tested in Problem 11.27. Using available computer software, again test this hypothesis at the 5 percent level of significance.

Ans. Reject H_0 ($P = 0.038$).

Chapter 12

The Chi-Square Test

12.1 THE CHI-SQUARE TEST AS A HYPOTHESIS-TESTING PROCEDURE

The procedures described in this chapter all involve comparing sample frequencies that have been entered into defined data categories with the expected pattern of frequencies that are based on a particular null hypothesis in each case. Thus the procedures are all hypothesis-testing procedures, and random-sample data are involved in the analyses.

The χ^2 (chi-square) probability distribution is described in Sections 9.6 and 11.8. The test statistic presented in the following section is distributed as the chi-square probability model, and since hypothesis testing is involved, the basic steps in hypothesis testing described in Section 10.1 apply in this chapter as well.

This chapter covers use of the chi-square test for *testing goodness of fit*, *testing the independence of two variables*, and *testing hypotheses concerning proportions*. One of the tests of proportions is that of testing the differences among *several* proportions, which is an extension of testing the difference between two proportions, as described in Section 11.7.

12.2 GOODNESS OF FIT TESTS

The null hypothesis in a goodness of fit test is a stipulation concerning the expected pattern of frequencies in a set of categories. The expected pattern may conform to the assumption of equal likelihood and may therefore be uniform, or the expected pattern may conform to such probability distributions as the binomial, Poisson, or normal.

EXAMPLE 1. A regional distributor of air-conditioning systems has subdivided his region into four territories. A prospective purchaser of the distributorship is told that installations of the equipment are about equally distributed among the four territories. The prospective purchaser takes a random sample of 40 installations performed during the past year from the company's files, and finds that the number installed in each of the four territories is as listed in the first row of Table 12.1 (where f_o means "observed frequency"). On the basis of the hypothesis that installations are equally distributed, the expected uniform distribution of the installations is given in the second row of Table 12.1 (where f_e means "expected frequency").

Table 12.1   Number of Installations of Air-Conditioning Systems According to Territory

	Territory				
	A	B	C	D	Total
Number installed in sample, f_o	6	12	14	8	40
Expected number of installations, f_e	10	10	10	10	40

For the null hypothesis to be accepted, the differences between observed and expected frequencies must be attributable to sampling variability at the designated level of significance. Thus, the chi-square test statistic is based on the magnitude of this difference for each category in the frequency distribution. The chi-square value for testing the difference between an obtained and expected pattern of frequencies is

$$\chi^2 = \sum \frac{(f_o - f_e)^2}{f_e} \qquad (12.1)$$

By formula (12.1) above, note that if the observed frequencies are very close to the expected frequencies, then the calculated value of chi-square will be close to zero. As the observed frequencies become increasingly different from the expected frequencies, the value of chi-square becomes larger. Therefore it follows that the chi-square test involves the use of just the *upper tail* of the chi-square distribution to determine whether an observed pattern of frequencies is different from an expected pattern.

EXAMPLE 2.   The calculation of the chi-square test statistic for the pattern of observed and expected frequencies in Table 12.1 is as follows:

$$\chi^2 = \sum \frac{(f_o - f_e)^2}{f_e} = \frac{(6-10)^2}{10} + \frac{(12-10)^2}{10} + \frac{(14-10)^2}{10} + \frac{(8-10)^2}{10} = \frac{40}{10} = 4.00$$

The required value of the chi-square test statistic to reject the null hypothesis depends on the level of significance which is specified and the degrees of freedom. In goodness of fit tests, the degrees of freedom df are equal to the number of categories minus the number of parameter estimators based on the sample and minus 1. Where k = number of categories of data and m = number of parameter values estimated on the basis of the sample, the degrees of freedom in a chi-square goodness of fit test are

$$df = k - m - 1 \qquad\qquad (12.2)$$

When the null hypothesis is that the frequencies are equally distributed, no parameter estimation is ever involved and $m = 0$. (Examples in which m is greater than zero are provided in Problems 12.6 and 12.8.) The subtraction of "1" is always included, because given a total number of observations, once observed frequencies have been entered in $k - 1$ categories of a table of frequencies, the last cell is in fact not "free" to vary. For instance, given that the first three categories in Table 12.1 have observed frequencies of 6, 12, and 14, respectively, it follows that the fourth category must have a frequency of 8 in order to cumulate to the designated sample size of $n = 40$.

EXAMPLE 3.   Following is a complete presentation of the hypothesis-testing procedure associated with the data in Table 12.1, with the null hypothesis being tested at the 5 percent level of significance.

H_0: The number of installations are uniformly distributed among the four territories.

H_1: The number of installations are not uniformly distributed among the four territories.

$$df = k - m - 1 = 4 - 0 - 1 = 3$$

Critical $\chi^2(df = 3, \alpha = 0.05) = 7.81$ (from Appendix 7)

Computed $\chi^2 = 4.00$ (from Example 2)

Because the calculated value of chi-square of 4.00 is not greater than the critical value of 7.81, the null hypothesis that the installations are equally distributed among the four territories cannot be rejected at the 5 percent level of significance.

Computed values of the chi-square test statistic are based on discrete counts, whereas the chi-square distribution is a continuous distribution. When the expected frequencies f_e for the cells are not small, this factor is not important in terms of the extent to which the distribution of the test statistic is approximated by the chi-square distribution. *A frequently used rule is that the expected frequency f_e for each cell, or category, should be at least 5.* Cells that do not meet this criterion should be combined with adjacent categories, when possible, so that this requirement is satisfied. The reduced number of categories then becomes the basis for determining the degrees of freedom df applicable in the test situation. See Problems 12.6 and 12.8. The expected frequencies for all cells of a data table also can be increased by increasing the overall sample size. Compare the expected frequencies in Problems 12.1 and 12.3.

The expected frequencies may be based on any assumption regarding the form of the population frequency distribution. If the assumption is simply based on the historical pattern of frequencies, then, as in the case of the equally likely hypothesis, no parameter estimation is involved, and $df = k - m - 1 = k - 0 - 1 = k - 1$.

EXAMPLE 4. Historically, a manufacturer of TV sets has had 40 percent of sales in small-screen sets (under 14 in.), 40 percent in the mid-range sizes (14 in. to 19 in.), and 20 percent in the large-screen category (21 in. and above). In order to ascertain appropriate production schedules for the next month, a random sample of 100 purchases during the current period is taken and it is found that 55 of the sets purchased were small, 35 were middle-sized, and 10 were large. Below we test the null hypothesis that the historical pattern of sales still prevails, using the 1 percent level of significance.

> H_0: The percentages of all purchases in the small-, medium-, and large-screen categories of TV sets
>   are 40 percent, 40 percent, and 20 percent, respectively.

> H_1: The present pattern of TV set purchases is different from the historical pattern in H_0.

$$df = k - m - 1 = 3 - 0 - 1 = 2$$

$$\text{Critical } \chi^2 (df = 2, \alpha = 0.01) = 9.21$$

Computed χ^2 (see Table 12.2 for observed and expected frequencies):

$$\chi^2 = \sum \frac{(f_o - f_e)^2}{f_e} = \frac{(55 - 40)^2}{40} + \frac{(35 - 40)^2}{40} + \frac{(10 - 20)^2}{20} = 11.25$$

The calculated chi-square statistic of 11.25 is greater than the critical value of 9.21. Therefore the null hypothesis is rejected at the 1 percent level of significance. Comparing the obtained and expected frequencies in Table 12.2, we find that the principal change involves more small sets and fewer large sets being sold, with some reduction in middle-sized sets also possibly occurring.

Table 12.2   Observed and Expected Purchases of TV Sets by Screen Size

	Screen size			
	Small	Medium	Large	Total
Observed frequency, f_o	55	35	10	100
Historical pattern, f_e	40	40	20	100

12.3. TESTS FOR INDEPENDENCE OF TWO CATEGORICAL VARIABLES (CONTINGENCY TABLE TESTS)

In the case of goodness of fit tests there is only one categorical variable, such as the screen size of TV sets which have been sold, and what is tested is the hypothesized pattern of frequencies, or the distribution, of the variable. The observed frequencies can be listed as a single row, or as a single column, of categories. *Tests for independence* involved two categorical variables, and what is tested is the assumption that the two variables are statistically independent. Independence implies that knowledge of the category in which an observation is classified with respect to one variable has no affect on the probability of being in one of the several categories of the other variables. Since two variables are involved, the observed frequencies are entered in a two-way classification table, or *contingency table* (see Section 5.8). The dimensions of such tables are defined by the expression $r \times k$, in which r indicates the number of rows and k indicates the number of columns.

EXAMPLE 5. Table 12.3 is repeated from Section 5.8 and is an example of the simplest possible format for a contingency table, in that each of the two variables (sex and age) has only two classification levels, or categories. Thus this is a 2×2 contingency table.

Table 12.3   Contingency Table for Stereo Shop Customers

	Sex		
Age	Male	Female	Total
Under 30	60	50	110
30 and over	80	10	90
Total	140	60	200

If the null hypothesis of independence is rejected for classified data such as in Table 12.3, this indicates that the two variables are *dependent* and that there is a *relationship* between them. For Table 12.3, for instance, this would indicate that there is a relationship between age and the sex of stereo shop customers.

Given the hypothesis of independence of the two variables, the expected frequency associated with each cell of a contingency table should be proportionate to the total observed frequencies included in the column and in the row in which the cell is located as related to the total sample size. Where f_r is the total frequency in a given row and f_k is the total frequency in a given column, then a convenient formula for determining the expected frequency for the cell of the contingency table that is located in that row and column is

$$f_e = \frac{f_r f_k}{n} \qquad (12.3)$$

The general formula for the degrees of freedom associated with a test for independence is

$$df = (r-1)(k-1) \qquad (12.4)$$

EXAMPLE 6.   The expected frequencies for the data of Table 12.3 are reported in Table 12.4. For row 1, column 1, for instance, the calculation of the expected frequency is

$$f_e = \frac{f_r f_k}{n} = \frac{(110)(140)}{200} = \frac{15,400}{200} = 77$$

Note that in this case the three remaining expected frequencies can be obtained by subtraction from row and column totals, as an alternative to using (12.3). This is a direct indication that there is one degree of freedom for a 2×2 contingency table, and that only one cell frequency is "free" to vary.

Table 12.4   Table of Expected Frequencies for the Observed Frequencies Reported in Table 12.3

	Sex		
Age	Male	Female	Total
Under 30	77	33	110
30 and over	63	27	90
Total	140	60	200

The chi-square test statistic for contingency tables is computed exactly as for the goodness of fit tests (see Section 12.2).

EXAMPLE 7.   Following is the test of the null hypothesis of independence for the data of Table 12.3, using the 1 percent level of significance.

H_0: Sex and age of stereo shop customers are independent.

H_1: Sex and age are dependent variables (there is a relationship between the variables sex and age).

$$df = (r-1)(k-1) = (2-1)(2-1) = 1$$

$$\text{Critical } \chi^2 (df = 1, \alpha = 0.01) = 6.63$$

$$\chi^2 = \sum \frac{(f_o - f_e)^2}{f_e} = \frac{(60-77)^2}{77} + \frac{(50-33)^2}{33} + \frac{(80-63)^2}{63} + \frac{(10-27)^2}{27} = 27.80$$

The calculated test statistic of 27.80 exceeds the required critical value of 6.63. Therefore the null hypothesis of independence is rejected at the 1 percent level of significance. Referring to Table 12.3, we see that male customers are more likely to be over 30 years of age while female customers are more likely to be under 30. The result of the chi-square test is that this observed relationship in the sample cannot be ascribed to chance at the 1 percent level of significance.

12.4.   TESTING HYPOTHESES CONCERNING PROPORTIONS

Testing a Hypothesized Value of the Proportion. Given a hypothesized population proportion and an observed proportion for a random sample taken from the population, in Section 11.5 we used the normal probability distribution as an approximation for the binomial process in order to test the hypothesized value. Mathematically it can be shown that such a two-tail test is equivalent to a chi-square goodness of fit test involving the one row of frequencies and two categories (a 1×2 table). Since the chi-square test involves an analysis of differences between obtained and expected frequencies regardless of the direction of the differences, there is no chi-square test procedure which is the equivalent of a one-tail test concerning a population proportion.

EXAMPLE 8.   A personnel department manager estimates that a proportion of $\pi = 0.40$ of the employees in a large firm will participate in a new stock investment program. A random sample of $n = 50$ employees are contacted and 10 indicate their intention to participate. The hypothesized value of the population proportion could be tested using the normal probability distribution, as described in Section 11.5. Following is the use of the chi-square test to accomplish the same objective, using the 5 percent level of significance.

$$H_0: \pi = 0.40 \qquad H_1: \pi \neq 0.40$$

$$df = k - m - 1 = 2 - 0 - 1 = 1$$

(There are two categories of observed frequencies, as indicated in Table 12.5.)

$$\text{Critical } \chi^2 (df = 1, \alpha = 0.05) = 3.84$$

Computed χ^2 (Table 12.5 indicates the observed and expected frequencies):

$$\chi^2 = \sum \frac{(f_o - f_e)^2}{f_e} = \frac{(10-20)^2}{20} + \frac{(40-30)^2}{30} = 8.33$$

Table 12.5   Observed and Expected Frequencies for Example 8

	Participation in program		Total
	Yes	No	
Number observed in sample, f_o	10	40	50
Number expected in sample, f_e	20	30	50

The calculated test statistic of 8.33 exceeds the critical value of 3.84. Therefore the null hypothesis is rejected at the 5 percent level of significance, and we conclude that the proportion of program participants in the entire firm is not 0.40.

Testing the Difference Between Two Proportions. A procedure for testing the difference between two proportions based on use of the normal probability distribution is presented in Section 11.7. Mathematically, it can be shown that such a two-tail test is equivalent to a chi-square contingency-table test in which the observed frequencies are entered in a 2×2 table. Again, there is no chi-square test equivalent to a one-tail test based on use of the normal probability distribution.

The sampling procedure used in conjunction with testing the difference between two proportions is that *two* random samples are collected, one for each of the two (k) categories. This contrasts with use of a 2×2 table for testing the independence of two variables in Section 12.3, in which *one* random sample is collected for the overall analysis.

EXAMPLE 9. Example 8 in Section 11.7 reports that 10 out of 50 households in one community watched a TV special on the national economy and 15 out of 50 households in a second community watched the TV special. In that example the null hypothesis H_0: $(\pi_1 - \pi_2) = 0$ or, equivalently, H_0: $\pi_1 = \pi_2$ is tested at the 1 percent level of significance. Below is the equivalent test using the chi-square test statistic.

$$H_0\text{: } \pi_1 = \pi_2 \qquad H_1\text{: } \pi_1 \neq \pi_2$$

$$df = (r-1)(k-1) = (2-1)(2-1) = 1$$

(The observed frequencies are entered in a 2×2 table, as indicated in Table 12.6.)

Table 12.6   Extent of TV Program Viewing in Two Communities

	Communities		
	Community 1	Community 2	Total
Number watching	10	15	25
Number not watching	40	35	75
Total	50	50	100

$$\text{Critical } \chi^2 (df = 1, \alpha = 0.01) = 6.63$$

Computed χ^2 (The observed frequencies are presented in Table 12.6 while the expected frequencies—calculated by formula (*12.3*)—are presented in Table 12.7):

$$\chi^2 = \sum \frac{(f_o - f_e)^2}{f_e} = \frac{(10-12.5)^2}{12.5} + \frac{(15-12.5)^2}{12.5} + \frac{(40-37.5)^2}{37.5} + \frac{(35-37.5)^2}{37.5} = 1.34$$

Table 12.7   Expected Frequencies for the Data of Table 12.6

	Communities		
	Community 1	Community 2	Total
Number watching	12.5	12.5	25
Number not watching	37.5	37.5	75
Total	50	50	100

The calculated test statistic of 1.34 is *not* greater than the critical value of 6.63. Therefore the null hypothesis cannot be rejected at the 1 percent level of significance, and we conclude that the proportion of viewers in the two communities does not differ.

Testing the Difference Among Several Population Proportions.   Given the basic approach in Example 9, the chi-square test can be used to test the difference among several (k) population proportions by using a $2 \times k$ tabular design for the analysis of the frequencies. In this case, there is no mathematically equivalent procedure based on the normal probability distribution. The null hypothesis in this case is that there is no difference among the several population proportions (or, that the several different sample proportions could have been obtained by chance from the same population). The sampling procedure followed is that several independent random samples are collected, one for each of the k data categories.

EXAMPLE 10.   From Example 9, suppose households in four communities are sampled in regard to the number viewing a TV special on the national economy. Table 12.8 presents the observed sample data while Table 12.9 presents the expected frequencies, based on formula (*12.3*). The test of the null hypothesis that there are no differences among the population proportions, using the 1 percent level of significance, follows.

$$H_0: \pi_1 = \pi_2 = \pi_3 = \pi_4 \qquad H_1: \text{Not all } \pi_1 = \pi_2 = \pi_3 = \pi_4.$$

(*Note:* Rejection of the null hypothesis does not indicate that *all* of the equalities are untrue, but only that *at least one* equality is untrue.)

$$df = (r-1)(k-1) = (2-1)(4-1) = 3$$

$$\text{Critical } \chi^2 (df = 3, \alpha = 0.01) = 11.35$$

$$\chi^2 = \sum \frac{(f_o - f_e)^2}{f_e} = \frac{(10-12)^2}{12} + \frac{(15-12)^2}{12} + \frac{(5-12)^2}{12} + \frac{(18-12)^2}{12} + \frac{(40-38)^2}{38} + \frac{(35-38)^2}{38} + \frac{(45-38)^2}{38} + \frac{(32-38)^2}{38}$$

$$= 0.33 + 0.75 + 4.08 + 3.0 + 0.11 + 0.24 + 1.29 + 0.95 = 10.75$$

Table 12.8   Extent of TV Program Viewing in Four Communities

	Communities				
	1	2	3	4	Total
Number watching	10	15	5	18	48
Number not watching	40	35	45	32	152
Total	50	50	50	50	200

Table 12.9   Expected Frequencies for the Data of Table 12.8

	Communities				
	1	2	3	4	Total
Number watching	12.0	12.0	12.0	12.0	48
Number not watching	38.0	38.0	38.0	38.0	152
Total	50	50	50	50	200

The calculated chi-square statistic of 10.75 is *not* greater than the critical value of 11.35. Therefore the differences in the proportion of viewers among the four sampled communities are not large enough to reject the null hypothesis at the 1 percent level of significance.

12.5 COMPUTER OUTPUT

Computer software for statistical analysis generally includes the capability of carrying out both goodness of fit tests and contingency table tests. For goodness of fit tests, both the observed frequencies and the expected frequencies usually have to be input for the analysis. Because the expected frequencies are determined directly from the sample data for contingency table tests, however, such tests require input only of the observed frequencies in order to carry out the test. As explained in Sections 12.3 and 12.4, contingency table tests can be used for testing the independence of two categorical variables or for testing hypotheses concerning the equality of two or more population proportions. The use of computer software for a contingency table test is illustrated in Problem 12.15.

Solved Problems

GOODNESS OF FIT TESTS

12.1 It is claimed that an equal number of men and women patronize a retail outlet specializing in the sale of jeans. A random sample of 40 customers are observed, and of these 25 are men and 15 are women. Test the null hypothesis that the overall number of men and women customers is equal by applying the chi-square test and using the 5 percent level of significance.

Table 12.10   Obtained and Expected Frequencies for Problem 12.1

	Customers		
	Men	Women	Total
Number in sample (f_o)	25	15	40
Number expected (f_e)	20	20	40

From Table 12.10,

H_0: The number of men and women customers is equal.

H_1: The number of men and women customers is not equal.

$$df = k - m - 1 = 2 - 0 - 1 = 1$$

Critical $\chi^2 (df = 1, \alpha = 0.05) = 3.84$

$$\chi^2 = \sum \frac{(f_o - f_e)^2}{f_e} = \frac{(25-20)^2}{20} + \frac{(15-20)^2}{20}$$

$$= \frac{(5)^2}{20} + \frac{(-5)^2}{20} = 2.50$$

The calculated test statistic of 2.50 is *not* greater than the critical value of 3.84. Therefore the null hypothesis cannot be rejected at the 5 percent level of significance.

12.2 With reference to Problem 12.1, suppose it had instead been claimed that twice as many men as compared with women are store customers. Using the observed data in Table 12.11, test this hypothesis using the 5 percent level of significance.

From Table 12.11,

H_0: There are twice as many men as there are women customers.

H_1: There are not twice as many men as women customers.

$$df = k - m - 1 = 2 - 0 - 1 = 1$$

Critical $\chi^2 (df = 1, \alpha = 0.05) = 3.84$

$$\chi^2 = \sum \frac{(f_o - f_e)^2}{f_e} = \frac{(25 - 26.67)^2}{26.67} + \frac{(15 - 13.33)^2}{13.33}$$

$$= \frac{(-1.67)^2}{26.67} + \frac{(1.67)^2}{13.33} = 0.10 + 0.21 = 0.31$$

The calculated chi-square statistic of 0.31 clearly does not exceed the critical value of 3.84. Therefore the null hypothesis cannot be rejected at the 5 percent level of significance. The fact that neither of the null hypotheses in Problems 12.1 and 12.2 could be rejected demonstrates the "benefit of the doubt" given to the null hypothesis in each case. However, the size of the sample also affects the probability of sample results (see Problem 12.3).

Table 12.11   Obtained and Expected Frequencies for Problem 12.2

	Customers		Total
	Men	Women	
Number in sample (f_o)	25	15	40
Number expected (f_e)	26.67	13.33	40

12.3 For the situation described in Problem 12.1, suppose the same null hypothesis is tested, but that the sample frequencies in each category are exactly doubled. That is, of 80 randomly selected customers 50 are men and 30 are women. Test the null hypothesis at the 5 percent level of significance and compare your decision with the one in Problem 12.1.

From Table 12.12,

H_0: The number of men and women customers is equal.

H_1: The number of men and women customers is not equal.

$$df = k - m - 1 = 2 - 0 - 1 = 1$$

Critical $\chi^2 (df = 1, \alpha = 0.05) = 3.84$

$$\chi^2 = \sum \frac{(f_o - f_e)^2}{f_e} = \frac{(50 - 40)^2}{40} + \frac{(30 - 40)^2}{40} = 5.00$$

The calculated chi-square value of 5.00 is greater than the critical value of 3.84. Therefore the null hypothesis is rejected at the 5 percent level of significance. Even though the sample data are proportionally the same as in Problem 12.1, the decision now is "reject H_0" instead of "accept H_0." This demonstrates the greater sensitivity of a statistical test associated with a larger sample size.

Table 12.12 Obtained and Expected Frequencies for Problem 12.3

	Customers		
	Men	Women	Total
Number in sample (f_o)	50	30	80
Number expected (f_e)	40	40	80

12.4 A manufacturer of refrigerators offers three basic product lines, which can be described as being "low," "intermediate," and "high" in terms of comparative price. Before a sales promotion aimed at highlighting the virtues of the high-priced refrigerators, the percentage sales in the three categories was 45, 30, and 25, respectively. Of a random sample of 50 refrigerators sold after the promotion, the number sold in the low-, intermediate-, and high-priced categories is 15, 15, and 20, respectively. Test the null hypothesis that the current pattern of sales does not differ from the historical pattern, using the 5 percent level of significance.

With reference to Table 12.13,

H_0: The present pattern of sales frequencies follows the historical pattern.

H_1: The present pattern of sales frequencies is different from the historical pattern.

$$df = k - m - 1 = 3 - 0 - 1 = 2$$

Critical $\chi^2 (df = 2, \alpha = 0.05) = 5.99$

$$\chi^2 = \sum \frac{(f_o - f_e)^2}{f_e} = \frac{(15 - 22.5)^2}{22.5} + \frac{(15 - 15)^2}{15} + \frac{(20 - 12.5)^2}{12.5}$$

$$= \frac{(-7.5)^2}{22.5} + \frac{(0)^2}{15} + \frac{(7.5)^2}{12.5} = 7.00$$

The calculated value of the test statistic of 7.00 is greater than the critical value of 5.99. Therefore the null hypothesis is rejected at the 5 percent level of significance. Although such rejection does not itself indicate in what respect the present pattern of sales differs from the historical pattern, a review of Table 12.13 indicates that more high-priced and fewer low-priced refrigerators were sold than would be expected in terms of the historical pattern of sales.

Table 12.13 Obtained and Expected Frequencies for Problem 12.4

	Price category of refrigerator			
	Low	Intermediate	High	Total
Number sold (f_o)	15	15	20	50
Number expected to be sold (f_e)	22.5	15	12.5	50

12.5 Any probability distribution can serve as the basis for determining the expected frequencies associated with a goodness of fit test (see Section 12.2). Suppose it is hypothesized that the distribution of machine breakdowns per hour in any assembly plant conforms to a Poisson probability distribution, as described in Section 6.6. However, the particular Poisson distribution as determined by the mean of the distribution, λ, is not specified. Table 12.14 presents the observed number of breakdowns during 40 sampled hours.

(a) Determine the value of λ to be used to test the hypothesis that the number of machine breakdowns conforms to a Poisson probability distribution.

(b) Construct the table of expected frequencies based on use of the Poisson distribution identified in (a) for a sample of $n = 40$ hr.

Table 12.14   Observed Number of Machine Breakdowns During 40 Sampled Hours and Worksheet for the Calculation of the Average Number of Breakdowns per Hour

Number of breakdowns (X)	Observed frequency (f_o)	$f_o(X)$
0	0	0
1	6	6
2	8	16
3	11	33
4	7	28
5	4	20
6	3	18
7	1	7
	$\Sigma f_o = 40$	$\Sigma[f_o(X)] = 128$

(a) $\bar{X} = \dfrac{\Sigma[f_o(X)]}{\Sigma f_o} = \dfrac{128}{40} = 3.2$ breakdowns per hour

Therefore, we set the mean of the Poisson distribution at $\lambda = 3.2$.

(b) The expected frequencies are determined by reference to Appendix 4 for the Poisson probability distribution. See Table 12.15.

Table 12.15   Determination of Expected Frequencies for the Machine-Breakdown Problem According to the Poisson Distribution with $\lambda = 3.2$ and $n = 40$

Number of breakdowns (X)	Probability (P)	Expected frequency $f_e (= nP)$
0	0.0408	1.6
1	0.1304	5.2
2	0.2087	8.3
3	0.2226	8.9
4	0.1781	7.1
5	0.1140	4.6
6	0.0608	2.4
7	0.0278	1.1
8	0.0111	0.4
9	0.0040	0.2
10	0.0013	0.1
11	0.0004	0.0
12	0.0001	0.0
13	0.0000	0.0
Total	1.0001	39.9

12.6 Given the information in Tables 12.14 and 12.15, test the null hypothesis that the distribution of machine breakdowns per hour conforms to a Poisson probability distribution at the 5 percent level of significance.

> H_0: The observed distribution of machine breakdowns per hour conforms to a Poisson-distributed variable.
>
> H_1: The distribution of machine breakdowns does not conform to a Poisson-distributed variable.
>
> Critical χ^2: Table 12.16 indicates the observed and expected frequencies to be compared. Note that in order to satisfy the requirement that each f_e be at least 5, a number of categories at each end of the frequency distribution had to be combined. Further, one parameter, λ, was estimated on the basis of the sample. Therefore, $df = k - m - 1 = 5 - 1 - 1 = 3$, and the critical $\chi^2 (df = 3, \alpha = 0.05) = 7.81$.
>
> Computed χ^2: As indicated in Table 12.16, the computed $\chi^2 = 0.675$.

Table 12.16   Observed and Expected Frequencies for the Machine-Breakdown Problem and the Calculation of the Chi-Square Value

Number of breakdowns	Observed frequency (f_o)	Expected frequency (f_e)	$\dfrac{(f_o - f_e)^2}{f_e}$
0	0 ⎫ 6	1.6 ⎫ 6.8	0.094
1	6 ⎭	5.2 ⎭	
2	8	8.3	0.011
3	11	8.9	0.496
4	7	7.1	0.001
5	4 ⎫	4.6 ⎫	
6	3	2.4	
7	1 ⎬ 8	1.1 ⎬ 8.8	0.073
8	0	0.4	
9	0	0.2	
10	0 ⎭	0.1 ⎭	
			$\chi^2 = \overline{0.675}$

Because the calculated test statistic of 0.675 clearly does not exceed the critical value of 7.81, the null hypothesis that the number of machine breakdowns per hour is a Poisson-distributed variable cannot be rejected at the 5 percent level of significance.

12.7 With respect to the sample data presented in Problem 12.5, suppose that in an established similar assembly plant machine breakdowns per hour follow a Poisson distribution with $\lambda = 2.5$. Determine if the breakdowns in the present plant differ significantly from such a pattern, using the 5 percent level of significance.

In this case, no parameter is estimated on the basis of the sample, and the expected frequencies are determined on the basis of the Poisson distribution with the mean $\lambda = 2.5$.

> H_0: The observed distribution of machine breakdowns per hour conforms to a Poisson-distributed variable with $\lambda = 2.5$.
>
> H_1: The observed distribution of machine breakdowns does not conform to a Poisson-distributed variable with $\lambda = 2.5$.

Table 12.17 illustrates the determination of the expected frequencies.

**Table 12.17   Determination of Expected Frequencies for the
Machine-Breakdown Problem According to the Poisson
Distribution with $\lambda = 2.5$ and with $n = 40$**

Number of breakdowns (X)	Probability (P)	Expected frequency $f_e (= nP)$
0	0.0821	3.3
1	0.2052	8.2
2	0.2565	10.3
3	0.2138	8.6
4	0.1336	5.3
5	0.0668	2.7
6	0.0278	1.1
7	0.0099	0.4
8	0.0031	0.1
9	0.0009	0.0
10	0.0002	0.0
Total	0.9999	40.0

Critical χ^2: Table 12.18 indicates the observed and expected frequencies to be compared. With the reduced number of categories being $k = 4$, and with no parameter estimated on the basis of the sample, $df = k - m - 1 = 4 - 0 - 1 = 3$, and critical $\chi^2 (df = 3, \alpha = 0.05) = 7.81$.

Computed χ^2: As indicated in Table 12.18, the computed $\chi^2 = 6.85$.

**Table 12.18   Observed and Expected Frequencies for the Machine-Breakdown
Problem and the Calculation of the Chi-Square Value**

Number of breakdowns (X)	Observed frequency (f_o)	Expected frequency (f_e)	$\dfrac{(f_o - f_e)^2}{f_e}$
0	0 ⎫ 6	3.3 ⎫ 11.5	2.63
1	6 ⎭	8.2 ⎭	
2	8	10.3	0.51
3	11	8.6	0.67
4	7 ⎫	5.3 ⎫	
5	4 ⎪	2.7 ⎪	
6	3 ⎬ 15	1.1 ⎬ 9.6	3.04
7	1 ⎪	0.4 ⎪	
8	0 ⎭	0.1 ⎭	
			$\chi^2 = \overline{6.85}$

The calculated chi-square statistic of 6.85 does not exceed the critical value of 7.81. Therefore the null hypothesis that the number of machine breakdowns per hour is distributed as a Poisson variable with $\lambda = 2.5$ cannot be rejected at the 5 percent level of significance. As would be expected, the χ^2 test statistic is larger in this problem than in Problem 12.6, where λ was based on the sample mean itself. However, the test statistic is still in the region of acceptance of the null hypothesis.

12.8   Table 12.19, taken from Problem 2.16, indicates the average number of injuries per thousand worker-hours in a sample of 50 firms taken from a particular industry. The mean for this distribution is $\bar{X} = 2.32$ in Problem 3.17; the sample standard deviation is $s = 0.42$ in Problem 4.23. Test the null hypothesis that the population pattern of frequencies is distributed as a normal distribution, using the 5 percent level of significance.

Table 12.19   Industrial Injuries in 50 Firms

Average number of injuries per thousand worker-hours	Number of firms
1.5–1.7	3
1.8–2.0	12
2.1–2.3	14
2.4–2.6	9
2.7–2.9	7
3.0–3.2	5
	50

H_0: The frequency distribution follows a normal distribution.

H_1: The frequency distribution does not follow a normal distribution.

The expected frequencies are determined in Table 12.20, based on use of Appendix 5 for the standard normal distribution and with use of the sample mean and sample standard deviation as estimators for the respective population parameters. Table 12.21 indicates the observed and expected frequencies to be compared.

Table 12.20   Determination of Expected Frequencies for the Industrial Injuries in 50 Firms

Average number of injuries per thousand worker-hours (class boundaries)	Class boundaries in standard-normal units (z)*	Probability of being in each category (P)†	Expected frequency $(= 50 \times P)$
1.45–1.75	−2.07 to −1.36	0.09	4.5
1.75–2.05	−1.36 to −0.64	0.17	8.5
2.05–2.35	−0.64 to 0.07	0.27	13.5
2.35–2.65	0.07 to 0.79	0.26	13.0
2.65–2.95	0.79 to 1.50	0.15	7.5
2.95–3.25	1.50 to 2.21	0.07	3.5
		1.01	50.5

*Based on $\hat{\mu} = \bar{X} = 2.32$ and $\hat{\sigma} = s = 0.42$; for example, for $X = 1.45$, $z = (X - \hat{\mu})/\hat{\sigma} = (1.45 - 2.32)/0.42 = -2.07$.
†The first probability value of 0.09 is the proportion of area in the entire "tail" to the left of $z = -1.36$ and the last probability value of 0.07 is the proportion of area in the entire "tail" to the right of $z = 1.50$. This procedure is necessary for the end classes so that the entire area under the normal curve is allocated in the frequency distribution.

Table 12.21   Observed and Expected Frequencies for the Industrial Injuries Data and the Determination of the χ^2 Value

Average number of injuries per thousand worker-hours	Observed frequency (f_o)	Expected frequency (f_e)	$\dfrac{(f_o - f_e)^2}{f_e}$
1.5–1.7	3 ⎫ 15	4.5 ⎫ 13.0	0.31
1.8–2.0	12 ⎭	8.5 ⎭	
2.1–2.3	14	13.5	0.02
2.4–2.6	9	13.0	1.23
2.7–2.9	7 ⎫ 12	7.5 ⎫ 11.0	0.09
3.0–3.2	5 ⎭	3.5 ⎭	
			$\chi^2 = 1.65$

$$df = k - m - 1 = 4 - 2 - 1 = 1$$

$$\text{Critical } \chi^2 (df = 1, \alpha = 0.05) = 3.84$$

$$\text{Computed } \chi^2 = 1.65 \qquad \text{(as indicated in Table 12.21)}$$

The calculated chi-square statistic of 1.65 is not greater than the critical value of 3.84. Therefore the null hypothesis that the frequency distribution of accident rates follows the normal distribution cannot be rejected at the 5 percent level of significance, and the hypothesis is therefore accepted.

TESTS FOR INDEPENDENCE OF TWO VARIABLES (CONTINGENCY TABLE TESTS)

12.9 Table 12.22 (a contingency table taken from Problem 5.21) presents voter reactions to a new property tax plan according to party affiliation. From these data, construct a table of the expected frequencies based on the assumption that there is no relationship between party affiliation and reaction to the tax plan.

Table 12.22 Contingency Table for Voter Reactions to a New Property Tax Plan

Party affiliation	Reaction			Total
	In favor	Neutral	Opposed	
Democratic	120	20	20	160
Republican	50	30	60	140
Independent	50	10	40	100
Total	220	60	120	400

The expected cell frequencies presented in Table 12.23 are determined by the formula $f_e = (f_r f_k)/n$ (see Section 12.3).

Table 12.23 Table of Expected Frequencies for the Observed Frequencies Reported in Table 12.22

Party affiliation	Reaction			Total
	In favor	Neutral	Opposed	
Democratic	88	24	48	160
Republican	77	21	42	140
Independent	55	15	30	100
Total	220	60	120	400

12.10 Referring to Tables 12.22 and 12.23, test the null hypothesis that there is no relationship between party affiliation and voter reaction, using the 1 percent level of significance.

H_0: Party affiliation and voter reaction are independent (there is no relationship).

H_1: Party affiliation and voter reaction are not independent.

$$df = (r-1)(k-1) = (3-1)(3-1) = 4$$

$$\text{Critical } \chi^2 (df = 4, \alpha = 0.01) = 13.28$$

$$\chi^2 = \sum \frac{(f_o - f_e)^2}{f_e} = \frac{(120-88)^2}{88} + \frac{(20-24)^2}{24} + \frac{(20-48)^2}{48} + \frac{(50-77)^2}{77}$$

$$+ \frac{(30-21)^2}{21} + \frac{(60-42)^2}{42} + \frac{(50-55)^2}{55} + \frac{(10-15)^2}{15} + \frac{(40-30)^2}{30}$$

$$= 11.64 + 0.67 + 16.33 + 9.47 + 3.86 + 7.71 + 0.45 + 1.67 + 3.33 = 55.13$$

The calculated test statistic of 55.13 clearly exceeds the critical value of 13.28. Therefore the null hypothesis is rejected at the 1 percent level of significance, and we conclude that there is a relationship between party affiliation and reaction to the new tax plan.

12.11 Table 12.24 indicates student reaction to expanding a college athletic program according to class standing, where "lower division" indicates freshman or sophomore class standing and "upper division" indicates junior or senior class standing. Test the null hypothesis that class standing and reaction to expanding the athletic program are independent variables, using the 5 percent level of significance.

Table 12.24   Student Reaction to Expanding the Athletic Program According to Class Standing

	Class standing		
Reaction	Lower division	Upper division	Total
In favor	20	19	39
Against	10	16	26
Total	30	35	65

H_0: Class standing and reaction to expanding the athletic program are independent.

H_1: Class standing and reaction to expanding the athletic program are not independent.

$$df = (r-1)(k-1) = (2-1)(2-1) = 1$$

$$\text{Critical } \chi^2 (df = 1, \alpha = 0.05) = 3.84$$

Computed χ^2 (The expected cell frequencies are presented in Table 12.25):

$$\chi^2 = \sum \frac{(f_o - f_e)^2}{f_e} = \frac{(20-18)^2}{18} + \frac{(19-21)^2}{21} + \frac{(10-12)^2}{12} + \frac{(16-14)^2}{14} = 1.03$$

The calculated test statistic of 1.03 is *not* greater than the critical value of 3.84. Therefore the null hypothesis cannot be rejected at the 5 percent level of significance, and the hypothesis that the two variables are independent is accepted.

Table 12.25   Table of Expected Frequencies for the Observed
Frequencies Reported in Table 12.24

Reaction	Class standing		Total
	Lower division	Upper division	
In favor	18	21	39
Against	12	14	26
Total	30	35	65

TESTING HYPOTHESES CONCERNING PROPORTIONS

12.12 Since Problem 12.1 involves a 1×2 table of observed frequencies, the procedure is equivalent to testing a hypothesized population proportion, as explained in Section 12.4.

(*a*)   Formulate the null hypothesis as a hypothesized proportion and interpret the result of the test carried out in Problem 12.1 from this standpoint.

(*b*)   Test the hypothesized proportion by using the normal probability distribution as the basis of the test and demonstrate that the result is equivalent to using the chi-square test.

(*a*) H_0: The proportion of men customers $\pi = 0.50$; H_1: $\pi \neq 0.50$

$$\text{Critical } \chi^2(df = 1, \alpha = 0.05) = 3.84$$

$$\chi^2 = 2.50 \quad \text{(from Problem 12.1)}$$

The calculated chi-square value of 2.50 is not greater than the critical value of 3.84. Therefore, we cannot reject the hypothesis that $\pi = 0.50$ at the 5 percent level of significance.

(*b*) H_0: $\pi = 0.50$ H_1: $\pi \neq 0.50$

$$\text{Critical } z \ (\alpha = 0.05) = \pm 1.96$$

Using formula (*11.12*),

$$\sigma_{\hat{p}} = \sqrt{\frac{\pi_0(1 - \pi_0)}{n}} = \sqrt{\frac{(0.50)(0.50)}{40}} = \sqrt{\frac{0.25}{40}} = \sqrt{0.00625} = 0.079$$

From formula (*11.14*),

$$z = \frac{\hat{p} - \pi_0}{\sigma_{\hat{p}}} = \frac{0.375 - 0.50}{0.079} = \frac{-0.125}{0.079} = -1.58$$

The calculated z value of -1.58 is not in a region of rejection of the null hypothesis. Therefore the null hypothesis cannot be rejected at the 5 percent level of significance, and we conclude that $\pi = 0.50$.

12.13 Because Problem 12.11 involves a 2×2 table of observed frequencies, the procedure is equivalent to testing the difference between two sample proportions. State the null and alternative hypothesis from this point of view and interpret the test carried out in Problem 12.11. In this case we assume that two independent random samples, one for each student class standing, were collected.

$$H_0: \pi_1 = \pi_2 \qquad H_1: \pi_1 \neq \pi_2$$

where $\pi_1 =$ proportion of lower-division students in favor of expanding the athletic program
$\pi_2 =$ proportion of upper-division students in favor of expanding the athletic program

$$\text{Critical } \chi^2(df = 1, \alpha = 0.05) = 3.84$$

$$\chi^2 = 1.03 \quad \text{(from Problem 12.11)}$$

The calculated test statistic of 1.03 is not greater than the critical value of 3.84. Therefore the null hypothesis cannot be rejected at the 5 percent level of significance, and the hypothesis that the proportion of lower-division students in favor of expanding the athletic program is equal to the proportion of upper-division students with this view is accepted.

12.14 Table 12.26 represents an extension of the study discussed in Problems 12.11 and 12.13. Formulate the null hypothesis from the viewpoint that a $2 \times k$ table of frequencies can be used to test the difference among k proportions, and carry out the test using the 5 percent level of significance. In this case we assume that three independent random samples were collected, one for each student class standing.

Table 12.26   Student Reaction to Expanding the Athletic Program According to Class Standing

Reaction	Class standing			Total
	Lower division	Upper division	Graduate	
In favor	20	19	15	54
Against	10	16	35	61
Total	30	35	50	115

$$H_0: \pi_1 = \pi_2 = \pi_3 \qquad H_1: \text{Not all } \pi_1 = \pi_2 = \pi_3$$

where π_1 = proportion of lower-division students in favor of expanding the athletic program
π_2 = proportion of upper-division students in favor of expanding the athletic program
π_3 = proportion of graduate students in favor of expanding the athletic program

$$df = (r-1)(k-1) = (2-1)(3-1) = 2$$

Critical $\chi^2 (df = 2, \alpha = 0.05) = 5.99$

Computed χ^2 (The expected cell frequencies are presented in Table 12.27):

$$\chi^2 = \sum \frac{(f_o - f_e)^2}{f_e} = \frac{(20-14.1)^2}{14.1} + \frac{(19-16.4)^2}{16.4} + \frac{(15-23.5)^2}{23.5} + \frac{(10-15.9)^2}{15.9} + \frac{(16-18.6)^2}{18.6} + \frac{(35-26.5)^2}{26.5} = 11.23$$

The calculated chi-square test statistic of 11.23 exceeds the critical value of 5.99. Therefore the null hypothesis is rejected at the 5 percent level of significance and we conclude that not all three population proportions are equal.

Table 12.27   Table of Expected Frequencies for the Observed Frequencies in Table 12.26

Reaction	Class standing			Total
	Lower division	Upper division	Graduate	
In favor	14.1	16.4	23.5	54
Against	15.9	18.6	26.5	61
Total	30	35	50	115

COMPUTER OUTPUT

12.15 Refer to the contingency table data in Table 12.22. Using computer software, test the null hypothesis that there is no relationship between party affiliation and voter reaction, using the 1 percent level of significance.

H_0: Party affiliation and voter reaction are independent (there is no relationship).

H_1: Party affiliation and voter reaction are not independent.

From the computer output in Fig. 12-1:

$$df = 4$$

$$\text{Computed } \chi^2 = 55.13$$

$$\text{Critical } \chi^2 (df = 4, \alpha = 0.01) = 13.28 \qquad \text{(from Appendix 7)}$$

Because the calculated test statistic of 55.13 exceeds the critical value of 13.28, the null hypothesis is rejected at the 1 percent level of significance, and we conclude that there is a relationship between party affiliation and reaction to the new tax plan. This result corresponds to the result arrived at by hand-calculation in Problem 12.10.

```
MTB ) READ CONTINGENCY-TABLE INTO C1-C3
DATA)   120      20      20
DATA)    50      30      60
DATA)    50      10      40
DATA) END
MTB ) CHISQUARE C1-C3
Expected counts are printed below observed counts

              C1       C2       C3     Total
     1       120       20       20      160
             88.0     24.0     48.0

     2        50       30       60      140
             77.0     21.0     42.0

     3        50       10       40      100
             55.0     15.0     30.0

Total        220       60      120      400

ChiSq =    11.64 +    0.67 +   16.33 +
            9.47 +    3.86 +    7.71 +
            0.45 +    1.67 +    3.33 = 55.13
df = 4
```

Fig. 12-1 Minitab output.

Supplementary Problems

GOODNESS OF FIT TESTS

12.16 Refer to Table 12.28 and test the null hypothesis that the consumer preferences for the four brands of wine are equal, using the 1 percent level of significance.

Ans. Reject H_0.

12.17 Using Table 12.28, test the hypothesis that Brand C is preferred by as many people as the other three brands combined at the 1 percent level of significance.

Table 12.28   Consumer Panel Preferences for Four
Brands of Rhine Wine

Brand				
A	B	C	D	Total
30	20	40	10	100

Ans. Accept H_0.

12.18 Table 12.29 reports the single most important safety feature desired by a random sample of car purchasers. Test the null hypothesis that the general population of car buyers is equally distributed in terms of primary preference for these safety features, using the (*a*) 5 percent and (*b*) 1 percent level of significance.

Table 12.29   Identification of Most Important Safety Feature Desired by Car Buyers

Safety feature					
Disk brakes	Modified suspension	Air bags	Automatic door locks	Cruise control	Total
20	10	30	25	15	100

Ans. (*a*) Reject H_0, (*b*) accept H_0.

12.19 In a college course in business statistics, the historical distribution of the A, B, C, D, and E grades has been 10, 30, 40, 10 and 10 percent, respectively. A particular class taught by a new instructor completes the semester with 8 students earning a grade of A, 17 with B, 20 with C, 3 with D, and 2 with E. Test the null hypothesis that this sample does not differ significantly from the historical pattern, using the 5 percent level of significance.

Ans. Accept H_0.

12.20 Table 12.30 reports the number of transistors that do not meet a stringent quality requirement in 20 samples of $n = 10$ each. Test the null hypothesis that this distribution is not significantly different from the binomial distribution with $n = 10$ and $p = 0.30$, using the 5 percent level of significance.

Table 12.30   Number of Defective Transistors in 20 Samples of Size $n = 10$ Each

Number defective per sample	0	1	2	3	4	5	6	7	8	9	10
Number of samples	0	1	2	4	5	5	2	1	0	0	0

Ans. Reject H_0.

12.21 Refer to Table 2.15 (page 25). For these grouped data, $\bar{X} = 28.95$ and $s = 2.52$ (see Problems 3.40 and 4.44). Test the null hypothesis that this distribution of frequencies conforms to a normal probability distribution at the 5 percent level of significance.

Ans. Cannot be tested, because $df = 0$.

12.22 Refer to Table 2.18 (page 27). For this sample of grouped data, $\bar{X} = 23.3$ years and $s = 3.4$ years (see Problems 3.46 and 4.52). Test the null hypothesis that this distribution of frequencies follows a normal probability distribution, using the 1 percent level of significance.

Ans. Reject H_0.

TESTS FOR INDEPENDENCE OF TWO VARIABLES (CONTINGENCY TABLE TESTS)

12.23 As an extension of Problem 12.18, the opinions of men and women were tallied separately (see Table 12.31). Test the hypothesis that there is no relationship between sex and which safety feature is preferred, using the 1 percent level of significance.

Table 12.31   Identification of Most Important Safety Feature Desired by Car Buyers, According to Sex

Respondents	Disk brakes	Modified suspension	Air bags	Automatic door locks	Cruise control	Total
Men	15	5	20	5	5	50
Women	5	5	10	20	10	50
Total	20	10	30	25	15	100

Ans. Reject H_0.

12.24 In order to investigate the relationship between employment status at the time a loan was arranged and whether or not the loan is now in default, a loan company manager chooses 100 accounts randomly, with the results indicated in Table 12.32. Test the null hypothesis that employment status and status of the loan are independent variables, using the 5 percent level of significance for the test.

Table 12.32   Employment Status and Loan Status for a Sample of 100 Accounts

Present status of loan	Employment status at time of loan		Total
	Employed	Unemployed	
In default	10	8	18
Not in default	60	22	82
Total	70	30	100

Ans. Accept H_0.

12.25 An elementary school principal categorizes parents into three income categories according to residential area, and into three levels of participation in school programs. From Table 12.33, test the hypothesis that there is no relationship between income and school program participation, using the 5 percent level of significance. Consider the meaning of the test results.

Ans. Reject H_0.

Table 12.33   Income Level and Participation in School Programs by Parents of Elementary School Students

Program participation	Income level			Total
	Low	Middle	High	
Never	28	48	16	92
Occasional	22	65	14	101
Regular	17	74	3	94
Total	67	187	33	287

TESTING HYPOTHESES CONCERNING PROPORTIONS

12.26   Refer to Problems 12.16 to 12.25 and identify those applications of the chi-square test which are equivalent to testing a hypothesized population proportion. For each problem you identify, formulate such a null hypothesis.

Ans.   Problem 12.17.

12.27   Refer to Problems 12.16 to 12.25 and identify those applications which are equivalent to testing the difference between two sample proportions, recognizing that two separate random samples are required. For each problem you identify, formulate such a null hypothesis.

Ans.   Problem 12.24.

12.28   Refer to Problems 12.16 to 12.25 and identify those applications which are equivalent to testing the differences among three or more proportions, recognizing that k separate random samples are required, and formulate the null hypothesis in each case.

Ans.   Problem 12.23.

COMPUTER OUTPUT

12.29   Refer to the contingency table data in Table 12.33. Use available computer software to test the null hypothesis that there is no relationship between income level and program participation, using the 5 percent level of significance.

Ans.   Reject H_0.

Chapter 13

Analysis of Variance

13.1 BASIC RATIONALE ASSOCIATED WITH TESTING THE DIFFERENCE AMONG SEVERAL MEANS

Whereas the chi-square test can be used to test the differences among several proportions (see Section 12.4), the analysis of variance (ANOVA) is used to test the differences among several means. A basic assumption underlying the analysis of variance is that the several sample means were obtained from normally distributed populations having the same variance σ^2. However, the test procedure has been found to be relatively unaffected by violations of the normality assumption when the populations are unimodal and the sample sizes are approximately equal. Because the null hypothesis is that the population means are equal, the assumption of equal variance (*homogeneity of variance*) also implies that for practical purposes the test is concerned with the hypothesis that the means came from the same population. This is so because any normally distributed population is defined by the mean and variance (or standard deviation) as the two parameters. (See Section 7.2 for a general description of the normal probability distribution.) All of the computational procedures presented in this chapter are for fixed-effects models as contrasted to random-effects models. This distinction is explained in Section 13.6.

The basic rationale underlying the analysis of variance was first developed by the British statistician Ronald A. Fisher, and the F distribution was named in his honor. The conceptual rationale is as follows:

(1) Compute the mean for each sample group and then determine the standard error of the mean $s_{\bar{x}}$ *based only on the several sample means.* Computationally, this is the standard deviation of these several mean values.

(2) Now, given the formula $s_{\bar{x}} = s/\sqrt{n}$, it follows that $s = \sqrt{n}\,s_{\bar{x}}$ and that $s^2 = ns_{\bar{x}}^2$. Therefore, the standard error of the mean computed in (1) can be used to estimate the variance of the (common) population from which the several samples were obtained. This estimate of the population variance is called the *mean square among treatment groups* (*MSTR*). Fisher called any variance estimate a "mean square" because computationally a variance is the mean of the squared deviations from the group mean (see Section 4.5).

(3) Compute the variance separately for each sample group and with respect to each group mean. Then pool these variance values by weighting them according to $n-1$ for each sample. This weighting procedure for the variance is an extension of the procedure for combining and weighting two sample variances (see Section 11.1). The resulting estimate of the population variance is called the *mean square error* (*MSE*) and is based on *within* group differences only. Again, it is called a "mean square" because it is a variance estimate. It is due to "error" because the deviations within each of the sample groups can be due only to random sampling error, and they cannot be due to any differences among the means of the population groups.

(4) If the null hypothesis that $\mu_1 = \mu_2 = \mu_3 = \cdots = \mu_k$ is true, then it follows that each of the two mean squares obtained in (2) and (3) is an unbiased and independent estimator of the same population variance σ^2. However, if the null hypothesis is false, then the expected value of *MSTR* is larger than *MSE*. Essentially, any differences among the population means will inflate *MSTR* while having no effect on *MSE*, which is based on *within* group differences only.

(5) Based on the observation in (4), the F distribution can be used to test the difference between the two variances, as described in Section 11.9. A one-tail test is involved, and the general form of the F test in the analysis of variance is

$$F_{df_1, df_2} = \frac{MSTR}{MSE} \qquad (13.1)$$

If the F ratio is in the region of rejection for the specified level of significance, then the hypothesis that the several sample means came from the same population is rejected.

Problem 13.1 illustrates the application of these five steps to a hypothesis-testing problem involving the difference among three means.

Although the above steps are useful for describing the conceptual approach underlying the analysis of variance, extension of this procedure for designs that are more complex than the simple comparison of k sample means is cumbersome. For this reason, in the sections which follow each design is described in terms of the linear model that identifies the components influencing the random variable. Also, a standard analysis-of-variance table which shows the calculation of the required mean square values is presented for each type of experimental design.

13.2 ONE-FACTOR COMPLETELY RANDOMIZED DESIGN (ONE-WAY ANOVA)

The one-way analysis of variance procedure is concerned with testing the difference among k sample means when the subjects are assigned randomly to each of the several treatment groups. Therefore, the general explanation in Section 13.1 concerns the one-way classification model.

The linear equation, or model, that represents the one-factor completely randomized design is

$$X_{ik} = \mu + \alpha_k + \varepsilon_{ik} \qquad (13.2)$$

where μ = the overall mean of all k treatment populations

α_k = effect of the treatment in the particular group k from which the value was sampled

ε_{ik} = the random error associated with the process of sampling (ε is the Greek "epsilon")

Table 13.1 is the summary table for the one-factor completely randomized design of the analysis of variance, including all computational formulas. The application of these formulas to sample data is illustrated in Problems 13.2 to 13.5. The symbol system used in this table is somewhat different from that used in Section 13.1 because of the need to use a system which can be extended logically to two-way analysis of variance. Thus, $MSTR$ becomes the *mean square among the A treatment groups* (MSA). Further, note that the definition of symbols in the context of analysis of variance is not necessarily consistent with the use of these symbols in general statistical analysis. For example, α_k in (13.2) is concerned with the effect on a randomly sampled value originating from the treatment group in which the value is located; it has nothing to do with the concept of α in general hypothesis-testing procedures as defined in Section 10.1. Similarly, N in Table 13.1 designates the total size of the sample for all treatment groups combined, rather than a population size. New symbols included in Table 13.1 are T_k, which represents the sum (total) of the values in a particular treatment group, and T, which represents the sum of the sampled values in all groups combined.

Table 13.1 Summary Table for One-Way Analysis of Variance (Treatment Groups Need Not Be Equal)

Source of variation	Degrees of freedom (df)	Sum of squares (SS)	Mean square (MS)	F ratio
Among treatment groups (A)	$K-1$	$SSA = \sum\limits_{k=1}^{K} \dfrac{T_k^2}{n_k} - \dfrac{T^2}{N}$	$MSA = \dfrac{SSA}{K-1}$	$F = \dfrac{MSA}{MSE}$
Sampling error (E)	$N-K$	$SSE = SST - SSA$	$MSE = \dfrac{SSE}{N-K}$	
Total (T)	$N-1$	$SST = \sum\limits_{i=1}^{n} \sum\limits_{k=1}^{K} X^2 - \dfrac{T^2}{N}$		

Instead of the form of the null hypothesis described in Section 13.1, the general form of the null hypothesis in the analysis of variance makes reference to the relevant component of the linear model. Thus, for the one-way analysis of variance the null and alternative hypotheses can be stated as

$H_0: \mu_1 = \mu_2 = \cdots = \mu_k$ or, equivalently, $H_0: \alpha_k = 0$ for all treatments (factor levels)

$H_1:$ not all $\mu_1 = \mu_2 = \cdots = \mu_k$ $H_1: \alpha_k \neq 0$ for some treatments

13.3 TWO-WAY ANALYSIS OF VARIANCE

Two-way analysis of variance is based on two dimensions of classifications, or treatments. For example, in analyzing the level of achievement in a training program we could consider both the effect of the method of instruction and the effect of prior school achievement. Similarly, we could investigate gasoline mileage according to the weight category of the car and according to the grade of gasoline. In data tables, the treatments identified in the column headings are typically called the *A* treatments; those in the row headings are called the *B* treatments.

Interaction in a two-factor experiment means that the two treatments are not independent, and that the effect of a particular treatment in one factor differs according to levels of the other factor. For example, in studying automobile mileage a higher-octane gasoline may improve mileage for certain types of cars but not for others. Similarly, the effectiveness of various methods of instruction may differ according to the ability levels of the students. In order to test for interaction, more than one observation or sampled measurement (i.e. *replication*) has to be included in each cell of the two-way data table. Section 13.4 presents the analytical procedure which is appropriate when there is only one observation per cell, and in which interaction between the two factors cannot be tested. The analytical procedure is extended to include replication and the analysis of interaction effects in Section 13.5.

13.4 THE RANDOMIZED BLOCK DESIGN (TWO-WAY ANOVA, ONE OBSERVATION PER CELL)

The two-way analysis of variance model in which there is only one observation per cell is generally referred to as the *randomized block design*, because of the principal use for this model. What if we extend the idea of using paired observations to compare two sample means (see Section 11.3) to the basic one-way analysis of variance model, and have groups of *k matched* individuals assigned randomly to each treatment level? In analysis of variance, such matched groups are called *blocks*, and because the individuals (or items) are randomly assigned based on the identified block membership, the design is referred to as the randomized block design. In such a design the "blocks" dimension is not a treatment dimension as such. The objective of using this design is not for the specific purpose of testing for a "blocks" effect. Rather, by being able to assign some of the variability among subjects to prior achievement, for example, the *MSE* can be reduced and the resulting test of the *A* treatments effect is more sensitive.

The linear model for the two-way analysis of variance model with one observation per cell (with no replication) is

$$X_{jk} = \mu + \beta_j + \alpha_k + \varepsilon_{jk} \qquad (13.3)$$

where μ = the overall mean regardless of any treatment
   β_j = effect of the treatment j or block j in the B dimension of classification
   α_k = effect of the treatment k in the A dimension of classification
   ε_{jk} = the random error associated with the process of sampling

Table 13.2 is the summary table for the two-way analysis of variance without replication. As compared with Table 13.1 for the one-way analysis of variance, the only new symbol in this table is T_j^2, which indicates that the total of each j group (for the B treatments, or blocks) is squared. See Problems 13.6 and 13.8 for application of these formulas.

Table 13.2   Summary Table for Two-Way Analysis of Variance with One Observation per Cell (Randomized Block Design)

Source of variation	Degrees of freedom (df)	Sum of squares (SS)	Mean square (MS)	F ratio
Among treatment groups (A)	$K-1$	$SSA = \sum\limits_{k=1}^{K} \dfrac{T_k^2}{n_k} - \dfrac{T^2}{N}$	$MSA = \dfrac{SSA}{K-1}$	$F = \dfrac{MSA}{MSE}$
Among treatment groups, or blocks (B)	$J-1$	$SSB = \dfrac{1}{K} \sum\limits_{j=1}^{J} T_j^2 - \dfrac{T^2}{N}$	$MSB = \dfrac{SSB}{J-1}$	$F = \dfrac{MSB}{MSE}$
Sampling error (E)	$(J-1)(K-1)$	$SSE = SST - SSA - SSB$	$MSE = \dfrac{SSE}{(J-1)(K-1)}$	
Total (T)	$N-1$	$SST = \sum\limits_{j=1}^{J} \sum\limits_{k=1}^{K} X^2 - \dfrac{T^2}{N}$		

13.5   TWO-FACTOR COMPLETELY RANDOMIZED DESIGN (TWO-WAY ANOVA, n OBSERVATIONS PER CELL)

As explained in Section 13.3, when replication is included within a two-way design, the interaction between the two factors can be tested. Thus, when such a design is used, three different null hypotheses can be tested by the analysis of variance: that there are no column effects (the column means are not significantly different), that there are no row effects (the row means are not significantly different), and that there is no interaction between the two factors (the two factors are independent). A significant interaction effect indicates that the effect of treatments for one factor varies according to levels of the other factor. In such a case, the existence of column and/or row effects may not be meaningful from the standpoint of the application of research results.

The linear model for the two-way analysis of variance when replication is included is

$$X_{ijk} = \mu + \beta_j + \alpha_k + \iota_{jk} + \varepsilon_{ijk} \tag{13.4}$$

where μ = the overall mean regardless of any treatment

β_j = effect of the treatment j in the B (row) dimension

α_k = effect of the treatment k in the A (column) dimension

ι_{jk} = effect of interaction between treatment j (of factor B) and treatment k (of factor A) (where ι is the Greek "iota")

ε_{ijk} = the random error associated with the process of sampling

Table 13.3 is the summary table for the two-way analysis of variance with replication. The formulas included in this table are based on the assumption that there are an equal number of observations in all of the cells. See Problem 13.9 for application of these formulas.

Table 13.3 Summary Table for Two-Way Analysis of Variance with More than One Observation per Cell

Source of variation	Degrees of freedom (df)	Sum of Squares (SS)	Mean square (MS)	F ratio
Among treatment groups (A)	$K-1$	$SSA = \sum\limits_{k=1}^{K} \dfrac{T_k^2}{nJ} - \dfrac{T^2}{N}$	$MSA = \dfrac{SSA}{K-1}$	$F = \dfrac{MSA}{MSE}$
Among treatment groups (B)	$J-1$	$SSB = \sum\limits_{j=1}^{J} \dfrac{T_j^2}{nK} - \dfrac{T^2}{N}$	$MSB = \dfrac{SSB}{J-1}$	$F = \dfrac{MSB}{MSE}$
Interaction (between) factors A and B) (I)	$(J-1)(K-1)$	$SSI = \dfrac{1}{n}\sum\limits_{j=1}^{J}\sum\limits_{k=1}^{K}\left(\sum\limits_{i=1}^{n} X\right)^2$ $- SSA - SSB - \dfrac{T^2}{N}$	$MSI = \dfrac{SSI}{(J-1)(K-1)}$	$F = \dfrac{MSI}{MSE}$
Sampling error (E)	$JK(n-1)$	$SSE = SST - SSA$ $- SSB - SSI$	$MSE = \dfrac{SSE}{JK(n-1)}$	
Total (T)	$N-1$	$SST = \sum\limits_{i=1}^{n}\sum\limits_{j=1}^{J}\sum\limits_{k=1}^{K} X^2 - \dfrac{T^2}{N}$		

13.6 ADDITIONAL CONSIDERATIONS

All of the computational procedures presented in this chapter are for fixed-effects models of the analysis of variance. In a *fixed-effects model*, all of the treatments of concern for a given factor are included in the experiment. For instance, in Problem 13.1 it is assumed that the only instructional methods of concern are the three methods included in the design. A *random-effects model*, however, includes only a random sample from all the possible treatments for the factor in the experiment. For instance, out of ten different instructional methods, three might have been randomly chosen. A different computational method is required in the latter case because the null hypothesis is that there are no differences among the various instructional methods in general, and not just among the particular instructional methods which were included in the experiment. In most experiments the fixed-effects model is appropriate, and therefore the presentation in this chapter has been limited to such models.

The concepts presented in this chapter can be extended to more than two factors. Designs involving three or more factors are called *factorial designs*, and in fact most statisticians include the two-way analysis of variance with replication in this category. Although a number of different null hypotheses can be tested with the same body of data by the use of factorial designs, the extension of such designs can lead to an extremely large number of categories (cells) in the data table, with related sampling problems. Because of such difficulties, designs have been developed which do not require that every possible combination of the treatment levels of every factor be included in the analysis. Such designs as the *Latin Square design* and *incomplete block designs* are examples of such developments and are described in specialized textbooks in the analysis of variance.

Whatever experimental design is used, rejection of a null hypothesis in the analysis of variance typically does not present the analyst with the basis for final decisions, because such rejection does not serve to pinpoint the exact differences among the treatments, or factor levels. For example, given that there is a significant difference in student achievement among three instructional methods, we would next want to determine which of the pairs of methods are different from one another. Various procedures have been developed for such pairwise comparisons carried out in conjunction with the analysis of variance.

13.7 COMPUTER APPLICATIONS

Computer software is available for all the designs of the analysis of variance described in this chapter, as well as for more complex designs that are outside the scope of this book. Uses of computer software for the one-factor completely randomized design, for the randomized-block design, and for the two-factor completely randomized design are illustrated in Problems 13.3, 13.7, and 13.10, respectively.

Solved Problems

ONE-FACTOR COMPLETELY RANDOMIZED DESIGN

13.1 Fifteen trainees in a technical program are randomly assigned to three different types of instructional approaches, all of which are concerned with developing a specified level of skill in computer-assisted design. The achievement test scores at the conclusion of the instructional unit are reported in Table 13.4, along with the mean performance score associated with each instructional approach. Use the analysis-of-variance procedure in Section 13.1 to test the null hypothesis that the three sample means were obtained from the same population, using the 5 percent level of significance for the test.

Table 13.4 Achievement Test Scores of Trainees under Three Methods of Instruction

Instructional method	Test scores					Total scores	Mean test scores
A_1	86	79	81	70	84	400	80
A_2	90	76	88	82	89	425	85
A_3	82	68	73	71	81	375	75

$$H_0: \mu_1 = \mu_2 = \mu_3 \quad \text{or, equivalently,} \quad H_0: \alpha_k = 0 \text{ for all treatments}$$

$$H_1: \text{not all } \mu_1 = \mu_2 = \mu_3 \qquad\qquad H_1: \alpha_{k \neq 0} \text{ for some treatments}$$

(1) The overall mean of all 15 test scores is

$$\bar{X}_T = \frac{\Sigma X}{n} = \frac{1,200}{15} = 80.0$$

The standard error of the mean, based on the three sample means reported, is

$$s_{\bar{x}} = \sqrt{\frac{\Sigma(\bar{X} - \bar{X}_T)^2}{\text{No. means} - 1}} = \sqrt{\frac{(80 - 80)^2 + (85 - 80)^2 + (75 - 80)^2}{3 - 1}} = \sqrt{\frac{50}{2}} = 5.0$$

(2) $MSTR = ns_{\bar{x}}^2 = 5(5.0)^2 = 5(25.0) = 125.0$

(3) From the general formula:

$$s^2 = \frac{\Sigma(X - \bar{X})^2}{n - 1}$$

the variance for each of the three samples is

$$s_1^2 = \frac{(86 - 80)^2 + (79 - 80)^2 + (81 - 80)^2 + (70 - 80)^2 + (84 - 80)^2}{5 - 1} = \frac{154}{4} = 38.5$$

$$s_2^2 = \frac{(90-85)^2+(76-85)^2+(88-85)^2+(82-85)^2+(89-85)^2}{5-1} = \frac{140}{4} = 35.0$$

$$s_3^2 = \frac{(82-75)^2+(68-75)^2+(73-75)^2+(71-75)^2+(81-75)^2}{5-1} = \frac{154}{4} = 38.5$$

Then,

$$\hat{\sigma}^2 \text{ (pooled)} = \frac{(n_1-1)s_1^2+(n_2-1)s_2^2+(n_3-1)s_3^2}{n_1+n_2+n_3-3} = \frac{(4)(38.5)+4(35.0)+4(38.5)}{5+5+5-3} = \frac{448.0}{12} = 37.3$$

Thus, $MSE = 37.3$.

(4) Since $MSTR$ is larger than MSE, a test of the null hypothesis is appropriate.

Critical F ($df = k-1$, $kn-k$; $\alpha = 0.05$) = F (2, 12; $\alpha = 0.05$) = 3.88

(5) $F = MSTR/MSE = 125/37.3 = 3.35$

Because the calculated F statistic of 3.35 is not greater than the critical F value of 3.88, the null hypothesis that the mean test scores for the three instructional methods in the population are all mutually equal cannot be rejected at the 5 percent level of significance.

13.2 Repeat the analysis of variance in Problem 13.1 (data in Table 13.4) by using the general procedure described in Section 13.2 with the accompanying formulas in Table 13.1.

The various quantities required for substitution in the formulas in Table 13.1 are

$$n_1 = 5 \qquad n_2 = 5 \qquad n_3 = 5 \qquad N = 15$$

$$T_1 = 400 \qquad T_2 = 425 \qquad T_3 = 375 \qquad T = 1,200$$

$$T_1^2 = 160,000 \qquad T_2^2 = 180,625 \qquad T_3^2 = 140,625 \qquad T^2 = 1,440,000$$

$$\frac{T^2}{N} = \frac{1,440,000}{15} = 96,000$$

$$\sum_{i=1}^{n}\sum_{k=1}^{K} X^2 = 86^2 + 79^2 + \cdots + 81^2 = 96,698$$

$$SST = \sum_{i=1}^{n}\sum_{k=1}^{K} X^2 - \frac{T^2}{N} = 96,698 - 96,000 = 698$$

$$SSA = \sum_{k=1}^{K} \frac{T_k^2}{n_k} - \frac{T^2}{N} = \frac{160,000}{5} + \frac{180,625}{5} + \frac{140,625}{5} - 96,000 = 250$$

$$SSE = SST - SSA = 698 - 250 = 448$$

Table 13.5 presents the analysis of variance (ANOVA) for the data in Table 13.4. As expected, the F ratio is identical to the one computed in Problem 13.1, and based on the 2 and 12 degrees of freedom, it

Table 13.5   ANOVA Table for Analysis of Three Methods of Instruction (Data in Table 13.4)

Source of variation	Degrees of freedom (df)	Sum of squares (SS)	Mean square (MS)	F ratio
Among treatment groups (A)	$3-1=2$	250	$\frac{250}{2} = 125$	$\frac{125}{37.33} = 3.35$
Sampling error (E)	$15-3=12$	448	$\frac{448}{12} = 37.33$	
Total (T)	$15-1=14$	698		

is less than the critical F of 3.88 required for significance at the 5 percent level. Thus, we conclude that there is no effect associated with the treatment levels (methods of instruction) and thereby also conclude that the differences among the means are not significant at the 5 percent levels.

13.3 Use available computer software to carry out the analysis in Problem 13.2 (data in Table 13.4) for the one-factor completely randomized design.

    The computer output in Fig. 13-1 corresponds to the results obtained by manual calculation in Table 13.5. As in Problem 13.2, we conclude that the differences among the means are not significant at the 5 percent level.

```
MTB > READ TEST SCORES BY METHOD INTO C1-C3
DATA> 86      90      82
DATA> 79      76      68
DATA> 81      88      73
DATA> 70      82      71
DATA> 84      89      81
DATA> END
MTB > NAME C1 = 'A-1', C2 = 'A-2', C3 = 'A-3'
MTB > AOVONEWAY FOR DATA IN C1-C3

ANALYSIS OF VARIANCE
SOURCE      DF        SS        MS        F
FACTOR       2     250.0     125.0     3.35
ERROR       12     448.0      37.3
TOTAL       14     698.0
                                         INDIVIDUAL 95 PCT CI'S FOR MEAN
                                         BASED ON POOLED STDEV
LEVEL        N      MEAN     STDEV   -----+---------+---------+---------+-
A-1          5    80.000     6.205            (----------*----------)
A-2          5    85.000     5.916                (----------*----------)
A-3          5    75.000     6.205   (----------*----------)
                                     -----+---------+---------+---------+-
POOLED STDEV =       6.110          72.0      78.0      84.0      90.0
```

Fig. 13-1   Minitab output for Problem 13.3.

ONE-WAY ANALYSIS OF VARIANCE WITH UNEQUAL GROUPS

13.4 Sometimes the data used as the basis for a one-way analysis of variance do not include equal group sizes for the several treatment levels. Table 13.6 reports the words per minute typed on different brands of electric typewriters by randomly assigned individuals with no prior experience on these machines, after the same amount of instruction. Test the null hypothesis

Table 13.6   **Average Words per Minute for Three Brands of Typewriters Based on a 15-min Test Period**

Typewriter brand	Number of words per minute					Total WPM	Mean WPM
A_1	79	83	62	51	77	352	70.4
A_2	74	85	72	—	—	231	77.0
A_3	81	65	79	55	—	280	70.0

that the mean words per minute achieved for the three machines is not different, using the 5 percent level of significance.

$$H_0: \mu_1 = \mu_2 = \mu_3 \qquad \text{or, equivalently,} \qquad H_0: \alpha_k = 0 \text{ for all treatments (typewriter brands)}$$

$$H_1: \text{not all } \mu_1 = \mu_2 = \mu_3 \qquad\qquad\qquad H_1: \alpha_k \neq 0 \text{ for some treatments}$$

The various quantities required for substitution in the standard formulas for the one-way analysis of variance are

$$n_1 = 5 \qquad n_2 = 3 \qquad n_3 = 4 \qquad N = 12$$

$$T_1 = 352 \qquad T_2 = 231 \qquad T_3 = 280 \qquad T = 863$$

$$T_1^2 = 123{,}904 \qquad T_2^2 = 53{,}361 \qquad T_3^2 = 78{,}400 \qquad T^2 = 744{,}769$$

$$\frac{T^2}{N} = \frac{744{,}769}{12} = 62{,}064.1$$

$$\sum_{i=1}^{n} \sum_{k=1}^{K} X^2 = 79^2 + 83^2 + \cdots + 55^2 = 63{,}441$$

$$SST = \sum_{i=1}^{n} \sum_{k=1}^{K} X^2 - \frac{T^2}{N} = 63{,}441 - 62{,}064.1 = 1{,}376.9$$

$$SSA = \sum_{k=1}^{K} \frac{T_k^2}{n_k} - \frac{T^2}{N} = \frac{123{,}904}{5} + \frac{53{,}361}{3} + \frac{78{,}400}{4} - 62{,}064.1 = 103.7$$

$$SSE = SST - SSA = 1{,}376.9 - 103.7 = 1{,}273.2$$

Table 13.7 presents the analysis of variance for this test. At the 5 percent level of significance, the critical value of F $(df = 2, 9)$ is 4.26. The computed F ratio of 0.37 is thus in the region of acceptance of the null hypothesis, and we conclude that based on these sample results there are no differences among the three brands of typewriters in terms of typing speed. In fact, because $MSTR$ is smaller than MSE we can observe that the variability among the three typewriters is less than the expected variability given that there are no differences among the typewriter brands.

Table 13.7   ANOVA Table for the Analysis of Typing Speed on Three Brands of Typewriters

Source of variation	Degrees of freedom (df)	Sum of squares (SS)	Mean square (MS)	F ratio
Among treatment groups (A) (brand of typewriter)	$3 - 1 = 2$	103.7	$\dfrac{103.7}{2} = 51.8$	$\dfrac{51.8}{141.5} = 0.37$
Sampling error (E)	$12 - 3 = 9$	1,273.2	$\dfrac{1{,}273.2}{9} = 141.5$	
Total (T)	$12 - 1 = 11$	1,376.9		

RELATIONSHIP OF THE ONE-FACTOR COMPLETELY RANDOMIZED DESIGN TO THE t TEST FOR TESTING THE DIFFERENCE BETWEEN THE MEANS OF TWO INDEPENDENT SAMPLES

13.5   The one-factor completely randomized design can be considered an extension to k groups of testing the difference between the means of two independent samples by use of Student's t distribution. In both types of applications a necessary assumption is that the samples have been obtained from the same normally distributed population for which the population variance σ^2

is unknown (see Section 11.2). Therefore, it follows that when one-way analysis of variance is applied to a design in which $k = 2$, the result is directly equivalent to using the t distribution to test the difference. In fact, for the analysis in which $k = 2$, $F = t^2$ both in terms of the respective critical values required for significance and the respective computed test-statistic values for the sample data. To demonstrate these observations, carry out (a) a t test and (b) the one-way analysis of variance for the difference between the means of the first two brands of typewriters reported in Table 13.6. Use the 5 percent level of significance to test the null hypothesis that there is no difference between the two means.

(a)   Using the t test for independent groups:

$$H_0: \mu_0 - \mu_2 = 0 \qquad (\text{or } \mu_1 = \mu_2) \qquad \bar{X}_1 = 70.4 \qquad \bar{X}_2 = 77.0$$

$$H_1: (\mu_1 - \mu_2 \neq 0 \qquad (\text{or } \mu_1 \neq \mu_2) \qquad n_1 = 5 \qquad n_2 = 3$$

$$\text{Critical } t \ (df = 6, \ \alpha = 0.05) = \pm 2.447$$

$$s_1^2 = \frac{\Sigma(X - \bar{X}_1)^2}{n_1 - 1} = \frac{(79 - 70.4)^2 + (83 - 70.4)^2 + (62 - 70.4)^2 + (51 - 70.4)^2 + (77 - 70.4)^2}{5 - 1}$$

$$= \frac{723.20}{4} = 180.8$$

$$s_2^2 = \frac{\Sigma(X - \bar{X}_2)^2}{n_2 - 1} = \frac{(74 - 77.0)^2 + (85 - 77.0)^2 + (72 - 77.0)^2}{3 - 1} = \frac{98.0}{2} = 49.0$$

$$\hat{\sigma}^2 = \frac{(n_1 - 1)s_1^2 + (n_2 - 1)s_2^2}{n_1 + n_2 - 2} = \frac{(4)180.8 + (2)49.0}{5 + 3 - 2} = \frac{821.2}{6} = 136.8667$$

$$\hat{\sigma}_{\bar{x}_1 - \bar{x}_2} = \sqrt{\frac{\hat{\sigma}^2}{n_1} + \frac{\hat{\sigma}^2}{n_2}} = \sqrt{\frac{136.8667}{5} + \frac{136.8667}{3}} = \sqrt{72.9955} = 8.54$$

$$t = \frac{\bar{X}_1 - \bar{X}_2}{\hat{\sigma}_{\bar{x}_1 - \bar{x}_2}} = \frac{70.4 - 77.0}{8.54} = \frac{-6.6}{8.54} = -0.77$$

The computed t statistic of -0.77 is in the region of acceptance of the null hypothesis, and therefore the null hypothesis cannot be rejected at the 5 percent level of significance.

(b)   For the two treatment levels (typewriter brands), based on the one-way analysis of variance:

$$H_0: \mu_1 = \mu_2 \qquad (\text{or } \alpha_k = 0) \qquad n_1 = 5 \qquad n_2 = 3 \qquad N = 8$$

$$H_1: \mu_1 \neq \mu_2 \qquad (\text{or } \alpha_k \neq 0) \qquad T_1 = 352 \qquad T_2 = 231 \qquad T = 583$$

$$T_1^2 = 123{,}904 \qquad T_2^2 = 53{,}361 \qquad T^2 = 339{,}889$$

$$\frac{T^2}{N} = \frac{339{,}889}{8} = 42{,}486.1$$

$$\sum_{i=1}^{n} \sum_{k=1}^{K} X^2 = 79^2 + 83^2 + \cdots + 72^2 = 43{,}389$$

$$SST = \sum_{i=1}^{n} \sum_{k=1}^{K} X^2 - \frac{T^2}{N} = 43{,}389 - 42{,}486.1 = 902.9$$

$$SSA = \sum_{k=1}^{K} \frac{T_k^2}{n_k} - \frac{T^2}{N} = \frac{123{,}904}{5} + \frac{53{,}361}{3} - 42{,}486.1 = 81.7$$

$$SSE = SST - SSA = 902.9 - 81.7 = 821.2$$

Table 13.8 presents the analysis of variance for this test. At the 5 percent level of significance, the critical value of F ($df = 1, 6$) is 5.99. Since the computed F statistic of 0.60 is in the region of acceptance of the null hypothesis, the hypothesis that there is no treatments effect cannot be rejected.

Table 13.8   ANOVA Table for the Analysis of Typing Speed on Two Brands of Typewriters

Source of variation	Degrees of freedom (df)	Sum of squares (SS)	Mean square (MS)	F ratio
Among treatment groups (A) (brand of typewriter)	$2-1=1$	81.7	$\dfrac{81.7}{1}=81.7$	$\dfrac{81.7}{136.9}=0.60$
Sampling error (E)	$8-2=6$	821.2	$\dfrac{821.2}{6}=136.9$	
Total (T)	$8-1=7$	902.9		

Thus, the observation about the comparability of the two testing procedures is supported. In terms of the critical values, critical $F=5.99$ and critical $t=\pm2.447$. Thus, $(\pm2.447)^2=5.99$. As for the computed test-statistic values of F and t, $F=0.60$ and $t=-0.77$; $(-0.77)^2=0.59\cong0.60$, with the slight difference being due solely to rounding error.

THE RANDOMIZED BLOCK DESIGN (TWO-WAY ANALYSIS WITHOUT INTERACTION)

13.6   For the data in Table 13.4, suppose a randomized block design was in fact used and trainees were matched before the experiment, with a trainee from each ability group (based on prior course achievement) assigned to each method of instruction. Table 13.9 is a revision of Table 13.4 in that the values reported are reorganized to reflect the randomized block design. Note, however, that the same values are included in each A treatment group as in Table 13.4, except that they are reported according to the B ability groups and therefore are arranged in a different order. Test the null hypothesis that there is no difference in the mean performance among the three methods of instruction, using the 5 percent level of significance.

Table 13.9   Achievement Test Scores of Trainees under Three Methods of Instruction, According to Ability Level

Level of ability (Block)	Method of instruction			Total (T_j)	Mean ($\bar{X}_j$)
	A_1	A_2	A_3		
B_1	86	90	82	258	86.0
B_2	84	89	81	254	84.7
B_3	81	88	73	242	80.7
B_4	79	76	68	223	74.3
B_5	70	82	71	223	74.3
Total (T_k)	400	425	375	Grand total $T=1,200$	
Mean ($\bar{X}_k$)	80.0	85.0	75.0		Grand mean $\bar{X}_T=80.0$

The various quantities required for the analysis of variance table are

For j:

$T_1 = 258$ $T_2 = 254$ $T_3 = 242$ $T_4 = 223$ $T_5 = 223$

$T_1^2 = 66,564$ $T_2^2 = 64,516$ $T_3^2 = 58,564$ $T_4^2 = 49,729$ $T_5^2 = 49,729$

For k:

$T_1 = 400$ $T_2 = 425$ $T_3 = 375$

$T_1^2 = 160,000$ $T_2^2 = 180,625$ $T_3^2 = 140,625$

$n_1 = 5$ $n_2 = 5$ $n_3 = 5$

Overall:

$T = 1,200$ $T^2 = 1,440,000$ $N = 15$

$$\frac{T^2}{N} = \frac{1,440,000}{15} = 96,000$$

$$\sum_{j=1}^{J} \sum_{k=1}^{K} X^2 = 86^2 + 84^2 + \cdots + 71^2 = 96,698$$

$$SST = \sum_{j=1}^{J} \sum_{k=1}^{K} X^2 - \frac{T^2}{N} = 96,698 - 96,000 = 698$$

$$SSA = \sum_{k=1}^{K} \frac{T_k^2}{n_k} - \frac{T^2}{N} = \frac{160,000}{5} + \frac{180,625}{5} + \frac{140,625}{5} - 96,000 = 250$$

$$SSB = \frac{1}{K} \sum_{j=1}^{J} T_j^2 - \frac{T^2}{N} = \frac{1}{3}(66,564 + 64,516 + 58,564 + 49,729 + 49,729) - 96,000 = 367.3$$

$$SSE = SST - SSA - SSB = 698 - 250 - 367.3 = 80.7$$

Table 13.10 presents the analysis of variance (ANOVA) for these data. With respect to the linear equation representing this model, the two F ratios in Table 13.10 are concerned with the tests of the following null and alternative hypotheses:

$H_0: \alpha_k = 0$ for all columns $\qquad H_0: \beta_j = 0$ for all rows

$H_1: \alpha_k \neq 0$ for some columns $\qquad H_1: \beta_j \neq 0$ for some rows

Table 13.10 ANOVA Table for Analysis of Three Methods of Instruction According to Ability Level

Source variation	Sum of squares (SS)	Degrees of freedom (df)	Mean square (MS)	F ratio
Among treatment groups (A) (method)	250.0	$3 - 1 = 2$	$\frac{250.0}{2} = 125.0$	$\frac{125}{10.1} = 12.4$
Among blocks (B) (ability level)	367.3	$5 - 1 = 4$	$\frac{367.3}{4} = 91.8$	$\frac{91.8}{10.1} = 9.1$
Sampling error (E)	80.7	$(5-1)(3-1) = 8$	$\frac{80.7}{8} = 10.1$	
Total (T)	698.0	$15 - 1 = 14$		

In terms of practical implications, the first null hypothesis is concerned with testing the difference among the column means, which is the basic purpose of the analysis. The second null hypothesis is concerned with testing the difference among the row means, which reflect the blocking that was done according to level of ability.

Using the 5 percent level of significance, the required F ratio for the rejection of the first null hypothesis $(df = 2, 8) = 4.46$ while the required F for the second null hypothesis $(df = 4, 8) = 3.84$. Thus, both of the calculated F ratios in Table 13.10 are in the region of rejection of the null hypothesis. We conclude that there is a significant difference in achievement test scores for the different methods of instruction, and also that there is a significant difference in achievement test scores for the different levels of ability. Note that this is in contrast to the result in Problem 13.2, where the data were treated as three independent samples. The MSE in the present analysis is substantially lower than that determined in Problem 13.2 because much of that variability could be identified as being due to differences in ability level.

13.7 Use available computer software to carry out the analysis in Problem 13.6 (data in Table 13.9) for the randomized block design.

The computer output in Fig. 13-2 corresponds to the results obtained by manual calculation in Table 13.10. For this particular software, however, note that the two F ratios are not reported, but rather they need to be calculated based on the mean squares that are included in the output. As in Problem 13.6, we conclude that there is a difference among the means according to method of instruction and there is a difference among the means according to the blocking factor of level of ability, both at the 5 percent level of significance.

```
MTB > READ METHOD IN C1, BLOCK IN C2, SCORE IN C3
DATA>   1     1      86
DATA>   1     2      84
DATA>   1     3      81
DATA>   1     4      79
DATA>   1     5      70
DATA>   2     1      90
DATA>   2     2      89
DATA>   2     3      88
DATA>   2     4      76
DATA>   2     5      82
DATA>   3     1      82
DATA>   3     2      81
DATA>   3     3      73
DATA>   3     4      68
DATA>   3     5      71
DATA> END
MTB > NAME C1 = 'METHOD', C2 = 'BLOCK', C3 = 'SCORE'
MTB > TWOWAY AOV FOR 'SCORE', DIMENSIONS 'METHOD', 'BLOCK'

ANALYSIS OF VARIANCE   SCORE

SOURCE        DF        SS         MS         F
METHOD        2       250.0      125.0      12.4
BLOCK         4       367.3       91.8       9.1
ERROR         8        80.7       10.1
TOTAL        14       698.0
```

Fig. 13-2 Minitab output for Problem 13.7.

RELATIONSHIP OF THE RANDOMIZED BLOCK DESIGN TO THE t TEST FOR TESTING THE DIFFERENCE BETWEEN TWO MEANS USING PAIRED OBSERVATIONS

13.8 When $k = 2$, the randomized block design is equivalent to the t test for the difference between the means of paired observations and $F = t^2$. This is similar to the case in Problem 13.5 for two independent samples. To demonstrate these points, apply the analysis of variance to the data in Table 13.11, which is taken from Example 4 in Chapter 11. In Example 4 the critical t ($df = 9$, $\alpha = 0.05$) is ± 2.262, the computed t statistic is $+1.59$ and thus the null hypothesis of no difference could not be rejected. Carry out the test at the 5 percent level of significance.

Table 13.11   Automobile Mileage Obtained with and without a Gasoline Additive for Ten Sampled Cars

Automobile	Mileage with additive (per gal)	Mileage without additive (per gal)	Total (T_j)	Mean ($\bar{X}_j$)
1	36.7	36.2	72.9	36.45
2	35.8	35.7	71.5	35.75
3	31.9	32.3	64.2	32.10
4	29.3	29.6	58.9	29.45
5	28.4	28.1	56.5	28.25
6	25.7	25.8	51.5	25.75
7	24.2	23.9	48.1	24.05
8	22.6	22.0	44.6	22.30
9	21.9	21.5	43.4	21.70
10	20.3	20.0	40.3	20.15
Total (T_k)	276.8	275.1	Grand total $T = 551.9$	
Mean ($\bar{X}_k$)	27.68	27.51		Grand mean $\bar{X}_T = 27.60$

The various quantities required for the analysis of variance table are

For j:

$T_1 = 72.9$	$T_2 = 71.5$	$T_3 = 64.2$	$T_4 = 58.9$	$T_5 = 56.5$
$T_1^2 = 5,314.41$	$T_2^2 = 5,112.25$	$T_3^2 = 4,121.64$	$T_4^2 = 3,469.21$	$T_5^2 = 3,192.25$
$T_6 = 51.5$	$T_7 = 48.1$	$T_8 = 44.6$	$T_9 = 43.4$	$T_{10} = 40.3$
$T_6^2 = 2,652.25$	$T_7^2 = 2,313.61$	$T_8^2 = 1,989.16$	$T_9^2 = 1,883.56$	$T_{10}^2 = 1,624.09$

For k:

$T_1 = 276.8$ $\qquad$ $T_2 = 275.1$

$T_1^2 = 76,618.24$ $\qquad$ $T_2^2 = 75,680.01$

$n_1 = 10$ $\qquad$ $n_2 = 10$

Overall:

$T = 551.9$ $\qquad$ $T^2 = 304,593.61$ $\qquad$ $N = 20$

$$\frac{T^2}{N} = \frac{304,593.61}{20} = 15,229.68$$

$$\sum_{j=1}^{J} \sum_{k=1}^{K} X^2 = 36.7^2 + 35.8^2 + \cdots + 20.0^2 = 6,798.87$$

$$SST = \sum_{j=1}^{J} \sum_{k=1}^{K} X^2 - \frac{T^2}{N} = 15,836.87 - 15,229.68 = 607.19$$

$$SSA = \sum_{k=1}^{K} \frac{T_k^2}{n_k} - \frac{T^2}{N} = \frac{76,618.24}{10} + \frac{75,680.01}{10} - 15,229.68 = 0.145$$

$$SSB = \frac{1}{K} \sum_{j=1}^{J} T_j^2 - \frac{T^2}{N} = \frac{1}{2}(5,314.41 + \cdots + 1,624.09) - 15,229.68 = 606.535$$

$$SSE = SST - SSA - SSB = 607.19 - 0.145 - 606.535 = 0.510$$

Table 13.12 presents the analysis of variance for the data of Table 13.11. The null hypothesis of interest, which is conceptually the same one tested in Example 4 of Chapter 11, is, for the treatment groups, H_0: $\alpha_k = 0$ and H_1: $\alpha_k \neq 0$. Critical F ($df = 1, 9$; $\alpha = 0.05$) = 5.12.

Since the computed value of F statistic of 2.54 is not greater than the critical value of 5.12, the hypothesis of no treatment effect (of no difference between the means) cannot be rejected at the 5 percent level of significance.

Note the very large value of the computed F for the blocks dimension. Referring to the data in Table 13.11 this result is not surprising, since the 10 automobiles were apparently in different weight categories and differed substantially in mileage achieved according to category.

Table 13.12   ANOVA Table for Analysis of Automobile Mileage with and without a Gasoline Additive

Source of variation	Degrees of freedom (df)	Sum of squares (SS)	Mean square (MS)	F ratio
Among treatment groups (A)	$2 - 1 = 1$	0.145	$\frac{0.145}{1} = 0.145$	$\frac{0.145}{0.057} = 2.54$
Among blocks (B) (different automobiles)	$10 - 1 = 9$	606.535	$\frac{606.535}{9} = 67.39$	$\frac{67.39}{0.057} = 1,182.28$
Sampling error (E)	$(2-1)(10-1) = 9$	0.510	$\frac{0.510}{9} = 0.057$	
Total (T)	$20 - 1 = 19$	607.190		

As expected, the critical value of F in this problem is equal to the square of the critical value used in the t test: $5.12 = (\pm 2.262)^2$. Also, except for the rounding error, the computed t^2 value in Example 4 of Chapter 11 is equal to the computed F of 2.54 here, with $t^2 = (1.59)^2 = 2.53$.

TWO-FACTOR COMPLETELY RANDOMIZED DESIGN

13.9   Nine trainees in each of four different subject areas were randomly assigned to three different methods of instruction. Three students were assigned to each instructional method. With reference to Table 13.13, test the various null hypotheses which are of interest with respect to such a design at the 5 percent level of significance.

The various quantities required for the analysis of variance table are

For J:

$T_1 = 717$	$T_2 = 709$	$T_3 = 722$	$T_4 = 732$
$T_1^2 = 514,089$	$T_2^2 = 502,681$	$T_3^2 = 521,284$	$T_4^2 = 535,824$

For k:

$T_1 = 960$	$T_2 = 1,020$	$T_3 = 900$
$T_1^2 = 921,600$	$T_2^2 = 1,040,400$	$T_3^2 = 810,000$

Table 13.13   Achievement Test Scores of Trainees under Three Methods of Instruction and for Four Subject Areas

Subject area	Method of instruction			Total (T_j)	Mean $(\bar{X}_j)$
	A_1	A_2	A_3		
B_1	70 79 72	83 89 78	81 86 79	717	79.7
B_2	77 81 79	77 87 88	74 69 77	709	78.8
B_3	82 78 80	94 83 79	72 79 75	722	80.2
B_4	85 90 87	84 90 88	68 71 69	732	81.3
Total (T_k)	960	1,020	900	Grand total $T = 2,880$	
Mean $(\bar{X}_k)$	80	85	75		Grand mean $\bar{X}_T = 80$

Overall:

$$T = 2,880 \qquad T^2 = 8,294,400 \qquad N = 36$$

$$\frac{T^2}{N} = \frac{8,294,400}{36} = 230,400$$

$$\sum_{j=1}^{J} \sum_{k=1}^{K} \left(\sum_{i=1}^{n} X \right)^2 = (70+79+72)^2 + (77+81+79)^2 + \cdots + (68+71+69)^2 = 694,694$$

$$\sum_{i=1}^{n} \sum_{j=1}^{J} \sum_{k=1}^{K} X^2 = 70^2 + 79^2 + 72^2 + 77^2 + \cdots + 69^2 = 232,000$$

$$SST = \sum_{i=1}^{n} \sum_{j=1}^{J} \sum_{k=1}^{K} X^2 - \frac{T^2}{N} = 232,000 - 230,400 = 1,600$$

$$SSA = \sum_{k=1}^{K} \frac{T_k^2}{nJ} - \frac{T^2}{N} = \frac{921,600}{(3)(4)} + \frac{1,040,400}{(3)(4)} + \frac{810,000}{(3)(4)} - 230,400 = 600$$

$$SSB = \sum_{j=1}^{J} \frac{T_j^2}{nK} - \frac{T^2}{N} = \frac{514,089}{(3)(3)} + \frac{502,681}{(3)(3)} + \frac{521,284}{(3)(3)} + \frac{535,824}{(3)(3)} - 230,400 = 30.8$$

$$SSI = \frac{1}{n} \sum_{j=1}^{J} \sum_{k=1}^{K} \left(\sum_{i=1}^{n} X \right)^2 - SSA - SSB - \frac{T^2}{N} = \frac{1}{3}(694,694) - 600 - 30.8 - 230,400 = 533.9$$

$$SSE = SST - SSA - SSB - SSI = 1,600.0 - 600.0 - 30.8 - 533.9 = 435.3$$

Table 13.14   ANOVA Table for Analysis of Three Methods of Instructions Applied in Four Subject Areas

Source of variation	Degrees of freedom (df)	Sum of squares (SS)	Mean square (MS)	F ratio
Between treatment groups (A) (method)	$3-1=2$	600.0	$\dfrac{600.0}{2}=300.0$	$\dfrac{300.0}{18.1}=16.57$
Between treatment groups (B) (subject)	$4-1=3$	30.8	$\dfrac{30.8}{3}=10.3$	$\dfrac{10.3}{18.1}=0.57$
Interaction between method and subject (I)	$(4-1)(3-1)=6$	533.9	$\dfrac{533.9}{6}=89.0$	$\dfrac{89.0}{18.1}=4.92$
Sampling error (E)	$(4)(3)(3-1)=24$	435.3	$\dfrac{435.3}{24}=18.1$	
Total (T)	$36-1=35$	1,600.0		

Table 13.14 presents the analysis of variance (ANOVA) for the data of Table 13.13. With respect to the linear model for two-way analysis of variance with replication, the three F ratios reported in Table 13.14 are concerned with the following null and alternative hypotheses:

$$H_0: \alpha_k = 0 \text{ for all columns; } H_1: \alpha_k \neq 0 \text{ for some columns}$$

$$H_0: \beta_j = 0 \text{ for all rows; } H_1: \beta_j \neq 0 \text{ for some rows}$$

$$H_0: \iota_{jk} = 0 \text{ for all cells; } H_1: \iota_{jk} \neq 0 \text{ for some cells}$$

Using the 5 percent level of significance, the required F ratio for the rejection of the first null hypothesis ($df = 2, 24$) is 3.40, for the second the required F ($df = 3, 24$) is 3.01, and for the third the required F ($df = 6, 24$) is 2.51. Thus, we conclude that there is a significant difference in test scores for the different methods of instruction, that there is no significant difference among the different subject areas, and that there is significant interaction between the two factors. The last conclusion indicates that the effectiveness of the three methods of instruction varies for the different subject areas. For example, in Table 13.13 notice that for subject area B_1 method A_1 was the least effective method while for subject area B_4, method A_3 was the least effective method. In reviewing this table, however, method A_2 is observed to be at least equal to the other methods for every subject area. Thus, the possibility of using different methods of instruction as being best for different subject areas does not appear to be the appropriate decision in this case, even though there was a significant interaction effect.

13.10   Use available computer software to carry out the analysis in Problem 13.9 (data in Table 13.13) for the two-factor completely randomized design.

   The computer output in Fig. 13-3 corresponds to the results obtained by manual calculation in Table 13.14. For this particular software, however, the three F ratios are not reported, but rather they need to be calculated based on the mean squares that are included in the output. The results of the tests of significance are the same as in Problem 13.9.

Supplementary Problems

ONE-FACTOR COMPLETELY RANDOMIZED DESIGN

13.11   Four types of advertising displays were set up in 12 retail outlets, with three outlets randomly assigned to each of the displays, for the purpose of studying the point-of-sale impact of the displays. With reference

```
MTB > READ METHOD IN C1, SUBJECT IN C2, SCORE IN C3
DATA>    1     1     70
DATA>    1     1     79
DATA>    1     1     72
DATA>    1     2     77
DATA>    1     2     81
DATA>    1     2     79
DATA>    1     3     82
DATA>    1     3     78
DATA>    1     3     80
DATA>    1     4     85
DATA>    1     4     90
DATA>    1     4     87
DATA>    2     1     83
DATA>    2     1     89
DATA>    2     1     78
DATA>    2     2     77
DATA>    2     2     87
DATA>    2     2     88
DATA>    2     3     94
DATA>    2     3     83
DATA>    2     3     79
DATA>    2     4     84
DATA>    2     4     90
DATA>    2     4     88
DATA>    3     1     81
DATA>    3     1     86
DATA>    3     1     79
DATA>    3     2     74
DATA>    3     2     69
DATA>    3     2     77
DATA>    3     3     72
DATA>    3     3     79
DATA>    3     3     75
DATA>    3     4     68
DATA>    3     4     71
DATA>    3     4     69
DATA> END
MTB > NAME C1 = 'METHOD', C2 = 'SUBJECT', C3 = 'SCORE'
MTB > TWOWAY AOV FOR 'SCORE', DIMENSIONS 'METHOD', 'SUBJECT'

ANALYSIS OF VARIANCE   SCORE
```

SOURCE	DF	SS	MS	F
METHOD	2	600.0	300.0	16.57
SUBJECT	3	30.9	10.3	0.57
INTERACTION	6	533.8	89.0	4.92
ERROR	24	435.3	18.1	
TOTAL	35	1600.0		

Fig. 13-3 Minitab output for Problem 13.10.

Table 13.15   Product Sales According to Advertising Display Used

Type display	Sales			Total sales	Mean sales
A_1	40	44	43	127	42.3
A_2	53	54	59	166	55.3
A_3	48	38	46	132	44.0
A_4	48	61	47	156	52.0

to Table 13.15, test the null hypothesis that there are no differences among the mean sales values for the four types of displays, using the 5 percent level of significance.

Ans.   Critical F ($df = 3, 8$) = 4.07. Computed $F = 4.53$. Therefore, reject the null hypothesis that $\alpha_k = 0$ for all treatment levels.

13.12   Use available computer software to carry out the analysis in Problem 13.11 (data in Table 13.15) for the one-factor completely randomized design.

Ans.   Same as for Problem 13.11.

THE RANDOMIZED BLOCK DESIGN (TWO-WAY ANALYSIS WITHOUT INTERACTION)

13.13   The designs produced by four automobile designers are evaluated by three product managers, as reported in Table 13.16. Test the null hypothesis that the average ratings of the designs do not differ, using the 1 percent level of significance.

Table 13.16   Ratings of Automobile Designs

Evaluator	Designer				Total (T_j)	Mean ($\bar{X}_j$)
	1	2	3	4		
A	87	79	83	92	341	85.25
B	83	73	85	89	330	82.50
C	91	85	90	92	358	89.50
Total (T_k)	261	237	258	273	Grand total $T = 1,029$	
Mean rating ($\bar{X}_k$)	87.0	79.0	86.0	91.0		Grand mean $\bar{X}_T = 85.75$

Ans.   Critical F ($df = 3, 6$) = 9.78. Computed $F = 12.29$. Therefore, reject the null hypothesis that $\alpha_k = 0$ for all treatment (column) effects.

13.14   Use available computer software to carry out the analysis in Problem 13.13 (data in Table 13.16) for the randomized block design.

Ans.   Same as for Problem 13.13.

TWO-FACTOR COMPLETELY RANDOMIZED DESIGN

13.15 The smallest data table for which two-way analysis of variance with interaction can be carried out is a 2×2 table with two observations per cell. Table 13.17, which presents sales data for a consumer product in eight randomly assigned regions, is such a table. Test the effect of the two factors and of the interaction between the two factors on the weekly sales levels, using the 1 percent level of significance. Consider the meaning of your test results.

Table 13.17   Weekly Sales in Thousands of Dollars with and without Advertising, and with and without Discount Pricing

Discount pricing	With advertising	Without advertising	Total (T_j)	Mean $(\bar{X}_j)$
With discounting	9.8 10.6	6.0 5.3	31.7	7.925
Without discounting	6.2 7.1	4.3 3.9	21.5	5.375
Total (T_k)	33.7	19.5	Grand total $T = 53.2$	
Mean $(\bar{X}_k)$	8.425	4.875		Grand mean $\bar{X}_T = 6.650$

Ans. At the 1 percent level of significance the critical values of F associated with the column, row, and interaction effects are, respectively, F $(df = 1, 4) = 21.20$; F $(df = 1, 4) = 21.20$; and F $(df = 1, 4) = 21.20$. The computed F ratios are, respectively, $F = 96.00$, $F = 49.52$, and $F = 7.66$. Therefore, there are significant column and row effects, but no significant interaction effects.

13.16 Use available computer software to carry out the analysis in Problem 13.15 (data in Table 13.17) for the two-factor completely randomized design.

Ans. Same as for Problem 13.15, except for slight rounding differences.

Chapter 14

Linear Regression and Correlation Analysis

14.1 OBJECTIVES AND ASSUMPTIONS OF REGRESSION ANALYSIS

The primary objective of regression analysis is to estimate the value of a random variable (the *dependent variable*) given that the value of an associated variable (the *independent variable*) is known. The *regression equation* is the algebraic formula by which the estimated value of the dependent variable is determined (see Section 14.3).

The term *simple regression analysis* indicates that the value of the dependent variable is estimated on the basis of one independent variable, whereas *multiple regression analysis* (covered in Chapter 15) is concerned with estimating the value of the dependent variable on the basis of two or more independent variables.

The general assumptions underlying the regression analysis model presented in this chapter are that (1) the dependent variable is a random variable, (2) the independent and dependent variables are linearly associated, and (3) the variances of the conditional distributions of the dependent variable, given different values of the independent variable, are equal (*homoscedasticity*). Assumption (1) indicates that although the values of the independent variable may be controlled, the values of the dependent variable must be obtained through the process of sampling.

If interval estimation is used in conjunction with regression analysis, an additional assumption is that (4) the conditional distributions of the dependent variable, given different values of the independent variable, are all normal distributions for the population of values.

EXAMPLE 1. An analyst wishes to estimate delivery time as the dependent variable based on distance as the independent variable for industrial parts shipped by truck. Suppose he chooses 10 recent shipments from the company's records such that the highway distances involved are about equally dispersed between 100 miles distance and 1,000 miles distance, and he records the delivery time for each shipment. Since the highway distance is to be used as the independent variable, his selection of trips of specific distances is acceptable. On the other hand, the dependent variable of delivery time is a random variable in this study, which conforms to the assumption underlying regression analysis. Whether or not the two variables have a linear relationship would generally be determined by constructing a scatter plot (see Section 14.2) or a residual plot (see Section 14.4). Such diagrams also are used to observe whether the vertical scatter (variance) is about equal along the regression line.

14.2 THE SCATTER PLOT

A *scatter plot* is a graph in which each plotted point represents an observed pair of values for the independent and dependent variables. The value of the independent variable X is plotted with respect to the horizontal axis, and the value of the dependent variable Y is plotted with respect to the vertical axis. (See Problem 14.1.)

The form of the relationship represented by the scatter diagram can be *curvilinear* rather than linear. While regression analysis for curvilinear relationships is outside of the scope of this book, there is a limited discussion of curvilinear trend analysis in Section 16.2. For relationships that are not linear, a frequent approach is to determine a method of transforming values of one or both variables so that the relationship of the transformed values is linear. Then linear regression analysis can be applied to the transformed values, and estimated values of the dependent variable can be transformed back to the original measurement scale.

EXAMPLE 2. An example of a curvilinear relationship would be the relationship between years since incorporation for a company and sales level, given that each year the sales level has increased by the same percentage over the preceding year. The resulting curve with an increasing slope would be indicative of a so-called exponential relationship.

If the scatter plot indicates a relationship that is generally linear, then a straight line is fitted to the data. The precise location of this line is determined by the method of least squares (see Section 14.3). As illustrated in Example 3, a regression line with a positive slope indicates a direct relationship between the variables, a negative slope indicates an inverse relationship between the variables, and a slope of zero indicates that the variables are unrelated. Further, the extent of vertical scatter of the plotted points with respect to the regression line indicates the degree of relationship between the two variables.

EXAMPLE 3. Figure 14-1 includes several scatter plots and associated regression lines demonstrating several types of relationships between the variables.

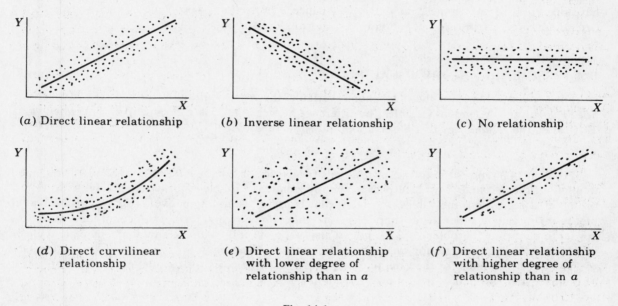

(a) Direct linear relationship

(b) Inverse linear relationship

(c) No relationship

(d) Direct curvilinear relationship

(e) Direct linear relationship with lower degree of relationship than in a

(f) Direct linear relationship with higher degree of relationship than in a

Fig. 14-1

14.3 THE METHOD OF LEAST SQUARES FOR FITTING A REGRESSION LINE

The linear model that represents the simple linear regression model is

$$Y_i = \beta_0 + \beta_1 X_i + \varepsilon_i \qquad (14.1)$$

where Y_i = value of the dependent variable in the ith trial, or observation
β_0 = first parameter of the regression equation, which indicates the value of Y when $X = 0$
β_1 = second parameter of the regression equation, which indicates the slope of the regression line
X_i = the specified value of the independent variable in the ith trial, or observation
ε_i = random-sampling error in the ith trial, or observation (ε is the Greek "epsilon")

The parameters β_0 and β_1 in the linear regression model are estimated by the values b_0 and b_1 based on sample data. Thus the linear regression equation based on sample data is

$$\hat{Y} = b_0 + b_1 X \qquad (14.2)$$

Depending on the mathematical criterion used, a number of different linear equations can be developed for a given scatter plot. By the *least-squares criterion* the best-fitting regression line (and equation) is that for which the sum of the squared deviations between the estimated and actual values of the dependent variable for the sample data is minimized. The computational formulas by which the values of b_0 and b_1 in the linear regression equation can be determined for the equation which satisfies the least-squares criterion are

$$b_1 = \frac{\Sigma XY - n\bar{X}\bar{Y}}{\Sigma X^2 - n\bar{X}^2} \qquad (14.3)$$

$$b_0 = \bar{Y} = b\bar{X} \qquad (14.4)$$

Once the regression equation is formulated, then this equation can be used to estimate the value of the dependent variable given the value of the independent variable. However, such estimation should be done only within the range of the values of the independent variable originally sampled, since there is no statistical basis to assume that the regression line is appropriate outside of these limits. Further, it should be determined whether the relationship expressed by the regression equation is real or could have occurred in the sample data purely by chance (see Section 14.6). Problem 14.2 illustrates the determination of a regression equation based on sample data.

14.4 RESIDUALS AND RESIDUAL PLOTS

For a given value X of the independent variable, the regression line value $\hat{Y}$ often is called the *fitted value* of the dependent variable. The difference between the observed value Y and the fitted value $\hat{Y}$ is called the *residual* for that observation and is denoted by e:

$$e = Y - \hat{Y} \qquad (14.5)$$

Problem 14.3 reports the full set of residuals for particular sample data. A *residual plot* is obtained by plotting the residuals e with respect to the independent variable X or, alternatively, with respect to the fitted regression line values $\hat{Y}$. Such a plot is useful as an alternative to the use of the scatter plot to investigate whether the assumptions concerning linearity and equality of conditional variances appear to be satisfied. Problem 14.3 includes a residual plot for the data. Residual plots are particularly important in multiple regression analysis, as described in Chapter 15.

The set of residuals for the sample data also serve as the basis for calculating the standard error of estimate, as described in the following section.

14.5 THE STANDARD ERROR OF ESTIMATE

The *standard error of estimate* is the conditional standard deviation of the dependent variable Y given a value of the independent variable X. For population data, the standard error of estimate is represented by the symbol $\sigma_{Y.X}$. The deviations formula by which this value is estimated on the basis of sample data is

$$s_{Y.X} = \sqrt{\frac{\Sigma(Y - \hat{Y})^2}{n-2}} = \sqrt{\frac{\Sigma e^2}{n-2}} \qquad (14.6)$$

Note that the numerator in formula (14.6) is the sum of the squares of the residuals discussed in the preceding section. Although formula (14.6) clearly reflects the idea that the standard error of estimate is the standard deviation with respect to the regression line (that is, it is the standard deviation of the vertical "scatter" about the line), computationally the formula requires that every fitted value $\hat{Y}$ be calculated for the sample data. The alternative computational formula that does not require determination of each fitted value and is therefore generally easier to use is

$$s_{Y.X} = \sqrt{\frac{\Sigma Y^2 - b_0 \Sigma Y - b_1 \Sigma XY}{n-2}} \qquad (14.7)$$

As will be seen in the following sections, the standard error of estimate serves as the cornerstone for the various standard errors used in the hypothesis-testing and interval-estimation procedures in regression analysis. Problems 14.5 and 14.6 illustrate the calculation of the standard error of estimate by the computational formula and deviations formula, respectively.

14.6 INFERENCES CONCERNING THE SLOPE

Before a regression equation is used for the purpose of estimation or prediction, we must first determine if a relationship in fact exists between the two variables in the population, or whether the observed relationship in the sample could have occurred by chance. In the absence of any relationship in the population, the slope of the population regression line would be zero, by definition: $\beta_1 = 0$. Therefore the usual null hypothesis tested is $H_0: \beta_1 = 0$. The null hypothesis can also be formulated as a one-tail test, in which case the alternative hypothesis is not simply that the two variables are related, but that the relationship is of a specific type (direct or inverse).

A hypothesized value of the slope is tested by computing a t statistic and using $n-2$ degrees of freedom. Two degrees of freedom are lost in the process of inference because *two* parameter estimates, b_0 and b_1, are included in the regression equation. The standard formula is

$$t = \frac{b_1 - (\beta_1)_0}{s_{b_1}} \quad (14.8)$$

where

$$s_{b_1} = \frac{s_{Y.X}}{\sqrt{\Sigma X^2 - n\bar{X}^2}} \quad (14.9)$$

However, when the null hypothesis is that the slope is zero, which generally is the case, formula (14.8) is simplified and is stated as

$$t = \frac{b_1}{s_{b_1}} \quad (14.10)$$

Problems 14.7 and 14.8 illustrate a two-tail test and a one-tail test for the slope, respectively.

The confidence interval for the population slope β_1 is constructed as follows, in which the degrees of freedom associated with the t are $n-2$:

$$b_1 \pm t s_{b_1} \quad (14.11)$$

Problem 14.9 illustrates the construction of a confidence interval for the population slope.

14.7 CONFIDENCE INTERVALS FOR THE CONDITIONAL MEAN

The point estimate for the conditional *mean* of the dependent variable, given a specific value of X, is the regression line value $\hat{Y}$. When we use the regression equation to estimate the conditional mean, the appropriate symbol for the conditional mean of Y is $\hat{\mu}_Y$:

$$\hat{\mu}_Y = b_0 + b_1 X \quad (14.12)$$

Based on sample data, the standard error of the conditional mean varies in value according to the designated value of X and is

$$s_{\hat{Y}.X} = s_{Y.X} \sqrt{\frac{1}{n} + \frac{(X - \bar{X})^2}{\Sigma X^2 - [(\Sigma X)^2/n]}} \quad (14.13)$$

Given the point estimate of the conditional mean and the standard error of the conditional mean, the confidence interval for the conditional mean, using $n-2$ degrees of freedom, is

$$\hat{\mu}_Y \pm t s_{\hat{Y}.X} \quad (14.14)$$

Again, it is $n-2$ degrees of freedom because the two parameter estimates b_0 and b_1 are required in the regression equation. Problem 14.10 illustrates the construction of a confidence interval for the conditional mean.

14.8 PREDICTION INTERVALS FOR INDIVIDUAL VALUES OF THE DEPENDENT VARIABLE

As contrasted to a confidence interval, which is concerned with estimating a population parameter, a *prediction interval* is concerned with estimating an *individual* value and is therefore a probability interval. It might seem that a prediction interval could be constructed by using the standard error of estimate defined in formulas (*14.6*) and (*14.7*). However, such an interval would be incomplete, because the standard error of estimate does not include the uncertainty associated with the fact that the regression line based on sample data also includes sampling error and generally is not identical to the population regression line. The complete standard error for a prediction interval is called the *standard error of forecast*, and it includes the uncertainty associated with the vertical "scatter" about the regression line *plus* the uncertainty associated with the position of the regression line value itself. The basic formula for the standard error of forecast is

$$s_{Y(\text{next})} = \sqrt{s_{Y.X}^2 + s_{\hat{Y}.X}^2} \qquad (14.15)$$

The computational version of the formula for the standard error of forecast is

$$s_{Y(\text{next})} = s_{Y.X} \sqrt{1 + \frac{1}{n} + \frac{(X - \bar{X})^2}{\Sigma X^2 - [(\Sigma X)^2/n]}} \qquad (14.16)$$

Finally, the prediction interval for an individual value of the dependent variable, using $n-2$ degrees of freedom, is

$$\hat{Y} \pm t s_{Y(\text{next})} \qquad (14.17)$$

The determination of the standard error of forecast and the construction of a prediction interval is illustrated in Problem 14.11.

14.9 OBJECTIVES AND ASSUMPTIONS OF CORRELATION ANALYSIS

In contrast to regression analysis, correlation analysis measures the degree of relationship between the variables. As was true in our coverage of regression analysis, in this chapter we restrict our coverage to *simple correlation analysis*, which is concerned with measuring the relationship between only one independent variable and the dependent variable. In Chapter 15, we describe multiple correlation analysis.

The population assumptions underlying simple correlation analysis are that (1) the relationship between the two variables is linear, (2) both of the variables are random variables, (3) for each variable the conditional variances given different values of the other variable are equal (homoscedasticity), and (4) for each variable the conditional distributions given different values of the other variable are all normal distributions. The last assumption is that of a *bivariate normal distribution*. Note that these assumptions are similar to the assumptions underlying interval estimate in regression analysis, except that in correlation analysis the assumptions apply to both variables whereas in regression analysis the independent variable can be set at various specific values and need not be a random variable.

EXAMPLE 4. Refer to the data collection procedure in Example 1, which concerns the variables of highway distance and delivery time for a sample of 10 recent shipments of industrial parts shipped by truck. Instead of the analyst choosing the 10 shipments so that they are about equally dispersed from 100 miles distance to 1,000 miles distance, the 10 shipments are chosen entirely randomly, without regard to either the highway distance or the delivery time included in each observation. Unlike Example 1, in which only the delivery time is a random variable, in the revised sampling plan both variables are random variables and therefore qualify for correlation analysis.

14.10   THE COEFFICIENT OF DETERMINATION

Consider that if an individual value of the dependent variable Y were estimated without knowledge of the value of any other variable, then the uncertainty associated with this estimate, and the basis for constructing a prediction interval, would be the variance σ_Y^2. Given a value of X, however, the uncertainty associated with the estimate is represented by $\sigma_{Y.X}^2$ (or $s_{Y.X}^2$ for sample data), as described in Section 14.5. If there is a relationship between the two variables, then $\sigma_{Y.X}^2$ will be smaller than σ_Y^2. For a perfect relationship, in which all values of the dependent variable are equal to the respective fitted regression line values, $\sigma_{Y.X}^2 = 0$. Therefore, in the absence of a perfect relationship, the value of $\sigma_{Y.X}^2$ indicates the uncertainty remaining *after* consideration of the value of the dependent variable. Or, we can say that the ratio of $\sigma_{Y.X}^2$ to σ_Y^2 indicates the proportion of variance (uncertainty) in the dependent variable which remains unexplained after a specific value of the dependent variable has been given:

$$\frac{\sigma_{Y.X}^2}{\sigma_Y^2} = \frac{\text{unexplained variance remaining in } Y}{\text{total variance in } Y} \qquad (14.18)$$

In (14.18) $\sigma_{Y.X}^2$ is determined by the procedure described in Section 14.5 (except that population data are assumed) and σ_Y^2 is calculated by the general formulas presented in Sections 4.5 and 4.6.

Given the proportion of unexplained variance, a useful measure of relationship is the *coefficient of determination*—the complement of the above ratio indicating the proportion of variance in the dependent variable which is statistically *explained* by the regression equation (i.e., by knowledge of the associated independent variable X). For population data the coefficient of determination is represented by the Greek ρ^2 ("rho squared") and is determined by

$$\rho^2 = 1 - \frac{\sigma_{Y.X}^2}{\sigma_Y^2} \qquad (14.19)$$

For sample data, the estimated value of the coefficient of determination can be obtained by the corresponding formula:

$$r^2 = 1 - \frac{s_{Y.X}^2}{s_Y^2} \qquad (14.20)$$

For computational purposes, the following formula for the sample coefficient of determination is convenient:

$$r^2 = \frac{b_0 \Sigma Y + b_1 \Sigma XY - n\bar{Y}^2}{\Sigma Y^2 - n\bar{Y}^2} \qquad (14.21)$$

Application of (14.21) is illustrated in Problem 14.12. Although this is a frequently used formula for computing the coefficient of determination for sample data, it does not incorporate any correction for biasedness and includes a slight positive bias. See the following section for the correction factor that can be used.

14.11   THE COEFFICIENT OF CORRELATION

Although the coefficient of determination r^2 is relatively easy to interpret, it does not lend itself to statistical testing. However, the square root of the coefficient of determination, which is called the *coefficient of correlation r*, does lend itself to statistical testing, because it can be used to define a test statistic which is distributed as the t distribution when the population correlation ρ equals 0. The value of the coefficient can range from -1.00 to $+1.00$. The arithmetic sign associated with the correlation coefficient, which is always the same as the sign associated with β_1 in the regression equation, indicates the direction of the relationship between X and Y (positive = direct; negative = inverse). The coefficient of correlation for population data, with the arithmetic sign being the same as that for β_1 in the regression equation, is

$$\rho = \sqrt{\rho^2} \qquad (14.22)$$

The coefficient of correlation for sample data is

$$r = \sqrt{r^2} \qquad (14.23)$$

In summary, the sign of the correlation coefficient indicates the direction of the relationship between the X and Y variables while the absolute value of the coefficient indicates the extent of relationship. The squared value of the correlation coefficient is the coefficient of determination and indicates the proportion of the variance in Y explained by knowledge of X (and vice versa).

EXAMPLE 5.   Figure 14-2 illustrates the general appearance of the scatter diagrams associated with several correlation values.

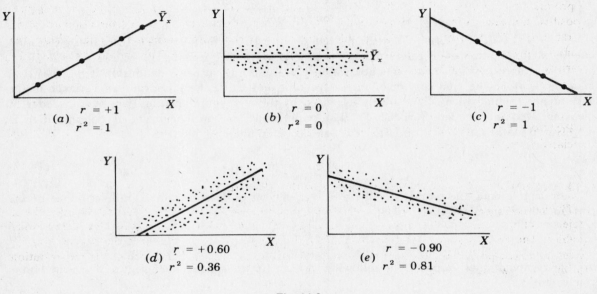

Fig. 14-2

As an alternative to (14.21), the following formula does not require prior determination of the regression values of b_0 and b_1. This formula would be used when the purpose of the analysis is to determine the extent and type of relationship between two variables, but without an accompanying interest in estimating Y given X. When this formula is used, the sign of the correlation coefficient is determined automatically, without the necessity of observing or calculating the slope of the regression line. The formula is

$$r = \frac{n\Sigma XY - \Sigma X \Sigma Y}{\sqrt{n\Sigma X^2 - (\Sigma X)^2}\sqrt{n\Sigma Y^2 - (\Sigma Y)^2}} \qquad (14.24)$$

Application of (14.24) is illustrated in Problem 14.13.

The sample coefficient of correlation r is somewhat biased as an estimator of ρ, with an absolute value which is too large. This factor is not mentioned in many textbooks because the amount of bias is slight, except for very small samples. An unbiased estimator for the coefficient of determination for the population can be obtained as follows:

$$\hat{\rho}^2 = 1 - (1 - r^2)\left(\frac{n-1}{n-2}\right) \qquad (14.25)$$

14.12 THE COVARIANCE APPROACH TO UNDERSTANDING THE CORRELATION COEFFICIENT

Another measure that can be used to express the relationship between two random variables is the sample *covariance*. The covariance measures the extent to which two variables "vary together" and is used in financial analyses concerned with determining the total risk associated with interrelated investments. Just as for the correlation coefficient, a positive sign indicates a *direct* relationship while a negative sign indicates an *inverse* relationship. The formula for the sample covariances is

$$\text{cov}(X, Y) = \frac{\Sigma[(X - \bar{X})(Y - \bar{Y})]}{n - 1} \tag{14.26}$$

By studying formula (*14.26*) we can conclude, for example, that when the observed value of Y tends to vary in the same direction from its mean as does the observed value of X from its mean, then the products of those deviations will tend to be positive values. The sum of such products will then be positive, indicating the direct, rather than inverse, relationship. Whereas the coefficient of correlation can range in value between only -1.00 and $+1.00$ and is a generalized measure of relationship, the covariance has no such limits in value and is not a generalized measure. The formula by which the covariance can be converted into the coefficient of correlation is

$$r = \frac{\text{cov}(X, Y)}{s_X s_Y} \tag{14.27}$$

Problem 14.14 illustrates the determination of a sample covariance and its conversion into the associated correlation coefficient.

14.13 SIGNIFICANCE OF THE CORRELATION COEFFICIENT

Typically, the null hypothesis of interest is that the population correlation $\rho = 0$, for if this hypothesis is rejected at a specified α level, we would conclude that there is an actual relationship between the variables. The hypothesis can also be formulated as a one-tail test. Given that the assumptions in Section 14.9 are satisfied, the following sampling statistic involving r is distributed as the t distribution with $df = n - 2$ when $\rho = 0$:

$$t = \frac{r}{\sqrt{\dfrac{1 - r^2}{n - 2}}} \tag{14.28}$$

Testing the null hypothesis that $\rho = 0$ is equivalent to testing the null hypothesis that $\beta = 0$ in the regression equation. (See Problem 14.15.)

14.14 PITFALLS AND LIMITATIONS ASSOCIATED WITH REGRESSION AND CORRELATION ANALYSIS

(1) In regression analysis a value of Y cannot be legitimately estimated if the value of X is outside of the range of values which served as the basis for the regression equation.

(2) If the estimate of Y involves the prediction of a result which has not yet occurred, the historical data which served as the basis for the regression equation may not be relevant for future events.

(3) The use of a prediction or a confidence interval is based on the assumption that the conditional distributions of Y are normal and have equal variances.

(4) A significant correlation coefficient does not necessarily indicate causation, but rather may indicate a common linkage to other events.

(5) A "significant" correlation is not necessarily an important correlation. Given a large sample, a correlation of, say, $r = +0.10$ can be significantly different from 0 at $\alpha = 0.05$. Yet the

coefficient of determination of $r^2 = 0.01$ for this example indicates that only 1 percent of the variance in Y is statistically explained by knowing X.

(6) The interpretation of the coefficients of correlation and determination is based on the assumption of a bivariate normal distribution for the population and, for each variable, equal conditional variances.

(7) For both regression and correlation analysis, a linear model is assumed. For a relationship which is curvilinear, a transformation to achieve linearity may be available. Another possibility is to restrict the analysis to the range of values within which the relationship is essentially linear.

14.15 COMPUTER OUTPUT

Computer software is available to carry out the various procedures of simple regression and correlation analysis that are covered in this chapter. Problem 14.17 illustrates the use of such software and compares various parts of the output to the results obtained by manual calculation in the preceding end-of-chapter problems.

Solved Problems

LINEAR REGRESSION ANALYSIS

14.1 Suppose an analyst takes a random sample of 10 recent truck shipments made by a company and records the distance in miles and delivery time to the nearest half-day from the time that the shipment was made available for pick-up. Construct the scatter plot for the data in Table 14.1 and consider whether linear regression analysis appears appropriate.

Table 14.1   Sample Observations of Trucking Distance and Delivery Time for 10 Randomly Selected Shipments

Sampled shipment	1	2	3	4	5	6	7	8	9	10
Distance (X), miles	825	215	1,070	550	480	920	1,350	325	670	1,215
Delivery time (Y), days	3.5	1.0	4.0	2.0	1.0	3.0	4.5	1.5	3.0	5.0

The scatter plot for these data is portrayed in Fig. 14-3. The first reported pair of values in the table is represented by the dot entered above 825 on the X axis and aligned with 3.5 with respect to the Y axis.

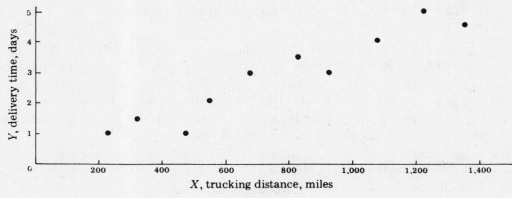

Fig. 14-3

The other nine points in the scatter plot were similarly entered. From the diagram, it appears that the plotted points generally follow a linear relationship and the vertical scatter at the line is about the same for low values and high values of X. Thus linear regression analysis appears appropriate.

14.2   Determine the least-squares regression equation for the data in Problem 14.1, and enter the regression line on the scatter plot for these data.

Referring to Table 14.2,

$$b_1 = \frac{\Sigma XY - n\bar{X}\bar{Y}}{\Sigma X^2 - n\bar{X}^2} = \frac{(26,370) - (10)(762)(2.85)}{7,104,300 - (10)(762)^2} = \frac{4,653}{1,297,860} = 0.0035851 \cong 0.0036$$

$$b_0 = \bar{Y} - b\bar{X} = 2.85 - (0.0036)(762) = 0.1068 \cong 0.11$$

Therefore, $$\hat{Y} = b_0 + b_1 X = 0.11 + 0.0036X$$

Table 14.2   Calculations Associated with Determining the Linear Regression Equation for Estimating Delivery Time on the Basis of Trucking Distance

Sampled shipment	Distance (X), miles	Delivery time (Y), days	XY	X^2	Y^2
1	825	3.5	2,887.5	680,625	12.25
2	215	1.0	215.0	46,225	1.00
3	1,070	4.0	4,280.0	1,144,900	16.00
4	550	2.0	1,100.0	302,500	4.00
5	480	1.0	480.0	230,400	1.00
6	920	3.0	2,760.0	846,400	9.00
7	1,350	4.5	6,075.0	1,822,500	20.25
8	325	1.5	487.5	105,625	2.25
9	670	3.0	2,010.0	448,900	9.00
10	1,215	5.0	6,075.0	1,476,225	25.00
Totals	7,620	28.5	26,370.0	7,104,300	99.75
Mean	$\bar{X} = \dfrac{\Sigma X}{n} = \dfrac{7,620}{10}$ $= 762$	$\bar{Y} = \dfrac{\Sigma Y}{n} = \dfrac{28.5}{10}$ $= 2.85$			

This estimated regression line based on sample data is entered in the scatter plot for these data in Fig. 14-4. Note the dashed lines indicating the amount of deviation between each sampled value of Y and the

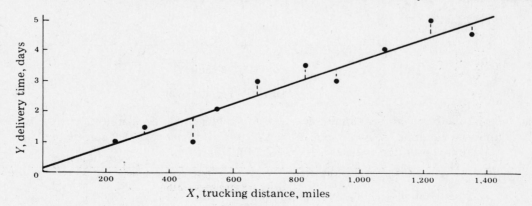

Fig. 14-4

corresponding estimated value, $\hat{Y}$. It is the sum of these squared deviations which is minimized by the linear regression line determined by the above procedure.

14.3 Determine the residuals and construct a residual plot with respect to the fitted values for the data in Table 14.2, using the regression equation developed in Problem 14.2. Compare the residual plot to the scatter plot in Fig. 14-4.

Table 14.3   Calculation of Residuals for the Delivery Time Problem

Sampled shipment	Distance (X), miles	Delivery time (Y), days	Fitted value $(\hat{Y})$	Residual $(e = Y - \hat{Y})$
1	825	3.50	3.08	0.42
2	215	1.00	0.88	0.12
3	1,070	4.00	3.96	0.04
4	550	2.00	2.09	−0.09
5	480	1.00	1.84	−0.84
6	920	3.00	3.42	−0.42
7	1,350	4.50	4.97	−0.47
8	325	1.50	1.28	0.22
9	670	3.00	2.52	0.48
10	1,215	5.00	4.48	0.52

Table 14.3 presents the calculation of the residuals, while Fig. 14-5 portrays the residual plot. Note that the general form of the scatter of the dependent-variable values Y with respect to the regression line in the scatter plot and the scatter of the residuals e with respect to the "0" line of deviation in the residual plot are similar, except for the "stretched" vertical scale in the residual plot.

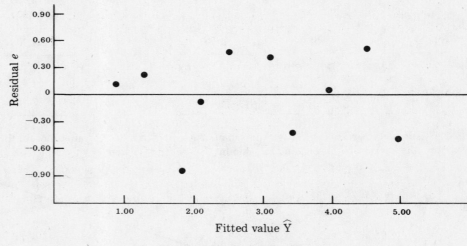

Fig. 14-5   Residual plot.

14.4 Using the regression equation developed in Problem 14.2, estimate the delivery time from the time that the shipment is available for pick-up for a shipment of 1,000 miles. Could this regression equation be used to estimate delivery time for a shipment of 2,500 miles?

$$\hat{Y} = 0.11 + 0.0036X = 0.11 + 0.0036(1,000) = 3.71 \text{ days}$$

It is not appropriate to use the above equation for a trip of 2,500 miles because the sample data for this estimated linear regression equation included trips up to 1,350 miles distance only.

14.5 Compute the standard error of estimate for the delivery time analysis problem, referring to values determined in the solution to Problem 14.2.

$$s_{Y.X} = \sqrt{\frac{\Sigma Y^2 - b_0 \Sigma Y - b_1 \Sigma XY}{n-2}} = \sqrt{\frac{99.75 - (0.11)(28.5) - (0.0036)(26,370)}{10-2}}$$

$$= \sqrt{\frac{1.683}{8}} = \sqrt{0.2104} = 0.4587 \cong 0.46$$

14.6 Compute the standard error of estimate by the basic formula that utilizes the residuals determined in Problem 14.3. Compare your answer to that obtained by the computational formula in Problem 14.5 above.

$$s_{Y.X} = \sqrt{\frac{\Sigma e^2}{n-2}} = \sqrt{\frac{1.8526}{10-2}} = 0.4812 \cong 0.48$$

Except for a slight rounding difference, this answer corresponds to the one obtained in Problem 14.5.

14.7 Using the standard error of estimate from Problem 14.5, test the null hypothesis $H_0: \beta_1 = 0$ for the trucking distance and delivery time data in Table 14.2, using the 5 percent level of significance.

Given $s_{Y.X} = 0.46$ and the values in Table 14.2,

$$s_{b_1} = \frac{s_{Y.X}}{\sqrt{\Sigma X^2 - n\bar{X}^2}} = \frac{0.46}{\sqrt{7,104,300 - 10(762)^2}} = \frac{0.46}{1,139.24} = 0.0004$$

$$H_0: \beta_1 = 0 \qquad H_1: \beta_1 \neq 0$$

$$\text{Critical } t(df = 8, \alpha = 0.05) = \pm 2.306$$

$$t = \frac{b_1}{s_{b_1}} = \frac{0.0036}{0.0004} = +9.00$$

Because the calculated t of $+9.00$ is in a region of rejection for this two-tail test, we conclude that there is a significant relationship between trucking distance and delivery time.

14.8 Referring to Problem 14.7, above, test the null hypothesis $H_0: \beta_1 \leq 0$ at the 5 percent level of significance.

$$H_0: \beta_1 \leq 0 \qquad H_1: \beta_1 > 0$$

$$\text{Critical } t(df = 8, \alpha = 0.05) = +1.860$$

$$t = +9.00 \qquad \text{(from Problem 14.7)}$$

Because the calculated t statistic of $+9.00$ exceeds the critical value of $+1.860$, the null hypothesis is rejected. For this one-tail test, we conclude that the slope of the population regression line is positive, and thus that there is a direct (rather than inverse) relationship between shipment distance and delivery time.

14.9 Determine the 95 percent confidence interval for β_1 for the trucking distance and delivery time data discussed in the preceding problems.

Since $b_1 = 0.0036$ (from Problem 14.2) and $df = n - 2 = 10 - 2 = 8$, the 95 percent confidence interval for β_1 is

$$b_1 \pm t s_{b_1} = 0.0036 \pm (2.306)(0.0004) = 0.0036 \pm 0.0009 = 0.0027 \text{ to } 0.0045$$

14.10 Using the values determined in the preceding problems, construct the 95 percent confidence interval for the *mean* delivery time for a trucking distance of 1,000 miles.

Given $\hat{\mu}_Y$ (for $X = 1,000$) = 3.71 days, $s_{Y.X} = 0.46$, and the values in Table 14.2,

$$s_{\hat{Y}.X} = s_{Y.X} \sqrt{\frac{1}{n} + \frac{(X - \bar{X})^2}{\Sigma X^2 - (\Sigma X)^2/n}} = 0.46 \sqrt{\frac{1}{10} + \frac{(1000 - 762)^2}{7,104,300 - (7,620)^2/10}} = 0.1748 \cong 0.17$$

The 95 percent confidence interval for the conditional mean (where $df = 10 - 2 = 8$) is

$$\hat{\mu}_Y \pm ts_{\hat{Y}.X} = 3.71 \pm (2.306)(0.17) = 3.71 \pm 0.39 = 3.32 \text{ to } 4.10 \text{ days}$$

Thus, for truck shipments of 1,000 miles, we estimate that the mean delivery time from the time that the shipment is available is between 3.32 and 4.10 days, with 95 percent confidence in this estimation interval.

14.11 Using the values determined in the preceding problems, determine the 95 percent prediction interval for the delivery time of a shipment given that a distance of 1,000 miles is involved. Compare this interval with the one constructed in Problem 14.10.

Since $\hat{Y}$ (for $X = 1,000$) = 3.71 days, $s_{Y.X} = 0.46$, and $s_{\hat{Y}.X} = 0.17$,

$$s_{Y \text{ next}} = \sqrt{s_{Y.X}^2 + s_{\hat{Y}.X}^2} = \sqrt{(0.46)^2 + (0.17)^2} = \sqrt{0.2405} = 0.4904 \cong 0.49$$

Where $df = 10 - 2 = 8$, the 95 percent prediction interval is

$$\hat{Y} \pm ts_{Y \text{ next}} = 3.71 \pm 2.306(0.49) = 3.71 \pm 1.13 = 2.58 \text{ to } 4.84 \text{ days}$$

As expected, this prediction interval is somewhat wider than the confidence interval for the conditional mean in Problem 14.10, since in the present application the interval concerns an *individual* value rather than a mean.

CORRELATION ANALYSIS

14.12 For the trucking distance and delivery time data, the sampling procedure described in Problem 14.1 indicates that both variables are in fact random variables. If we further assume a bivariate normal distribution for the population and, for each variable, equal conditional variances, then correlation analysis can be applied to the sample data. Using values calculated in Problem 14.2, calculate the coefficient of determination for the sample data (ignore the slight biasedness associated with this coefficient).

$$r^2 = \frac{b_0 \Sigma Y + b_1 \Sigma XY - n\bar{Y}^2}{\Sigma Y^2 - n\bar{Y}^2}$$

$$= \frac{(0.11)(28.5) + (0.0036)(26,370) - (10)(2.85)^2}{99.75 - (10)(2.85)^2}$$

$$= \frac{16.842}{18.525} = 0.9091 \cong 0.91$$

Thus, as a point estimate we can conclude that about 91 percent of the variance in delivery time is statistically explained by the trucking distance involved. Further, given the trucking distance involved, we can also conclude that only about 9 percent of the variance remains unexplained.

14.13 For the trucking distance and delivery time data, (*a*) calculate the coefficient of correlation by reference to the coefficient of determination in Problem 14.12, and (*b*) determine the coefficient of correlation by using the alternative computational formula for *r*.

(*a*) $r = \sqrt{r^2} = \sqrt{0.9091} = +0.9535 \cong +0.95$

The positive value for the correlation value is based on the observation that the slope b_1 of the regression line is positive, as determined in Problem 14.2.

(b) $\quad r = \dfrac{n\Sigma XY - \Sigma X \Sigma Y}{\sqrt{n\Sigma X^2 - (\Sigma X)^2}\sqrt{n\Sigma Y^2 - (\Sigma Y)^2}} = \dfrac{(10)(26,370) - (7,620)(28.5)}{\sqrt{(10)(7,104,300) - (7,620)^2}\sqrt{(10)(99.75) - (28.5)^2}}$

$\qquad = \dfrac{46,530}{(3,602.5824)(13.6107)} = \dfrac{46,530}{49,033.668} = +0.9489 \cong +0.95$

Except for a slight difference due to rounding, the two values are the same.

14.14 For the trucking distance and delivery time data, (a) calculate the sample covariance, and (b) convert the covariance value into the coefficient of correlation. Compare your answer in (b) to the answers for the coefficient of correlation obtained in Problem 14.13 above.

(a) $\quad \text{cov}(X, Y) = \dfrac{\Sigma[(X - \bar{X})(Y - \bar{Y})]}{n - 1}$

$\qquad = \dfrac{4,653}{9} = +517.00 \qquad \text{(from Table 14.4)}$

(b) $\quad r = \dfrac{\text{cov}(X, Y)}{s_X s_Y}$

$\quad$ where $\quad s_X = \sqrt{\dfrac{\Sigma X^2 - n\bar{X}^2}{n - 1}} \qquad$ [from formula (*4.11*)]

$\qquad = \sqrt{\dfrac{7,104,300 - 10(762)^2}{10 - 1}} \qquad \text{(from Table 14.2)}$

$\qquad = \sqrt{144,206.66} = 379.7455$

$\quad s_Y = \sqrt{\dfrac{\Sigma Y^2 - n\bar{Y}^2}{n - 1}} \qquad$ [from formula (*4.11*)]

$\qquad = \sqrt{\dfrac{99.75 - 10(2.85)^2}{10 - 1}} \qquad \text{(from Table 14.2)}$

$\qquad = \sqrt{2.058333} = 1.4347$

$\quad r = \dfrac{517.00}{(379.7455)(1.4347)} = +0.9489 \cong +0.95$

The correlation coefficient matches those calculated in Problem 14.13 above.

Table 14.4   Calculation of the Covariance

Sampled shipment	Distance, (X), miles	Delivery time (Y), days	$X - \bar{X}$	$Y - \bar{Y}$	$(X - \bar{X})(Y - \bar{Y})$
1	825	3.5	63	0.65	40.95
2	215	1.0	−547	−1.85	1,011.95
3	1,070	4.0	308	1.15	354.20
4	550	2.0	−212	−0.85	180.20
5	480	1.0	−282	−1.85	521.70
6	920	3.0	158	0.15	23.70
7	1,350	4.5	588	1.65	970.20
8	325	1.5	−437	−1.35	589.95
9	670	3.0	−92	0.15	−13.80
10	1,215	5.0	453	2.15	973.95
	$\bar{X} = 762$	$\bar{Y} = 2.85$			$\Sigma = 4,653.00$

14.15 Determine whether the correlation value computed in Problem 14.14(b) is significantly different from zero at the 5 percent level of significance.

$$H_0: \rho = 0 \qquad H_1: \rho \neq 0$$

$$\text{Critical } t(df = 8, \alpha = 0.05) = \pm 2.306$$

$$t = \frac{r}{\sqrt{\dfrac{1-r^2}{n-2}}} = \frac{0.9489}{\sqrt{\dfrac{1-0.9004}{10-2}}} = \frac{0.9489}{0.1116} = +8.50$$

Because the test statistic of $t = +8.50$ is in a region of rejection, the null hypothesis that there is no relationship between the two variables is rejected, and we conclude that there is a significant relationship between trucking distance and delivery time. Note that this conclusion coincides with the test of the null hypothesis that $\beta_1 = 0$ in Problem 14.7. The difference in the calculated values of t in these two tests is due to rounding of values in previous calculations.

14.16 For a sample of $n = 10$ loan recipients at a finance company, the correlation coefficient between household income and amount of outstanding short-term debt is found to be $r = +0.50$.

(a) Test the hypothesis that there is no correlation between these two variables for the entire population of loan recipients, using the 5 percent level of significance.

(b) Interpret the meaning of the correlation coefficient which was computed.

(a)

$$H_0: \rho = 0 \qquad H_1: \rho \neq 0$$

$$\text{Critical } t(df = 10 - 2 = 8, \alpha = 0.05) = \pm 2.306$$

$$t = \frac{r}{\sqrt{\dfrac{1-r^2}{n-2}}} = \frac{0.50}{\sqrt{\dfrac{1-(0.50)^2}{8}}} = \frac{0.50}{0.306} = +1.634$$

Because the computed t statistic of $+1.634$ is not in a region of rejection, the null hypothesis cannot be rejected, and we continue to accept the assumption that there is no relationship between the two variables. The observed sample relationship can be ascribed to chance at the 5 percent level of significance.

(b) Based on the correlation coefficient of $r = +0.50$, we might be tempted to conclude that because $r^2 = 0.25$, approximately 25 percent of the variance in short-term debt is explained statistically by the amount of household income. *However*, because the null hypothesis in part (a) above was not rejected, a more appropriate interpretation is that none of the variance in Y is associated with changes in X. By this approach, it is appropriate to consider the interpretation of r^2 only if the null hypothesis that there is no relationship has been rejected.

COMPUTER OUTPUT

14.17 Use available computer software to carry out the regression and correlation analysis for the data on distance and delivery time in Table 14.2. Compare particular results to the results of the manual calculations in the preceding exercises.

The computer input and output is presented in Fig. 14-6.
The following items are identified in the output:

① The scatter plot, which corresponds to the manually prepared plot in Fig. 14-3.

② The regression equation, which corresponds to the solution in Problem 14.2, except for the effect of the rounding of values in the manual calculations.

③ The residual plot, which corresponds to the manually prepared plot in Fig. 14-5.

```
MTB > NAME C1 = 'DISTANCE', C2 = 'DELTIME'
MTB > READ 'DISTANCE', 'DELTIME'
DATA >      825      3.5
DATA >      215      1.0
DATA >     1070      4.0
DATA >      550      2.0
DATA >      480      1.0
DATA >      920      3.0
DATA >     1350      4.5
DATA >      325      1.5
DATA >      670      3.0
DATA >     1215      5.0
DATA > END
MTB > PLOT 'DELTIME' VS. 'DISTANCE'
```

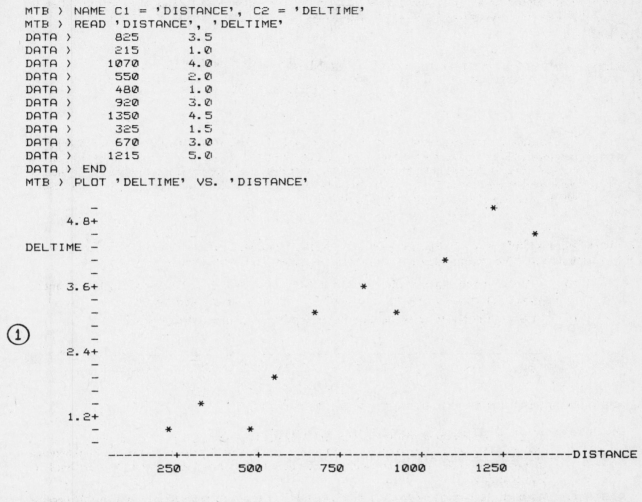

```
MTB > REGRESS 'DELTIME' USING 1 PREDICTOR 'DISTANCE', C3, PUT FITS IN C4;
SUBC) RESIDUALS IN C5;
SUBC) PREDICT 1000.

The regression equation is
DELTIME = 0.118 + 0.00359 DISTANCE

Predictor          Coef          Stdev      t-ratio
Constant          0.1181        0.3551        0.33
DISTANCE        0.0035851     0.0004214        8.51

s = 0.4800      R-sq = 90.0%     R-sq(adj) = 88.8%

Analysis of Variance

SOURCE         DF          SS          MS
Regression      1       16.682      16.682
Error           8        1.843       0.230
Total           9       18.525
```

Fig. 14-6 Minitab output (numbered labels are defined in the solution to Problem 14.17).
(*Continued on p. 258.*)

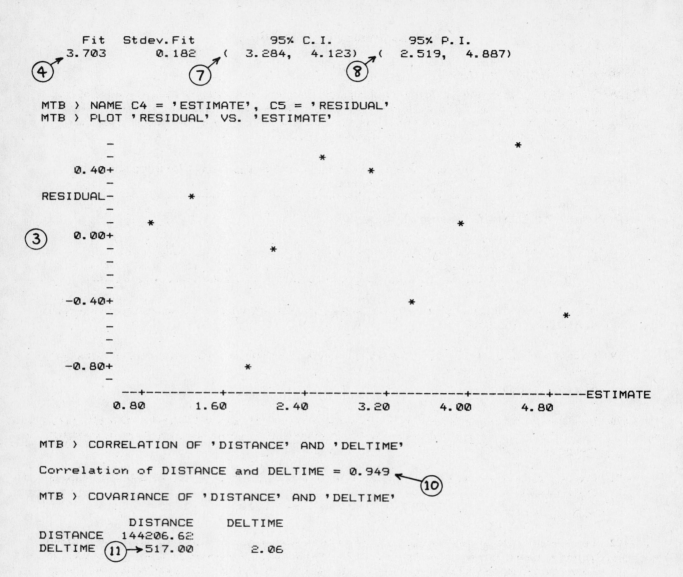

```
        Fit   Stdev.Fit        95% C. I.        95% P. I.
④→     3.703     0.182    ⑦→( 3.284,  4.123)⑧→( 2.519,  4.887)

MTB > NAME C4 = 'ESTIMATE', C5 = 'RESIDUAL'
MTB > PLOT 'RESIDUAL' VS. 'ESTIMATE'

           -                                                    *
           -
     0.40+                          *
           -                                   *
RESIDUAL-          *
           -
③    -        *                              *
     0.00+
           -                   *
           -
           -
     -0.40+                              *
           -                                                *
           -
           -
     -0.80+            *
           -
        --+---------+---------+---------+---------+---------+----ESTIMATE
         0.80      1.60      2.40      3.20      4.00      4.80

MTB > CORRELATION OF 'DISTANCE' AND 'DELTIME'

Correlation of DISTANCE and DELTIME = 0.949 ←⑩

MTB > COVARIANCE OF 'DISTANCE' AND 'DELTIME'

              DISTANCE    DELTIME
DISTANCE    144206.62
DELTIME  ⑪→517.00       2.06
```

Fig. 14-6 (*continued*)

④ The estimated delivery time for a shipment of 1,000 miles, which corresponds to the solution in Problem 14.4.

⑤ The standard error of estimate, which corresponds to the solution in Problem 14.6.

⑥ The t test for the null hypothesis, $H_0: \beta_1 = 0$, which corresponds to the solution in Problem 14.7.

⑦ The 95 percent confidence interval for the mean delivery time for all shipments of 1,000 miles, which corresponds to the solution in Problem 14.10.

⑧ The 95 percent prediction interval for the delivery time for one particular shipment of 1,000 miles, which corresponds to the solution in Problem 14.11.

⑨ The coefficient of determination between distance and delivery time, which corresponds to the solution in Problem 14.12.

⑩ The coefficient of correlation between distance and delivery time, which corresponds to the solution in Problem 14.13.

⑪ Covariance of distance and delivery time, which corresponds to the solution in Problem 14.14(*a*).

Supplementary Problems

LINEAR REGRESSION ANALYSIS

14.18 Table 14.5 presents sample data relating the number of study hours spent by students outside of class during a three-week period for a course in business statistics and their scores in an examination given at the end of that period. Prepare a scatter plot for these data and observe if the assumptions of linearity and equality of the conditional variances seem to be satisfied.

Table 14.5   Hours Spent on a Statistics Course and Examination Grades for a Sample of $n = 8$ Students

Sampled student	1	2	3	4	5	6	7	8
Hours of study (X)	20	16	34	23	27	32	18	22
Examination grade (Y)	64	61	84	70	88	92	72	77

Ans. Based on the scatter plot, the assumptions of linearity and equality of conditional variances seem to be reasonably satisfied.

14.19 Determine the least-squares regression line for the data in Table 14.5 and enter the line on the plot constructed in Problem 14.18.

Ans. $\hat{Y} = 40 + 1.5X$

14.20 Determine the residuals and construct a residual plot with respect to the fitted values for the data in Table 14.5, using the regression equation developed in Problem 14.19. Compare the residual plot to the scatter plot prepared in Problem 14.18.

14.21 Calculate the standard error of estimate for the data in Table 14.5 by using the residuals determined in Problem 14.20.

Ans. $s_{Y.X} = 6.16$

14.22 For the sample information presented in Problem 14.18, (*a*) test the null hypothesis that the slope of the regression line is zero, using the 1 percent level of significance, and interpret the result of your test. (*b*) Repeat the test for the null hypothesis that the true regression coefficient is equal to *or less than* zero, using the 1 percent level of significance.

Ans. (*a*) Reject H_0 and conclude that there is a significant relationship. (*b*) Reject H_0 and conclude that there is a significant positive relationship.

14.23 Use the regression equation determined in Problem 14.19 to estimate the examination grade of a student who devoted 30 hr of study to the course material.

Ans. $Y = 85$

14.24 Refer to Problems 14.21 and 14.23. Construct the 90 percent confidence interval for estimating the mean exam grade for students who devote 30 hr to course preparation.

Ans. 79.02 to 90.98

14.25 Referring to the preceding problems, construct the 90 percent prediction interval for the examination grade of an individual student who devoted 30 hr to course preparation.

Ans. 71.59 to 98.41

14.26 Table 14.6 presents data relating the number of weeks of experience in a job involving the wiring of miniature electronic components and the number of components which were rejected during the past week for 12 randomly selected workers. Plot these sample data on a scatter plot.

Table 14.6   Weeks of Experience and Number of Components Rejected During a Sampled Week for 12 Assembly Workers

Sampled worker	1	2	3	4	5	6	7	8	9	10	11	12
Weeks of experience (X)	7	9	6	14	8	12	10	4	2	11	1	8
Number of rejects (Y)	26	20	28	16	23	18	24	26	38	22	32	25

14.27 From Table 14.6 determine the regression equation for predicting the number of components rejected given the number of weeks experience, and enter the regression line on the scatter plot. Comment on the nature of the relationship as indicated by the regression equation.

Ans. $\hat{Y} = 35.57 - 1.40X$

14.28 Test the null hypothesis that there is no relationship between the variables in Table 14.6, and that the slope of the population regression line is zero, using the 5 percent level of significance for the test.

Ans. Reject H_0 and conclude that there is a relationship.

14.29 For the sample information in Problems 14.26 and 14.27, construct the 95 percent confidence interval for estimating the value of the population regression coefficient β_1, and interpret the value of this coefficient.

Ans. -1.85 to -0.95

14.30 Using the regression equation developed in Problem 14.27, estimate the number of components rejected for an employee with 3 weeks of experience in the job.

Ans. $\hat{Y} = 31.37$

14.31 Continuing with Problem 14.30, construct the 95 percent confidence interval for estimating the mean number of rejects for employees with 3 weeks experience in the operation.

Ans: 28.74 to 34.00

14.32 Refer to Problems 14.30 and 14.31. Construct the 95 percent prediction interval for the number of components rejected for an employee with 3 weeks experience in the job.

Ans. 25.09 to 37.65

CORRELATION ANALYSIS

14.33 Compute the coefficient of determination and the coefficient of correlation for the data in Table 14.5 and analyzed in Problems 14.18 to 14.25, taking advantage of the fact that the values of b_0 and b_1 for the regression equation were calculated in Problem 14.19. Interpret the computed coefficients.

Ans. $r^2 = 0.7449 \cong 0.74$, $r = +0.863 \cong +0.86$

14.34 For the sample correlation value determined in Problem 14.33, test the null hypothesis that (*a*) $\rho = 0$ and (*b*) $\rho \leq 0$, using the 1 percent level of significance with respect to each test. Interpret your results.

Ans. (*a*) Reject H_0 and conclude that there is a significant relationship. (*b*) Reject H_0 and conclude that there is a significant positive relationship.

14.35 For the data in Table 14.5, (*a*) calculate the sample covariance and (*b*) convert the covariance value into the coefficient of correlation. Compare your answer in (*b*) to the correlation coefficient obtained in Problem 14.33.

Ans. (*a*) $\text{cov}(X, Y) = +62.86$, (*b*) $r = +0.8622 \cong +0.86$

14.36 For the sample data reported in Table 14.6 and analyzed in Problems 14.26 to 14.32, determine the value of the correlation coefficient by the formula which is not based on the use of the estimated regression line values of b_0 and b_1. Interpret the meaning of this value by computing the coefficient of determination.

Ans. $r = -0.908 \cong -0.91$, $r^2 = 0.8245 \cong 0.82$

COMPUTER OUTPUT

14.37 Use available computer software to carry out the regression and correlation analysis for the data in Table 14.6. As part of the output, include a scatter plot, determination of the regression equation, the 95 percent confidence interval for the mean number of rejects for employees with 3 weeks experience, the 95 percent prediction interval for an individual employee with 3 weeks experience, and the sample correlation coefficient. Compare the computer output to the results of the manual calculations in the preceding exercises.

Ans. Except for minor differences due to rounding, the results are the same as for the manual calculations.

Chapter 15

Multiple Regression and Correlation

15.1 OBJECTIVES AND ASSUMPTIONS OF MULTIPLE LINEAR REGRESSION ANALYSIS

Multiple regression analysis is an extension of simple regression analysis, as described in Chapter 14, to applications involving the use of two or more independent variables to estimate the value of the dependent variable. In the case of two independent variables, denoted by X_1 and X_2, the linear algebraic model is

$$Y_i = \beta_0 + \beta_1 X_{i,1} + \beta_2 X_{i,2} + \varepsilon_i \qquad (15.1)$$

The definitions of the above terms are equivalent to the definitions in Section 14.3 for simple regression analysis, except that more than one independent variable is involved in the present case. Based on sample data, the linear regression equation for the case of two independent variables is

$$\hat{Y} = b_0 + b_1 X_1 + b_2 X_2 \qquad (15.2)$$

(*Note:* In some textbooks and computer software the dependent variable is designated X_1, with the several independent variables then identified sequentially beginning with X_2.)

The multiple regression equation identifies the best fitting line based on the method of least squares, as described in Section 14.3. In the case of multiple regression analysis, the best fitting line is a line through n-dimensional space (3-dimensional in the case of two independent variables). The calculations required for determining the values of the parameter estimates in a multiple regression equation and the associated standard error values are quite complex and generally involve matrix algebra. However, computer software is widely available for carrying out such calculations, and the solved problems at the end of this chapter are referenced to the use of such software. Specialized textbooks in regression and correlation analysis include complete descriptions of the mathematical analyses involved.

The assumptions of multiple linear regression analysis are similar to those of the simple case involving only one independent variable. For point estimation, the principal assumptions are that (1) the dependent variable is a random variable whereas the independent variables need not be random variables, (2) the relationship between the several independent variables and the one dependent variable is linear, and (3) the variances of the conditional distributions of the dependent variable, given various combinations of values of the independent variables, are all equal (homoscedasticity). For interval estimation, an additional assumption is that (4) the conditional distributions for the dependent variable follow the normal probability distribution.

15.2 ADDITIONAL CONCEPTS IN MULTIPLE REGRESSION ANALYSIS

Constant (in the regression equation): Although the b_0 and the several b_i values are all estimates of parameters in the regression equation, in most computer output the term "constant" refers to the value of the b_0 intercept. In multiple regression analysis, this is the regression-equation value of the dependent variable Y given that all of the independent variables are equal to zero.

Partial regression coefficient (or *net regression coefficient*): Each of the b_i regression coefficients is in fact a partial regression coefficient. A partial regression coefficient is the conditional coefficient given that one or more other independent variables (and their coefficients) are also included in the regression equation. Conceptually, a partial regression coefficient represents the slope of the regression line between the independent variable of interest and the dependent variable *given* that the other independent variables are included in the model and are thereby "held constant." The symbol $b_{Y1.2}$ (or $b_{12.3}$ when the dependent variable is designated by X_1) is the partial regression coefficient for the first independent variable given that a second independent variable is also included in the regression equation. For simplicity, when the entire regression equation is presented, this coefficient usually is designated by b_1.

Use of the F test: As described in Section 15.5, the analysis of variance is used in regression analysis to test for the significance of the overall model, before considering the significance of the individual independent variables.

Use of t tests: As illustrated in the Solved Problems at the end of the chapter, t tests are used to determine if the partial regression coefficient for each independent variable represents a significant contribution to the overall model.

Confidence interval for the conditional mean: The determination of such a confidence interval is illustrated in Problem 15.9. Where the standard error of the conditional mean in the case of two independent variables is designated by $s_{\bar{Y}.12}$, the formula for the confidence interval is

$$\hat{\mu}_Y \pm t s_{\bar{Y}.12} \qquad\qquad (15.3)$$

Prediction intervals: The prediction interval for estimating the value of an individual observation of the dependent variable, given the values of the several independent variables, is similar to the prediction interval in simple regression analysis as described in Section 14.8. Determination of such an interval is illustrated in Problem 15.10. The general formula for the prediction interval is

$$\hat{Y} \pm t s_{Y(\text{next})} \qquad\qquad (15.4)$$

Stepwise regression analysis: In forward stepwise regression analysis, one independent variable is added to the model in each step of selecting such variables. In backward stepwise regression analysis, we begin with all the variables under consideration being included in the model, and then (possibly) we remove one variable in each step. These are two of several available approaches to choosing the "best" set of independent variables for the model, as discussed in specialized books in linear algebraic models and multiple regression analysis. Computer software is available to do either forward or backward stepwise regression analysis. The end-of-chapter Solved Problems use a backward stepwise procedure.

15.3 THE USE OF INDICATOR (DUMMY) VARIABLES

Although the linear regression model is based on the independent variables being on a quantitative measurement scale, it is possible to include a qualitative (categorical) variable in a multiple regression model. Examples of such variables are the sex of an employee in a study of salary levels and the location codes in a real estate appraisal model.

The indicator variable utilizes a binary 0,1 code. Where k designates the number of categories that exist for the qualitative variable, $k-1$ indicator variables are required to code the qualitative variable. Thus, the sex of an employee can be coded by one indicator variable, because $k=2$ in this case, and $2-1=1$. The code system then can be 0 = female and 1 = male (or vice versa). For a real estate appraisal model in which there are three types of locations, labeled A, B, and C, $k=3$ and therefore $3-1=2$ indicator variables are required. These could be coded as follows, assuming that two other independent variables, X_1 and X_2, are already in the model:

Location	X_3	X_4
A	0	0
B	1	0
C	0	1

(*Note:* In the binary code scheme as illustrated above, there can be no more than one "1" in each row, and there must be exactly one "1" in each column.)

The difficulty associated with qualitative variables that have more than two categories is that more than one indicator variable is used to represent the same qualitative variable. Specifically, indicator variables X_3 and X_4, above, taken *together*, represent the one variable of location. In stepwise regression, the question then arises as to whether, and how, such indicator variables appropriately can be selected only singly for inclusion in a model if they are a part of a set that is used to code one qualitative variable. This issue is treated in advanced textbooks in regression analysis. The end-of-chapter Solved Problems include consideration of a two-category qualitative variable.

15.4 RESIDUALS AND RESIDUAL PLOTS

As originally defined in Section 14.4, a residual in regression analysis is

$$e = Y - \hat{Y} \tag{15.5}$$

In Section 14.4 the *residual plot* was defined as a plot of the residuals e with respect to the independent variable X or with respect to the fitted value $\hat{Y}$. In simple regression analysis, either a scatter plot or a residual plot can be used to observe if the assumptions of linearity and equality of conditional variances appear to be satisfied. In multiple regression analysis, however, the only type of plot that addresses these issues for the overall model is the residual plot with respect to the fitted value $\hat{Y}$, because this is the only two-dimensional plot that can incorporate the use of *several* independent variables. If a problem with respect to linearity or equality of conditional variances is observed to exist on such a plot, then individual scatter plots or residual plots can be prepared for each independent variable in the model in order to trace the source of the problem. The computer output associated with the Solved Problems includes the residual plot for the overall model as well as residual plots for the individual independent variables.

15.5 ANALYSIS OF VARIANCE IN LINEAR REGRESSION ANALYSIS

As indicated in Section 15.2, an F test is used to test for the significance of the overall model. That is, it is used to test the null hypothesis that there is no relationship in the population between the (several) independent variables taken as a group and the one dependent variable. If there is only one independent variable in the regression model, then the F test is equivalent to a two-tailed t test directed at the slope b_1. Therefore use of the F test is not required in simple regression analysis. For clarity, however, we focus on the simple regression model to explain the rationale of using the analysis of variance. The explanation applies similarly to the case of multiple regression, except that the estimated $\hat{Y}$ values are based on several independent variables rather than only one.

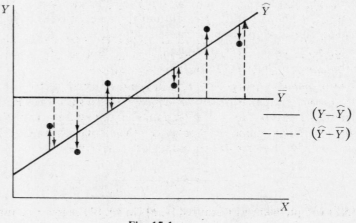

Fig. 15-1

Consider the scatter plot in Fig. 15-1. If there is no regression effect in the population, then the $\hat{Y}$ line differs from the $\bar{Y}$ line purely by chance. It follows that the variance estimate based on the differences ($\hat{Y} - \bar{Y}$), called *mean square regression* (*MSR*), would be different only by chance from the variance estimate based on the residuals ($Y - \hat{Y}$), called *mean square error* (*MSE*). On the other hand, if there is a regression effect, then the mean square regression is inflated in value as compared with the mean square error. Table 15.1 presents the standard format for the analysis of variance table that is used to test for the significance of an overall regression effect. The degrees of freedom k associated with MSR in the table is the number of independent variables in the multiple regression equation. As indicated in the table, the test statistic is

$$F = \frac{MSR}{MSE} \tag{15.6}$$

Problems 15.1 and 15.3 illustrate the use of the F test in multiple regression analysis.

Table 15.1   Analysis of Variance Table for Testing the Significance of the Regression Effect

Source of variation	Degrees of freedom (*df*)	Sum of squares (*SS*)	Mean square (*MS*)	*F* ratio
Regression (*R*)	k	SSR	$MSR = \dfrac{SSR}{k}$	$F = \dfrac{MSR}{MSE}$
Sampling error (*E*)	$n - k - 1$	SSE	$MSE = \dfrac{SSE}{n - k - 1}$	
Total (*T*)	$n - 1$	SST		

15.6   OBJECTIVES AND ASSUMPTIONS OF MULTIPLE CORRELATION ANALYSIS

Multiple correlation analysis is an extension of simple correlation analysis, as described in Chapter 14, to the situations involving two or more independent variables and their degree of association with the dependent variable. As is the case for multiple regression analysis described in Section 15.1, the dependent variable is designated by Y while the several independent variables are designated sequentially beginning with X_1. (*Note:* In some textbooks and computer software the dependent variable is designated by X_1, in which case the independent variables are designated sequentially beginning with X_2.)

The coefficient of multiple correlation, which is designated by $R_{Y.12}$ for the case of two independent variables, is indicative of the extent of relationship between two independent variables taken as a group and the dependent variable. Because it is possible for one of the independent variables to have a positive relationship with the dependent variable while the other independent variable has a negative relationship with the dependent variable, all R values are reported as absolute values, without an arithmetic sign. Problem 15.12 includes determination of a coefficient of multiple correlation and consideration of how this coefficient is tested for significance.

The *coefficient of multiple determination* is designated by $R_{Y.12}^2$ for the case of two independent variables. Similar to the interpretation of the simple coefficient of determination (see Section 14.10), this coefficient indicates the proportion of variance in the dependent variable which is statistically accounted for by knowledge of the two (or more) independent variables. The sample coefficient of multiple determination for the case of two independent variables is

$$R_{Y.12}^2 = 1 - \frac{s_{Y.12}^2}{s_Y^2} \tag{15.7}$$

Formula (*15.7*) is equivalent to formula (*14.20*) in simple regression analysis and is presented for conceptual purposes, rather than for computational application. Because it is the orientation in this chapter that computer software should be used for multiple regression and correlation analysis, computational procedures are not included. Problem 15.12 includes consideration of a coefficient of multiple determination included in computer output and the interpretation of this coefficient.

The assumptions of multiple correlation analysis are similar to those of the simple case involving only one independent variable. These are that (1) all variables involved in the analysis are random variables, (2) the relationships are all linear, (3) the conditional variances are all equal (homoscedasticity), and (4) the conditional distributions are all normal. These requirements are quite stringent and are seldom completely satisfied in real data situations. However, multiple correlation analysis is quite robust in the sense that some of these assumptions, and particularly the assumption about all the conditional distributions being normally distributed, can be violated without serious consequences in terms of the validity of the results.

15.7   ADDITIONAL CONCEPTS IN MULTIPLE CORRELATION ANALYSIS

In addition to the coefficient of multiple correlation and the coefficient of multiple determination described in the preceding section, the following concepts or procedures are unique to multiple correlation analysis.

Coefficient of partial correlation: Indicates the correlation between one of the independent variables in the multiple correlation analysis and the dependent variable, with the other independent variable(s) held constant statistically. The partial correlation with the first of two independent variables would be designated by $r_{Y1.2}$, while the partial correlation with the second of two independent variables would be designated by $r_{Y2.1}$. (If the dependent variable is designated by X_1, then these two coefficients would be designated by $r_{12.3}$ and $r_{13.2}$, respectively.) The partial correlation value is different from a simple correlation value because for the latter case other independent variables are not statistically controlled.

Coefficient of partial determination: This is the squared value of the coefficient of partial correlation described above. The coefficient indicates the proportion of variance statistically accounted for by one particular independent variable, with the other independent variable(s) held constant statistically.

15.8   PITFALLS AND LIMITATIONS ASSOCIATED WITH MULTIPLE REGRESSION AND CORRELATION ANALYSIS

Two principal areas of difficulty are those associated with colinearity and autocorrelation. These are described briefly below. Detailed discussions of these problems and what can be done about them are included in specialized textbooks in regression and correlation analysis.

Colinearity (or *multicolinearity*): When the independent variables in a multiple regression analysis are highly correlated with one another, the partial (or net) regression coefficients are unreliable in terms of meaning. Similarly, the practical meaning of the coefficients of partial correlation may be questionable. It is possible, for example, that the partial correlation for a given independent variable will be highly negative even though the simple correlation is highly positive. In general, therefore, care should be taken in interpreting partial regression coefficients and partial correlation coefficients when there are independent variables that have a high positive or negative correlation with one another.

Autocorrelation: Refers to the absence of independence in the sampling of the dependent variable Y. The situation is likely to exist when the Y values are time-series values, in which case the value of the dependent variable in one time period is almost invariably related to values in adjoining time periods. In such a case, the standard error associated with each partial regression coefficient b_i is understated, as is the value of the standard error of estimate. The result is that any prediction or confidence intervals are narrower (more precise) than they should be, and null hypotheses concerning the absence of relationship are rejected too frequently.

15.9 COMPUTER OUTPUT

The Solved Problems at the end of this chapter all make reference to the Minitab analysis in Fig. 15-2, applied to the data in Table 15.2. As is illustrated in the problems, a backward stepwise procedure is used (see Section 15.2). However, instead of using the automatic backward stepwise procedure that is available in Minitab, we use separate regression commands at each step in order to highlight the logic associated with such a procedure for selecting the independent variables for inclusion in the multiple regression model.

Table 15.2　Annual Salaries for a Sample of Supervisors

Sampled individual	Annual salary (Y)	Years of experience (X_1)	Years of postsecondary education (X_2)	Sex (X_3)
1	$34,900	5.5	4.0	F
2	40,500	9.0	4.0	M
3	38,900	4.0	5.0	F
4	39,000	8.0	4.0	M
5	37,500	9.5	5.0	M
6	35,500	3.0	4.0	F
7	36,000	7.0	3.0	F
8	32,700	1.5	4.5	F
9	45,000	8.5	5.0	M
10	40,000	7.5	6.0	F
11	36,000	9.5	2.0	M
12	33,600	6.0	2.0	F
13	35,000	2.5	4.0	M
14	32,500	1.5	4.5	M

```
MTB >  NAME C1 = 'SALARY',  C2 = 'EXPER',  C3 = 'EDUC',  C4 = 'SEX'
MTB >  READ 'SALARY',  'EXPER',  'EDUC',  'SEX'
DATA >  34900     5.5     4.0        0
DATA >  40500     9.0     4.0        1
DATA >  38900     4.0     5.0        0
DATA >  39000     8.0     4.0        1
DATA >  37500     9.5     5.0        1
DATA >  35500     3.0     4.0        0
DATA >  36000     7.0     3.0        0
DATA >  32700     1.5     4.5        0
DATA >  45000     8.5     5.0        1
DATA >  40000     7.5     6.0        0
DATA >  36000     9.5     2.0        1
DATA >  33600     6.0     2.0        0
DATA >  35000     2.5     4.0        1
DATA >  32500     1.5     4.5        1
DATA >  END
MTB >  REGRESS C1 ON 3 PREDICTORS C2-C4
```

(1) The regression equation is
SALARY = 25495 + 802 EXPER + 1596 EDUC + 383 SEX

Fig. 15-2　Minitab output (numbered labels are referenced in the solutions). (*Continued on pp. 268–270.*)

```
Predictor         Coef              Stdev       t-ratio
Constant         25495              2810           9.07
EXPER            801.6              228.5          3.51
EDUC             1595.7             560.6          2.85
SEX                383              1287           0.30
```

s = 2252   R-sq = 67.5%   R-sq(adj) = 57.8%

Analysis of Variance

```
SOURCE          DF             SS                 MS              F
Regression       3        105309888            35103296         6.92
Error           10         50702240             5070224
Total           13        156012128
```

```
SOURCE          DF          SEQ SS
EXPER            1        63281472
EDUC             1        41580688
SEX              1          447733
```

Unusual Observations

```
Obs.    EXPER      SALARY        Fit  Stdev.Fit    Residual    St.Resid
  5      9.50       37500       41472     1193        -3972      -2.08R
  9      8.50       45000       40670     1084         4330       2.19R
```

R denotes an obs. with a large st. resid.

MTB) REGRESS C1 ON 2 PREDICTORS C2-C3, C5, PUT FITS IN C6;
SUBC) RESIDUALS IN C7;
SUBC) PREDICT 5.5 4.0.

The regression equation is
SALARY = 25511 + 826 EXPER + 1604 EDUC

```
Predictor         Coef              Stdev       t-ratio
Constant         25511              2690           9.48
EXPER            825.7              204.6          4.04
EDUC             1603.7             536.3          2.99
```

s = 2156   R-sq = 67.2%   R-sq(adj) = 61.3%

Analysis of Variance

```
SOURCE          DF             SS                 MS              F
Regression       2        104862160            52431072        11.28
Error           11         51149968             4649997
Total           13        156012128
```

```
SOURCE          DF          SEQ SS
EXPER            1        63281472
EDUC             1        41580688
```

Unusual Observations

```
Obs.    EXPER      SALARY        Fit  Stdev.Fit    Residual    St.Resid
  5      9.50       37500       41374     1098        -3874      -2.09R
  9      8.50       45000       40548      961         4452       2.31R
```

R denotes an obs. with a large st. resid.

Fig. 15-2 (*continued*)

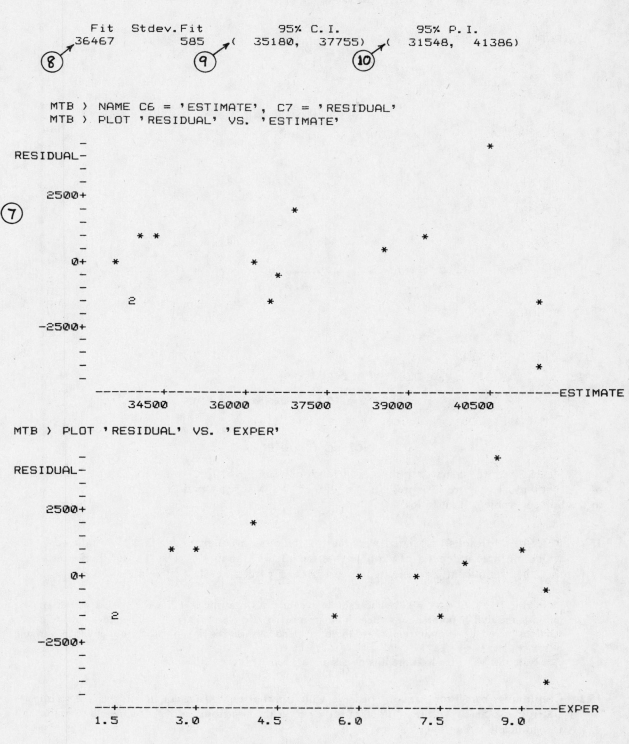

```
        Fit    Stdev.Fit        95% C.I.          95% P.I.
       36467          585    (  35180,   37755)  (  31548,   41386)
  (8)                  (9)                        (10)


      MTB > NAME C6 = 'ESTIMATE', C7 = 'RESIDUAL'
      MTB > PLOT 'RESIDUAL' VS. 'ESTIMATE'

           -                                                        *
RESIDUAL-
           -
   2500+
(7)        -                                    *
           -
           -               *   *                             *
     0+        *                        *               *
           -                               *
           -
           -        2                   *
  -2500+                                                           *
           -
           -
           -
           --------+---------+---------+---------+---------+---------+---------ESTIMATE
              34500     36000     37500     39000     40500

MTB > PLOT 'RESIDUAL' VS. 'EXPER'

           -                                                        *
RESIDUAL-
           -
   2500+
           -                           *
           -           *   *                             *
     0+                                      *        *
           -                                                  *
           -        2                 *           *
  -2500+
           -
           -                                               *
           -
           --+---------+---------+---------+---------+---------+----EXPER
            1.5       3.0       4.5       6.0       7.5       9.0
```

Fig. 15-2   (*continued*)

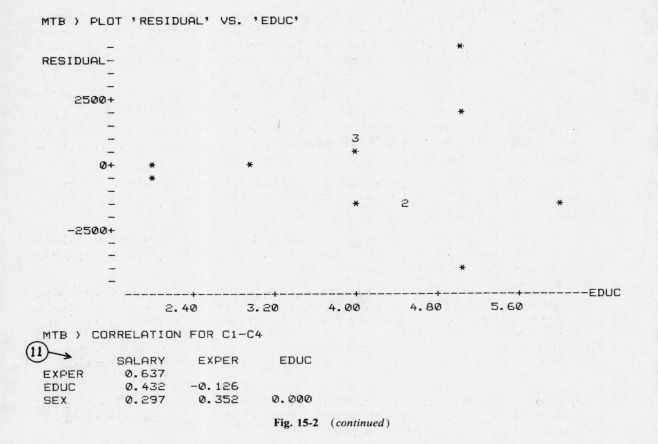

Fig. 15-2   (*continued*)

Solved Problems

Table 15.2 reports salary data for $n = 14$ randomly sampled supervisors from a large population of such individuals. Figure 15-2 presents the Minitab analysis that serves as the basis for the solutions to the Solved Problems that follow.

15.1   Refer to the results associated with the first regression command in Fig. 15-2, labeled ①, in which all three independent variables are included in the model. Test the null hypothesis that there is a significant regression effect, using the 5 percent level of significance.

Critical $F(3, 10, 0.05) = 3.71$. Because the obtained test statistic is $F = 6.92$ (labeled with ② in the output), the null hypothesis that there is no regression effect is rejected. We conclude that there *is* a relationship in the population among the three independent variables as a group and the dependent variable of salary.

Note that with this software it is necessary to calculate the F statistic manually.

15.2   Continuing from Problem 15.1, observe which partial regression coefficient has the smallest t statistic associated with it, and determine whether the contribution of that variable to the model is significant at the 5 percent level.

Note that the degrees of freedom for the t test are the same as for MSE in the analysis of variance and are equal to the sample size minus the number of parameters estimated in the multiple regression equation or, in this case, $df = 14 - 4 = 10$.

The variable for which the t statistic is smallest is the indicator variable for sex (X_3), with $t = 0.30$ (labeled ③ in Fig. 15-2). Because the critical t (10, 0.05) = ±2.228, the null hypothesis that $\beta_3 = 0$ cannot be rejected. Therefore we conclude that sex is *not* a significant contributor to the model and should be removed from the multiple regression equation. As discussed in Section 15.2, such a procedure of removing one variable at a time is called *backward stepwise regression analysis.*

15.3 Refer to the output of the second regression command in Fig. 15-2, labeled ④, which is based on using years of experience and years of education as independent variables, but not the indicator variable for sex. Test the null hypothesis that there is no regression effect, using the 5 percent level of significance.

Critical $F(2, 11, 0.05) = 3.98$. Because the calculated F statistic is $F = 11.28$ (labeled ⑤ in the output), the null hypothesis is rejected and we conclude that there *is* a significant regression effect.

15.4 Continuing from Problem 15.3, observe which partial regression coefficient has the smallest t statistic associated with it, and determine whether the contribution of this variable to the multiple regression model is significant at the 5 percent level.

The variable for which the t statistic is smallest is education (X_2), with $t = 2.99$ (labeled ⑥ in the output). Because critical $t(11, 0.05) = ±2.201$, the null hypothesis that $\beta_2 = 0$ is rejected at the 5 percent level of significance, and we conclude that this variable *does* contribute significantly to the multiple regression model.

The fact that no variable is removed in this step of the backward stepwise procedure means that the independent variables to be included in the multiple regression model have now been determined.

15.5 Refer to the residual plot labeled with a ⑦ in Fig. 15-2, which plots the residuals with respect to the fitted $\hat{Y}$ values, and observe whether the requirements of linearity and equality of conditional variances appear to be satisfied.

The assumption of linearity appears to be satisfied. With respect to the equality of the conditional variances, however, it appears that the conditional variances may be somewhat greater at large values of the estimated salary, beyond $40,000 on the horizontal scale of the residual plot.

15.6 Continuing from Problem 15.5, refer to the next two residual plots in Fig. 15-2. Comment on the purpose for studying these two plots and on the results of your observations.

Such individual plots can be used to attempt to identify the source of any problems observed on the residual plot for the overall model. The individual residual plots for both years of experience and years of education are linear, but both exhibit higher conditional variances at higher numbers of years, thus identifying the source of the possible difficulty identified in the solution to Problem 15.5 above. Although we proceed with further calculations in these solved problems, the possible nonequality of the conditional variances would have to be considered in the interpretation of confidence intervals and prediction intervals.

15.7 Referring to the multiple regression equation based on using the two independent variables in Fig. 15-2 (labeled with ④ in the output), estimate the annual salary for an individual with 5.5 years of experience and 4.0 years of postsecondary education.

$\hat{Y} = b_0 + b_1 X_1 + b_2 X_2$

$\quad = 25,511 + 826 X_1 + 1,604 X_2$

$\quad = 25,511 + 826(5.5) + 1,604(4.0)$

$\quad = \$36,470 \qquad$ (or from the computer output based on less rounding of values during calculations, and labeled ⑧ in Fig. 15-2, $36,467)

15.8 Interpret the meanings of the several b_i values included in the regression equation in Problem 15.7 above.

  The quantity $b_0 = 25,511$ is the intercept point with the Y axis when $X_1 = 0$ and $X_2 = 0$. Literally, it would be the estimated annual salary for someone with no experience and no postsecondary education. It has no such practical meaning, however, because these values for the independent variables were not included in the range of sampled values.

  The quantity $b_1 = 826$ indicates that, on the average, an increase of 1 year in years of experience is associated with an increase of \$826 in annual salary, *given that the variable of years of education also is included in the regression model* and is thereby statistically controlled, or "held constant."

  The quantity $b_2 = 1,604$ indicates that, on the average, an increase of 1 year in years of education is associated with an increase of \$1,604 in annual salary, given that the variable of years of experience also is included in the regression model.

15.9 Refer to the appropriate portion of the output and determine the 95 percent confidence interval for the mean annual salary of all individuals in the population with 5.5 years of experience and 4.0 years of postsecondary education.

$$95\% \text{ C.I.} = \$35,180 \text{ to } \$37,755 \qquad (\text{labeled } ⑨ \text{ in the output})$$

15.10 Refer to the appropriate portion of the output in Fig. 15-2 and determine the 95 percent prediction interval for a particular individual who has 5.5 years of education and 4.0 years of postsecondary education. Compare this interval with the one obtained in Problem 15.9 and interpret the meaning of this prediction interval.

$$95\% \text{ P.I.} = \$31,548 \text{ to } \$41,386 \qquad (\text{labeled } ⑩ \text{ in the output})$$

  As expected, this interval is wider than the confidence interval in Problem 15.9. A prediction interval is a probability interval for the value of an *individual* value of Y, rather than for a mean. The prediction interval indicates that the probability is 0.95 that an individual with 5.5 years of experience and 4.0 years of postsecondary education will have an annual salary between \$31,548 and \$41,386.

15.11 Refer to the correlation table in the last portion of the output, labeled ⑪. (*a*) Which variable has the highest correlation with the annual salary, and what is that correlation? (*b*) Which two variables other than salary have the highest correlation with each other, and what is that correlation?

  (*Note:* The significance of these simple coefficients of correlation would be determined by using the t test described in Section 14.13. In fact, the only simple correlation value in the table that is significant at the 5 percent level is the coefficient of 0.637.)

  (*a*) The highest correlation is between experience and salary, with $r = 0.637$.

  (*b*) The highest correlation is between experience and sex, with $r = 0.352$. In the computer input, notice that the code used in the indicator variable for sex is female = 0 and male = 1. The fact that the correlation between experience and sex appears to be positive indicates that years of experience for sex code "1" (male) tends to be greater than the years of experience for sex code "0" (female).

15.12 From the output in Fig. 15-2, determine the coefficient of multiple determination and the coefficient of multiple correlation for the reduced model with two independent variables. Interpret the coefficient of determination and indicate how the coefficient of correlation would be tested for significance.

  The coefficient of multiple determination is $R^2 = 0.675$ (not adjusted for degrees of freedom), labeled by ⑫ in the output. This indicates that approximately 67.5 percent of the variance in the supervisors' salaries is statistically accounted for by their years of experience and years of postsecondary education.

The coefficient of multiple correlation is $R = \sqrt{0.675} = 0.82$, reported as an absolute value. The null hypothesis that this population coefficient of multiple correlation is equal to zero is tested by the F test applied to the associated multiple regression model in Problem 15.3, in which we rejected the null hypothesis that there is no relationship between the independent variables of experience and education, taken as a group, with the dependent variable of salary.

Supplementary Problems

Table 15.3 reports single-family house prices, with a random sample of 10 houses taken from each of three housing subdivisions. As indicated, in addition to the subdivision and the price, the square footage of each house and each lot have been collected. With price being the dependent variable, carry out a backward stepwise regression analysis using computer software. Also obtain a residual plot for the final regression model and a correlation matrix of all the simple correlation coefficients as the basis for the solutions to the supplementary problems that follow. Use the binary coding scheme in Section 15.3 for the two indicator variables that are required for coding the housing subdivision.

Table 15.3   Single-Family House Prices in Three Subdivisions

Sampled house	Price	Living area sq. ft	Lot size, sq. ft	Subdivision
1	102,200	1,500	12,000	A
2	103,950	1,200	10,000	A
3	87,900	1,200	10,000	A
4	110,000	1,600	15,000	A
5	97,000	1,400	12,000	A
6	95,700	1,200	10,000	A
7	113,600	1,600	15,000	A
8	109,600	1,500	12,000	A
9	110,800	1,500	12,000	A
10	90,600	1,300	12,000	A
11	109,000	1,600	13,000	B
12	133,000	1,900	15,000	B
13	134,000	1,800	15,000	B
14	120,300	2,000	17,000	B
15	137,000	2,000	17,000	B
16	122,400	1,700	15,000	B
17	121,700	1,800	15,000	B
18	126,000	1,900	16,000	B
19	128,000	2,000	16,000	B
20	117,500	1,600	13,000	B
21	158,700	2,400	18,000	C
22	186,800	2,600	18,000	C
23	172,400	2,300	16,000	C
24	151,200	2,200	16,000	C
25	179,100	2,800	20,000	C
26	182,300	2,700	20,000	C
27	195,850	3,000	22,000	C
28	168,000	2,400	18,000	C
29	199,400	2,500	20,000	C
30	163,000	2,400	18,000	C

15.13 Obtain the multiple regression equation for estimating house price based on all three variables of house size, lot size, and location. Test the multiple regression model for significance at the 5 percent level.

> *Ans.* $\hat{Y} = 40,462 + 36.0X_1 + 0.94X_2 + 4,267X_3 + 26,648X_4$. The regression model is significant at the 5 percent level.

15.14 Observe which variable has the partial regression coefficient with the lowest reported t statistic, and determine whether the contribution of that variable is significant at the 5 percent level. (*Note:* For our solution we take the view that given that the two indicator variables represent the *one* qualitative variable of location, either both should be removed or neither should be removed from the model. Therefore we will consider only the *higher* of the two t ratios associated with the indicator variables.)

> *Ans.* Lot size has the smallest test-statistic value ($t = 0.44$) and should be removed from the model.

15.15 Continuing from Problem 15.14, obtain the multiple regression equation for the reduced model and test the model for significance at the 5 percent level.

> *Ans.* $\hat{Y} = 41,153 + 43.6X_1 + 4,025X_3 + 24,319X_4$. The regression model is significant at the 5 percent level.

15.16 As in Problem 15.14, observe which variable has the partial regression coefficient with the lowest reported t statistic and determine whether the contribution of that variable to the multiple regression model is significant at the 5 percent level.

> *Ans.* The lowest test statistic is for the first indicator variable ($t = 0.79$), which we ignore based on the approach to the indicator variables described in Problem 15.14. The next-lowest test statistic is for the second indicator variable ($t = 2.44$), and it is significant at the 5 percent level. Therefore, no variable should be removed from the model.

15.17 Observe the residual plot for the multiple regression model in Problem 15.15. Do the assumptions of linearity and equality of conditional variances appear to be satisfied?

> *Ans.* The assumptions appear to be reasonably satisfied. There is one "outlier" in the upper-right portion of the plot for which one should check the accuracy of the original data for that sampled house (house no. 29).

15.18 Use the regression model in Problem 15.15 and estimate the price of a house with (*a*) 1,200 sq. ft. and in subdivision A, (*b*) 1,800 sq. ft. and in subdivision B, (*c*) 2,400 sq. ft. and in subdivision C.

> *Ans.* (*a*) \$93,423, (*b*) \$123,583, (*c*) \$170,012

15.19 Use the regression model in Problem 15.15 to determine the 95 percent confidence interval for the conditional mean price of all houses with (*a*) 1,200 sq. ft. and in subdivision A, (*b*) 1,800 sq. ft. and in subdivision B, (*c*) 2,400 sq. ft. and in subdivision C.

> *Ans.* (*a*) \$87,118 to \$99,728, (*b*) \$118,229 to \$128,937, (*c*) \$164,250 to \$175,774

15.20 Continuing with Problem 15.19, determine the 95 percent prediction interval for the price of an individual house in each of the three categories identified above.

> *Ans.* (*a*) \$75,428 to \$111,419, (*b*) \$105,898 to \$141,268, (*c*) \$152,199 to \$187,825

15.21 Refer to the matrix of simple correlation coefficients for all the variables included in the study. (*a*) Which variable has the highest correlation with house price? (*b*) Which two of the "candidate" variables to be independent variables have the highest correlation with one another?

> *Ans.* (*a*) Square footage of the house, with $r = 0.962$, (*b*) square footage of house and lot, with $r = 0.961$.

15.22 Determine (*a*) the coefficient of multiple determination (uncorrected for degrees of freedom) and (*b*) the coefficient of multiple correlation for the final multiple regression model.

> *Ans.* (*a*) $R^2 = 0.945$, (*b*) $R = \sqrt{0.945} = 0.972$

Chapter 16

Time Series Analysis and Business Forecasting

16.1 THE CLASSICAL TIME SERIES MODEL

A *time series* is a set of observed values, such as production or sales data, for a sequentially ordered series of time periods. Examples of such data are sales of a particular product for a series of months and the number of workers employed in a particular industry for a series of years. A time series is portrayed graphically by a line graph (see Section 2.8), with the time periods represented on the horizontal axis and time series values represented on the vertical axis.

EXAMPLE 1. Figure 16-1 is a line graph which portrays the annual dollar sales for a computer software company (fictional) that was incorporated in 1980. As can be observed, a *peak* in annual sales was achieved in 1985, followed by two years of declining sales that culminated in the *trough* ("trŏf") in 1987, which was then again followed by increasing levels of sales during the final three years of the reported time series values.

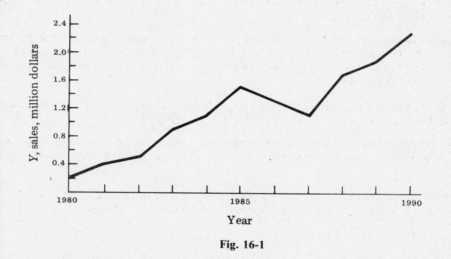

Fig. 16-1

Time series analysis is the procedure by which the time-related factors that influence the values observed in the time series are identified and segregated. Once identified, they can be used to aid in the interpretation of historical time series values and to forecast future time series values. The classical approach to time series analysis identifies four such influences, or *components*:

(1) *Trend (T):* The general long-term movement in the time series values (Y) over an extended period of years.

(2) *Cyclical fluctuations (C):* Recurring up and down movements with respect to trend that have a duration of several years.

(3) *Seasonal variations (S):* Up and down movements with respect to trend that are completed within a year and recur annually. Such variations typically are identified on the basis of monthly or quarterly data.

(4) *Irregular variations (I):* The erratic variations from trend that cannot be ascribed to the cyclical or seasonal influences.

The model underlying classical time series analysis is based on the assumption that for any designated period in the time series the value of the variable is determined by the influences of the four components

275

defined above, and, furthermore, that the components have a multiplicative relationship. Thus, where Y represents the observed time series value,

$$Y = T \times C \times S \times I \qquad (16.1)$$

The model represented by (16.1) is used as the basis for separating the influences of the various components influencing time series values, as described in the following sections of this chapter.

16.2 TREND ANALYSIS

Because trend analysis is concerned with the long-term direction of movement in the time series, such analysis is performed using annual data. Typically, at least 15 or 20 years of data should be used, so that cyclical movements involving several years duration are not taken to be indicative of the overall trend of the time series values.

The method of least squares (see Section 14.3) is the most frequent basis used for identifying the trend component of the time series by determining the equation for the best fitting trend line. Note that statistically speaking, a trend line is not a regression line because the dependent variable Y is not a random variable, but, rather, an accumulated historical value. Further, there can be only one historical value for any given time period (not a distribution of values) and the values associated with adjoining time periods are dependent, rather than independent. Nevertheless, the least squares method is a convenient basis for determining the trend component of a time series. When the long-term increase or decrease appears to follow a linear trend, the equation for the trend line values, with X representing the year, is

$$Y_T = b_0 + b_1 X \qquad (16.2)$$

As explained in Section 14.3, the b_0 in (16.2) represents the point of intersection of the trend line with the Y axis, whereas the b_1 represents the slope of the trend line. Where X is the year and Y is the observed time-series value, the formulas for determining the values of b_0 and b_1 for the linear trend equation are

$$b_1 = \frac{\Sigma XY - n\bar{X}\bar{Y}}{\Sigma X^2 - n\bar{X}^2} \qquad (16.3)$$

$$b_0 = \bar{Y} - b_1\bar{X} \qquad (16.4)$$

See Problem 16.1 for the determination of a linear trend equation.

In the case of nonlinear trend, two types of trend curves often used as the basis of trend analysis are the exponential trend curve and the parabolic trend curve. A typical *exponential trend curve* is one that reflects a constant rate of growth during a period of years, as might apply to the sales of personal computers during the 1980s. See Fig. 16-2(a). An exponential curve is so named because the independent variable X is the exponent of b_1 in the general equation:

$$Y_T = b_0 b_1^X \qquad (16.5)$$

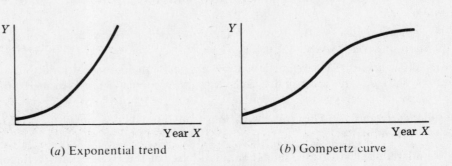

(a) Exponential trend   (b) Gompertz curve

Fig. 16-2

Taking the logarithm of both sides of (16.5) results in a linear logarithmic trend equation,

$$\log Y_T = \log b_0 + X \log b_1 \tag{16.6}$$

The advantage of the transformation into logarithms is that the linear equation for trend analysis can be applied to the logs of the values when the time series follows an exponential curve. The forecasted log values for Y_T can then be reconverted to the original measurement units by taking the antilog of the values. We do not demonstrate such analysis in this book.

Many time series for the sales of a particular product can be observed to include three stages: an *introductory stage* of slow growth in sales, a *middle stage* of rapid sales increases, and a *final stage* of slow growth as market saturation is reached. For some products, such as structural steel, the complete set of three stages may encompass many years. For other products, such as citizen band radios, the stage of saturation may be reached relatively quickly. The particular trend curve that includes the three stages just described is the *Gompertz curve*, as portrayed in Fig. 16-2(b). The equation for the Gompertz trend curve is

$$Y_T = b_0 b_1^{(b_2)^X} \tag{16.7}$$

The values of b_0, b_1, and b_2 are determined by first taking the logarithm of both sides of the equation, as follows:

$$\log Y_T = \log b_0 + (\log b_1) b_2^X \tag{16.8}$$

Finally, the values to form the trend curve are calculated by taking the antilog of the values calculated by formula (16.8). The details of such calculations are included in specialized books in time series analysis.

16.3 ANALYSIS OF CYCLICAL VARIATIONS

Annual time-series values represent the effects of only the trend and cyclical components, because the seasonal and irregular components are defined as short-run influences. Therefore, for annual data the cyclical component can be identified by dividing the observed values by the associated trend value, as follows:

$$\frac{Y}{Y_T} = \frac{T \times C}{T} = C \tag{16.9}$$

The ratio in (16.9) is multiplied by 100 so that the mean cyclical relative will be 100.0. A cyclical relative of 100 would indicate the absence of any cyclical influence on the annual time-series value. See Problem 16.2.

In order to aid in the interpretation of cyclical relatives, a *cycle chart* which portrays the cyclical relatives according to year is often prepared. The peaks and troughs associated with the cyclical component of the time series can be made more apparent by the construction of such a chart. See Problem 16.3.

16.4 MEASUREMENT OF SEASONAL VARIATIONS

The influence of the seasonal component on time series values is identified by determining the seasonal index number associated with each month (or quarter) of the year. The arithmetic mean of all 12 monthly index numbers (or four quarterly index numbers) is 100. The identification of positive and negative seasonal influences is important for production and inventory planning.

EXAMPLE 2. An index number of 110 associated with a given month indicates that the time-series values for that month have averaged 10 percent higher than for other months because of some positive seasonal factor. For instance, the unit sales of men's shavers might be 10 percent higher in June as compared with other months because of Father's Day.

The procedure most frequently used to determine seasonal index numbers is the *ratio-to-moving-average method*. By this method, the ratio of each monthly value to the moving average centered at that month is first determined. Because a moving average based on monthly (or quarterly) data for an entire year would "average out" the seasonal and irregular fluctuations, but not the longer-term trend and cyclical influences, the ratio of a monthly (or quarterly) value to a moving average can be represented symbolically by

$$\frac{Y}{\text{Moving average}} = \frac{T \times C \times S \times I}{T \times C} = S \times I \qquad (16.10)$$

The second step in the ratio-to-moving-average method is to average out the irregular component. This is typically done by listing the several ratios applicable to the same month (or quarter) for the several years, eliminating the highest and lowest values, and computing the mean of the remaining ratios. The resulting mean is called a *modified mean*, because of the elimination of the two extreme values.

The final step in the ratio-to-moving-average method is to adjust the modified mean ratios by a correction factor so that the sum of the 12 monthly ratios is 1,200 (or 400 for four quarterly ratios). See Problem 16.4.

16.5 APPLYING SEASONAL ADJUSTMENTS

One frequent application of seasonal indexes is that of adjusting observed time-series data by removing the influence of the seasonal component from the data. Such adjusted data are called *seasonally adjusted data*, or *deseasonalized data*. Seasonal adjustments are particularly relevant if we wish to compare data for different months to determine if an increase (or decrease) relative to seasonal expectations has taken place.

EXAMPLE 3. An increase in lawn fertilizer sales of 10 percent from April to May of a given year represents a relative *decrease* if the seasonal index number for May is 20 percent above the index number for April. In other words, if an increase occurs but is not as large as expected based on historical data, then relative to these expectations a decline in demand has occurred.

The observed monthly (or quarterly) time-series values are adjusted for seasonal influences by dividing each value by the monthly (or quarterly) index for that month. The result is then multiplied by 100 to maintain the decimal position of the original data. The process of adjusting data for the influence of seasonal variations can be represented by

$$\frac{Y}{S} = \frac{T \times C \times S \times I}{S} = T \times C \times I \qquad (16.11)$$

Although the resulting values after the application of (16.11) are in the same measurement units as the original data, they do not represent actual occurrences. Rather, they are relative values and are meaningful for comparative purposes only. See Problem 16.5.

16.6 FORECASTING BASED ON TREND AND SEASONAL FACTORS

A beginning point for long-term forecasting of annual values is provided by use of the trend line (16.2) equation. However, a particularly important consideration in long-term forecasting is the cyclical component of the time series. There is no standard method by which the cyclical component can be forecast based on historical time-series values alone, but certain economic indicators (see Section 16.7) are useful for anticipating cyclical turning points.

For short-term forecasting, the beginning point is the projected trend value which is then adjusted for the seasonal component. Because the equation for the trend line is normally based on the analysis of annual values, the first step required is to "step down" this equation so that it is expressed in terms of months (or quarters). A trend equation based on annual data is modified to obtain projected monthly values as follows:

$$Y_T = \frac{b_0}{12} + \left(\frac{b_1}{12}\right)\left(\frac{X}{12}\right) = \frac{b_0}{12} + \frac{b_1}{144} X \qquad (16.12)$$

A trend equation based on annual dates is modified to obtain projected quarterly values as follows:

$$Y_T = \frac{b_0}{4} + \left(\frac{b_1}{4}\right)\left(\frac{X}{4}\right) = \frac{b_0}{4} + \frac{b_1}{16} X \qquad (16.13)$$

The basis for the above modifications is not obvious if one overlooks the fact that trend values are not associated with points in time, but rather, with periods of time. Because of this consideration, all *three* elements in the equation for annual trend (b_0, b_1, and X) have to be stepped down.

By the transformation for monthly data in (16.12), the base point in the year which was formerly coded $X = 0$ would be at the middle of the year, or July 1. Because it is necessary that the base point be at the middle of the first month of the base year, or January 15, the intercept $b_0/12$ in the modified equation is then reduced by 5.5 times the modified slope. A similar adjustment is made for quarterly data. Thus, a trend equation which is modified to obtain projected monthly values *and* with $X = 0$ placed at January 15 of the base year is

$$Y_T = \frac{b_0}{12} - (5.5)\left(\frac{b_1}{144}\right) + \frac{b_1}{144} X \qquad (16.14)$$

Similarly, a trend equation which is modified to obtain projected quarterly values *and* with $X = 0$ placed at the middle of the first quarter of the base year is

$$Y_T = \frac{b_0}{4} - (1.5)\left(\frac{b_1}{16}\right) + \frac{b_1}{16} X \qquad (16.15)$$

Problem 16.6 illustrates the process of stepping down a trend equation. After monthly (or quarterly) trend values have been determined, each value can be multiplied by the appropriate seasonal index (and divided by 100 to preserve the decimal location in the values) to establish a beginning point for short-term forecasting. See Problem 16.7.

16.7   CYCLICAL FORECASTING AND BUSINESS INDICATORS

As indicated in Section 16.6, forecasting based on the trend and seasonal components of a time series is considered only a beginning point in economic forecasting. One reason is the necessity to consider the likely effect of the cyclical component during the forecast period, while the second is the importance of identifying the specific causative factors which have influenced the time series variables.

For short-term forecasting, the effect of the cyclical component is often assumed to be the same as included in recent time-series values. However, for longer periods, or even for short periods during economic instability, the identification of the *cyclical turning points* for the national economy is important. Of course, the cyclical variations associated with a particular product may or may not coincide with the general business cycle.

EXAMPLE 4.   Historically, factory sales of passenger cars have coincided closely with the general business cycle for the national economy. On the other hand, sales of automobile repair parts tend to be counter-cyclical with respect to the general business cycle.

The National Bureau of Economic Research has identified a number of published time series that historically have been indicators of cyclical revivals and recessions with respect to the general business cycle. One group, called *leading indicators*, have usually reached cyclical turning points prior to the corresponding change in general economic activity. The leading indicators include such measures as layoff rate in manufacturing, value of new orders in durable goods industries, and a common stock price index. A second group, called *coinciding indicators*, are time series which have generally had turning points coinciding with the general business cycle. Coinciding indicators include such measures as the employment rate and the industrial production index. The third group, called *lagging indicators*, are those time series for which the peaks and troughs usually lag behind those of the general business cycle. Lagging indicators include such measures as manufacturing and trade inventories and the average prime rate charged by banks.

In addition to considering the effect of cyclical fluctuations and forecasting such fluctuations, specific causative variables that have influenced the time series values historically should also be studied. Regression and correlation analysis (see Chapters 14 and 15) are particularly applicable for such studies as the relationship between pricing strategy and sales volume. Beyond the historical analyses, the possible implications of new products and of changes in the marketing environment are also areas of required attention.

16.8   EXPONENTIAL SMOOTHING AS A FORECASTING METHOD

Exponential smoothing is a forecasting method not based on the analysis of the historical time series components as such. Rather, a weighted moving average is used as the forecast, with the assigned weights decreasing exponentially for periods farther in the past. There are, in fact, several types of exponential smoothing models, as described in specialized books in business forecasting. The method presented in this section is *single exponential smoothing*.

The following algebraic model serves to represent how the exponentially decreasing weights are determined. Specifically, where α is a smoothing constant discussed below, the most recent value of the time series is weighted by α, the next most recent value is weighted by $\alpha(1-\alpha)$, the next value by $\alpha(1-\alpha)^2$, and so forth, and all the weighted values are then summed to determine the forecast:

$$\hat{Y}_{t+1} = \alpha Y_t + \alpha(1-\alpha)Y_{t-1} + \alpha(1-\alpha)^2 Y_{t-2} + \cdots + \alpha(1-\alpha)^k Y_{t-k} \qquad (16.16)$$

where $\hat{Y}_{t+1}$ = forecast for the next period
   α = smoothing constant $(0 \leq \alpha \leq 1)$
   Y_t = actual value for the most recent period
   Y_{t-1} = actual value for the period preceding the most recent period
   Y_{t-k} = actual value for k periods preceding the most recent period

Although the above formula serves to present the rationale of exponential smoothing, its use is quite cumbersome. A simplified procedure that requires an initial "seed" forecast but does not require the determination of weights generally is used instead. The formula for determining the forecast by the simplified method of exponential smoothing is

$$\hat{Y}_{t+1} = \hat{Y}_t + \alpha(Y_t - \hat{Y}_t) \qquad (16.17)$$

where $\hat{Y}_{t+1}$ = forecast for the next period
   $\hat{Y}_t$ = forecast for the most recent period
   α = smoothing constant $(0 \leq \alpha \leq 1)$
   Y_t = actual value for the most recent period

Because the most recent time series value must be available to determine a forecast for the following period, exponential smoothing can be used only to forecast the value for the *next* period in the time series, not for several periods into the future. The closer the value of the smoothing constant is set to 1.0, the more heavily is the forecast weighted by the most recent results. See Problem 16.8.

Single exponential smoothing is most effective as a forecasting method when cyclical and irregular influences comprise the main effects on time series values. When there is a strong linear trend, double exponential smoothing yields a better forecast. Triple exponential smoothing is appropriate in the case of a strong nonlinear trend, while other methods described in advanced textbooks in forecasting can be used to incorporate the existence of a strong seasonal influence.

16.9 COMPUTER OUTPUT

A wide variety of software is available in time series analysis and business forecasting, with much of it incorporating more sophisticated techniques than can be included in our limited exposure to this specialized topic. Problem 16.9 illustrates the use of Minitab to determine a linear trend equation, while Problem 16.10 illustrates the use of a program called SEASON for determining seasonal indexes and for seasonally adjusting the input data.

Solved Problems

TREND ANALYSIS

16.1 Table 16.1 presents sales data for an 11-year period for a computer software company (fictional) incorporated in 1980, as described in Example 1, and for which the time series data are portrayed by the line graph in Fig. 16-1. Included also are worktable calculations needed to determine the equation for the trend line. (*a*) Determine the linear trend equation for these data by the least squares method, coding 1980 as 0 and carrying all values to two places beyond the decimal point. (*b*) Enter the trend line on the line graph in Fig. 16.1.

Table 16.1 Annual Sales for a Computer Software Firm, with Worktable to Determine the Equation for the Trend Line

Year	Coded year (X)	Sales, in millions (Y)	XY	X^2
1980	0	$ 0.2	0	0
1981	1	0.4	0.4	1
1982	2	0.5	1.0	4
1983	3	0.9	2.7	9
1984	4	1.1	4.4	16
1985	5	1.5	7.5	25
1986	6	1.3	7.8	36
1987	7	1.1	7.7	49
1988	8	1.7	13.6	64
1989	9	1.9	17.1	81
1990	10	2.3	23.0	100
Totals	55	12.9	85.2	385

(*a*)
$$Y_T = b_0 + b_1 X$$

where
$$\bar{X} = \frac{\Sigma X}{n} = \frac{55}{11} = 5.00$$

$$\bar{Y} = \frac{\Sigma Y}{n} = \frac{12.9}{11} = 1.17$$

$$b_1 = \frac{\Sigma XY - n\bar{X}\bar{Y}}{\Sigma X^2 - n\bar{X}^2} = \frac{85.2 - 11(5.00)(1.17)}{385 - 11(5.00)^2}$$

$$= \frac{20.85}{110} = 0.19$$

$$b_0 = \bar{Y} - b_1\bar{X} = 1.17 - 0.19(5.00) = 0.22$$

$$Y_T = 0.22 + 0.19X \quad (\text{with } X = 0 \text{ at } 1980)$$

This equation can be used as a beginning point for forecasting, as described in Section 16.6. The slope of 0.19 indicates that during the 11-year existence of the company there has been an average increase in sales of 0.19 million dollars ($190,000) annually.

(b)   Figure 16-3 repeats the line graph in Fig. 16-1, but with the trend line entered in the graph. The peak and trough in the time series that were briefly discussed in Example 1 are now more clearly visible.

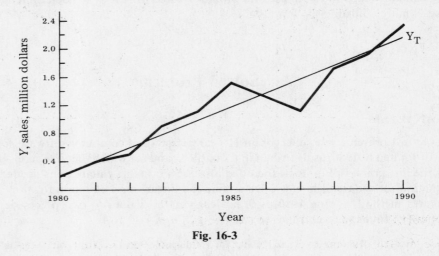

Fig. 16-3

ANALYSIS OF CYCLICAL VARIATIONS

16.2   Determine the cyclical component for each of the time series values reported in Table 16.1, utilizing the trend equation determined in Problem 16.1.

Table 16.2 presents the determination of the cyclical relatives. As indicated in the last column of the table, each cyclical relative is determined by multiplying the observed time series value by 100 and dividing by the trend value. Thus, the cyclical relative of 90.0 for 1980 was determined by calculating $100(0.20)/0.22$.

Table 16.2   Determination of Cyclical Relatives

Year	Coded year (X)	Sales, in $ millions Actual (Y)	Sales, in $ millions Expected (Y)	Cyclical relative $100 Y / Y$
1980	0	0.20	0.22	90.9
1981	1	0.40	0.41	97.6
1982	2	0.50	0.60	83.3
1983	3	0.90	0.79	113.9
1984	4	1.10	0.98	112.2
1985	5	1.50	1.17	128.2
1986	6	1.30	1.36	95.6
1987	7	1.10	1.55	71.0
1988	8	1.70	1.74	97.7
1989	9	1.90	1.93	98.4
1990	10	2.30	2.12	108.5

16.3   Construct a cycle chart for the sales data reported in Table 16.1, based on the cyclical relatives determined in Table 16.2.

The cycle chart is presented in Fig. 16-4, and includes the peak and trough observed previously in Fig. 16-3.

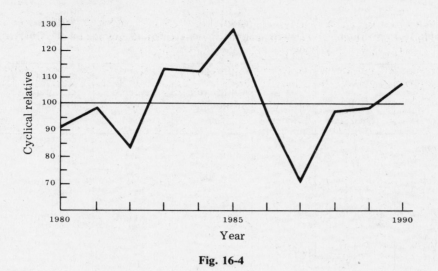

Fig. 16-4

MEASUREMENT OF SEASONAL VARIATIONS

16.4   Table 16.3 presents quarterly sales data for the computer software company for which annual data are reported in Table 16.1. Determine the seasonal indexes by the ratio-to-moving-average method.

Table 16.3   Quarterly Sales for the Computer Software Firm

Quarter	Quarterly sales, in $1,000s					
	1985	1986	1987	1988	1989	1990
1	500	450	350	550	550	750
2	350	350	200	350	400	500
3	250	200	150	250	350	400
4	400	300	400	550	600	650

Table 16.4 is concerned with the first step in the ratio-to-moving-average method, that of computing the ratio of each quarterly value to the 4-quarter moving average centered at that quarter.

The 4-quarter moving totals are centered between the quarters in the table because as a moving total of an even number of quarters, the total would always fall between two quarters. For example, the first listed total of 1,500 (in $1,000s) is the total sales amount for the first through the fourth quarters of 1985. Since four quarters are involved, the total is centered vertically in the table between the second and third quarter.

Because it is desired that the moving average be centered at each quarter, instead of between quarters, adjoining 4-quarter moving totals are combined to form the 2-year centered moving totals. Note that this type of total does *not* include 2 years of data as such. Rather, two overlapping 4-quarter periods are included in the total. For example, the first centered moving total of 2,950 (in $1000s) includes the 4-quarter total of 1,500 for the four quarters of 1985 plus the 4-quarter total of 1,450 for the second quarter of 1985 through the first quarter of 1986.

The 4-quarter centered moving average simply is the 2-year centered moving total divided by 8.

Finally, the ratio to moving average in the last column of Table 16.4 is the ratio of each quarterly sales value to the centered moving average for that quarter. This ratio is multiplied by 100 and in effect is reported as a percentage.

Table 16.4   Worktable for Determining the Ratios to Moving Average for the Quarterly Sales Data (Sales in $1,000s)

Year	Quarter	Sales	4-quarter moving total	2-year centered moving total	4-quarter centered moving average	Ratio to moving average (percent)
1985	I	500				
	II	350				
			1,500			
	III	250		2,950	368.75	67.8
			1,450			
	IV	400		2,900	362.50	110.3
			1,450			
1986	I	450		2,850	356.25	126.3
			1,400			
	II	350		2,700	337.50	103.7
			1,300			
	III	200		2,500	312.50	64.0
			1,200			
	IV	300		2,250	281.25	106.7
			1,050			
1987	I	350		2,050	256.25	136.6
			1,000			
	II	200		2,100	262.50	76.2
			1,100			
	II	150		2,400	300.00	50.0
			1,300			
	IV	400		2,750	343.75	116.4
			1,450			
1988	I	550		3,000	375.00	146.7
			1,550			
	II	350		3,250	406.25	86.2
			1,700			
	III	250		3,400	425.00	58.8
			1,700			
	IV	550		3,450	431.25	127.5
			1,750			
1989	I	550		3,600	450.00	122.2
			1,850			
	II	400		3,750	468.75	85.3
			1,900			
	III	350		4,000	500.00	70.0
			2,100			
	IV	600		4,300	537.50	111.6
			2,200			
1990	I	750		4,450	556.25	134.8
			2,250			
	II	500		4,550	568.75	87.9
			2,300			
	II	400				
	IV	650				

Table 16.5 incorporates the second and third steps in determining the seasonal indexes. The modified mean for each quarter is the mean of the three remaining ratios for each quarter *after* elimination of the highest and lowest ratios. For example, for Quarter 1 the two extreme ratios 122.2 and 146.7 are eliminated from consideration, with the mean of the remaining three ratios being 132.6.

Finally, the modified means are multiplied by an adjustment factor so that the sum of the indexes is approximately 400 (1,200 for monthly indexes). The adjustment factor used with modified mean quarterly indexes to obtain the quarterly indexes is

$$\text{Qtrly. Adj. Factor} = \frac{400}{\text{sum of quarterly means}} \qquad (16.18)$$

The adjustment factor used with modified mean monthly indexes to obtain the monthly indexes is

$$\text{Mo. Adj. Factor} = \frac{1,200}{\text{sum of monthly means}} \qquad (16.19)$$

For the ratios in Table 16.5, the adjustment factor is 400/395.4 or 1.0116. Each modified mean is multiplied by this adjustment factor to obtain the quarterly indexes. Overall, we can observe that the quarter with the greatest "positive" seasonal effect is Quarter 1, with sales generally being about 34.1 percent higher than the typical quarter. On the other hand, sales in Quarter 3 generally are just 64.2 percent of the sales in the typical quarter. By being aware of such seasonal variations, a firm can take them into consideration when planning work schedules and in making forecasts of sales.

Table 16.5   Calculation of the Seasonal Indexes for the Quarterly Data

Quarter	1985	1986	1987	1988	1989	1990	Modified mean	Seasonal index: mean × 1.0116*
1		126.3	136.6	146.7	122.2	134.8	132.6	134.1
2		103.7	76.2	86.2	85.3	87.9	86.5	87.5
3	67.8	64.0	50.0	58.8	70.0		63.5	64.2
4	110.3	106.7	116.4	127.5	111.6		112.8	114.1
							395.4	399.9

*Adjustment factor = 400/395.4 = 1.0116

APPLYING SEASONAL ADJUSTMENTS

16.5   Refer to the seasonal indexes determined in Table 16.5.

(a)   Compute the seasonally adjusted value for each quarter, rounding the values to whole thousand dollar amounts.

(b)   Compare the results for the third and fourth quarters of 1990 based on (1) the reported level of actual sales, and (2) the seasonally adjusted values.

(a)   The seasonally adjusted values are reported in Table 16.6. Each deseasonalized value was determined by dividing the quarterly value in Table 16.3 by the seasonal index for that quarter (from Table 16.5), and multiplying by 100 to preserve the location of the decimal point. For example, the seasonally adjusted value of 373 (in $1,000s) for Quarter I of 1985 was obtained by dividing 500 (from Table 16.3) by 134.1 (from Table 16.5) and multiplying by 100.

(b)   The reported levels of sales for the third and fourth quarters indicate a substantial increase in sales between the two quarters—from $400,000 to $650,000. However, on a seasonally adjusted basis there is a decline of 8.5 percent, from $623,000 to $570,000. Thus, although sales increased in the fourth quarter, the increase was not as large as it should have been, based on the different historical seasonal influences for these two quarters.

Table 16.6   Seasonally Adjusted Values for the Quarterly Data

Quarter	1985	1986	1987	1988	1989	1990
1	$373	$336	$261	$410	$410	$559
2	400	400	229	400	457	571
3	389	312	234	389	545	623
4	351	263	351	482	526	570

FORECASTING BASED ON TREND AND SEASONAL FACTORS

16.6   Step down the trend equation developed in Problem 16.1 so that it is expressed in terms of quarters rather than years. Also, adjust the equation so that the trend values are in thousands of dollars instead of millions of dollars. Carry the final values to the first place beyond the decimal point.

$$Y_T \text{ (quarterly)} = \frac{b_0}{4} - 1.5\left(\frac{b_1}{16}\right) + \left(\frac{b_1}{16}\right)X$$

$$= \frac{0.22}{4} - 1.5\left(\frac{0.19}{16}\right) + \left(\frac{0.19}{16}\right)X$$

$$= 0.0550 - 0.0178 + 0.0119X$$

$$= 0.0372 + 0.0119X$$

The quarterly trend equation in units of thousands of dollars is:

$$Y_T \text{ (quarterly)} = 1,000\,(0.0372 + 0.0119X) = 37.2 + 11.9X$$

For the above equation, $X = 0$ at the first quarter of 1980.

16.7   Forecast the level of quarterly sales for each quarter of 1991 based on the quarterly trend equation determined in Problem 16.6 and the seasonal indexes determined in Table 16.5.

The forecasted values based on the quarterly trend equation, and then adjusted for the quarterly seasonal indexes, are

$$\text{First quarter, 1991} = [37.2 + 11.9(44)] \times \frac{134.1}{100} = 752.0$$

$$\text{Second quarter, 1991} = [37.2 + 11.9(45)] \times \frac{87.5}{100} = 501.1$$

$$\text{Third quarter, 1991} = [37.2 + 11.9(46)] \times \frac{64.2}{100} = 375.3$$

$$\text{Fourth quarter, 1991} = [37.2 + 11.9(47)] \times \frac{114.1}{100} = 680.6$$

EXPONENTIAL SMOOTHING AS A FORECASTING METHOD

16.8   Refer to the annual time series data in Table 16.1. Using the actual level of sales for 1984 of 1.1 million dollars as the "seed" forecast for 1985, determine the forecast for each annual sales amount by the method of single exponential smoothing, rounding your forecast to one place beyond the decimal point. First use a smoothing constant of $\alpha = 0.80$, then use a smoothing constant of $\alpha = 0.20$, and compare the two sets of forecasts.

Table 16.7 is the worktable that reports the two sets of forecasts. For example, the forecast amount for 1986 based on $\alpha = 0.20$ was determined as follows:

$$\hat{Y}_{t+1} = \hat{Y}_t + \alpha(Y_t - \hat{Y}_t)$$
$$\hat{Y}_{1986} = \hat{Y}_{1985} + \alpha(Y_{1985} - \hat{Y}_{1985})$$
$$= \$1.1 + 0.20(0.4) = 1.1 + 0.08 = 1.18 \cong \$1.2$$

The forecast errors are generally lower for $\alpha = 0.80$. Thus, the greater weight given to the forecast errors leads to better forecasts for these data.

Table 16.7   Year-by-Year Forecasts by the Method of Exponential Smoothing

Year (t)	Sales in millions (Y_t)	$\alpha = 0.20$		$\alpha = 0.80$	
		Forecast ($\hat{Y}_t$)	Forecast error ($Y_t - \hat{Y}_t$)	Forecast ($\hat{Y}_t$)	Forecast error ($Y_t - \hat{Y}_t$)
1985	$1.5	$1.1	$0.4	$1.1	$0.4
1986	1.3	1.2	0.1	1.4	−0.1
1987	1.1	1.2	−0.1	1.3	−0.2
1988	1.7	1.2	0.5	1.1	0.6
1989	1.9	1.3	0.6	1.6	0.3
1990	2.3	1.4	0.9	1.8	0.5
1991		1.6		2.2	

COMPUTER OUTPUT

16.9   Apply computer software to obtain the linear trend equation for the annual sales data in Table 16.1. Use the coded years in Table 16.1.

Figure 16-5 includes the linear equation, which differs only slightly from the equation determined manually in Problem 16.1 because of the rounding of values. As explained in Section 16.2, note that this equation is *not* a regression equation as such, even though such a command was used to obtain the linear equation.

```
MTB ) NAME C1 = 'YEAR', C2 = 'SALES'
MTB ) READ 'YEAR', 'SALES'
DATA)   0    0.2
DATA)   1    0.4
DATA)   2    0.5
DATA)   3    0.9
DATA)   4    1.1
DATA)   5    1.5
DATA)   6    1.3
DATA)   7    1.1
DATA)   8    1.7
DATA)   9    1.9
DATA)  10    2.3
DATA) END
MTB ) REGRESS 'SALES' USING 1 PREDICTOR 'YEAR'
```

Fig. 16-5   Minitab output for Problem 16.9. (*Continued on p. 288.*)

```
The regression equation is
SALES = 0.232 + 0.188 YEAR

Predictor        Coef        Stdev      t-ratio
Constant        0.2318      0.1169       1.98
YEAR            0.18818     0.01976      9.52

s = 0.2072      R-sq = 91.0%      R-sq(adj) = 90.0%

Analysis of Variance

SOURCE          DF          SS          MS
Regression       1        3.8954      3.8954
Error            9        0.3865      0.0429
Total           10        4.2818

Unusual Observations
Obs.     YEAR     SALES      Fit  Stdev.Fit  Residual   St.Resid
  8       7.0    1.1000    1.5491   0.0739    -0.4491     -2.32R

R denotes an obs. with a large st. resid.
```

Fig. 16-5 *(continued)*

INPUT VALUES

PERIOD	1985	1986	1987	1988	1989	1990	1991	1992
I	500.00	450.00	350.00	550.00	550.00	750.00	0.00	0.00
II	350.00	350.00	200.00	350.00	400.00	500.00	0.00	0.00
III	250.00	200.00	150.00	250.00	350.00	400.00	0.00	0.00
IV	400.00	300.00	400.00	550.00	600.00	650.00	0.00	0.00

SEASONAL INDEXES

(USING PERCENT OF FOUR-QUARTER CENTERED MOVING AVERAGES)

PERIOD	1985	1986	1987	1988	1989	1990	1991	1992	SEASONAL INDEX
I	0.00	126.32	136.59	146.67	122.22	134.33	0.00	0.00	134.13
II	0.00	103.70	76.19	86.15	85.33	87.91	0.00	0.00	87.48
III	67.80	64.00	50.00	58.32	70.00	0.00	0.00	0.00	64.29
IV	110.34	106.57	116.36	127.54	111.63	0.00	0.00	0.00	114.10
									400.00

SEASONALLY ADJUSTED DATA

PERIOD	1985	1986	1987	1988	1989	1990	1991	1992
I	372.77	335.49	260.94	410.04	410.04	559.15	0.00	0.00
II	400.09	400.09	228.62	400.09	457.24	571.55	0.00	0.00
III	388.89	311.11	233.33	388.39	544.45	622.23	0.00	0.00
IV	350.56	262.92	350.56	482.03	525.85	569.67	0.00	0.00

Fig. 16-6 Computer output for Problem 16.10. (*SEASON, copyright Leonard J. Kazmier.*)

16.10 Use computer software to determine the seasonal indexes based on the quarterly data in Table 16.3 and to seasonally adjust the data.

Figure 16-6 includes the quarterly indexes and the deseasonalized data. Except for slight differences due to rounding, the results correspond to the solutions in Problems 16.4 and 16.5.

Supplementary Problems

TREND ANALYSIS

16.11 Given the data in Table 16.8, determine the linear trend equation for construction contracts for commercial and industrial buildings in the United States, designating 1975 as the base year for the purpose of coding the years.

Ans. $Y_T = 269.401 + 4.274X$ (in mil. sq. ft)

Table 16.8   Construction Contracts for Commercial and Industrial Buildings, million sq. ft

Year	Construction contracts
1975	291.60
1976	205.72
1977	251.84
1978	322.92
1979	361.36
1980	311.24
1981	310.88
1982	229.52
1983	255.44
1984	312.28
1985	345.68

Source: U.S. Department of Commerce, *Business Conditions Digest.*

16.12 Construct the line graph for the data in Table 16.8 and enter the trend line on this graph.

ANALYSIS OF CYCLICAL VARIATIONS

16.13 Determine the cyclical relatives for the annual time series data reported in Table 16.8.

16.14 Prepare a cycle chart to portray the cyclical relatives determined in Problem 16.13 above.

Ans. The cyclical relatives reflect peaks (e.g., 1979) and troughs (e.g., 1982) in the national economy. Incidentally, these construction contracts are considered to be a *leading economic indicator*, as described in Section 16.7.

MEASUREMENT OF SEASONAL VARIATIONS

16.15 Table 16.9 presents quarterly data for the number of customer orders received in a mail order firm (fictional). Determine the seasonal indexes based on these data, carrying three places beyond the decimal point during calculations and rounding the indexes to two places.

Table 16.9 Number of Customer Orders in a Mail-Order Firm (in 1,000s)

Quarter	Year				
	1986	1987	1988	1989	1990
I	9	10	13	11	14
II	16	15	22	17	18
III	18	18	17	25	25
IV	21	20	24	21	26

Ans. I = 68.19, II = 93.76, III = 113.42, IV = 124.64

APPLYING SEASONAL ADJUSTMENTS

16.16 Continuing with Problem 16.15, seasonally adjust the data and comment on the comparison of the resulting values for the third and fourth quarters of 1990.

Ans. Although there was an increase in orders from the third to the fourth quarters in the original data, on a seasonally adjusted basis there was a decline in orders.

FORECASTING BASED ON TREND AND SEASONAL FACTORS

16.17 Suppose that the trend equation for quarterly sales orders (in thousands) is $Y_T = 13.91 + 0.43X$, with $X = 0$ at the first quarter of 1986. Forecast the number of orders for each of the four quarters of 1991 based on the trend and seasonal components of the time series.

Ans. I = 15.34, II = 21.51, III = 26.51, IV = 29.66

EXPONENTIAL SMOOTHING AS A FORECASTING METHOD

16.18 Refer to the construction data in Table 16.8. Using the 1982 construction amount of 229.52 (mil. sq. ft.) as the "seed" forecast for 1983, use the method of single exponential smoothing with a smoothing constant of $\alpha = 0.70$ to determine year-by-year forecasts for 1983 through 1986.

Ans. 229.52, 273.58, 339.37, 350.10

COMPUTER OUTPUT

16.19 Use available computer software to determine the linear trend equation for the construction data in Table 16.8, designating 1975 as the coded year $X = 0$.

16.20 Use available computer software to (a) determine the quarterly seasonal indexes and (b) deseasonalize the data on the number of customer orders in Table 16.9.

Chapter 17

Index Numbers for Business and Economic Data

17.1 INTRODUCTION

An *index number* is a percentage relative by which a measurement in a *given period* is expressed as a ratio to the measurement in a designated *base period*. The measurements can be concerned with *quantity, price,* or *value.*

EXAMPLE 1. The Consumer Price Index (CPI) prepared by the U.S. Department of Labor is an example of a price index, whereas the Federal Reserve Board Index of Industrial Production is an example of a quantity index.

When the index number represents a comparison for an *individual* product or commodity, it is a *simple index number.* In contrast, when the index number has been constructed for a *group* of items or commodities, it is an *aggregate index number* or *composite index number.*

EXAMPLE 2. The price index of, say, 130 for butter is a simple price index indicating that the price of butter in the given period was 30 percent higher than in the base period, for which the price index is 100 by definition. The same price index of 130 for the Consumer Price Index (an aggregate price index) would indicate that the average price of a "market basket" of about 400 goods and services was 30 percent higher in the given period as compared with the base period.

17.2 CONSTRUCTION OF SIMPLE INDEXES

Where p_n indicates the price of a commodity in the given period and p_0 indicates the price in the base period, the general formula for the simple price index, or *price relative,* is

$$I_p = \frac{p_n}{p_0} \times 100 \qquad \text{(See Problem 17.1)} \qquad (17.1)$$

Similarly, where q_n indicates the quantity of an item produced or sold in the given period and q_0 indicates the quantity in the base period, the general formula for the simple quantity index, or *quantity relative,* is

$$I_q = \frac{q_n}{q_0} \times 100 \qquad \text{(See Problem 17.2)} \qquad (17.2)$$

Finally, the *value* of a commodity in a designated period is equal to the price of the commodity multiplied by the quantity produced (or sold). Therefore $p_n q_n$ indicates the value of a commodity in the given period, and $p_0 q_0$ indicates the value of the commodity in the base period. The general formula for a simple value index, or *value relative,* is

$$I_p = \frac{p_n q_n}{p_0 q_0} \times 100 \qquad \text{(See Problem 17.3)} \qquad (17.3)$$

17.3 CONSTRUCTION OF AGGREGATE PRICE INDEXES

To obtain an aggregate price index, the prices of the several items or commodities could simply be summed for the given period and for the base period, respectively, and then compared. Such an index would be an *unweighted* aggregate price index. An unweighted index generally is not very useful because the implicit weight of each item in the index depends on the units upon which the prices are based.

EXAMPLE 3.   If the price of milk is reported "per gallon" as contrasted to "per quart," then the price would make a much greater contribution to an unweighted price index for a group of commodities which includes milk.

Because of the difficulty described above, aggregate price indexes generally are weighted according to the quantities q of the commodities. The question as to which period quantities should be used is the issue that serves as the basis for different types of aggregate price relatives. One of the more popular aggregate price indexes is *Laspeyres' index*, in which the prices are weighted by the quantities associated with the *base* year before being summed. The formula is

$$I(L) = \frac{\Sigma p_n q_0}{\Sigma p_0 q_0} \times 100 \qquad \text{(See Problem 17.4)} \qquad\qquad (17.4)$$

Instead of using the base-year quantities as weights, the *given*-year quantities could be used. This is *Paasche's index*. The formula is

$$I(P) = \frac{\Sigma p_n q_n}{\Sigma p_0 q_n} \times 100 \qquad \text{(See Problem 17.5)} \qquad\qquad (17.5)$$

Both the Laspeyres and Paasche methods of constructing an aggregate price index can be described as following the *weighted-aggregate-of-prices* approach. An alternative is the *weighted-average-of-price-relatives* approach, by which the simple price index for each individual commodity is weighted by a value figure pq. The values used may either be for the base year, $p_0 q_0$, or for the given year, $p_n q_n$. Typically, the base-year values are used as weights, resulting in the following formula for the weighted-average-of-price-relatives:

$$I_p = \frac{\Sigma[(p_0 q_0)(p_n/p_0 \times 100)]}{\Sigma p_0 q_0} \qquad \text{(See Problem 17.6)} \qquad\qquad (17.6)$$

Algebraically, (*17.6*) is equivalent to Laspeyres' index, while use of given period values as weights would result in an index equivalent to Paasche's index. The reason for the popularity of the weighted-average-of-price-relatives method is that by this procedure the simple indexes for each commodity are computed first, and these simple price indexes are often desired for analytical purposes in addition to the aggregate price index itself.

17.4   LINK RELATIVES

Link relatives are indexes for which the base is always the preceding period. Therefore, for a set of link relatives for annual value of sales, each index number represents a percentage comparison with the preceding year. Such relatives are useful for highlighting year-to-year comparisons, but are not convenient as the basis for making long-run comparisons. See Problem 17.7.

17.5   SHIFTING THE BASE PERIOD

The base of an established index number series is often shifted to a more recent year so that current comparisons are more meaningful. Assuming that the original quantities underlying the index number series are not available, the base period for an index number can be shifted by dividing each (original) index by the index of the newly designated base year and multiplying the result by 100:

$$I_{n(\text{shifted})} = \frac{I_{n(\text{old})}}{\text{old index of new base}} \times 100 \qquad \text{(See Problem 17.9)} \qquad\qquad (17.7)$$

17.6 SPLICING TWO SERIES OF INDEX NUMBERS

An index number may often undergo change by addition of certain new products or by exclusion of certain old products, as well as by changes in the base year. Yet, for the purpose of historical continuity it is desirable to have a unified series of index numbers available. In order to *splice* two such separate time series to form one continuous series of index numbers, there must be one year of overlap for the two series such that both types of index numbers have been calculated for that year. Generally, that year of overlap also is the new base, because it is the year at which new products have been added to and/or removed from the aggregate index. The index numbers that have to be changed in the process of splicing are the indexes of the old series. This change is achieved by dividing the new index number for the overlap year (100.0 if this is the new base) by the old index for that year and then multiplying each of the index numbers of the old index number series by this quotient. See Problem 17.10.

17.7 THE CONSUMER PRICE INDEX (CPI)

The *Consumer Price Index* is the most widely known of the published indexes because of its use as an indicator of the cost of living. It is published monthly by the Bureau of Labor Statistics of the U.S. Department of Labor and is an aggregate price index for a "market basket" of several hundred goods and services. Two types of indexes have been published since 1978: the CPI-W, which is a continuation of the historical CPI and focuses on urban wage and clerical workers, and the broader CPI-U, for all urban consumers. The newer CPI-U is now regarded as the general CPI. Whereas the base year for this index is 1967, the quantity weights (beginning in January 1987) are based on a 1982–1984 survey of consumer expenditures.

17.8 PURCHASING POWER AND THE DEFLATION OF TIME SERIES VALUES

Whereas the Consumer Price Index is indicative of the price of a market basket of goods and services compared to the 1967 base year, the reciprocal of the CPI indicates the *purchasing power* of the dollar relative to the base year (i.e., in 1967 dollars):

$$\text{Value of dollar} = \frac{1}{\text{CPI}} \times 100 \qquad \text{(See Problem 17.11)} \qquad (17.8)$$

Deflation of a time series is the process by which a series of current-year values are converted to constant-dollar values. Using formula (*17.8*), restatement of such values into the constant-dollar values of the base year is accomplished by

$$\text{Deflated amount} = \frac{\text{reported amount}}{\text{CPI}} \times 100 \qquad (17.9)$$

See Problem 17.12.

17.9 OTHER PUBLISHED INDEXES

The *Producer Price Indexes* include separate indexes for *crude materials*, *intermediate materials*, and *finished goods* published by the Bureau of Labor Statistics. The price index for finished goods often is referred to as the "wholesale price index" in the business press. Finally, the *Index of Industrial Production* is an aggregate quantity index published monthly by the Federal Reserve Board that measures the output of the nation's factories, mines, and electric and gas utilities. It is a weighted average of quantity relatives, with values of output for the 1967 base period used as weights.

Solved Problems

SIMPLE INDEXES

17.1 Referring to Table 17.1, determine the simple price indexes for 1990 for the three commodities, using 1985 as the base year.

Table 17.1   Prices and Consumption of Three Commodities in a Particular Metropolitan Area, 1985 and 1990

Commodity	Unit quotation	Average price		Per capita consumption (per month)	
		1985 (p_0)	1990 (p_n)	1985 (q_0)	1990 (q_n)
Milk	Half-gallon	$0.99	$1.29	15.0	18.0
Bread	$1\frac{1}{2}$-lb loaf	1.10	1.20	3.8	3.7
Eggs	Dozen	0.80	1.20	1.0	1.2

For milk:
$$I_p = \frac{p_n}{p_0} \times 100 = \frac{1.29}{0.99} \times 100 = 130.3$$

For bread:
$$I_p = \frac{p_n}{p_0} \times 100 = \frac{1.20}{1.10} \times 100 = 109.1$$

For eggs:
$$I_p = \frac{p_n}{p_0} \times 100 = \frac{1.20}{0.80} \times 100 = 150.0$$

17.2 Referring to Table 17.1, determine the simple quantity indexes for the three commodities for 1990, using 1985 as the base year.

For milk:
$$I_q = \frac{q_n}{q_0} \times 100 = \frac{18.0}{15.0} \times 100 = 120.0$$

For bread:
$$I_q = \frac{q_n}{q_0} \times 100 = \frac{3.7}{3.8} \times 100 = 97.4$$

For eggs:
$$I_q = \frac{q_n}{q_0} \times 100 = \frac{1.2}{1.1} \times 100 = 109.1$$

17.3 Compute the simple value relatives for 1990 for the three commodities in Table 17.1, using 1985 as the base year.

For milk:
$$I_v = \frac{p_n q_n}{p_0 q_0} \times 100 = \frac{(1.29)(18.0)}{(0.99)(15.0)} \times 100 = 156.4$$

For bread:
$$I_v = \frac{p_n q_n}{p_0 q_0} \times 100 = \frac{(1.20)(3.7)}{(1.10)(3.8)} \times 100 = 106.2$$

For eggs:
$$I_v = \frac{p_n q_n}{p_0 q_0} \times 100 = \frac{(1.20)(1.2)}{(0.80)(1.0)} \times 100 = 180.0$$

AGGREGATE PRICE INDEXES

17.4 Compute Laspeyres' aggregate price index for 1990 for the three commodities in Table 17.1, using 1985 as the base year.

Referring to Table 17.2, the index is determined as follows:

$$I(L) = \frac{\Sigma p_n q_0}{\Sigma p_0 q_0} \times 100 = \frac{25.11}{19.83} \times 100 = 126.7$$

Table 17.2 Worksheet for the Calculation of Laspeyres' Index for the Data in Table 17.1

Commodity	$p_n q_0$	$p_0 q_0$
Milk	$19.35	$14.85
Bread	4.56	4.18
Eggs	1.20	0.80
Total	$\Sigma p_n p_0 = \$25.11$	$\Sigma p_0 q_0 = \$19.83$

17.5 Compute Paasche's aggregate price index for 1990 for the three commodities in Table 17.1, using 1985 as the base year.

Using Table 17.3, we calculate the index as follows:

$$I(P) = \frac{\Sigma p_n q_n}{\Sigma p_0 q_n} \times 100 = \frac{29.10}{22.85} \times 100 = 127.4$$

Table 17.3 Worksheet for the Calculation of Paasche's Index for the Data in Table 17.1

Commodity	$p_n q_n$	$p_0 q_n$
Milk	$23.22	$17.82
Bread	4.44	4.07
Eggs	1.44	0.96
Total	$\Sigma p_n q_n = \$29.10$	$\Sigma p_0 q_n = \$22.85$

17.6 Compute the price index by the weighted-average-of-price-relatives method for the three commodities in Table 17.1, using 1985 as the base year.

With reference to Table 17.4,

$$I_p = \frac{\Sigma[(p_0 q_0)(p_n/p_0 \times 100)]}{\Sigma p_0 q_0} = \frac{2,510.46}{19.83} = 126.6$$

This answer corresponds to Laspeyres' index, calculated in Problem 17.4.

Table 17.4   Worksheet for the Computation of the Weighted Average of Price Relatives for the Data in Table 17.1

Commodity	Price relative ($p_n/p_0 \times 100$)	Value weight ($p_0 q_0$)	Weighted relative $[(p_0 q_0)(p_n/p_0 \times 100)]$
Milk	130.30	$14.85	1,934.96
Bread	109.09	4.18	456.00
Eggs	150.00	0.80	120.00
Total		$19.83	2,510.96

LINK RELATIVES

17.7   The dollar value of sales (in millions) for the Ford Motor Company between 1980 and 1985 is as follows: 1980, 37,086; 1981, 38,247; 1982, 37,067; 1983, 44,455; 1984, 52,366; 1985, 52,774. Determine the link relatives for these data.

The link relatives are reported in Table 17.5. For example, the link relative of 103.1 for 1981 indicates that the dollar sales for 1981 were 3.1 percent higher than dollar sales in the preceding year, 1980.

Table 17.5   Ford Motor Company Sales and Link Relatives, 1980–1985

Year	1980	1981	1982	1983	1984	1985
Sales, millions of dollars	37,086	38,247	37,067	44,455	52,366	52,774
Link relative	—	103.1	96.9	119.9	117.8	100.8

Source: Ford Motor Company, *Annual Report 1985.*

17.8   Referring to the sales amounts in Table 17.5, compute the value indexes for these six years using 1980 as the base year.

Table 17.6 reports the value indexes.

Table 17.6   Ford Motor Company Value Indexes, 1980–1985

Year	1980	1981	1982	1983	1984	1985
Value index (1980 = 100)	100.0	103.1	99.9	119.9	141.2	142.3

Source: Table 17.5.

SHIFTING THE BASE PERIOD

17.9   Shift the value indexes reported in Table 17.6 from the base year of 1980 to 1985 as the common base year for the indexes, using only the indexes reported in that table.

Table 17.7 reports the indexes computed with 1985 as the base year. As one example, the new value index for 1980 was determined as follows:

$$I_{N(\text{shifted})} = \frac{I_{n(\text{old})}}{\text{old index of new base year}} \times 100 = \frac{I_{1980(\text{old})}}{\text{old index for 1985}} \times 100$$

$$= \frac{100.0}{142.3} \times 100 = 70.3$$

Table 17.7   Value Indexes Using Base Years of 1980 and 1985 for Ford Motor Company Sales

Year	1980	1981	1982	1983	1984	1985
Value index (1980 = 100)	100.0	103.1	99.9	119.9	141.2	142.3
Value index (1985 = 100)	70.3	72.5	70.2	84.3	99.2	100.0

Source: Table 17.6.

SPLICING TWO SERIES OF INDEX NUMBERS

17.10   Table 17.8 presents two hypothetical series of price indexes: one computed for a group of commodities for the years 1975 to 1980 with the base year being 1975, and the other computed beginning in 1980 for a revised group of commodities. Thus, 1980 is the overlap year for which both indexes were determined. Splice these two series to form one continuous series of index numbers with 1980 as the base year.

The spliced series of indexes is reported in the last column of Table 17.8. The quotient which results from dividing the new index number for 1980 (100.0) by the index number for 1980 in the old series (119.2) is 0.839, and this value is used as the multiplication factor to convert each index number in the old series.

Table 17.8   Splicing Two Index Number Series

Year	Old price index (1975 = 100)	Revised price index (1980 = 100)	Spliced price index (1980 = 100)
1975	100.0		83.9
1976	103.1		86.5
1977	106.9		89.7
1978	110.0		92.3
1979	114.1		95.7
1980	119.2	100.0	100.0
1981		105.2	105.2
1982		111.3	111.3
1983		117.5	117.5
1984		124.8	124.8
1985		129.9	129.9
1986		137.7	137.7

THE CONSUMER PRICE INDEX AND DEFLATION OF TIME SERIES VALUES

17.11   Table 17.9 reports values of the Consumer Price Index for 1980 through 1986. Determine the purchasing power of the dollar for each of these years in terms of the value of the dollar in the base year 1967.

The last column of Table 17.9 reports the value of the dollar in each year. For example, the value for 1980 was determined as follows:

$$\text{Value of dollar} = \frac{1}{\text{CPI}} \times 100 = \frac{1}{246.8} \times 100 = \$0.41$$

Thus, in 1980 the dollar was worth 41¢ in terms of 1967 dollars, on the average.

Table 17.9   Consumer Price Indexes and the Value of the U.S. Dollar, 1980–1986 (Base Year = 1967)

Year	Consumer Price Index (CPI-U)	Value of dollar
1980	246.8	$0.41
1981	272.4	0.37
1982	289.1	0.35
1983	298.4	0.34
1984	311.1	0.32
1985	322.2	0.31
1986	328.4	0.30

Source: U.S. Department of Commerce, *Survey of Current Business.*

17.12 The first part of Table 17.10 reports the annual salary amounts for an accountant who began employment with a business firm in 1980. (*a*) Deflate these time series values so that the amounts are expressed in 1967 dollars. (*b*) Restate the salary amounts in 1980 dollars. (*c*) What was the percent salary increase between 1980 and 1986 in terms of stated (current) dollar amounts? (*d*) What was the percent salary increase between 1980 and 1986 in terms of constant dollars?

Table 17.10   Annual Salary Amounts

Year	Salary	Deflated salary (1967 dollars)*	Restated salary (1980 dollars)
1980	$16,000	$6,483	$16,000
1981	17,000	6,241	15,403
1982	19,000	6,572	16,220
1983	19,500	6,535	16,128
1984	23,000	7,393	18,246
1985	25,000	7,759	19,149
1986	27,000	8,222	20,291

*Using CPI-U indexes in Table 17.9.

(*a*) The third column of Table 17.10 reports the deflated amounts. For example, the deflated amount for 1980 was determined by:

$$\text{Deflated amount} = \frac{\text{reported amount}}{\text{CPI}} \times 100$$

$$= \frac{\$16,000}{246.8} \times 100 = \$6,483$$

(*b*) The last column of Table 17.10 reports the salary amounts in terms of 1980 dollars. These amounts can be determined by multiplying each current amount by the ratio of the CPI for 1980 by the CPI for each respective year. However, given that the deflated amounts in 1967 dollars have been determined, a simpler procedure is to multiply each deflated amount by the CPI for 1980. For instance, the salary amount in 1981 stated in 1980 dollars is

$$1981 \text{ salary (1980 dollars)} = 1981 \text{ salary (1967 dollars)} \times \frac{\text{CPI (1980)}}{100}$$

$$= \$6,241 \times \frac{246.8}{100} = \$15,403$$

(c) $$\text{Increase (stated dollars)} = \frac{\text{1986 salary} - \text{1980 salary}}{\text{1980 salary}} \times 100$$

$$= \frac{\$27,000 - \$16,000}{\$16,000} \times 100 = 68.75\%$$

(d) $$\text{Increase (1967 dollars)} = \frac{\$8,222 - \$6,483}{\$6,483} \times 100 = 26.82\%$$

Supplementary Problems

SIMPLE INDEXES

17.13 Referring to Table 17.11 and using 1985 as the base year, compute the simple price relatives for the (a) copier paper, (b) lined pads, (c) marker pens, and (d) paper clips.

Ans. (a) 111.7, (b) 121.6, (c) 109.5, (d) 109.1

Table 17.11   Average Prices and Monthly Consumption of a Selected Sample of Office Supplies in a Departmental Office, 1985 and 1990

Item	Unit of quotation	Average price		Monthly consumption	
		1985	1990	1985	1990
Copier paper	Ream	$2.64	$2.95	4.5	8.0
Lined pads	Ream	0.37	0.45	10.0	16.0
Marker pens	Each	0.21	0.23	8.0	6.0
Paper clips	Box	0.11	0.12	2.0	2.0

17.14 Referring to Table 17.11 and using 1985 as the base year, compute the simple quantity relatives for the (a) copier paper, (b) lined pads, (c) marker pens, and (d) paper clips.

Ans. (a) 177.8, (b) 160.0, (c) 75.0, (d) 100.0

17.15 Referring to Table 17.11 and using 1985 as the base year, compute the simple value relatives for the (a) copier paper, (b) lined pads, (c) marker pens, and (d) paper clips.

Ans. (a) 198.7, (b) 194.6, (c) 82.1, (d) 109.1

AGGREGATE PRICE INDEXES

17.16 Determine the Laspeyres' price index for the 1990 prices of the office supplies in Table 17.11.

Ans. $I(L) = 113.6$

17.17 Determine the Paasche price index for the 1990 prices of the office supplies in Table 17.11.

Ans. $I(P) = 113.7$

17.18 Determine the aggregate price index for the 1990 prices in Table 17.11 by using the weighted-average-of-relatives method.

Ans. $I_p = 113.6$

LINK RELATIVES

17.19   Table 17.12 reports the worldwide unit sales of motor vehicles (cars and trucks) by the Ford Motor Company during the years 1980 to 1985. Determine the link relatives for these data.

   Ans.   Beginning with 1981: 101.8, 99.0, 115.6, 113.2, 98.5

Table 17.12   Worldwide Motor Vehicle Sales of the Ford Motor Company, 1980–1985 (in 1,000s)

Year	1980	1981	1982	1983	1984	1985
Unit sales	4,328	4,313	4,268	4,934	5,585	5,550

Source: Ford Motor Company, *Annual Report 1985.*

17.20   Referring to the unit sales in Table 17.12, compute the quantity relatives for 1980 to 1985, using 1980 as the base year.

   Ans.   Beginning with 1980: 100.0, 101.8, 100.7, 116.4, 131.8, 129.8

SHIFTING THE BASE PERIOD

17.21   Shift the quantity indexes computed in Problem 17.20 from the base year of 1980 to the year 1985 by using the indexes determined in that problem, rather than the original quantities in Table 17.12.

   Ans.   Beginning with 1980: 77.0, 78.4, 77.6, 89.7, 101.5, 100.0

SPLICING TWO SERIES OF INDEX NUMBERS

17.22   The base period used for the Consumer Price Index prior to the 1967 base year was the 1957–1959 period. Table 17.13 reports the Consumer Price Index for the years 1960 through 1969 using the base period of

Table 17.13   Consumer Price Indexes for the Old Series (1957–1959 = 100) and New Series (1967 = 100)

Year	Old price index (1957–1959 = 100)	New price index (1967 = 100)
1960	103.1	
1961	104.2	
1962	105.4	
1963	106.7	
1964	108.1	
1965	109.9	
1966	113.1	
1967	116.3	100.0
1968	121.2	104.2
1969	127.7	109.8
1970		116.3
1971		121.3
1972		125.3
1973		133.1
1974		147.7
1975		161.2

Source: U.S. Department of Commerce, *Survey of Current Business.*

1957 through 1959, and the revised Consumer Price Index for 1967 through 1975 using the base year 1967. An economist wishes to develop a unified series of indexes with a common base year of 1967. Splice the two historic series of indexes to satisfy this objective.

THE CONSUMER PRICE INDEX AND DEFLATION OF TIME SERIES VALUES

17.23 Table 17.14 reports beginning wage rates paid to employees in a transit company between 1980 and 1986. Using the Consumer Price Indexes reported in Table 17.9, (*a*) deflate the wage rates and express them in 1967 dollars and (*b*) restate the wage rates in terms of constant 1980 dollars.

Ans. Beginning with 1980: (*a*) $3.22, 3.05, 2.99, 3.01, 3.35, 3.23, 3.11; (*b*) $7.94, 7.53, 7.37, 7.43, 8.27, 7.97, 7.67

Table 17.14 Average Wage Rates in a Transit Company, 1980–1986

Year	1980	1981	1982	1983	1984	1985	1986
Hourly rate	$7.94	$8.30	$8.63	$8.97	$10.42	$10.42	$10.21

Chapter 18

Bayesian Decision Analysis: Payoff Tables and Decision Trees

18.1 THE STRUCTURE OF PAYOFF TABLES

From the standpoint of statistical decision theory, a decision situation under conditions of uncertainty can be represented by certain common ingredients which are included in the structure of the *payoff table* for the situation. Essentially, a payoff table identifies the conditional gain (or loss) associated with every possible combination of decision acts and events; it also typically indicates the probability of occurrence for each of the mutually exclusive events.

Table 18.1   General Structure of a Payoff Table

Events	Probability	Acts				
		A_1	A_2	A_3	$\ldots$	A_n
E_1	P_1	X_{11}	X_{12}	X_{13}	$\ldots$	X_{1n}
E_2	P_2	X_{21}	X_{22}	X_{23}	$\ldots$	X_{2n}
E_3	P_3	X_{31}	X_{32}	X_{33}	$\ldots$	X_{3n}
$\ldots$	$\ldots$	$\ldots$	$\ldots$	$\ldots$	$\ldots$	$\ldots$
E_m	P_m	X_{m1}	X_{m2}	X_{m3}	$\ldots$	X_{mn}

In Table 18.1, the *acts* are the alternative courses of action, or strategies, that are available to the decision maker. As the result of the analysis, one of these acts is chosen as being the best act. The basis for this choice is the subject matter of this chapter. As a minimum, there must be at least two possible acts available, so that the opportunity for choice in fact exists. An example of an act is the number of units of a particular item to be ordered for stock.

The *events* identify the occurrences which are outside of the decision maker's control and which determine the level of success for a given act. These events are often called "states of nature," "states," or "outcomes." We are concerned only with discrete events in this chapter. An example of an event is the level of market demand for a particular item during a stipulated time period.

The *probability* of each event is included as part of the general format of a decision table when such probability values are in fact available. However, one characteristic of Bayesian decision analysis is that such probabilities should always be available since they can be based on either objective data or be determined subjectively on the basis of judgment. Because the events in the payoff table are mutually exclusive and exhaustive, the sum of the probability values should be 1.0.

Finally, the cell entries are the conditional values, or conditional economic consequences. These values are usually called *payoffs* in the literature, and they are conditional in the sense that the economic result which is experienced depends on the decision act that is chosen and the event that occurs.

EXAMPLE 1. A heating and air-conditioning contractor must commit himself to the purchase of central air-conditioning units as of April 1 for resale and installation during the following summer season. Based on demand during the previous summer, current economic conditions, and competitive factors in the market, he estimates that there is a 0.10 probability of selling only 5 units, a 0.30 probability of selling 10 units, a 0.40 probability of selling 15 units, and a 0.20 probability of selling 20 units. The air-conditioning units can be ordered only in groups of five, with the cost per unit being $1,000 and the retail price being $1,300 (plus installation charges). Any unsold units at the end of the season are returned to the manufacturer for a net credit of $800, after deduction of shipping charges.

Table 18.2 is the payoff table for this situation. Note that because we estimate that there will be a market demand for at least 5 units but not more than 20 units, these are logically also the limits for our possible acts (units ordered). Our stipulation that units can only be ordered in groups of five serves to simplify the problem and reduce the size of the payoff table. The payoffs in Table 18.2 are based on a markup of $300 per unit sold and a loss of $200 for each unsold unit. Thus, for example, if 15 units are ordered for stock and only 10 are demanded, the economic result is a gain of $3,000 on the 10 units sold less a loss of $1,000 on the 5 units returned to the manufacturer, for a resulting payoff of $2,000 at A_3, E_2 in the table.

Table 18.2   Payoff Table for the Number of Air-Conditioning Units to Be Ordered

Market demand	Probability	Order quantity			
		A_1: 5	A_2: 10	A_3: 15	A_4: 20
E_1: 5	0.10	$1,500	$ 500	−$ 500	−$1,500
E_2: 10	0.30	1,500	3,000	2,000	1,000
E_3: 15	0.40	1,500	3,000	4,500	3,500
E_4: 20	0.20	1,500	3,000	4,500	6,000
	1.00				

In the following sections we refer to Example 1 in order to demonstrate the application of different decision criteria, or standards, that can be used to identify the decision act which is considered to be the best act. The methods in this chapter which involve the use of the probability values associated with each event are concerned only with the probabilities formulated during the initial structuring of the payoff table. Because these values are formulated before the collection of any additional information, they are called *prior probabilities* in Bayesian decision analysis. See Chapters 19 and 20 for other concepts in Bayesian decision analysis.

18.2   DECISION MAKING BASED UPON PROBABILITIES ALONE

A complete application of Bayesian decision analysis involves use of all of the information included in a payoff table. However, in this section we briefly consider the criteria that would be used if the economic consequences were ignored (or not determined) and if the decision was based entirely on the probabilities associated with the possible events.

In such cases, one decision criterion which might be used is to identify the event with the *maximum probability* of occurrence and to choose the decision act corresponding with that event. Another basis for choosing the best act would be to calculate the *expectation* of the event and to choose the act accordingly. However, because neither of these criteria make reference to the economic consequences associated with the various decision acts and events, they represent an incomplete basis for choosing the best decision.

EXAMPLE 2.   Table 18.3 presents the probability distribution for the market demand of central air-conditioning units in Example 1. The event with the maximum probability is $E_3 = 15$, for which $P = 0.40$. On the basis of the criterion of highest probability, the number of units ordered would be 15.

The calculation of the expected demand $E(D)$ is included in Table 18.3 (also see Section 6.2). Since the air-conditioning units can be ordered only as whole units, and further, only in groups of five, the expected demand level of 13.5 units cannot be ordered. Either 10 units would be ordered with the expectation of being 3.5 units short (as a long-run average), or 15 units would be ordered with the expectation of having an excess of 1.5 units, on the average.

Table 18.3   Probability Distribution of Market Demand for Air-Conditioning Units and the Calculation of the Expected Demand

Market demand (D)	Probability [$P(D)$]	$(D)P(D)$
E_1: 5	0.10	0.5
E_2: 10	0.30	3.0
E_3: 15	0.40	6.0
E_4: 20	0.20	4.0
	1.00	$E(D) = 13.5$

One difficulty associated with the two criteria described in Example 2 is that their long-run success cannot really be evaluated without some reference to economic consequences. For example, suppose the contractor can order air conditioners for stock without any prepayment and with the opportunity to return unsold units at the manufacturer's expense. In such a circumstance there would be no risk associated with overstocking, and the best decision would be to order 20 units so that the inventory would be adequate for the highest possible demand level.

18.3   DECISION MAKING BASED UPON ECONOMIC CONSEQUENCES ALONE

The payoff matrix that is used in conjunction with decision making based only upon economic consequences is similar to Table 18.1, except for the absence of the probability distribution associated with the possible events. Three criteria that have been described and used in conjunction with such a decision matrix are the maximin, maximax, and minimax regret criteria.

The *maximin criterion* is the standard by which the best act is the one for which the minimum value is larger than the minimum for any other decision act. Use of this criterion leads to a highly conservative decision strategy, in that the decision maker is particularly concerned about the "worst that can happen" with respect to each act. Computationally, the minimum value in each column of the payoff table is determined, and the best act is the one for which the resulting value is largest.

EXAMPLE 3.   Table 18.4 presents the economic consequences associated with the various acts and events for the problem described in Example 1. The minimum value associated with each decision act is listed along the bottom of this table. Of these values, the largest economic result (the maximum of the four minima) is $1,500. Since the act "$A_1$: Order 5 air-conditioning units" is associated with this outcome, this is the best decision act from the standpoint of the maximin criterion.

Table 18.4   Number of Air-Conditioning Units to Be Ordered According to the Maximin Criterion

Market demand	A_1: 5	A_2: 10	A_3: 15	A_4: 20
E_1: 5	$1,500	$ 500	−$ 500	−$1,500
E_2: 10	1,500	3,000	2,000	1,000
E_3: 15	1,500	3,000	4,500	3,500
E_4: 20	1,500	3,000	4,500	6,000
Minimum	$1,500	$ 500	−$ 500	−$1,500

The *maximax criterion* is the standard by which the best act is the one for which the maximum value is larger than the maximum for any other decision act. This criterion is philosophically the opposite of the maximin criterion, since the decision maker is particularly oriented toward the "best that can happen" with respect to each act. Computationally, the maximum value in each column of the payoff table is determined, and the best act is the one for which the resulting value is largest.

EXAMPLE 4. In Table 18.5 the *maximum* value associated with each decision act is listed along the bottom. The largest of these maximum values (the maximum of the maxima) is $6,000, and thus the associated act "A_4: Order 20 air-conditioning units" would be chosen as the best act from the standpoint of the maximax criterion. Rather than going through the two-step procedure of identifying column maxima and then determining the maximum of these several values, a shortcut which can be used is simply to locate the largest value in the table.

Table 18.5   Number of Air-Conditioning Units to Be Ordered According to the Maximax Criterion

Market demand	A_1: 5	A_2: 10	A_3: 15	A_4: 20
E_1: 5	$1,500	$ 500	−$ 500	−$1,500
E_2: 10	1,500	3,000	2,000	1,000
E_3: 15	1,500	3,000	4,500	3,500
E_4: 20	1,500	3,000	4,500	6,000
Maximum	$1,500	$3,000	$4,500	$6,000

Analysis by the *minimax regret* criterion is based on so-called regrets rather than on conditional values as such. A *regret*, or conditional *opportunity loss*, for each act is the difference between the economic outcome for the act and the economic outcome of the best act *given that a particular event has occurred*. Thus, the "best" or most desirable regret value is "0," which indicates that the act is perfectly matched with the given event. Also, note that even when there is an economic gain associated with a particular act and a given event, there could also be an opportunity loss, because some other act could result in a higher payoff with the given event.

The construction of a table of opportunity losses, or regrets, is illustrated in Example 5. The best act is identified as being the one for which the maximum possible regret is smallest. Philosophically, the minimax regret criterion is similar to the maximin criterion in terms of "assuming the worst." However, use of the concept of opportunity loss results in a broader criterion, in that the failure to improve a payoff is considered to be a type of loss.

EXAMPLE 5. Table 18.6 is the table of opportunity losses associated with the conditional values in Table 18.5. Computationally, the regret values are determined given each event in turn, that is, according to rows. For example,

Table 18.6   Opportunity Loss Table for the Number of Air-Conditioning Units to Be Ordered and Application of the Minimax Regret Criterion

Market demand	A_1: 5	A_2: 10	A_3: 15	A_4: 20
E_1: 5	$ 0	$1,000 (= $1,500 − 500)	$2,000 [= $1,500 − (−500)]	$3,000 [= $1,500 − (−1,500)]
E_2: 10	1,500 (= 3,000 − 1,500)	0	1,000 (= 3,000 − 2,000)	2,000 (= 3,000 − 1,000)
E_3: 15	3,000 (= 4,500 − 1,500)	1,500 (= 4,500 − 3,000)	0	1,000 (= 4,500 − 3,500)
E_4: 20	4,500 (= 6,000 − 1,500)	3,000 (= 6,000 − 3,000)	1,500 (= 6,000 − 4,500)	0
Maximum regret	$4,500	$3,000	$2,000	$3,000

given that E_1 occurs, the best act (by reference to the payoff matrix in Table 18.2) is A_1, with a value of \$1,500. If A_2 is chosen, the result is \$500, which differs from the best act by \$1,000, and which is then the regret value for A_2. If A_3 is chosen, regret $= \$1,500 - (-500) = \$2,000$. If A_4 is chosen, regret $= \$1,500 - (-1,500) = \$3,000$. The remaining opportunity loss values are similarly determined, with the best conditional value in each row serving as the basis for determining regret values in that row. For instance, in row 2, for E_2: 10, the best act is A_2: 10 with a payoff value of \$3,000 in Table 18.5.

The maximum regret which can occur in conjunction with each decision act is listed along the bottom of Table 18.6. The smallest of these maxima (the minimum of the maximum regrets) is \$2,000; "A_3: Order 15 units" would thus be chosen as the best act from the standpoint of the minimax regret criterion.

18.4 DECISION MAKING BASED UPON BOTH PROBABILITIES AND ECONOMIC CONSEQUENCES: THE EXPECTED PAYOFF CRITERION

The methods presented in this section utilize all the information contained in the basic payoff table (see Section 18.1). Thus, we consider both the probabilities associated with the possible events and the economic consequences for all combinations of the several acts and several events.

The *expected payoff* (*EP*) criterion is the standard by which the best act is the one for which the expected economic outcome is the highest, as a long-run average. Note that in the present case we are concerned about the long-run average economic result, and not simply the long-run average event value (demand level) discussed in Section 18.2. Computationally, the expected payoff for each act is determined by multiplying the conditional payoff for each event/act combination by the probability of the event and summing these products for each act.

EXAMPLE 6.   Table 18.7 repeats the information from Table 18.2, except that the expected payoffs have been identified along the bottom. Table 18.8 illustrates the procedure by which such expected payoffs are calculated. In Table 18.7, the largest expected payoff is \$3,250, and thus the associated act "A_3: Order 15 units" is the best act from the standpoint of the expected payoff criterion.

Table 18.7   Payoff Table for the Number of Air-Conditioning Units to Be Ordered and Determination of the Best Act According to the Expected Payoff Criterion

Market demand	Probability	Order quantity			
		A_1: 5	A_2: 10	A_3: 15	A_4: 20
E_1: 5	0.10	\$1,500	\$ 500	-\$ 500	-\$1,500
E_2: 10	0.30	1,500	3,000	2,000	1,000
E_3: 15	0.40	1,500	3,000	4,500	3,500
E_4: 20	0.20	1,500	3,000	4,500	6,000
Expected payoff (*EP*)		\$1,500	\$2,750	\$3,250	\$2,750

Table 18.8   Determination of the Expected Payoff for Decision A_4 in Table 18.7

Market demand	Payoff for A_4: X	$P(X)$	$XP(X)$
E_1: 5	-\$1,500	0.10	-\$ 150
E_2: 10	1,000	0.30	300
E_3: 15	3,500	0.40	1,400
E_4: 20	6,000	0.20	1,200
			$\Sigma XP(X) = \$2,750$

The expected payoff criterion often is referred to as the *Bayesian criterion*. Use of the adjective "Bayesian" here is distinct from the use of Bayes' theorem for revising a prior probability value (see Sections 5.7 and 19.2). Thus, a reference to "Bayesian procedures" in decision analysis can involve use of the expected payoff criterion, revision of prior probability values, or both.

The best act identified by the expected payoff criterion can also be determined by identifying the act with the minimum expected opportunity loss (EOL), or expected regret. This is so because the act with the largest expected gain logically would have the smallest expected regret. In textbooks which refer to opportunity losses simply as *losses*, *expected losses* are understood to mean expected opportunity losses.

EXAMPLE 7.   Table 18.9 repeats the opportunity loss values from Table 18.6. It also shows the probability of each event and the expected opportunity loss associated with each decision act. Table 18.10 illustrates the procedure by which each expected opportunity loss was calculated. As indicated in Table 18.9, the minimum expected opportunity loss is $800, and thus the associated act "Order 15 units" is the best act from the standpoint of minimizing the expected opportunity loss. Note that this is the same act as identified by use of the expected payoff criterion in Example 6.

Table 18.9   Opportunity Loss Table for the Number of Air-Conditioning Units to Be Ordered and Computation of Expected Opportunity Losses

Market demand	Probability	Order quantity			
		A_1: 5	A_2: 10	A_3: 15	A_4: 20
E_1: 5	0.10	$ 0	$1,000	$2,000	$3,000
E_2: 10	0.30	1,500	0	1,000	2,000
E_3: 15	0.40	3,000	1,500	0	1,000
E_4: 20	0.20	4,500	3,000	1,500	0
Expected opportunity loss (EOL)		$2,550	$1,300	$ 800	$1,300

Table 18.10   Determination of the Expected Opportunity Loss for Decision A_4 in Table 18.9

Market demand	Conditional opportunity loss for A_4: OL	$P(OL)$	$(OL)P(OL)$
E_1: 5	$3,000	0.10	$ 300
E_2: 10	2,000	0.30	600
E_3: 15	1,000	0.40	400
E_4: 20	0	0.20	0
			$EOL = \$1,300$

18.5   DECISION TREE ANALYSIS

Frequently a decision problem is complicated by the fact that the payoffs are not only associated with an initial decision, but rather, that subsequent events lead to the need for additional decisions at each step of a sequential process. The evaluation of the alternative decision acts in the first step of such a sequential process must of necessity be based on an evaluation of the events and decisions in the overall process. *Decision tree analysis* is the method which can be used to identify the best initial act, as well as the best subsequent acts. The decision criterion satisfied is the Bayesian expected payoff criterion (see Section 18.4).

The first step in decision tree analysis is to construct the decision tree that corresponds to a sequential decision situation. The tree is constructed from left to right with appropriate identification of *decision points* (sequential points at which a choice has to be made) and *chance events* (sequential points at which a probabilistic event will occur). The construction of a valid decision tree for a sequential decision situation is particularly dependent on appropriate analysis of the overall decision situation. See Problem 18.12.

After a decision tree is constructed, the probability values associated with the chance events and the payoffs that can occur are entered in the diagram. Many of the payoffs are several steps removed from the initial decision point. In order to determine the expected payoffs of the alternative acts at the initial decision point, expected payoffs are systematically calculated from right to left in the decision tree. This process is sometimes called "folding back" (see Problem 18.13). As the result of applying this analytical process, the best act at the initial decision point can be identified.

18.6 EXPECTED UTILITY AS THE DECISION CRITERION

The expected payoff criterion is typically used in conjunction with both payoff table analysis and decision tree analysis (see Section 18.4). However, when the decision maker perceives one or more of the economic consequences as being unusually large or small, the expected payoff criterion does not necessarily provide the basis for identifying the "best" decision. This is particularly likely for unique, rather than repetitive, situations.

EXAMPLE 8. Consider the decision acts in Table 18.11 and assume that the choice with respect to each pair will be made only once. Although it can easily be demonstrated that the expected value of A_2 is greater than that for A_1 in every case, most people would choose A_1 in preference to A_2 for each of these pairs. In the context of each choice being a one-time act, the choice of A_1 is rational even though the expected payoff is not maximized. The implication of such a conclusion is that monetary values may not adequately represent true values to the decision maker in decision situations that include the possibility of exceptional losses and/or exceptional gains.

Table 18.11   Three Pairs of Alternative Decision Acts with Associated Consequences

A_1: Receive $1,000,000 for certain.	A_2: Receive $2,000,000 with a probability of 0.50 or receive $100 with a probability of 0.50.
A_1: Pay $10.	A_2: Experience a loss of $8,000 with a probability of 0.001 or experience no loss with a probability of 0.999.
A_1: Receive $15,000 with a probability of 0.50 or receive $5,000 with a probability of 0.50.	A_2: Receive $50,000 with a probability of 0.50 or experience a loss of $25,000 with a probability of 0.50.

Utility is a measure of value which expresses the true relative value of various outcomes, including economic consequences, for a decision maker. This book deals only with the utility of economic consequences. Any given utility scale can begin at an arbitrary minimum value and have an arbitrarily assigned maximum value. However, it is convenient to have utility values begin at a minimum of 0 and extend to a maximum of 1.00, and this is the scale most frequently used. With such a scale, an outcome with a utility of 0.60 is understood to be twice as desirable as one with a utility of 0.30. On the other hand, note that an economic outcome of $60,000 is not necessarily twice as desirable as an outcome of $30,000 for a decision maker with limited resources.

Using a *reference contract*, you can determine an individual's utility values for different monetary values. By this approach, the individual is asked to designate an *amount certain* that he would accept, or pay, as being equivalent to each of a series of uncertain situations involving risk. The first risk situation portrayed always includes the two extreme limits of the range of monetary values of interest, i.e., lower and upper limits having utilities of 0 and 1.0, respectively. See Problem 18.14.

Once a reference-contract procedure has been established, it can be continued by changing either the designated probabilities or one or more of the economic consequences in the risk situation. In this way, the set of utility values corresponding to a range of monetary values can be determined. See Problem 18.15.

After the utility values have been determined, the paired values may be plotted on a graph. A best-fitting smooth line can be drawn through the plotted points as an approximation of the decision maker's *utility function* for various payoffs. The standard convention is to plot the monetary values with respect to the horizontal axis and the utility values with respect to the vertical axis. This graph can be used as the basis for estimating the utility value of any monetary outcome between the designated limits of the function. In turn, this makes it possible to substitute utility values for monetary values in a payoff table, and to determine the best act by identifying that act for which the *expected utility* (EU) is maximized. See Problems 18.16 and 18.17.

The form of the utility function indicates whether a decision maker is a risk averter or a risk seeker (see Fig. 18-1). For the risk averter, each additional dollar along the horizontal axis is associated with a declining slope of the utility function. That is, one might say that each additional increment has positive value to the decision maker, but not as much as the preceding increments (the curve is concave). Conversely, Fig. 18-1(c) indicates that for the risk seeker each additional monetary increment has increasing value to the decision maker (the curve is convex). It can also be shown that the risk averter designates an amount certain that is consistently *less* than the expected payoff for the risk situation, while the risk seeker designates an amount certain that is consistently *greater* than the expected payoff for the risk situation. See Problem 18.16.

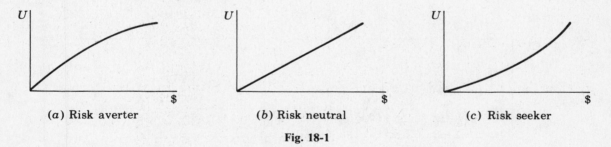

(a) Risk averter   (b) Risk neutral   (c) Risk seeker

Fig. 18-1

Overall, the form of the utility curve reflects a decision maker's attitude toward risk, and therefore is an important factor in determining which act is best *for the decision maker* at a particular time. For a firm in financial jeopardy, a particular contract opportunity that includes the possibility of a high loss and subsequent business failure may not represent a good opportunity even though the expected payoff associated with the contract is positive and represents a substantial return. The concept of utility provides the basis for demonstrating why both parties to a contract (such as insurer and insured) can experience a positive utility and why a risk situation which is "right" for one firm may not be right for another. However, if the utility function is linear or approximately so, as in Fig. 18-1(b), then the expected payoff criterion is equivalent to the expected utility criterion for that decision maker.

Solved Problems

PAYOFF TABLES

18.1   Based on a new technological approach, a manufacturer has developed a color TV set with a 45-in. picture tube. The owner of a small retail store estimates that at the selling price of $2,800 the probability values associated with his selling 2, 3, 4, or 5 sets during the 3 months of concern are 0.30, 0.40, 0.20 and 0.10, respectively. Based only on these probability values, what number of sets should the retailer order for stock, assuming no reorders are possible during the period?

Based on the criterion of maximum probability, three sets would be ordered, since the probability of 0.40 associated with three sets being sold is higher than the probability of any other event.

On the other hand, the expectation of the demand level is

$$2(0.30) + 3(0.40) + 4(0.20) + 5(0.10) = 3.1$$

Based on this expectation of the event, the act which comes closest to corresponding with it is also that of ordering three sets.

18.2 For the inventory decision situation in Problem 18.1, the profit margin for each set sold is $200. If any sets are not sold during the three months, the total loss per set to the retailer will be $300. Based on these economic consequences alone, and ignoring the probability values identified in Problem 18.1, determine the best decision acts from the standpoint of the maximin and maximax criteria.

With reference to Table 18.12, for the maximin criterion the best act is A_1: Order two sets. For the maximax criterion, the best act is A_4: Order five sets.

Table 18.12   Number of TV Sets to Be Ordered According to the Maximin and Maximax Criteria

Market demand	Order quantity			
	A_1: 2	A_2: 3	A_3: 4	A_4: 5
E_1: 2	$400	$100	−$200	−$ 500
E_2: 3	400	600	300	0
E_3: 4	400	600	800	500
E_4: 5	400	600	800	1,000
Minimum	$400	$100	−$200	−$ 500
Maximum	$400	$600	$800	$1,000

18.3 Determine the best decision act from the standpoint of the minimax regret criterion for the decision situation described in Problems 18.1 and 18.2.

Table 18.13   Opportunity Loss Table for the Number of TV Sets to Be Ordered and Application of the Minimax Regret Criterion

Market demand	Order quantity			
	A_1: 2	A_2: 3	A_3: 4	A_4: 5
E_1: 2	$ 0	$300	$600	$900
E_2: 3	200	0	300	600
E_3: 4	400	200	0	300
E_4: 5	600	400	200	0
Maximum regret	$600	$400	$600	$900

Table 18.13 indicates the opportunity losses (regrets) for this decision situation. From the standpoint of the minimax regret criterion, the best act is A_2: Order three sets.

18.4   With reference to Problems 18.1 and 18.2, determine the best act from the standpoint of the expected payoff criterion.

Table 18.14 is the complete decision table for this problem and reports the expected payoffs associated with the possible decision acts. As indicated in the table, the best act from the standpoint of the expected payoff criterion is A_2: Order three sets.

18.5   Using Table 18.14, determine the best act for the Bayesian criterion by identifying the act with the lowest expected opportunity loss.

Table 18.15 repeats the opportunity loss values from Table 18.13 and identifies the expected opportunity loss associated with each act. As indicated in the table, the best act from the standpoint of minimizing the expected opportunity loss is A_2: Order three sets. This result was expected, since the act which satisfies the Bayesian criterion of maximizing the expected payoff will also be the act which has the minimum expected opportunity loss.

Table 18.14   Payoff Table for the Number of TV Sets to Be Ordered and Application of the Expected Payoff Criterion

Market demand	Probability	Order quantity			
		A_1: 2	A_2: 3	A_3: 4	A_4: 5
E_1: 2	0.30	$400	$100	−$200	−$ 500
E_2: 3	0.40	400	600	300	0
E_3: 4	0.20	400	600	800	500
E_4: 5	0.10	400	600	800	1,000
Expected payoff (EP)		$400	$450	$300	−$ 50

Table 18.15   Opportunity Loss Table for the Number of TV Sets to Be Ordered and Computation of Expected Opportunity Losses

Market demand	Probability	Order quantity			
		A_1: 2	A_2: 3	A_3: 4	A_5: 5
E_1: 2	0.30	$ 0	$300	$600	$900
E_2: 3	0.40	200	0	300	600
E_3: 4	0.20	400	200	0	300
E_4: 5	0.10	600	400	200	0
Expected opportunity loss (EOL)		$220	$170	$320	$570

18.6   Table 18.16 presents the payoffs (returns) associated with five alternative types of investment decisions for a 1-year period. Given that the probabilities associated with the possible states are not available, determine the best acts from the standpoint of the maximin and maximax criteria.

Table 18.16   Monetary Returns Associated with Several Investment Alternatives for a $10,000 Fund

State of economy	Investment decision				
	A_1 Savings account	A_2 Corporate bonds	A_3 Blue chip stocks	A_4 Speculative stocks	A_5 Stock options
E_1: Recession	$600	$500	−$2,500	−$ 5,000	−$10,000
E_2: Stable	600	900	800	400	−5,000
E_3: Expansion	600	900	4,000	10,000	20,000

Table 18.17 identifies the minimum and maximum values for each act in Table 18.16. As would be expected, the maximin criterion leads to the selection of the very conservative decision A_1: Invest in a savings account. The maximax criterion, on the other hand, leads to the selection of the decision act at the other extreme, A_5: Invest in stock options.

Table 18.17   Investment Decision to Be Made According to the Maximin and Maximax Criteria

Economic consequence	Investment decision				
	A_1 Savings account	A_2 Corporate bonds	A_3 Blue chip stocks	A_4 Speculative stocks	A_5 Stock options
Minimum	$600	$500	−$2,500	−$ 5,000	−$10,000
Maximum	$600	$900	$4,000	$10,000	$20,000

18.7   Determine the best decision act from the standpoint of the minimax regret criterion for the situation described in Problem 18.6.

Table 18.18 indicates that the best act in this case is A_4: Invest in speculative stocks. Note the extent to which the maximum regret values for the first four acts are influenced by the possibility of the large gain with A_5, and that there is no consideration of probability values using the minimax regret criterion.

Table 18.18   Opportunity Loss Table for the Investment Decision Problem and Application of the Minimax Regret Criterion

State of economy	Investment decision				
	A_1 Savings account	A_2 Corporate bonds	A_3 Blue chip stocks	A_4 Speculative stocks	A_5 Stock options
E_1: Recession	$ 0	$ 100	$ 3,100	$ 5,600	$10,600
E_2: Stable	300	0	100	500	5,900
E_3: Expansion	19,400	19,100	16,000	10,000	0
Maximum regret	$19,400	$19,100	$16,000	$10,000	$10,600

18.8   For Problem 18.6, suppose the probabilities associated with a recession, economic stability, and with an expansion are 0.30, 0.50, and 0.20, respectively. Determine the best act from the standpoint of the Bayesian criterion of maximizing the expected payoff.

Table 18.19 is the complete payoff table for this problem and also indicates the expected payoff associated with each decision act. The best act from the standpoint of the expected payoff criterion is A_2: Invest in corporate bonds.

Table 18.19   Payoff Table for the Investment Decision Problem and the Determination of the Best Act According to Expected Payoff Criterion

State of economy	Probability	Investment decision				
		A_1 Savings account	A_2 Corporate bonds	A_3 Blue Chip stocks	A_4 Speculative stocks	A_5 Stock options
E_1: Recession	0.30	$600	$500	−$2,500	−$ 5,000	−$10,000
E_2: Stable	0.50	600	900	800	400	−5,000
E_3: Expansion	0.20	600	900	4,000	10,000	20,000
Expected payoff (*EP*)		$600	$780	$ 450	$ 700	−$ 1,500

18.9   Thus far, we have been concerned with situations in which positive as well as negative economic consequences can occur. When the consequences are all *cost* values, the same techniques can be applied *provided that all cost figures are identified as being negative values* (payouts). However, cost figures are frequently identified as "costs" but are reported as absolute values without the negative signs. In such a case, keep in mind that the minimum cost, rather than the maximum, is the "best" cost, and adjust the application of the various decision criteria accordingly. Using Table 18.20, determine the best acts from the standpoint of the maximin and maximax criteria by modifying the application of these criteria appropriately.

Table 18.20   Cost Table Associated with Three Product Inspection Plans

Proportion defect in shipment	Inspection plan		
	A_1 100% inspection	A_2 5% inspection	A_3 No inspection
E_1: 0.01	$100	$ 20	$ 0
E_2: 0.05	100	150	400

Since the lowest cost is the best cost and the highest cost is the worst cost, when the maximin criterion is applied, the objective is to choose that act whose maximum cost is the smallest. In this context, the maximin criterion could be called the "minimax" criterion. As indicated in Table 18.21, the decision act which satisfies this criterion is A_1: 100% inspection. By using this criterion, the maximum cost which can occur is minimized.

In the context of a table of costs, using the maximax criterion involves choosing that act whose minimum value is the smallest. Thus, the criterion could be called the "minimin" criterion in this case. As indicated in Table 18.21, the criterion which satisfies the optimistic orientation represented by the maximax criterion (i.e. which minimizes the minimum cost) is A_3: No inspection.

Table 18.21 Quality Inspection Plan to Be Chosen by the Maximin and Maximax Criteria

Proportion defective in shipment	Inspection plan		
	A_1 100% inspection	A_2 5% inspection	A_3 No inspection
E_1: 0.01 E_2: 0.05	$100 100	$ 20 150	$ 0 400
Maximum cost	$100	$150	$400
Minimum cost	$100	$ 20	$ 0

18.10 Determine the best decision act from the standpoint of the minimax regret criterion for the situation in Problem 18.9.

Refer to Table 18.22. Because the best act for a given state is the one which results in the *lowest* cost, that is the act which is assigned an opportunity loss of zero. Thus, given E_1, A_3 is the best act in terms of having the lowest cost, it thus has $0 regret. The act which minimizes the maximum regret which can occur is A_2: 5% inspection.

Table 18.22 Opportunity Loss Table for the Quality Inspection Plan Problem and Application of the Minimax Regret Criterion

Proportion defective in shipment	Inspection plan		
	A_1 100% inspection	A_2 5% inspection	A_3 No inspection
E_1: 0.01 E_2: 0.05	$100 0	$20 50	$ 0 300
Maximum regret	$100	$50	$300

18.11 From Problem 18.9, determine the best act from the standpoint of the general Bayesian criterion of maximizing the expected payoff, given that the probability is 0.80 that a shipment will contain a proportion of 0.01 defectives and the probability is 0.20 that a shipment will contain a proportion of 0.05 defectives.

Table 18.23 Cost Table for the Quality Inspection Plan Problem and Application of the Bayesian Criterion

Proportion defective in shipment	Probability	Inspection plan		
		A_1 100% inspection	A_2 5% inspection	A_3 No inspection
E_1: 0.01 E_2: 0.05	0.80 0.20	$100 100	$ 20 150	$ 0 400
Expected cost		$100	$ 46	$ 80

When dealing with cost figures, the Bayesian criterion is satisfied by choosing that act for which the expected cost is minimized. Table 18.23 is a complete decision table for this problem and indicates that the best act from this standpoint is A_2: 5% inspection. Note that in terms of expected (long-run) cost, the "conservative" strategy of 100% inspection is the least preferred act.

DECISION TREE ANALYSIS

18.12 A manufacturer has been presented with a proposal for a new product and must decide whether or not to develop it. The cost of the development project is $200,000; the probability of success is 0.70. If development is unsuccessful, the project will be terminated. If it is successful, the manufacturer must then decide whether to begin manufacturing the product at a high level or at a low level. If demand is high, the incremental profit given a high level of manufacturing is $700,000; given a low level, it is $150,000. If demand is low, the incremental profit given a high level of manufacturing is $100,000; given a low level, it is $150,000. All of these incremental profit values are gross figures (i.e., *before* subtraction of the $200,000 development cost). The probability of high demand is estimated as $P = 0.40$, and of low demand as $P = 0.60$. Construct the decision tree for this situation.

Figure 18-2 is the decision tree for this problem. As is typical for such diagrams, each decision point is identified by a square and each chance event is identified by a circle.

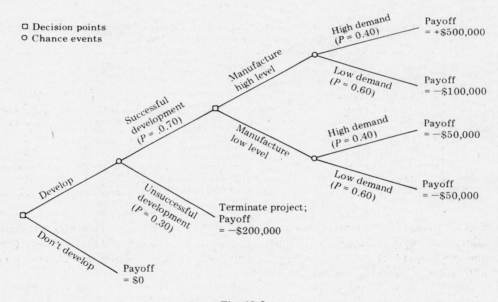

Fig. 18-2

18.13 Referring to Fig. 18-2, determine whether or not the manufacturer should proceed with the attempt to develop this product by determining the expected payoff associated with the alternative acts "Develop" and "Don't develop."

Figure 18-3 repeats the decision tree presented in Fig. 18-2, except that the expected payoffs associated with each possible decision in the sequential process have been entered, and the nonpreferred act at each decision point has been eliminated from further consideration in each case by superimposing a double bar (‖) at that branch. Working from right to left, the expected payoff of the act "High-level manufacturing" is $140,000, which is determined as follows:

$$EP(\text{High-level manufacturing}) = (0.40)(500,000) + (0.60)(-100,000) = \$140,000$$

316   BAYESIAN ANALYSIS: PAYOFF TABLES, DECISION TREES   [CHAP. 18

Similarly,

$$EP(\text{Low-level manufacturing}) = (0.40)(-50,000) + (0.60)(-50,000) = -\$50,000$$

Comparing the two expected payoffs, the best act is "high-level manufacturing"; the other act possibility is eliminated. Moving leftward to the next decision point (which is also the initial decision point in this case), the expected payoffs for the two possible decision acts are

$$EP(\text{Develop}) = (0.70)(140,000) + (0.30)(-200,000) = \$38,000$$

$$EP(\text{Don't develop}) = \$0$$

Comparing the two expected payoffs, the best act at the initial decision point is "Develop." In the calculation of the expected payoff for "Develop," note that the probability of 0.70 (for "Successful development") is multiplied by \$140,000, with the −\$50,000 in the adjoining branch being ignored because it is associated with a decision act eliminated in the previous step of analysis.

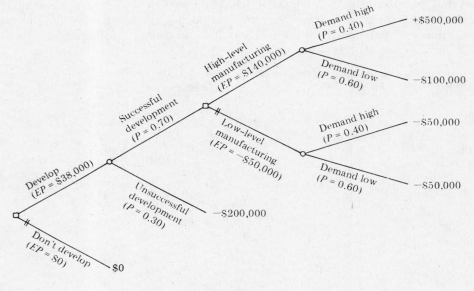

Fig. 18-3

UTILITY FUNCTIONS AND EXPECTED UTILITY

18.14 Referring to Table 18.2, we see that the two extreme monetary outcomes are −\$1,500 and +\$6,000. Suppose that a decision maker indicates that he would be indifferent to receiving an amount certain of \$1,200 in lieu of a risk situation in which there is a 50 percent chance of gaining \$6,000 and a 50 percent chance of a \$1,500 loss. Following the convention of assigning the extreme lower and upper monetary values with utility values of "0" and "1.0," respectively, determine the utility values associated with −\$1,500, \$1,200, and \$6,000.

We begin by assigning utility values to the two extreme monetary consequences:

$$U(-\$5,000) = 0 \qquad U(\$6,000) = 1.0$$

The utility associated with the amount certain of \$1,200 is determined as follows:

$$U(\text{amount certain}) = P(U \text{ of high outcome}) + (1 - P)(U \text{ of low outcome})$$

$$U(\$1,200) = 0.50(1.0) + 0.50(0)$$

$$U(\$1,200) = 0.50$$

18.15 Continuing with Problem 18.14, suppose that four additional reference contracts are presented to the decision maker (see Table 18.24). Determine the utility values associated with each amount certain.

Table 18.24   Four Reference Contracts with Associated Amounts Certain

Contract number	Probability of $6,000 gain	Probability of $1,500 loss	Equivalent amount certain
1	0.10	0.90	−$1,000
2	0.30	0.70	0
3	0.70	0.30	3,000
4	0.90	0.10	5,000

Contract No. 1:

$$\text{Amount certain} = 0.10(+\$6,000) \text{ vs. } 0.90(-\$1,500)$$
$$-\$1,000 = 0.10(+\$6,000) \text{ vs. } 0.90(-\$1,500)$$
$$U(-\$1,000) = 0.10(1.0) + 0.90(0) = 0.10$$

Contract No. 2:

$$\text{Amount certain} = 0.30(+\$6,000) \text{ vs. } 0.70(-\$1,500)$$
$$\$0 = 0.30(+\$6,000) \text{ vs. } 0.70(-\$1,500)$$
$$U(\$0) = 0.30(1.0) + 0.70(0) = 0.30$$

Contract No. 3:

$$\text{Amount certain} = 0.70(+\$6,000) \text{ vs. } 0.30(-\$1,500)$$
$$\$3,000 = 0.70(+\$6,000) \text{ vs. } 0.30(-\$1,500)$$
$$U(\$3,000) = 0.70(1.0) + 0.30(0) = 0.70$$

Contract No. 4:

$$\text{Amount certain} = 0.90(+\$6,000) \text{ vs. } 0.10(-\$1,500)$$
$$\$5,000 = 0.90(+\$6,000) \text{ vs. } 0.10(-\$1,500)$$
$$U(\$5,000) = 0.90(1.0) + 0.10(0) = 0.90$$

Table 18.25 summarizes the utility values which are equivalent to the several monetary values determined above and in Problem 18.14. Note that when the risk situation in the reference contract involves the two extreme outcomes with the associated utility values of "1.0" and "0," the utility values of the amount certain which is designated by the decision maker is always equal to the probability of the outcome which has the assigned utility of 1.0.

Table 18.25   Utility Values and Equivalent Monetary Values

Monetary value, $	−1,500	−1,000	0	1,200	3,000	5,000	6,000
Utility value	0	0.10	0.30	0.50	0.70	0.90	1.00

18.16 Referring to Table 18.25, construct a graph to portray the utility function for this decision maker. Describe his attitude toward the risks inherent in the reference contracts which were presented to him.

Figure 18-4 shows that this decision maker is a risk averter in the range of monetary consequences included in this problem. See Fig. 18-1. Basically, this indicates that the amount certain which he designates as being equivalent to the designated risk situation in the reference contract is always less than the associated expected payoff, except for the two extreme end points of the utility curve. For instance, in Problem 18.14 the decision maker stated that for him a certain $1,200 was equivalent to a probability of 0.50 of gaining $6,000 and a probability of 0.50 of losing $1,500. However, the payoff of this risk situation is $0.50(\$6,000) + 0.50(-\$1,500) = \$3,000 + (-\$750) = \$2,250$.

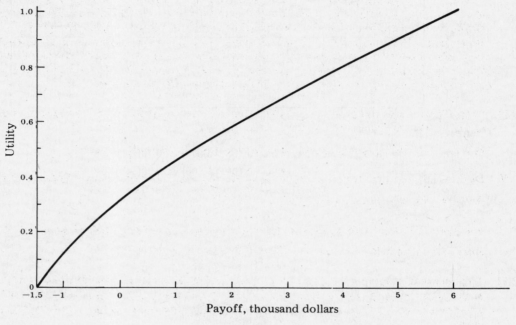

Fig. 18-4

18.17 Referring to the utility function in Fig. 18-4, determine the approximate utility values corresponding to each payoff in Table 18-2, and determine the best act from the standpoint of maximizing the expected utility.

Table 18.26 identifies the utility values which are equivalent to the payoffs in Table 18.2. Using the criterion of maximizing the expected utility, the best act is A_3, as it was for the expected payoff criterion (see Example 6). However, note that whereas the expected payoffs for acts A_2 and A_4 were equal in Table 18.7, the expected utility values of these two acts are not equal.

Table 18.26 Utility Table for the Number of Air-Conditioning Units to Be Ordered and Determination of the Best Act According to the Expected Utility Criterion

Market demand	Probability	Order quantity			
		A_1: 5	A_2: 10	A_3: 15	A_4: 20
E_1: 5	0.10	0.55	0.38	0.21	0
E_2: 10	0.30	0.55	0.70	0.61	0.47
E_3: 15	0.40	0.55	0.70	0.86	0.78
E_4: 20	0.20	0.55	0.70	0.86	1.00
Expected utility (EU)		0.550	0.668	0.720	0.653

Supplementary Problems

PAYOFF TABLES

18.18 An investment analyst estimates that there is about a 50 percent chance of an "upturn" in the chemical industry during the first quarter of the year, with the probabilities of "no change" and a "downturn" being about equal. A client is considering either the investment of $10,000 in a mutual fund specializing in chemical industry common stocks or investing in corporate AAA-rated bonds yielding 8.0 percent per year. If the chemical industry experiences an upturn during the first quarter, the value of the mutual fund shares (including dividends) will increase by 15.0 percent during the next 12 months. If there is no change, the value will increase by 3.0 percent. If there is a downturn, the value will *decrease* by 10.0 percent. Ignoring any commission costs, construct a payoff table for this investment problem.

18.19 For the investment decision described in Problem 18.18, determine the best decision acts from the standpoint of the (*a*) maximin and (*b*) maximax criteria.

   Ans.   (*a*) A_2: Invest in AAA bonds, (*b*) A_1: Invest in mutual fund

18.20 Determine the best act from the standpoint of the minimax regret criterion for Problem 18.18.

   Ans. A_2: Invest in AAA bonds

18.21 For Problem 18.18, determine the best act from the standpoint of the expected payoff criterion.

   Ans. A_2: Invest in AAA bonds

18.22 A retailer buys a certain product for $3 per case and sells it for $5 per case. The high markup is reflective of the perishability of the product, since it has no value after 5 days. Based on experience with similar products, the retailer is confident that the demand for the item will be somewhere between 9 and 12 cases, inclusive.

   Construct an appropriate payoff table. Determine the best act from the standpoint of (*a*) the maximin criterion and (*b*) the maximax criterion.

   Ans.   (*a*) A_1: Order 9 cases, (*b*) A_4: Order 12 cases

18.23 Construct the table of opportunity losses (regrets) for Problem 18.22, and determine the best act from the standpoint of the minimax regret criterion.

   Ans. A_2: Order 10 cases

18.24 Continuing with Problem 18.22, the retailer further estimates that the probability values associated with selling 9 to 12 cases of the item are 0.30, 0.40, 0.20, and 0.10, respectively. Determine the best decision acts from the standpoint of the (*a*) maximum probability criterion and (*b*) expectation of the event.

   Ans.   (*a*) A_2: Order 10 cases, (*b*) A_2: Order 10 cases

18.25 Referring to Problems 18.22 and 18.23, determine the best order quantity from the standpoint of the (*a*) expected payoff criterion and (*b*) the criterion of minimizing the expected opportunity loss. Demonstrate that these are equivalent criteria by also identifying the "second best" and "worst" acts with respect to each criterion.

   Ans.   (*a*) A_2: Order 10 cases, (*b*) A_2: Order 10 cases

18.26 In conjunction with the installation of a new marine engine, a charter vessel owner has the opportunity to buy spare propellers at $200 each for use during the coming cruise season. The propellers are custom fit for the vessel, and because the owner has already contracted to sell the boat after the cruise season, there is no value associated with having spare propellers left over after the season. If a spare propeller is not immediately available when needed, the cost of buying a needed spare propeller, including lost cruise time, is $400. Based on previous experience, the vessel owner estimates that the probability values associated with needing 0 to 3 propellers during the coming cruise season are 0.30, 0.30, 0.30, and 0.10, respectively. Construct a cost table for this problem and determine the number of propellers that should be ordered from the standpoint of (a) the maximin criterion, (b) the maximax criterion.

Ans. (a) A_4: Order three spare propellers, (b) A_1: Order no spare propellers

18.27 For the situation in Problem 18.26, determine the best act from the standpoint of the (a) maximum probability criterion and (b) the expectation of the event.

Ans. (a) Indifferent between 0, 1, and 2 propellers; (b) A_2: Order 1 propeller

18.28 For Problem 18.26, determine the best act from the standpoint of the expected payoff criterion (in this case, expected cost).

Ans. A_2: Order 1 propeller

18.29 Several weeks before an annual "Water Carnival," a college social group must decide to order either blankets, beach umbrellas, or neither for resale at the event. The monetary success of the decision to sell one of the two items is dependent on the weather conditions, as indicated in Table 18.27. Determine the best acts from the standpoint of the (a) maximin and (b) maximax criteria.

Ans. (a) A_3: Neither, (b) A_2: Beach umbrellas

Table 18.27 Payoffs for the "Water Carnival" Decision Problem

Weather	A_1: Blankets	A_2: Beach umbrellas	A_3: Neither
Cool	$100	−$ 80	$0
Hot	−50	150	0

18.30 Determine the best act from the standpoint of the minimax regret criterion for Problem 18.29.

Ans. A_3: Neither

18.31 Continuing with Problem 18.29, one of the club members calls the weather bureau and learns that during the past 10 years the weather has been "hot" on 6 of the 10 days on the date when the Water Carnival is to be held. Use this information to estimate the probabilities associated with the two weather conditions. Then determine the best acts from the standpoint of the (a) maximum probability and (b) expected payoff criteria.

Ans. (a) A_2: Beach umbrellas, (b) A_2: Beach umbrellas

DECISION TREE ANALYSIS

18.32 An investor is considering placing a $10,000 deposit to reserve a franchise opportunity in a new residential area for 1 year. There are two areas of uncertainty associated with this sequential decision situation: whether or not a prime franchise competitor will decide to locate an outlet in the same area and whether or not the residential area will develop to be a moderate or large market. The investor estimates that there is a 50-50 chance that the competing franchise system will develop an outlet. Thus the investor must first decide whether to make the initial $10,000 down payment. After the decision of the competing system is known the investor must then decide whether or not to proceed with constructing the franchise outlet. If there is competition and the market is large, the net gain during the relevant period is estimated as being $15,000; if the market is moderate, there will be a net loss of $10,000. If there is no competition and the market is large, the net gain will be $30,000; if the market is moderate, there will be a net gain of $10,000. The investor estimates that there is about a 40 percent chance that the market will be large. Using decision tree analysis, determine whether or not the initial deposit of $10,000 should be made.

Ans. Make the deposit ($EP = $9,000)

UTILITY FUNCTIONS AND EXPECTED UTILITY

18.33 For Problems 18.6 to 18.8, the extreme points of the possible monetary consequences are −$10,000 and $20,000. Describe how you would go about developing a utility function for the individual who is involved in this investment decision.

18.34 Based on your answer to Problem 18.33, suppose the set of utility values reported in Table 18.28 has been developed. Construct the appropriate utility curve and use it to determine whether the investor can be described as a risk averter, a risk seeker, or neutral with respect to risk. Illustrate the meaning of your conclusion in the context of this decision problem.

Ans. The decision maker is a risk seeker.

Table 18.28   Utility Values Equivalent to Monetary Values for the Investment Decision Problem

Monetary value, $	−10,000	−2,500	2,000	10,000	14,000	18,000	20,000
Utility value	0	0.1	0.2	0.4	0.6	0.8	1.0

18.35 Determine the best act for the investment decision described in Problems 18.6 to 18.8, based on the expected utility criterion and using the utility function developed in Problem 18.34.

Ans. A_5: Invest in stock options ($EU \cong 0.230$)

Chapter 19

Bayesian Decision Analysis: The Use of Sample Information

19.1 THE EXPECTED VALUE OF PERFECT INFORMATION (*EVPI*)

Given that the several possible events, or states, are uncertain and are associated with a probability distribution, the availability of *perfect information* indicates that the decision maker knows which event will occur with respect to each individual decision opportunity. The *expected value of perfect information* (*EVPI*) is the difference between the (long-run) expected payoff given such information minus the expected payoff associated with the best act under conditions of uncertainty. Although perfect information as such is seldom available, the *EVPI* serves as an indication of the maximum value that any sample can have to the decision maker. In determining the *EVPI*, note that under conditions of uncertainty a "best act" is identified as the consistent strategy for each and every decision opportunity. On the other hand, the availability of perfect information leads to a mixed strategy in which the decision act is perfectly matched to the event, or state, for each decision opportunity. The formula for determining the expected value of perfect information, where *EPPI* designates the *expected payoff with perfect information*, is

$$EVPI = EPPI - EP \text{ (under conditions of uncertainty)} \qquad (19.1)$$

EXAMPLE 1. Refer to Table 18.7 (page 306), where the best act under conditions of uncertainty is A_3: Order 15 units, with an expected payoff of \$3,250. If perfect information were available, act A_1 would be chosen 10 percent of the time, act A_2 30 percent of the time, and so forth corresponding to the relative frequencies of the associated demand levels. See Table 19.1. As indicated, the expected payoff with perfect information is \$4,050. Therefore, $EVPI = 4,050 - 3,250 = \$800$. This is the maximum amount that any sample information could be worth on the average, because it is the maximum amount by which the expected payoff could be increased with all uncertainty removed.

Table 19.1 Determination of the Expected Payoff with Perfect Information for the Decision Situation in Table 18.7

Market demand	Probability	Best act	Conditional value of best act	Expected payoff
E_1: 5	0.10	A_1	\$1,500	\$ 150
E_2: 10	0.30	A_2	3,000	900
E_3: 15	0.40	A_3	4,500	1,800
E_4: 20	0.20	A_4	6,000	1,200
				\$4,050

Consider now the relationship between opportunity loss and the value of perfect information. By definition, *EOL* of the best act is the average amount of regret associated with the optimum act, as a long-run average. See Section 18.4. Given that the *EOL* of the best act is the average amount by which the best possible gain given a perfect (mixed) strategy is missed, it follows that this should also be the value of perfect information for a given decision situation. Therefore, an alternative basis for determining the expected value of perfect information is

$$EVPI = EOL \text{ (best act)} \qquad (19.2)$$

322

EXAMPLE 2.   In Example 7 of Chapter 18 (see page 307), we observed that the same decision act (A_3) was identified as being the best act whether the act with the highest *EP* or the lowest *EOL* were identified. Now, we can further observe that the *EOL* value of the best act ($800) also indicates the *EVPI* for this situation. This value corresponds with the value obtained in Example 1, above.

In the above examples, note the difference between the expected payoff *with* perfect information and the expected value *of* perfect information. The expected payoff *with* perfect information (*EPPI*) indicates the long-run average result with all uncertainty removed, whereas the expected value *of* perfect information (*EVPI*) is the average *incremental* value of removing uncertainty in the decision situation.

As was the case in Chapter 18, in this chapter we restrict our coverage to events that follow a discrete probability distribution. The determination of the value of perfect information and the value of sample information for events that follow the normal probability distribution is covered in Chapter 20.

19.2   PRIOR AND POSTERIOR PROBABILITY DISTRIBUTIONS

In Section 5.7 Bayes' theorem is used to determine the revised probability of a first event given that a designated second event has occurred. This section presents the application of Bayes' theorem for the purpose of revising the *several* probabilities associated with the entire set of possible events, or states, in a decision situation. In this context, the *prior probability distribution* is the probability distribution which is applicable before the collection of any sample information. In Bayesian decision analysis such a probability distribution is often subjective in that it is based on judgments, although it could also be based on historical information. The *posterior probability distribution* is the probability distribution after sample information has been observed and has been used to revise the prior probability distribution by application of Bayes' theorem.

As illustrated in Section 5.7, in order to apply Bayes' theorem the prior probability of the uncertain event and the conditional probability of the sample result must be known. Typically, the conditional probabilities are determined by the use of some standard probability distribution according to the nature of the sampling situation.

EXAMPLE 3.   When a manufacturing process is in control, the proportion defective is only 0.01; when it is out of control, the proportion defective is 0.10. Historically, the process has been found to be out of control 5 percent of the time, and this is the basis for the probability values indicated in Table 19.2. A sample of $n = 10$ items is

Table 19.2   Prior Probability Values Associated with Two Possible Levels of Fraction Defective in a Manufacturing Process

Fraction defective	Prior probability
0.01	0.95
0.10	0.05

inspected and one item is found to be defective. In order to use Bayes' formula to revise the prior probability values, the binomial probability distribution (see Section 6.3) serves as the basis for determining the conditional probability of the sample result given each fraction defective in turn. For instance, the probability that $x = 1$ item will be defective, given a sample $n = 10$ and the fraction defective $p = 0.01$, is 0.0914 by reference to Appendix 2. Note that in the present context the value of "p" in the binomial table is the fraction defective, and *not* the prior probability value associated with that fraction defective. The tree diagram for this sampling situation (Fig. 19-1) includes all of the conditional probabilities as well as the prior probabilities.

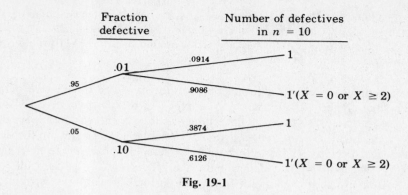

Fig. 19-1

Using formula (*5.16*), we determine the posterior probability that the true fraction defective is 0.01, given that one item was found to be defective in a sample of $n = 10$, as follows:

$$P(A \mid B) = \frac{P(A)P(B \mid A)}{P(A)P(B \mid A) + P(A')P(B \mid A')}$$

$$P(p = 0.01 \mid 1 \text{ def.}) = \frac{P(p = 0.01) \times P(1 \text{ def} \mid p = 0.01)}{P(p = 0.01) \times P(1 \text{ def} \mid p = 0.01) + P(p = 0.10) \times P(1 \text{ def} \mid p = 0.10)}$$

$$= \frac{(0.95)(0.0914)}{(0.95)(0.0914) + (0.05)(0.3874)} = \frac{0.08683}{0.10620} = 0.81761 \cong 0.82$$

We could now also apply Bayes' formula to determine the probability that the true fraction defective is 0.10 given the sample result. However, since only two levels of fraction defective are possible, this value obviously is the complement of the 0.82 determined above, or 0.18. The prior and posterior probability values for this example are summarized in Table 19.3. Note that the shift in the probability values after the sample reflects the fact that observing one defective item in a random sample of $n = 10$ is more likely with the fraction defective of 0.10 than with 0.01.

Table 19.3   Prior and Posterior Probability Distributions

Fraction defective	Prior probability	Posterior probability
0.01 0.10	0.95 0.05	0.82 0.18

Example 3 presents a direct application of the methodology in Section 5.7. When an entire set of probability values is being revised, however, a *tabular approach* for developing the posterior probability distribution is more convenient than the repeated application of Bayes' formula for determining each posterior probability value.

Table 19.4   Revision of Probabilities for the Fraction-Defective Problem

(1) Event	(2) Prior P	(3) Conditional probability of sample result	(4) Joint probability, Col. (2) × Col. (3)	(5) Posterior P Col. (4) ÷ sum
0.01 0.10	0.95 0.05	0.0914 0.3874	0.08683 0.01937	$0.81761 \cong 0.82$ $0.18239 \cong 0.18$
Total	1.00		0.10620	1.00

EXAMPLE 4. Table 19.4 illustrates the use of the tabular approach for developing the posterior probability distribution for the situation described in Example 3. For each event, the posterior probability value is determined by dividing the joint probability value in that row of column (4) by the sum of column (4). By this procedure, each entry in column (4) is equivalent to the numerator in Bayes' formula while the sum of column (4) is equivalent to the denominator. As indicated, these are the same posterior probability values as in Example 3.

19.3 BAYESIAN POSTERIOR ANALYSIS AND THE VALUE OF SAMPLE INFORMATION (AFTER SAMPLING)

The procedure by which sample information can be used to revise a prior probability distribution can now be applied to a decision problem concerned with whether or not to accept a shipment from a vendor. The best act is first determined based on the prior distribution and then on the basis of a posterior distribution. Finally, we determine the value of the particular sample result based on whether the identification of the best act was changed by the revision of the prior probability values.

EXAMPLE 5. Table 19.5 shows that the fraction (proportion) defective for a shipment of 1,000 items from a vendor can be at one of four levels: 0.01, 0.05, 0.10, and 0.20. The prior probability values based on historical experience with this vendor are included in the table. The costs associated with A_1: Accept are based on the fact that identification and removal of a defective item which becomes part of an assembled component costs $1.00. Thus, for a fraction defective of 0.01 there are $0.01 \times 1,000 = 10$ defective items, which involves a cost of $10 \times \$1.00 = \$10.00$ for later removal. The costs associated with A_2: Reject are based on the information that if the true fraction defective is 0.05 or less, our company is contractually obligated to accept the shipment, and thus must pay for the additional shipment expenses when the vendor returns the same shipment back to the company. Based on the prior probability distribution, the act which has the lowest expected cost is A_1: Accept, and thus in the absence of any sample information such a shipment should be accepted routinely. Indeed, if the best act were otherwise, it would indicate that we have contracted with the wrong vendor!

Table 19.5   Decision Table for the Shipment Problem Based on Use of the Prior Distribution

Fraction defective (p_i)	Prior (P)	Conditional cost of A_1: Accept	Conditional cost of A_2: Reject	Expected cost of A_1: Accept	Expected cost of A_2: Reject
0.01	0.50	$ 10.00	$200.00	$ 5.00	$100.00
0.05	0.30	50.00	200.00	15.00	60.00
0.10	0.10	100.00	0	10.00	0
0.20	0.10	200.00	0	20.00	0
Total	1.00			$50.00	$160.00

Whereas the analysis in Example 5 is based on the prior probability distribution, in Example 6, we assume that sample information is available as a basis for revising the prior probability distribution.

EXAMPLE 6. For the situation described in Example 5, suppose a random sample of $n = 10$ items is selected and two of the items are found to be defective. Using the tabular procedure described in Section 19.2, the prior probability distribution is revised in Table 19.6. As expected, the posterior probability values associated with the 0.10 and 0.20 fraction defective states are larger than the respective prior probability values, because of the sample result. The conditional probabilities in Table 19.6 are based on the binomial probability distribution. As in Section 19.2, note that, when the binomial table is used, the value of "p_i" is each fraction defective in turn, and *not* the prior probability value.

Table 19.6   Revision of Probabilities for the Shipment Problem

Fraction defective (p_i)	Prior (P)	Conditional probability of sample result $[P(X = 2 \mid n = 10, p_i)]$	Joint probability	Posterior (P)
0.01	0.50	0.0042	0.00210	$0.0284 \cong 0.03$
0.05	0.30	0.0746	0.02238	$0.3022 \cong 0.30$
0.10	0.10	0.1937	0.01937	$0.2616 \cong 0.26$
0.20	0.10	0.3020	0.03020	$0.4078 \cong 0.41$
Total	1.00		0.07405	1.00

Bayesian posterior analysis is the process of determining a best act by revising a prior probability distribution on the basis of sample data, and then using the resulting posterior probability distribution to determine the best decision act. With the posterior probability distribution available from Example 6, in Example 7 we complete the Bayesian posterior analysis.

EXAMPLE 7.   In Table 19.7, the expected cost for A_2: Reject ($66.00) is lower than the expected cost associated with A_1: Accept. Therefore, given that a sample of $n = 10$ contained two defective items, the best act from the standpoint of the Bayesian criterion of minimizing the expected cost is to reject the shipment and return it, rather than to enter the items into the manufacturing process.

Table 19.7   Cost Table for the Shipment Problem Based on Use of the Posterior Probability Distribution

Fraction defective (p_i)	Posterior (P)	Conditional cost of A_1: Accept	Conditional cost of A_2: Reject	Expected cost of A_1: Accept	Expected cost of A_2: Reject
0.01	0.03	$ 10.00	$200.00	$ 0.30	$ 6.00
0.05	0.30	50.00	200.00	15.00	60.00
0.10	0.26	100.00	0	26.00	0
0.20	0.41	200.00	0	82.00	0
Total	1.00			$123.30	$66.00

Having illustrated the process of Bayesian posterior analysis in Examples 5 through 7, we can now consider the value associated with a sample that has already been taken. The estimated *value of sample information* (*VSI*) for a sample which has already been observed is based on the difference between the *posterior expected payoffs* (or costs) of the best acts identified before and after sampling, or

$$VSI = \text{(posterior expected payoff of best posterior act)}$$

$$- \text{(posterior expected payoff of best prior act)} \qquad (19.3)$$

Thus, if the identification of the best act has not been changed as a result of the posterior analysis, then the value of the sample information is $0, even though the expected payoff associated with that act would generally have changed.

EXAMPLE 8.   In order to estimate the value of the sample information included in Example 7, we first note that as a result of the posterior analysis the best decision act was changed to A_2: Reject, from the best act of A_1: Accept, which was applicable without the sample. The estimated value of the sample information which was obtained is

$$VSI = -66.00 - (-123.30) = -66.00 + 123.30 = \$57.30$$

(*Note:* Since both of the expected values are costs in this case, they are entered as negative values above so that the *VSI* is appropriately positive. In the long run, the sample information served to revise the prior probabilities toward a more accurate assessment. Even though we thought that the expected cost of A_1: Accept was $50.00 based on the prior analysis, with the sample information considered we recognize that a more reliable assessment of this expected cost is $123.30. The sample information caused us to change our decision from this act to act A_2: Reject, with the (posterior) expected cost of $66.00, thus reducing the expected cost by $57.30 from what we would have experienced had act A_1 been chosen.)

19.4 PREPOSTERIOR ANALYSIS: THE EXPECTED VALUE OF SAMPLE INFORMATION (*EVSI*) PRIOR TO SAMPLING

Preposterior analysis is the process by which the value of sample information is estimated before the sample is collected. The basic procedure is to consider all of the possible sample outcomes, determine the estimated value in the decision process of each sample outcome, and then determine the *expected values of the sample information* (*EVSI*) by weighting each of these different values by the probability that the associated sample outcome will occur.

Example 9 illustrates the process of preposterior analysis for a sample of only $n = 1$. Because all possible sample outcomes have to be considered, the computation of *EVSI* for samples that are not small is tedious. However, the procedure described and illustrated below can be applied for larger samples by the use of a computer.

EXAMPLE 9.   Suppose we wish to determine the expected value of sample information associated with a sample of $n = 1$ for the decision problem described in Example 5. Table 19.5 is the cost table for this problem, and on the basis of the prior distribution, the best act is A_1: Accept the shipment, with an expected cost of $50.00 (as contrasted to A_2: Reject the shipment, with an expected cost of $160.00). We proceed by determining the *VSI* associated with each possible sample outcome, and then combine these values to compute the *EVSI* for the sample of $n = 1$.

Step 1: Determine the posterior distribution if zero items are defective in the sample of $n = 1$. The revision procedure and the posterior probability distribution are presented in Table 19.8. Note that the probability values have "shifted" somewhat in the expected direction. Also notice that the sum of the "joint probability" column is identified as indicating the overall (unconditional) probability of the sample result being observed. This probability value is used in the final step of determining the *EVSI*.

Table 19.8   Revision of Probabilities for the Shipment Problem, Given No Defective in a Sample of $n = 1$

Fraction defective (p_i)	Prior (P)	Conditional probability of sample result $[P(X = 0 \mid n = 1, p_i)]$	Joint probability	Posterior (P)
0.01	0.50	0.99	0.495	$0.5211 \cong 0.521$
0.05	0.30	0.95	0.285	$0.3000 = 0.300$
0.10	0.10	0.90	0.090	$0.0947 \cong 0.095$
0.20	0.10	0.80	0.080	$0.0842 \cong 0.084$
Total	1.00		$P(X = 0) = 0.950$	1.000

Step 2: Determine the best act if zero items are defective in the sample of $n = 1$ and determine the expected payoffs (costs) associated with the alternative acts. As indicated in Table 19.9, the best act is A_1: Accept the shipment, which is the same act which was determined best on the basis of the prior analysis (before considering any sample information) in Example 5.

Table 19.9   Determination of the Best Act for the Shipment Problem, Given No Defective in a Sample of
n = 1

Fraction defective (p_i)	Posterior (P)	Conditional cost of A_1: Accept	Conditional cost of A_2: Reject	Expected cost of A_1: Accept	Expected cost of A_2: Reject
0.01	0.521	$ 10.00	$200.00	$ 5.21	$104.20
0.05	0.300	50.00	200.00	15.00	60.00
0.10	0.095	100.00	0	9.50	0
0.20	0.084	200.00	0	16.80	0
				Total: $46.51	Total: $164.20

Step 3: Determine the estimated value of the information that zero items are defective in the sample of $n = 1$. Since the best act is not changed if this sample result is observed, the value of the sample is $0. Or, using formula (*19.3*),

$$VSI = \text{(posterior expected payoff of best posterior act)} - \text{(posterior expected payoff of best prior act)}$$
$$= -46.51 - (-46.51) = -46.51 + 46.51 = \$0$$

Step 4: Determine the posterior distribution if one item is defective in the sample of $n = 1$. The revision procedure and the posterior probability distribution are presented in Table 19.10. Again, note that the probability values have shifted in the expected direction. Also, the sum of the "joint probability" column again indicates the overall probability of the sample result being considered, in this case that $X = 1$ given $n = 1$ and given the prior probability distribution.

Table 19.10   Revision of Probabilities for the Shipment Problem, Given One Defective in a Sample of
n = 1

Fraction defective (p_i)	Prior (P)	Conditional probability of sample result $[P(X = 1 \mid n = 1, p_i)]$	Joint probability	Posterior (P)
0.01	0.50	0.01	0.005	0.10
0.05	0.30	0.05	0.015	0.30
0.10	0.10	0.10	0.010	0.20
0.20	0.10	0.20	0.020	0.40
			Total $P(X = 1) = \overline{0.050}$	Total: $\overline{1.00}$

Step 5: Determine the best act if one item is defective in the sample of $n = 1$ and determine the expected payoffs (costs) associated with the alternative acts. As indicated in Table 19.11, the best act is A_2: Reject the shipment, which is a change from the best act based on the prior probability distribution.

Table 19.11   Determination of the Best Act for the Shipment Problem, Given One Defective in a Sample
of n = 1

Fraction defective (p_i)	Posterior (P)	Conditional cost of A_1: Accept	Conditional cost of A_2: Reject	Expected cost of A_1: Accept	Expected cost of A_2: Reject
0.01	0.10	$ 10.00	$200.00	$ 1.00	$20.00
0.05	0.30	50.00	200.00	15.00	60.00
0.10	0.20	100.00	0	20.00	0
0.20	0.40	200.00	0	80.00	0
				Total: $116.00	Total: $80.00

Step 6: Determine the estimated value of the information that one item is defective in the sample of $n = 1$. Since the best act is changed if this sample result is observed, the value is determined using formula (*19.3*):

$$VSI = (\text{Posterior expected payoff of best posterior act}) - (\text{posterior expected payoff of best prior act})$$

$$= -80.00 - (-116.00) = -80.00 + 116.00 = \$36.00$$

Step 7: Since the *VSI* associated with each possible sample outcome has now been determined, combine these values to obtain the expected value of sample information by means of the following formula:

$$EVSI = \sum (VSI_i) P(X_i) \qquad\qquad (19.4)$$

Applying formula (*19.4*),

$$EVSI = 0(0.950) + 36.00(0.050) = 0 + 1.80 = \$1.80$$

In the above calculation, each possible *VSI* amount is multiplied by the probability of the associated sample result occurring, and the resulting products are summed. As indicated in Steps 1 and 4 above, the probability values of the two different sample results occurring for the present problem are indicated by the sums of the "joint probability" columns in Tables 19.8 and 19.10, respectively.

19.5 EXPECTED NET GAIN FROM SAMPLING (*ENGS*) AND OPTIMUM SAMPLE SIZE

For each sample size considered, the *expected net gain from sampling* (*ENGS*) is defined as the difference between the expected value of sample information and the cost of the sample (*CS*):

$$ENGS = EVSI - CS \qquad\qquad (19.5)$$

If the *ENGS* of a sample being considered is greater than zero in value, then the sample should be taken unless some other sample has a higher *ENGS*. This matter is covered later in this section, with respect to optimum sample size.

EXAMPLE 10.   For the sample size of $n = 1$ which was considered in Example 9, $EVSI = \$1.80$. Suppose the cost of inspection is $1.20 per item sampled. We determine the expected net gain from taking a sample of $n = 1$ in this decision situation as follows:

$$ENGS = EVSI - CS = 1.80 - 1.20 = \$0.60$$

Therefore, it is better to take a sample of one item before deciding whether or not to accept the shipment, rather than to make the decision without the sample.

As the sample size which is being considered is increased in preposterior analysis, the *EVSI* becomes progressively larger and approaches the value of *EVPI* for the decision situation. The *optimum sample size* is the size such that the *ENGS* is maximized. Figure 19-2 illustrates the relationship among *EVPI*, *EVSI*, and *CS* as the sample size is increased. The general relationship between sample size and *ENGS* is indicated in Fig. 19-3.

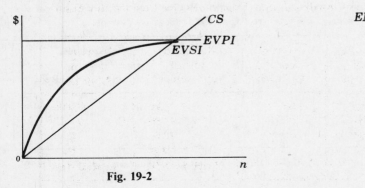

Fig. 19-2

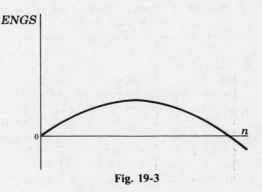

Fig. 19-3

EXAMPLE 11.　The values of *EVSI*, *CS*, and *ENGS* for the shipment inspection problem discussed in Examples 9 and 10 are presented in Table 19.12, in order of sample size. The values for $n = 0$ are based on the absence of a sample while the values for $n = 1$ were determined in Examples 9 and 10. The values for $n = 2$ were determined by a procedure similar to that followed for $n = 1$, except that three possible sample results had to be considered ($X = 0$, $X = 1$, and $X = 2$). These calculations are not presented in this chapter. Since *ENGS* always "peaks" at some sample size (including possibly at $n = 0$) and then declines, we conclude from the information in Table 19.12 that the optimum sample size for this simplified problem is $n = 1$. For more realistic problems involving the consideration of larger sample sizes, the use of a computer would be required.

Table 19.12　*EVSI*, *CS*, and *ENGS* for the Shipment Problem According to Sample Size

Sample size (n)	EVSI	CS	ENGS
0	$0	$0	$0
1	1.80	1.20	0.60
2	2.83	2.40	0.43

In some texts in the area of Bayesian decision analysis, attention is given to determining the *overall terminal expected payoff* (*OTEP*) and *net overall terminal expected payoff* (*NOTEP*) in conjunction with preposterior analysis. For each possible sample size, the overall terminal expected payoff is the expected value of the best act before sampling plus the expected value of the sample information. This sum is reduced by the cost of sampling to determine the net overall expected payoff. Thus, the overall terminal expected payoff is

$$OTEP = EP \text{ (best prior act)} + EVSI \qquad (19.6)$$

The net overall terminal expected payoff is

$$NOTEP = OTEP - CS \qquad (19.7)$$

Finally, we should observe that if a sample is to be taken, then the best decision will depend on the observed sample result. Therefore, a complete Bayesian decision rule which involves sampling requires that the optimum sample size be identified, and further, that the best act according to various sample results be specified.

EXAMPLE 12.　Table 19.13 is a summary table for the shipment inspection problem. In addition to values which were previously reported, the *OTEP* and *NOTEP* values are indicated for each of the three sample sizes considered. The costs are expressed as negative values so that the subtraction leading to the determination of the *NOTEP* values would be algebraically consistent with the general formula. Therefore, the optimum sample size in Table 19.13 is the one with the largest (algebraic) value of *NOTEP*, or $n = 1$. Note that this conclusion is consistent with

Table 19.13　A Summary Table of the Preposterior Analysis for the Shipment Problem and the Designation of the Best Bayesian Decision Rule (the Inspection Costs Are Expressed as Negative Numbers)

Sample size (n)	EVSI	OTEP	CS	ENGS	NOTEP	Decision rule	
						A_1: Accept	A_2: Reject
0	$0	−$50.00*	$0	$0	−$50.00	Always	Never
1†	1.80	−48.20	1.20	0.60	−49.40	$X = 0$	$X = 1$
2	2.83	−47.17	2.40	0.43	−49.57	$X = 0$	$X \geq 1$

*This is the expected payoff (cost) of the best act based on the prior probability distribution and without sampling. See Table 19.7.
†The optimum sample size, based either on maximizing *ENGS* or maximizing *NOTEP*.

the result indicated in Example 11 and in Table 19.12. Given that a sample of $n = 1$ should be taken, in Table 19.13 we also observe that if the sampled item is not defective ($X = 0$), the best act is A_1: Accept. If the sampled item is defective ($X = 1$), the best act is A_2: Reject.

Solved Problems

THE EXPECTED VALUE OF PERFECT INFORMATION

19.1   Determine the *EVPI* for the investment decision situation in Problems 18.6 to 18.8 by determining the *EOL* associated with the best act based on the Bayesian criterion of expected payoff.

As indicated in Table 18.19, the best act from the standpoint of the expected payoff criterion is A_2: Corporate bonds. The opportunity-loss values for all acts are identified in Table 18.18. Thus,

$$EVPI = EOL(A_2) = 100(0.30) + 0(0.50) + 19,100(0.20) = \$3,850$$

19.2   Determine the *EVPI* for the shipment problem described in Example 5 by first determining the expected cost with perfect information.

Because this decision problem is concerned with costs, note that *EVPI* in this case is determined by subtracting the expected cost with perfect information from the expected cost of the best act under conditions of uncertainty. Referring to Table 19.5, under perfect information we would accept the shipment whenever the fraction defective is either 0.01 or 0.05, and we would reject it whenever it is 0.10 or 0.20. On the basis of this decision rule, the expected cost and then *EVPI* are calculated as follows:

$$EC \text{ (with perfect information)} = 10(0.50) + 50(0.30) + 0(0.10) + 0(0.10) = \$20.00$$
$$EC \text{ (under conditions of uncertainty)} = \$50.00 \qquad \text{(from Example 5)}$$
$$EVPI = EC \text{ (under conditions of uncertainty)} - EC \text{ (with perfect information)}$$
$$= 50.00 - 20.00 = \$30.00$$

19.3   In Problem 18.5 the best act with respect to how many large-screen television sets are to be ordered is determined by identifying the act with the lowest *EOL*. The best act is A_2: Order three sets, with $EOL(A_2) = \$170$. Referring to Table 18.14, demonstrate that this *EOL* is equal to the *EVPI* value which can be calculated by first determining the expected payoff with perfect information.

$$EPPI = 400(0.30) + 600(0.40) + 800(0.20) + 1,000(0.10) = \$620$$
$$EP \text{ (under conditions of uncertainty)} = \$450 \qquad \text{(from Problem 18.4)}$$
$$EVPI = EPPI - EP \text{ (under conditions of uncertainty)}$$
$$= 620 - 450 = \$170$$

PRIOR AND POSTERIOR DECISION ANALYSIS

19.4   A new product being considered by our firm will have either a low, moderate, or high level of sales. We have the choice of either marketing or not marketing the product. Table 19.14 is the payoff table which indicates the probability values associated with the three possible sales levels, based on our best judgment, and the economic consequences for this decision situation.

(*a*)   Determine the best act from the standpoint of the Bayesian criterion of maximizing expected payoff.

(*b*)   Determine the maximum value that any further information can have.

Table 19.14   Payoff Table for Marketing a New Product

Event	Probability	Decision act	
		A_1: Market	A_2: Don't market
E_1: Low sales	0.40	−$30,000	$0
E_2: Moderate sales	0.40	10,000	0
E_3: High sales	0.20	50,000	0

(*a*)
$$EP(A_1) = -30,000(0.40) + 10,000(0.40) + 50,000(0.20) = \$2,000$$
$$EP(A_2) = \$0$$

Therefore, the best act is A_1: Market the product.

(*b*)   Table 19.15 portrays the table of opportunity losses, or regrets, for this decision situation. Since the *EVPI* is equal to the *EOL* of the best act,

$$EVPI = EOL(A_1) = 30,000(0.40) + 0(0.40) + 0(0.20) = \$12,000$$

Alternatively, we can determine the *EVPI* by first determining the expected payoff with perfect information and then subtracting the expected payoff of the best act (under uncertainty) from this value:

$$EPPI = 0(0.40) + 10,000(0.40) + 50,000(0.20) = \$14,000$$
$$EVPI = EPPI - EP \text{ (under conditions of uncertainty)}$$
$$= 14,000 - 2,000 = \$12,000$$

Table 19.15   Opportunity Loss Table for Marketing a New Product

Event	Probability	Decision act	
		A_1: Market	A_2: Don't market
E_1: Low sales	0.40	$30,000	$ 0
E_2: Moderate sales	0.40	0	10,000
E_3: High sales	0.20	0	50,000

19.5   For the decision situation in Problem 19.4, test-marketing of the product in one region results in a moderate level of sales in that region. A review of historical sales patterns in that region indicates that for products with low sales nationally, sales in the region were moderate 30 percent of the time. For products with a moderate level of sales nationally, sales in the region were also moderate 70 percent of the time. For products with a high level of sales nationally, sales in the region were moderate only 10 percent of the time.

(*a*)   Revise the prior probabilities used in Problem 19.4, on the basis of the sample result that sales were at a moderate level in the test-market region.

(*b*)   Determine the best decision act on the basis of posterior decision analysis.

(*a*)   The posterior probability distribution is determined in Table 19.16. As would be expected, the probability that the true level of demand is moderate is increased after taking the sample results into consideration.

(b)   The best act after the sample (the Bayesian posterior act) is determined by using the posterior probabilities determined in Table 19.16 in conjunction with the conditional values presented in Table 19.14. The expected payoffs associated with acts A_1 (Market) and A_2 (Don't market) are

$$EP(A_1) = -30,000(0.29) + 10,000(0.67) + 50,000(0.05) = \$500$$

$$EP(A_2) = \$0$$

Therefore, as in the case of the prior analysis, the best act is A_1: Market the product.

Table 19.16   Revision of Probabilities for Marketing a New Product Based on a Moderate Level of Sales in the Test Market

Event	Prior (P)	Conditional probability of sample result	Joint probability	Posterior (P)
E_1: Low sales	0.40	0.30	0.12	$0.2857 \cong 0.29$
E_2: Moderate sales	0.40	0.70	0.28	$0.6666 \cong 0.67$
E_3: High sales	0.20	0.10	0.02	$0.0476 \cong 0.05$
Total	1.00		0.42	1.01

19.6   What is the estimated value of the sample information that was obtained in Problem 19.5? How do you reconcile this value with the observed difference in the expected payoff in this risk situation after the sample as compared with before the sample?

Because the optimal act was not changed by the sample result, the sample had no value as such. More specifically,

$$VSI = \text{(posterior expected payoff of the best posterior act)}$$
$$- \text{(posterior expected payoff of the best prior act)}$$
$$= 500 - 500 = \$0$$

The difference between the posterior and prior values of the optimal act is $500 - 2,000 = -\$1,500$. In other words, the posterior expected payoff is \$1,500 less than the prior expected payoff for the same act. However, this change took place because the sample provided more information about the uncertain states, and in this sense the sample information did not cause a change as such. Without the sample we determined that the expected payoff of the best act was \$2,000, but after the sample we recognize that for this particular decision situation the prior expectation should be changed. However, because we would still choose the same act as the optimal act, this knowledge does not cause us to change the eventual expected monetary result from that which would have occurred without the sample information.

19.7   The owner of a small manufacturing concern has the opportunity to purchase five used lathes as a group from excess equipment being disposed of by a large firm in the area. The uncertain factor is the number of machines that will require a major overhaul before being put to productive use in his factory. The decision to purchase them must take into account the cost of the machines to him, their value if in good condition, and the cost of overhaul. The various economic consequences for this decision are given in Table 19.17. Determine the best act for this decision situation and the expected payoff associated with this act.

Table 19.17 Decision Table for the Purchase of the Used Lathes

Number of defective lathes	Probability	Decision act	
		A_1: Purchase	A_2: Don't purchase
0	0.05	$5,000	$0
1	0.10	3,000	0
2	0.20	1,000	0
3	0.30	−1,000	0
4	0.30	−3,000	0
5	0.05	−5,000	0

$$EP(A_1) = 5,000(0.05) + 3,000(0.10) + 1,000(0.20) + (-1,000)(0.30) + (-3,000)(0.30) + (-5,000)(0.05)$$
$$= -\$700$$
$$EP(A_2) = \$0$$

Therefore, the best act is A_2: Don't purchase the machines, with an expected payoff of $0.

19.8 Determine the expected value of perfect information for the decision situation described in Problem 19.7.

$$EPPI = 5,000(0.05) + 3,000(0.10) + 1,000(0.20) + 0(0.30) + 0(0.30) + 0(0.05) = \$750$$
$$EP \text{ (under conditions of uncertainty)} = \$0 \qquad \text{(from Problem 19.10)}$$
$$EVPI = EPPI - EP \text{ (under conditions of uncertainty)} = 750 - 0 = \$750$$

19.9 Referring to Problem 19.7, suppose that a thorough inspection of each machine is permitted, but such inspection costs $50 per machine. The owner of the manufacturing company chooses two machines at random, has them inspected, and learns that neither one of these machines is in need of an overhaul. Determine the best act based on Bayesian posterior analysis.

Because the population (of five machines) is finite and relatively small in size, the probability distribution which is used to determine the conditional probabilities of the sample result must take into consideration the fact that sampling without replacement is involved in this decision situation. The required probabilities can be determined by the use of tree diagrams or by the use of the hypergeometric distribution (see Section 6.5). For the population of five machines which includes two defective machines, for example, Fig. 19-4 can be used in conjunction with the multiplication rule for dependent events to obtain the required conditional probability (see Section 5.6). In this figure, D stands for a defective machine being observed

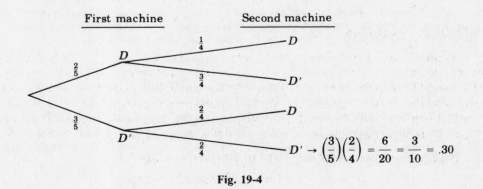

Fig. 19-4

and D' stands for the observed machine being nondefective. As indicated, the probability of observing no defective item in a sample of $n = 2$ for the given population is 0.30. Conditional probabilities for other populations can be similarly determined.

The revision of the prior probability distribution based on the result that two sampled machines were both found to be nondefective is presented in Table 19.18, with the conditional probability values in the third column having been determined by either of the two procedures described above. Using the posterior probability distribution, we compute the posterior expected payoffs as follows:

$$EP(A_1) = 5,000(0.25) + 3,000(0.30) + 1,000(0.30) + (-1,000)(0.15) + (-3,000)(0.00) + (-5,000)(0.00)$$
$$= \$2,300$$
$$EP(A_2) = \$0$$

Therefore, the best Bayesian posterior act is A_1: Purchase the machines, with an expected payoff of $2,300.

Table 19.18   Revision of Probabilities for the Purchase of the Used Lathes Based on No Machines Being Defective for a Sample of $n = 2$

Number of defective lathes	Prior (P)	Conditional probability of sample result	Joint probability	Posterior (P)
0	0.05	1.00	0.05	0.25
1	0.10	0.60	0.06	0.30
2	0.20	0.30	0.06	0.30
3	0.30	0.10	0.03	0.15
4	0.30	0	0	0
5	0.05	0	0	0
Total	1.00		$P(X = 0) = 0.20$	1.00

19.10   What is the estimated net value of the sample information obtained in Problem 19.9?

$$VSI = (\text{posterior expected value of the best posterior act})$$
$$- (\text{posterior expected value of the best prior act})$$
$$= 2,300 - 0 = \$2,300$$
$$\text{Net } VSI = VSI - CS = 2,300 - 100 = \$2,200$$

Note that the net value above is *not ENGS* as described in Section 19.5, because the above value is the estimated net gain associated with a particular sample that has already been taken, not the expected value of a sample of a given size prior to sampling.

PREPOSTERIOR ANALYSIS

19.11   In Section 19.5 we indicated that the overall terminal expected payoff (*OTEP*) can be determined by adding the *EVSI* for the given sample size to the *EP* (prior). Logically, *OTEP* can also be determined by identifying the *EP* (posterior) that would be associated with each possible sample result and then weighting these payoffs by their respective probabilities to determine the overall terminal expected payoff. Apply this procedure to the preposterior analysis in Example 9, and demonstrate that the resulting value for *OTEP* is the same as reported in Table 19.13.

The alternative formula for the overall terminal expected payoff is

$$OTEP = \sum (\text{posterior } EP_i) P(X_i) \qquad (19.8)$$

The possible posterior *EP* amounts for the sample of $n = 1$ are reported in Tables 19.9 and 19.11, while the respective probabilities of the associated sample results are reported in Tables 19.8 and 19.10. Substituting these values in the above equation,

$$OTEP = (-46.51)(0.95) + (-80.00)(0.05) = -\$48.18$$

Except for rounding error, this is the same negative payoff (cost) as the $-\$48.20$ reported in Table 19.13.

Supplementary Problems

THE EXPECTED VALUE OF PERFECT INFORMATION

19.12   For Problems 18.22 to 18.25, determine the expected value of perfect information by subtracting the expected payoffs of the best act under uncertainty from the expected payoff with perfect information. Compare your answer with the expected opportunity loss of the best act, as determined in Problem 18.25.

Ans. $EVPI = 20.20 - 18.50 = \$1.70$

19.13   For Problems 18.26 to 18.28, determine the expected value of perfect information regarding the number of spare propellers that will be required.

Ans. $EVPI = \$160$

PRIOR AND POSTERIOR DECISION ANALYSIS

19.14   A jobber has the opportunity to purchase a shipment of 10 high-quality stereo systems manufactured abroad for \$1,000. However, the equipment has been in transit by ocean freighter for some time, and there is a distinct possibility of moisture damage having occurred. Based on his previous experience with the shipping company involved, the jobber estimates that there is a 20 percent chance the shipment is damaged. If the shipment is damaged, the jobber can sell the stereo sets for just \$500. If no moisture damage has occurred, he can resell the entire shipment for a net profit of \$300. Determine whether or not the jobber should purchase the shipment from the standpoint of the expected payoff criterion.

Ans.   Purchase, with $EP = \$140$.

19.15   With respect to the jobber's decision in Problem 19.14, for practical purposes "damaged" means that one-half of the stereo sets are affected, because more extensive damage would be visually obvious. The jobber pays \$10 to have one randomly selected stereo set uncrated and tested, and it is found to perform adequately. Determine the best act given this sample information, and the expected payoff associated with this act.

Ans.   Purchase, $EP = \$212$.

19.16   What is the estimated value of the sample information which was collected in Problem 19.15?

Ans.   \$0, or considering the cost of the sample, $-\$10$.

19.17   Refer to the investment decision described in Problems 18.18 to 18.21. Determine (*a*) the expected payoff with perfect information (*EPPI*), and (*b*) the *EVPI* for this decision problem.

Ans.   (*a*) \$1,150, (*b*) \$350

19.18   For Problems 18.18 to 18.21, by a somewhat simplified approach the investment analyst defines "upturn" to mean that (at least) 70 percent of the usual purchasers of chemical products increase their order amounts, "no change" to mean that (about) 50 percent of the purchasers increase their order amounts, and "downturn" to mean that 30 percent (or fewer) of the purchasers increase their order amounts. He contacts a random sample of 20 purchasers of chemical products and learns that 14 of them are increasing their order amounts over previous periods.

(a) Revise the prior probability distribution regarding the three possible states of the chemical industry, as given in Problem 18.18, by taking this sample result into consideration.

(b) Using the posterior probability distribution, determine the best act from the standpoint of the expected payoff criterion and compare it with your answer in Problem 18.21.

Ans. (a) P (upturn) $\cong 0.91$, P (no change) $\cong 0.09$, P (downturn) $\cong 0.00$;

   (b) Invest in mutual fund ($EP = \$1,392$).

19.19 Estimate the value of the sample information which was obtained in Problem 19.18.

Ans. $592

19.20 Using the posterior probability distribution developed in Problem 19.18, determine (a) the expected payoff with perfect information and (b) the $EVPI$ after the sample has been taken. Compare your results with those in Problem 19.17.

Ans. (a) $1,437, (b) $45

19.21 Referring to Problem 18.29, suppose that the club need not commit itself to buying the blankets or beach umbrellas until 3 days before the event. Therefore, the members decide to add the weatherman's forecast as additional information. Looking at the weatherman's record, for days that are in fact cool he has correctly forecast the weather 90 percent of the time. For days that are in fact hot, he has correctly forecast the weather 70 percent of the time. For the day of the "Water Carnival" the weatherman forecasts cool weather. Revise the prior probability distribution given in Problem 18.31 based on this forecast.

Ans. P (Cool) $\cong 0.67$, P (Hot) $\cong 0.33$

19.22 Taking the weather forecast given in Problem 19.21 into consideration, determine the best decision act for the situation described in Problem 18.29.

Ans. A_1: Order blankets.

19.23 Estimate the value of the weather forecast given in Problem 19.21.

Ans. $54.60

PREPOSTERIOR ANALYSIS

19.24 In Example 9 the $EVSI$ for a sample of $n = 1$ was determined to be $1.80. Using the computational procedure illustrated in Example 9, determine the VSI associated with a sample of $n = 2$ and with the number of defective items found being $X = 0$, $X = 1$, and $X = 2$, respectively.

Ans. $0, $23.22, and $134.14

19.25 Referring to the conditional VSI figures calculated in Problem 19.24, determine the $EVSI$ associated with a sample of $n = 2$.

Ans. $2.83

Chapter 20

Bayesian Decision Analysis: Application of the Normal Distribution

20.1 INTRODUCTION

This chapter deals with Bayesian decision analysis for an event (state) that is a continuous random variable and is normally distributed, rather than being a discrete random variable. The methodology by which the mean and standard deviation for such a prior probability distribution are determined is illustrated in Section 20.2. The techniques of analysis in this case are further based on the requirement that only two decision acts are being evaluated, or compared, and that the payoff functions associated with these acts are linear payoff functions (see Section 20.3). Although these requirements may appear to be quite restrictive, a broad range of decision problems can be analyzed by these techniques.

EXAMPLE 1. An electronics firm has decided to assemble a telephone paging device for subsequent distribution by the company. The specific decision which has to be made now is whether Type 1 or Type 2 equipment is to be installed in the assembly plant. A larger capital investment is required for the Type 2 equipment, but the variable manufacturing cost is lower, which in turn leads to a higher incremental profit after the fixed (capital) cost has been covered (see Example 5). However, the Type 2 equipment would be more profitable than Type 1 equipment only at relatively higher levels of sales, which is the uncertain event in this problem. It is assumed that the decision maker's estimate of the uncertain sales level follows the normal distribution. Because the sales level estimate is the key factor in determining which type of equipment will yield the best return, it is the first area of interest in the prior analysis. In this example we have indicated that (1) our uncertainty in regard to the event follows the normal distribution; (2) two acts are being considered—purchasing Type 1 or Type 2 equipment; and (3) there is a constant incremental profit over variable cost per unit sold, which is indicative of the existence of linear payoff functions for the two acts.

20.2 DETERMINING THE PARAMETERS OF THE NORMAL PRIOR PROBABILITY DISTRIBUTION

The prior probability distribution is descriptive of the uncertainty which is associated with the decision maker's estimate of the random event. It is not the event which follows the probability distribution, but rather, the estimate of the event. Since the estimate is based on a judgment, there is no mathematical theorem which would justify the use of the normal distribution with respect to such a judgment in any specific instance. However, for judgment situations in which an informed decision maker is aware of a number of uncertain factors which could influence the value of the eventual outcome in either one direction or the other, the use of the normal distribution has been found to be a satisfactory approximation of the uncertainty inherent in the estimate.

As described in Section 7.2, a normal distribution is defined by identifying the mean and the standard deviation of the distribution. The mean of the prior distribution can be obtained by asking the decision maker to identify the "most likely" value of the random event or by asking him for that value such that there is a 50 percent chance that the actual value will be lower and a 50 percent chance that the actual value will be higher. Note that the first approach in fact is a request for the mode of the probability distribution while the second approach asks for the median of the distribution. As indicated in Section 3.6, the mean, median, and mode are all at the same point for a normally distributed variable. The mean of the prior probability distribution is designated by M_0 in this text.

EXAMPLE 2. For the decision problem introduced in Example 1, suppose the decision maker says that the most likely level of sales nationally for the telephone paging device is 40,000 units during the relevant period. On the basis of this judgment we designate the value of the mean of the prior distribution as $M_0 = 40,000$ units.

The prior mean is obtained by describing either the mode or median because these measures are easily described and conceptualized in a nonmathematical fashion. In the case of the standard deviation, it would be even more difficult to obtain such a prior estimate directly. Instead, the decision maker provides the boundaries for a stated probability interval, and the standard deviation of the prior distribution is determined on the basis of the observed size of the interval. As contrasted to a classical confidence interval based on sample information, an interval based on judgment is often called a *credibility interval*. The "middle 50 percent" interval is the one which is most frequently used, since this possible range of outcomes is relatively the easiest to conceptualize. Then the standard deviation is determined by observing that the middle 50 percent of the normal distribution is contained within approximately $\pm 2/3\sigma$ units (the specific value is 0.67σ). Where the prior standard deviation is designated by S_0, the computational formula to determine the prior standard deviation by use of the middle 50 percent credibility interval is

$$\frac{2}{3} S_0 = \frac{\text{middle 50\% interval}}{2} \tag{20.1}$$

or

$$S_0 = \frac{3(\text{middle 50\% interval})}{4} \tag{20.2}$$

EXAMPLE 3.   For the decision problem discussed in Examples 1 and 2, if the decision maker states that he has 50 percent confidence that the actual sales level will be somewhere between 35,000 and 45,000 units, the standard deviation of the prior probability distribution is

$$\frac{2}{3} S_0 = \frac{45,000 - 35,000}{2} = 5,000$$

$$S_0 = 7,500 \text{ units}$$

By the procedures illustrated in Examples 2 and 3 we have determined the values of M_0 and S_0 for total market demand. However, it is often the case that the estimate for an event such as sales is analyzed on the basis of mean results per outlet, rather than for the overall market. There are two reasons for this. First, a manager may find this easier. Second, and more important from the standpoint of analysis, if the prior probability distribution is to be revised on the basis of sample information, then that sample information will be obtained from randomly chosen outlets. In order to use such sample information, the prior distribution must itself be formulated on a per-outlet basis.

EXAMPLE 4.   If we anticipate collecting sample information as well as managerial judgment, the estimates obtained in Examples 2 and 3 can be converted to a per-outlet basis. Alternatively, we could have asked for per-outlet estimates in the first place. Suppose the paging device is to be distributed through a total of 500 retail outlets in the country. If we ask the decision maker for the most likely level of sales per outlet, the answer should be $M_0 = 80$ to be consistent with his estimate of 40,000 units for the total market in Example 2. Similarly, he should state that there is a 50 percent chance that average sales per outlet will be between 70 and 90 units to be consistent with the interval for total sales given in Example 3. Thus, on a per-outlet basis the prior standard deviation is

$$\frac{2}{3} S_0 = \frac{90 - 70}{2} = 10$$

$$S_0 = 15$$

We have now demonstrated how the parameters of a prior probability distribution can be obtained given that the distribution is assumed to be normally distributed. The student should be particularly alert to the fact that the prior mean and prior standard deviation are descriptive of the decision maker's estimate of the random event and his uncertainty with respect to this event. Given that the estimate is assumed to be unbiased, it follows that M_0 is an estimate of the overall population mean μ (for example, the actual mean level of sales for the 500 retail outlets). However, it does *not* follow that S_0 is an estimator of σ, the population standard deviation. In this regard, it is more appropriate to think of M_0 and S_0 as being similar to $\bar{X}$ and $\sigma_{\bar{x}}$, respectively, but being based on judgment rather than on a sample.

20.3 DEFINING THE LINEAR PAYOFF FUNCTIONS AND DETERMINING THE BEST ACT

The existence of a linear payoff function indicates that the expected payoff associated with an act is a linear function with respect to the uncertain level of the state. Algebraically,

$$\text{Payoff}\,(A_1) = k_1 + b_1 X \tag{20.3}$$

$$\text{Payoff}\,(A_2) = k_2 + b_2 X \tag{20.4}$$

where k is the constant factor (a capital expenditure would be a negative constant), b is the incremental revenue (or cost) per unit of the random variable and thus indicates the slope of the line, and X indicates the value of the random variable.

EXAMPLE 5.   For the decision situation described in Example 1, suppose that the capital expenditure required is \$350,000 for Type 1 equipment and \$450,000 for Type 2 equipment. Because of the lower variable manufacturing cost associated with the Type 2 equipment, the incremental profit (over variable cost) is \$12 per unit when this equipment is used as compared with a \$10 per unit incremental profit for the Type 1 equipment. The payoff functions are represented algebraically below, and graphically in Fig. 20-1.

$$\text{Payoff}\,(A_1) = -350{,}000 + 10X$$

$$\text{Payoff}\,(A_2) = -450{,}000 + 12X$$

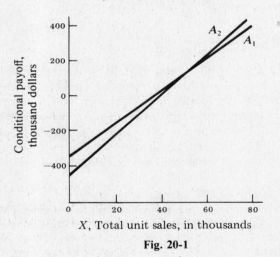

Fig. 20-1

The point at which the two linear payoff functions cross indicates the point of indifference with respect to choosing between the two acts, because the conditional payoffs of the acts are equal at that point. This is generally called the *breakeven point* in statistical decision analysis. The specific value of the random variable at the breakeven point can be algebraically determined by setting the two equations equal to one another and solving for X_b:

$$k_1 + b_1 X_b = k_2 + b_2 X_b$$

$$b_1 X_b - b_2 X_b = k_2 - k_1$$

$$X_b(b_1 - b_2) = k_2 - k_1$$

$$X_b = \frac{k_2 - k_1}{b_1 - b_2} \tag{20.5}$$

Having established the breakeven point, the best act can be determined graphically by observing whether the prior mean is above or below the breakeven point, and choosing the act which is optimal on that side of the breakeven point. The best act can also be identified by substituting the value of the prior mean in each payoff function and choosing the act with the highest expected payoff. The expected payoffs for two decision acts which have linear payoff functions are

$$EP(A_1) = k_1 + b_1 M_0 \qquad\qquad (20.6)$$

$$EP(A_2) = k_2 + b_2 M_0 \qquad\qquad (20.7)$$

EXAMPLE 6. For the acts and payoff functions described in Example 5, the breakeven point is

$$X_b = \frac{k_2 - k_1}{b_1 - b_2} = \frac{-450,000 - (-350,000)}{10 - 12} = \frac{-100,000}{-2} = 50,000 \text{ units}$$

The prior mean was determined to be $M_0 = 40,000$ units in Example 2. Referring to Fig. 20-1, we can observe that this value is to the left of the breakeven point, and that act A_1 (purchase Type 1 equipment) is the best act. Alternatively, we can determine the expected payoff associated with each act:

$$EP(A_1) = k_1 + b_1 M_0 = -350,000 + 10(40,000) = \$50,000$$

$$EP(A_2) = k_2 + b_2 M_0 = -450,000 + 12(40,000) = \$30,000$$

Therefore, the best act is A_1 (purchase Type 1 equipment), with an associated expected payoff of $50,000. Note that in determining the best act, only M_0 is required for the prior probability distribution.

The analysis presented thus far in this section has been concerned with the total market demand. As indicated in Section 20.2, however, if we anticipate the collection of sample information, an analysis on a per-outlet basis is more convenient for subsequent use of the sample results. In per-outlet analysis, the linear payoff functions are based on mean sales per outlet as the variable:

$$\text{Payoff } (A_1) = k_1 + N b_1 \mu \qquad\qquad (20.8)$$

$$\text{Payoff } (A_2) = k_2 + N b_2 \mu \qquad\qquad (20.9)$$

In the equations above, N is the number of outlets and μ is the average sales level per outlet. Similarly, the breakeven point in terms of mean sales per outlet must include consideration of the number of outlets involved:

$$\mu_b = \frac{k_2 - k_1}{N b_1 - N b_2} \qquad\qquad (20.10)$$

Finally, the expected payoffs associated with the two acts when the linear payoff functions are based on mean sales per outlet are determined as follows:

$$EP(A_1) = k_1 + N b_1 M_0 \qquad\qquad (20.11)$$

$$EP(A_2) = k_2 + N b_2 M_0 \qquad\qquad (20.12)$$

EXAMPLE 7. Given that there are 500 retail outlets, then each additional unit of mean sales per outlet represents 500 additional units in total. In place of the payoff functions determined in Example 5, we have:

$$\text{Payoff } (A_1) = k_1 + N b_1 \mu = -350,000 + (500)(10)\mu = -350,000 + 5,000\mu$$

$$\text{Payoff } (A_2) = k_2 + N b_2 \mu = -450,000 + (500)(12)\mu = -450,000 + 6,000\mu$$

In place of the breakeven point for total sales reported in Example 6, on a per-outlet basis the mean breakeven point is

$$\mu_b = \frac{k_2 - k_1}{N b_1 - N b_2} = \frac{-450,000 - (-350,000)}{500(10) - 500(12)} = 100 \text{ units per outlet}$$

From Example 4, $M_0 = 80$ on a per-outlet basis. Therefore, the expected payoff associated with each act based on a per-outlet analysis is

$$EP(A_1) = k_1 + Nb_1 M_0 = -350,000 + (500)(10)(80) = \$50,000$$

$$EP(A_2) = k_2 + Nb_2 M_0 = -450,000 + (500)(12)(80) = \$30,000$$

These values are identical to those obtained by the total-market analysis in Example 6.

If the linear functions are concerned with costs as opposed to revenues, the best act is the one with the lowest expected cost. See Problems 20.5 to 20.8.

20.4 LINEAR PIECEWISE LOSS FUNCTIONS AND THE EXPECTED VALUE OF PERFECT INFORMATION (*EVPI*)

In the context of statistical decision analysis, the term *loss function* always refers to an opportunity loss function. For the two-action problem with linear payoff functions, it follows that the loss function associated with each act will be a linear piecewise function made up of two linear pieces, or segments. This is true because on one side of the breakeven point the opportunity loss associated with a given act will be zero, while on the other side the opportunity loss increases linearly with each additional unit from the breakeven point. The process of determining linear piecewise loss functions is best presented by an example.

EXAMPLE 8.   In Examples 4 through 6 the following payoff functions and breakeven point were determined:

$$\text{Payoff } (A_1) = k_1 + b_1 X = -\$350,000 + 10X$$

$$\text{Payoff } (A_2) = k_2 - b_2 X = -\$450,000 + 12X$$

$$X_b = 50,000$$

We observed in Example 6 that act A_1 is the optimum act (with no opportunity loss) when the level of sales is below the breakeven point and act A_2 is the optimum act when the sales level is above the breakeven point. For act A_1, if the actual level of sales is greater than the breakeven point, then the opportunity loss is the difference between the two incremental revenue amounts ($12 - \$10$) multiplied by the number of units by which the sales exceed the breakeven point. Specifically,

$$OL(A_1, X) \begin{cases} = \$2.00(X - 50,000) & \text{for } X > X_b (= 50,000) \\ = 0 & \text{for } X \le X_b (= 50,000) \end{cases}$$

Conversely, for act A_2 there is an increment of $2.00 in the opportunity loss for each unit that the actual sales level is below the breakeven value:

$$OL(A_2, X) \begin{cases} = 0 & \text{for } X \ge X_b (= 50,000) \\ = \$2.00(50,000 - X) & \text{for } X < X_b (= 50,000) \end{cases}$$

The piecewise linear loss functions developed for A_1 and A_2 above are graphically portrayed in Figs. 20-2 and 20-3, respectively.

We can observe from Example 8 that the per-unit opportunity loss for act A_1 is equal to the per-unit opportunity loss for act A_2 when each type of loss occurs. In other words, the slope of the opportunity loss function in terms of absolute value is the same for both acts for that piecewise portion that is not equal to zero. It is convenient to designate the absolute value of the slope of the loss function as b. Computationally,

$$b = |b_1 - b_2| \qquad (20.13)$$

As indicated in Section 19.1, the expected value of perfect information is the expected opportunity loss of the best act. Thus, for the two-act situation with piecewise linear loss functions, only the opportunity loss function of the best act is relevant for computing *EOL*. The formula to determine the

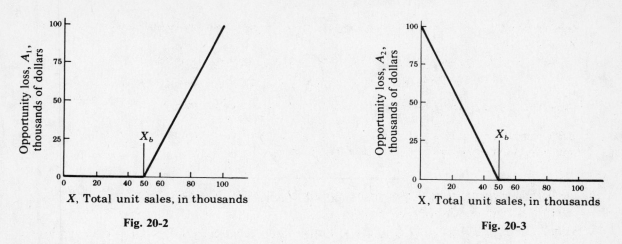

Fig. 20-2   Fig. 20-3

expected value of perfect information for a two-action problem with piecewise linear loss functions and a normal prior probability distribution is

$$EVPI = bS_0L(D) \qquad (20.14)$$

where $b = |b_1 - b_2|$

$S_0 =$ standard deviation of the prior probability distribution

$$D = \left| \frac{X_b - M_0}{S_0} \right|$$

$L(D) =$ the unit normal loss function for D (see Appendix 9)

EXAMPLE 9.   In Example 6, the best act for the equipment purchase problem was determined to be act A_1: Purchase Type 1 equipment with an expected payoff of $50,000. Using information from the preceding examples, we can determine the $EVPI$ as follows:

$$EVPI = bS_0L(D) = (2)(7,500)(0.04270) = \$640.50$$

where $b = |b_1 - b_2| = |10 - 12| = 2$

$S_0 = 7,500$ (from Example 3)

$$D = \left| \frac{X_b - M_0}{S_0} \right| = \left| \frac{50,000 - 40,000}{7,500} \right| = 1.33$$

$L(D) = 0.04270$ (from Appendix 9)

The maximum amount by which the expected payoff in this decision situation can be increased in the long run, given the removal of uncertainty, is $640.50. Therefore, no sample information could be worth more than this amount, as a long-run average.

In order to understand the computational procedure inherent in the formula for $EVPI$, it is useful to superimpose the normal probability function on the opportunity loss function of the best act. In Fig. 20-4, for the $EVPI$ calculation in Example 9, opportunity losses occur to the right of $X_b = 50$. Note that the greater the difference between M_0 and X_b, the lower the probability of experiencing an opportunity loss, as would be indicated by the proportion of the probability distribution that lies to the right of X_b. Therefore, the larger the value of D in the $EVPI$ formula, the smaller the value of $L(D)$. Conceptually, $L(D)$ represents the product of the opportunity loss function and the probability function to the right of the breakeven point in Fig. 20-4, and can be thought of as being the EOL of the best act *given a loss function with a slope of 1.0 and a normal distribution with a standard deviation of 1.0*. Therefore, the multiplication by b and by S_0 in the $EVPI$ formula serves to transform the $L(D)$ value given in Appendix 9 for the unit normal loss function to the appropriate value for the given application.

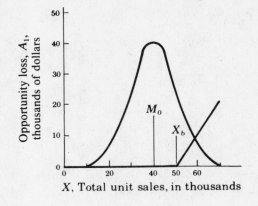

Fig. 20-4

As indicated in Sections 20.2 and 20.3, an analysis of data on a per-outlet basis is more appropriate than the total-market analysis if we anticipate subsequent collection of sample data from randomly chosen outlets. In such a per-outlet analysis, the calculation of *EVPI* has to include consideration of the fact that the per-unit opportunity loss should represent the opportunity loss per unit change in the *mean* value per outlet. The slope of the payoff functions of the two acts with respect to changes in the mean sales per outlet is designated by Nb_1 and Nb_2. Therefore, the absolute value of the slope of the opportunity loss function for the per-outlet analysis is determined by

$$b = |Nb_1 - Nb_2| \qquad (20.15)$$

EXAMPLE 10. In Example 7, the payoff functions for the two acts under consideration in terms of mean sales per outlet were identified as

$$\text{Payoff}\,(A_1) = -350{,}000 + 5{,}000\mu$$

$$\text{Payoff}\,(A_2) = -450{,}000 + 6{,}000\mu$$

The breakeven point in terms of mean sales per unit was identified as $\mu_b = 100$ units per outlet.

In Example 4 the prior mean and standard deviation on a per-outlet basis were identified as $M_0 = 80$ and $S_0 = 15$, respectively.

With the information summarized above, we determine *EVPI* for this decision problem as follows:

$$EVPI = bS_0L(D) = (1{,}000)(15)(0.04270) = \$640.50$$

where
$$b = |Nb_1 - Nb_2| = |5{,}000 - 6{,}000| = 1{,}000$$

$$S_0 = 15$$

$$D = \left| \frac{\mu_b - M_0}{S_0} \right| = \left| \frac{100 - 80}{15} \right| = 1.33$$

$$L(D) = 0.04270 \qquad \text{(from Appendix 9)}$$

Therefore, the *EVPI* determined on the basis of mean sales per outlet corresponds with the result in Example 9 based on the total-market analysis.

20.5 BAYESIAN POSTERIOR ANALYSIS

When sample information has been collected after a prior probability distribution has been formulated, a posterior probability distribution can be determined by computing the values of the posterior mean and the posterior variance. The posterior mean is designated by M_1 and is computed by either of the following formulas:

$$M_1 = \frac{(1/S_0^2)M_0 + (1/\sigma_{\bar{x}}^2)\bar{X}}{(1/S_0^2) + (1/\sigma_{\bar{x}}^2)} \qquad (20.16)$$

or

$$M_1 = \frac{M_0\sigma_{\bar{x}}^2 + \bar{X}S_0^2}{S_0^2 + \sigma_{\bar{x}}^2} \qquad (20.17)$$

Formula (20.16) indicates the conceptual basis for determining the posterior mean. Essentially, the posterior mean is a weighted mean, with the prior and sample means being weighted by the reciprocal of their respective variances. However, for computational purposes formula (20.17) is more convenient. If the population standard deviation σ is unknown, then it is generally considered acceptable to use the sample standard deviation s and to substitute $s_{\bar{x}}$ for $\sigma_{\bar{x}}$ in the above formulas.

EXAMPLE 11. In Example 4 the prior mean and standard deviation for the equipment decision problem were determined to be $M_0 = 80$ and $S_0 = 15$. Suppose the paging device is manufactured on a pilot basis and distributed through nine randomly selected retail outlets. The mean sales level per outlet is found to be $\bar{X} = 110$ units with a standard deviation of $s = 18$ units. With σ unknown, we use the estimator of $\sigma_{\bar{x}}^2$: $s_{\bar{x}}^2 = s^2/n = 324/9 = 36$ and (20.17) to calculate the posterior mean:

$$M_1 = \frac{M_0\sigma_{\bar{x}}^2 + \bar{X}S_0^2}{S_0^2 + \sigma_{\bar{x}}^2} = \frac{(80)(36) + (110)(225)}{225 + 36} = \frac{27,630}{261} = 105.9 \text{ units}$$

The variance of the posterior distribution is designated S_1^2, with the square root of this value being the posterior standard deviation S_1. The conceptual basis by which the posterior variance is determined is indicated by the equation

$$\frac{1}{S_1^2} = \frac{1}{S_0^2} + \frac{1}{\sigma_{\bar{x}}^2} \qquad (20.18)$$

Formula (20.18) indicates that the reciprocal of the posterior variance is equal to the reciprocal of the prior variance plus the reciprocal of the variance of the mean. Note that from the standpoint of decision analysis a larger variance indicates greater uncertainty, and that the larger the variance, the smaller its reciprocal.

The computational formula derived from (20.18) is

$$S_1^2 = \frac{S_0^2\sigma_{\bar{x}}^2}{S_0^2 + \sigma_{\bar{x}}^2} \qquad (20.19)$$

As was the case in determining the posterior mean, if the population standard deviation σ is unknown, the sample deviation s is generally used to compute $s_{\bar{x}}^2$ as an estimator of $\sigma_{\bar{x}}^2$.

EXAMPLE 12. Given the information in Example 11, we compute the variance and standard deviation of the posterior distribution as follows.

$$S_1^2 = \frac{S_0^2\sigma_{\bar{x}}^2}{S_0^2 + \sigma_{\bar{x}}^2} \cong \frac{S_0^2 s_{\bar{x}}^2}{S_0^2 + s_{\bar{x}}^2} = \frac{(225)(36)}{225 + 36} = \frac{8,100}{261} = 31.03$$

$$S_1 = 5.57 \cong 5.6$$

After the posterior probability distribution is formulated, the best act in the decision situation is determined on the basis of this revised probability distribution and the payoff functions previously established. As explained in Section 19.3, the sample information only has value if there is a change in the identification of the best act as the result of the prior distribution having been revised. As in Chapter 19, the estimated value of sample information for a sample that has already been taken is

$$VSI = \text{(posterior expected payoff of best posterior act)}$$

$$-\text{(posterior expected payoff of best prior act)} \qquad (20.20)$$

EXAMPLE 13. In Example 7, the payoff functions for the equipment decision problem based on mean sales per outlet were identified as:

$$\text{Payoff}(A_1) = -350,000 + 5,000\mu \qquad \text{Payoff}(A_2) = -450,000 + 6,000\mu$$

Using the posterior mean value of $M_1 = 105.9$, we determine the expected payoff associated with each act:

$$\text{Payoff}(A_1) = -350,000 + 5,000(105.9) = \$179,500 \qquad \text{Payoff}(A_2) = -450,000 + 6,000(105.9) = \$185,400$$

Therefore, on the basis of the Bayesian posterior analysis act A_2 is chosen, as contrasted to act A_1 being chosen on the basis of the prior analysis. The estimated value of the sample which was obtained in this case is

$$VSI = (\text{posterior expected payoff of best posterior act}) - (\text{posterior expected payoff of the best prior act})$$
$$= 185,400 - 179,500 = \$5,900$$

20.6 PREPOSTERIOR ANALYSIS AND THE EXPECTED VALUE OF SAMPLE INFORMATION (*EVSI*)

Because the prior mean is considered to be an unbiased estimator of the population mean, there is no basis for anticipating a difference between the values of the prior mean and the posterior mean before a sample is taken. However, we can anticipate that the posterior variance will be smaller in value than the prior variance. From Section 20.4, recall that the expected opportunity loss of the best act is dependent on the difference between the mean of the probability distribution and the breakeven point. When the variance is reduced, the *EOL* (and thus the *EVPI*) is reduced because the difference between the mean and breakeven point is then greater in units of the standard deviation. Conceptually, the *expected value of sample information* is the expected difference between the *EVPI* before the sample and the anticipated *EVPI* after the sample. Thus the *EVSI* is the expected reduction in the expected opportunity loss associated with the best act as the result of taking a sample of a specified size.

The reduction in the variance between the prior and posterior probability distributions is designated S_*^2, and can be determined by either of the following formulas:

$$S_*^2 = S_0^2 - S_1^2 \tag{20.21}$$

$$S_*^2 = \frac{S_0^4}{S_0^2 + \sigma_{\bar{x}}^2} \tag{20.22}$$

In the computational formula (20.22), the value of the population standard deviation is required in order to determine the variance of the mean $\sigma_{\bar{x}}^2$. However, this value is generally not known. Further, a sample estimator is not available because the sample has not yet been taken. Therefore, the value of σ has to be estimated by reference to other similar decision situations.

Once the value of S_*^2 is determined, the expected value of sample information is calculated by the formula

$$EVSI = bS_* L(D_*) \tag{20.23}$$

where $b = |b_1 - b_2|$

$$D_* = \left| \frac{\mu_b - M_0}{S_*} \right|$$

$L(D_*) = $ the unit normal loss function for D_* (see Appendix 9)

In (20.23), b is the slope of the loss function associated with the best act (see Section 20.4) and D_* is similar to D in the formula for *EVPI* presented in Section 20.4, except that S_* is in the denominator of the formula instead of S_0.

EXAMPLE 14. In Example 4 the prior mean and standard deviation for the equipment decision problem on a per-outlet basis were determined to be $M_0 = 80$ and $S_0 = 15$. For paging devices manufactured by other companies we have information which indicates that the standard deviation of the sales per outlet for the time period of concern is approximately $\sigma = 20$. We can determine the expected value of the information from a market study involving nine retail outlets as follows:

$$S_*^2 = \frac{S_0^4}{S_0^2 + \sigma_{\bar{x}}^2} = \frac{(15)^4}{(15)^2 + 44.44} = \frac{50,625}{225 + 44.44} = \frac{50,625}{269.44} = 187.89$$

where Est. $\sigma_{\bar{x}}^2 = \frac{\text{Est } \sigma^2}{n} = \frac{(20)^2}{9} = \frac{400}{9} = 44.44$

$$EVSI = bS_*L(D_*) = (1,000)(13.71)(0.03208) = \$439.82$$

where $b = |Nb_1 - Nb_2| = |5,000 - 6,000| = 1,000$ (from Example 10)

$$D_* = \left|\frac{\mu_b - M_0}{S_*}\right| = \left|\frac{100 - 80}{13.71}\right| = 1.46$$

$L(D_*) = 0.03208$

Recall that in Example 13 the estimated value of sample information for a particular sample of $n = 9$ which had *already* been collected was found to be \$5,900. The *EVSI* of \$439.82 computed in Example 14 indicates that this would be the long-run average value for a sample of size $n = 9$. In many specific instances a sample will have a value of \$0, because the identification of the best act will not have been affected by the sample information.

20.7   EXPECTED NET GAIN FROM SAMPLING (*ENGS*) AND OPTIMUM SAMPLE SIZE

The expected net gain from sampling and the optimum sampling size for our equipment decision problem are determined in Example 15. See Section 19.5 for a discussion of these concepts and procedures.

EXAMPLE 15.   In Example 14 we found that the *EVSI* for a sample of $n = 9$ is \$439.82. Suppose that the cost of designing the marketing study is \$300 and the cost of obtaining the sales data from each sampled retail outlet is \$10. Given the prior probability distribution with $M_0 = 80$, $S_0 = 15$, and estimated $\sigma = 20$, we can determine the *ENGS* associated with the alternative sample sizes of $n_1 = 9$, $n_2 = 12$, and $n_3 = 15$ as follows:

For $n = 9$:

$$EVSI = \$439.82 \qquad \text{(from Example 14)}$$

$$ENGS = EVSI - CS = 439.82 - 390.00 = \$49.82$$

For $n = 12$:

$$S_*^2 = \frac{S_0^4}{S_0^2 + \sigma_{\bar{x}}^2} = \frac{(15)^4}{(15)^2 + 33.33} = \frac{50,625}{258.33} = 195.97$$

where Est. $\sigma_{\bar{x}}^2 = \frac{\text{Est } \sigma^2}{n} = \frac{(20)^2}{12} = \frac{400}{12} = 33.33$

$$EVSI = bX_*L(D_*) = (1,000)(14.00)(0.03431) = \$480.34$$

where $b = |b_1 - b_2| = |5,000 - 6,000| = 1,000$

$$D_* = \left|\frac{\mu_b - M_0}{S_*}\right| = \left|\frac{100 - 80}{14.00}\right| = 1.43$$

$$ENGS = EVSI - CS = 480.34 - 420.00 = \$60.34$$

For $n = 15$:

$$S_*^2 = \frac{S_0^4}{S_0^2 + \sigma_{\bar{x}}^2} = \frac{(15)^4}{(15)^2 + 26.67} = \frac{50,625}{251.67} = 201.16$$

where $\sigma_{\bar{x}}^2 = \frac{\text{Est } \sigma^2}{n} = \frac{(20)^2}{15} = \frac{400}{15} = 26.67$

$$EVSI = bS_* L(D_*) = (1,000)(14.18)(0.03587) = \$508.64$$

where $\quad b = |b_1 - b_2| = |5,000 - 6,000| = 1,000$

$$D_* = \left| \frac{\mu_b - M_0}{S_*} \right| = \left| \frac{100 - 80}{14.18} \right| = 1.41$$

$$ENGS = EVSI - CS = 508.64 - 450.00 = \$58.64$$

Refer to the summarized information in Table 20.1. Of the sample sizes considered, the optimum sample size is at $n = 12$, with the $ENGS = \$60.34$. Although all possible sample sizes were not considered, because the $ENGS$ associated with $n = 9$ and $n = 15$ are both smaller than the $ENGS$ for $n = 12$, it is obvious that by the sample size of $n = 15$ the point of the optimum sample size has been passed, and that the optimum is somewhere between $n = 9$ and $n = 15$.

Table 20.1 $\quad$ **$EVSI$, CS, and $ENGS$ for the Equipment Purchase Decision According to Sample Size**

Sample size (n)	$EVSI$	CS	$ENGS$
0	\$ 0	\$ 0	\$ 0
9	439.82	390.00	49.82
12	480.34	420.00	60.34
15	508.64	450.00	58.64

Finally, the *overall terminal expected payoff* (*OTEP*) and *net overall terminal expected payoff* (*NOTEP*) can be determined in conjunction with preposterior analysis. These concepts are explained in Section 19.5, where the calculations are illustrated in Example 12.

20.8 BAYESIAN DECISION ANALYSIS VS. CLASSICAL DECISION PROCEDURES

This chapter and the preceding two chapters have been concerned with Bayesian decision analysis, as contrasted to the coverage of classical decision procedures in most of the book. The principal techniques of classical inference are interval estimation and hypothesis testing. The principal concern of Bayesian decision analysis is the choice of a decision act. Although the classical techniques are directly concerned with estimating or testing hypotheses concerning population parameters, the results of these procedures relate to alternative courses of action, or decisions. For example, the acceptance of the null hypothesis that the average sales level for a product will be below the breakeven point would be associated with the decision not to market the product. Thus, both classical and Bayesian procedures can be concerned with the process of choosing best decision acts.

The main difference between classical and Bayesian procedures is the use of subjective (prior) information in Bayesian decision analysis and the evaluation of alternative decision acts in terms of economic consequences (or possibly utilities). The economic consequences can be formulated in terms of either payoffs or conditional opportunity losses (regrets). Essentially, the choice of α and β levels for the probabilities of Type I and Type II error in hypothesis testing is the basis by which the importance of the two alternative types of mistakes is assessed. The use of opportunity losses in Bayesian analysis represents a similar evaluation in a more explicit way. Whereas classical decision procedures are based entirely on the analysis of data collected through random sampling, Bayesian procedures can also include the analysis of sample data (through posterior analysis), but are not dependent on the availability of such data.

From the standpoint of practical considerations, an important factor associated with the development of Bayesian decision analysis is that such analysis begins with an identification of managerial judgments, and such judgments are included in the analysis. This means that the statistical analyst is required to work closely with managerial personnel. In contrast, the exclusive orientation toward sample data in classical decision procedures may not give managers the opportunity to input their judgments into the decision analysis, and to feel that their judgments are considered to be important in the decision process.

Solved Problems

DECISION ANALYSIS PRIOR TO ANY SAMPLING

20.1   The owner of a small manufacturing company is considering the addition of electrical generators to the line of automotive electrical equipment being manufactured. The capital investment required to manufacture the generators is $150,000, and there is a profit of $2.00 per generator sold through the established distribution system of 100 retail outlets. During the total relevant period of time, the owner estimates that the average sales of the electrical generators per retail outlet will be 700 units, and that there is a 50 percent chance that the average sales level per outlet will be between 600 and 800 units.

(a)   Assuming a normal prior probability distribution, determine the mean and standard deviation of this distribution.

(b)   The two possible decision acts are A_1: Manufacture the generators and A_2: Don't manufacture the generators. Formulate the linear payoff functions associated with these acts and portray them on a common graph.

(a) $$M_0 = 700 \text{ units}$$

$$\frac{2}{3} S_0 = \frac{\text{middle 50\% interval}}{2} = \frac{800 - 600}{2} = 100$$

$$S_0 = 150$$

(b) $$\text{Payoff } (A_1) = k_1 + Nb_1\mu = -150{,}000 + (100)(2)\mu = -150{,}000 + 200\mu$$

$$\text{Payoff } (A_2) = k_2 + Nb_2\mu = 0 + (100)(0)\mu = 0$$

Figure 20-5 portrays these two linear payoff functions.

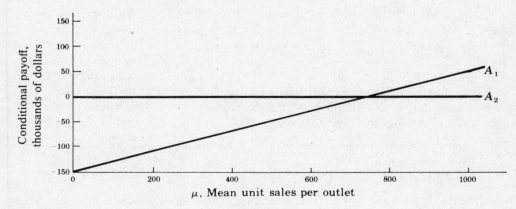

Fig. 20-5

20.2 Determine the best act for the decision situation in Problem 20.1 by calculating the expected values associated with the two possible acts.

$$EP(A_1) = k_1 + Nb_1 M_0 = -150{,}000 + (100)(2)(700) = -\$10{,}000$$

$$EP(A_2) = k_2 + Nb_2 M_0 = 0 + 100(0)700 = 0$$

Therefore, the best act is A_2: Don't manufacture the generators.

20.3 Referring to Problem 20.1, (a) determine the breakeven value in terms of the mean unit sales per outlet and (b) formulate the linear piecewise loss functions for the two possible acts. Portray the loss function of act A_2: Don't manufacture on a graph with the normal prior probability distribution superimposed on this graph.

(a)
$$\mu_b = \frac{k_2 - k_1}{Nb_1 - Nb_2} = \frac{0 - (-150{,}000)}{100(2) - 100(0)} = \frac{150{,}000}{200} = \$750$$

Note that this is the mean sales volume at which the two linear payoff functions cross in Fig. 20-5.

(b)
$$OL(A_1, \mu) \begin{cases} =0 & \text{for } \mu \ge \mu_b \ (=750) \\ =200(750 - \mu) & \text{for } \mu < \mu_b \ (=750) \end{cases}$$

$$OL(A_2, \mu) \begin{cases} =200(\mu - 750) & \text{for } \mu > \mu_b \ (=750) \\ =0 & \text{for } \mu \le \mu_b \ (=750) \end{cases}$$

See Fig. 20-6.

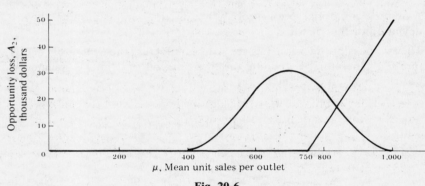

Fig. 20-6

20.4 For Problems 20.1 to 20.3, (a) determine the expected value of perfect information. (b) Suppose the owner of this manufacturing firm chooses the best act as identified by the Bayesian analysis. What is the probability that his decision will turn out to be wrong?

(a) $$EVPI = bS_0 L(D) = (200)(150)(0.2555) = \$7{,}665$$

where $b = |Nb_1 - Nb_2| = |200 - 0| = 200$

$S_0 = 150$

$$D = \left| \frac{\mu_b - \mu_0}{S_0} \right| = \left| \frac{750 - 700}{150} \right| = \frac{50}{150} = 0.33$$

$L(D) = 0.2555$ (from Appendix 9)

(b)   Refer to Fig. 20-6. The probability that the decision not to market the generators will turn out to be wrong is equivalent to the probability that the mean sales per outlet will exceed the breakeven of 750 units. Converting the breakeven value into a unit-normal z value, we have

$$z = \frac{\mu_b - M_0}{S_0} = \frac{750 - 700}{150} = 0.33$$

Therefore

$$P(\mu > 750) = P(z > 0.33) = 0.5000 - P(0 \le z \le 0.33)$$
$$= 0.5000 - 0.1293 = 0.3707 \cong 0.37$$

20.5   An office manager can purchase a specialty photocopy machine for reproduction of blueprints from one of two manufacturers. The brand Y machine costs \$8,000 and involves a variable cost of 5¢ per page of copy produced. Brand Z costs \$9,000 but involves a variable cost of 4.5¢ per page of copy produced.

   (a)   Formulate the linear cost function associated with act A_1: Buy brand Y and act A_2: Buy brand Z.

   (b)   The manager estimates the useful life of each machine as being 500 weeks. Reformulate the linear cost functions determined in (a) in terms of the average pages of copy produced per week.

   (a)
$$C(A_1) = k_1 + b_1 X = 8,000 + 0.05X$$
$$C(A_2) = k_2 + b_2 X = 9,000 + 0.045X$$

   Note: These cost functions could be expressed as payoff functions by attaching negative signs to both the capital investment and the incremental costs in the above equations. However, in comparative cost studies it is generally considered more convenient to consider the values as positive values but to label them as representing costs.

   (b)
$$C(A_1) = k_1 + Nb_1\mu = 8,000 + (500)(0.05)\mu = 8,000 + 25\mu$$
$$C(A_2) = k_2 + Nb_2\mu = 9,000 + (500)(0.045)\mu = 9,000 + 22.5\mu$$

20.6   (a)   The volume of work during the period of time that the photocopy machine in Problem 20.5 is to be used is not expected to change (an obvious simplifying assumption!). The manager estimates that an average of about 500 copies of blueprints will be produced per week on the machine. Determine which brand of machine should be purchased based on the overall expected cost associated with each of the two brands of copy equipment.

   (b)   Determine the breakeven volume in terms of the mean number of copies produced per week which would result in the manager being indifferent between the two brands of equipment in terms of overall expected cost.

   (a)
$$EC(A_1) = k_1 + Nb_1 M_0 = 8,000 + 25(500) = \$20,500$$
$$EC(A_2) = k_2 + Nb_2 M_0 = 9,000 + 22.5(500) = \$20,250$$

   Therefore, the best act is A_2: Buy brand Z, because this brand has a lower expected cost associated with it. (*Note:* If the analysis had been done in terms of expected values, the two values above would have negative signs and thus the largest value would be associated with act A_2.)

   (b)
$$\mu_b = \frac{k_2 - k_1}{Nb_1 - Nb_2} = \frac{9,000 - 8,000}{500(0.05) - 500(0.045)} = \frac{1,000}{25 - 22.5} = 400 \text{ copies per week}$$

20.7   Refer to Problems 20.5 and 20.6. (a) Assuming a normal prior probability distribution, the standard deviation of such a distribution can be determined based on the manager estimating any two percentile points in this distribution, and not necessarily by his identifying a middle 50 percent credibility interval. The established prior mean of 500 copies per week is of course at the 50th percentile point of the distribution. In addition, suppose the manager estimates that there is only a 10 percent chance that the average number of copies produced per week will exceed 700. Determine the standard deviation of the prior probability distribution. (b) Formulate the linear piecewise loss functions for the two acts on the basis of the mean number of copies produced per week.

(a) By reference to the standard normal distribution (Appendix 5), we observe that the value associated with the 90th percentile point is approximately $z = +1.28$. We can use a general formula for z and solve for the unknown S_0 as follows:

$$z = \frac{\mu - M_0}{S_0} \qquad S_0 = \frac{\mu - M_0}{z} = \frac{700 - 500}{1.28} = 156.25$$

(b)

$$OL(A_1, \mu) \begin{cases} = 2.5(400 - \mu) & \text{for } \mu > \mu_b \ (=400) \\ = 0 & \text{for } \mu \leq \mu_b \ (=400) \end{cases}$$

$$OL(A_2, \mu) \begin{cases} = 0 & \text{for } \mu \geq \mu_b \ (=400) \\ = 2.5(\mu - 400) & \text{for } \mu < \mu_b \ (=400) \end{cases}$$

20.8 Determine the *EVPI* for the photocopy machine decision described in Problems 20.5 to 20.7.

$$EVPI = bS_0 L(D) = (2.5)(156.25)(0.1580) = \$61.71875 \cong \$61.72$$

where $b = |Nb_1 - Nb_2| = |25 - 22.5| = 2.5$

$S_0 = 156.25$

$$D = \left| \frac{\mu_b - M_0}{S_0} \right| = \left| \frac{400 - 500}{156.25} \right| = \frac{100}{156.25} = 0.64$$

$L(D) = 0.1580$ (from Appendix 9)

The low value of the *EVPI* indicates that there is neither a great deal of financial risk associated with this decision-making situation, nor much expected economic gain possibility associated with sampling.

BAYESIAN POSTERIOR ANALYSIS

20.9 With reference to Problems 20.1 to 20.4, a generator similar to the one which is to be manufactured is test-marketed in 10 randomly selected outlets. The mean sales per outlet is $\bar{X} = 800$ with the sample standard deviation $s = 110$. Determine the mean and standard deviation of the posterior distribution.

$$M_1 = \frac{M_0 \sigma_{\bar{x}}^2 + \bar{X} S_0^2}{S_0^2 + \sigma_{\bar{x}}^2} = \frac{(700)(1,100) + (800)(22,500)}{22,500 + 1,100} = \frac{18,770,000}{23,600} = 795.3 \text{ units}$$

where Est. $\sigma_{\bar{x}}^2 = \frac{s^2}{n} \left(\frac{N - n}{N - 1} \right) = \frac{12,100}{10} \left(\frac{100 - 10}{100 - 1} \right) = (1,210)(0.909) = 1,099.999 \cong 1,100$

(The finite correction factor is used in this case because $n > 0.05 N$. See Section 8.2.)

$$S_1^2 = \frac{S_0^2 \sigma_{\bar{x}}^2}{S_0^2 + \sigma_{\bar{x}}^2} = \frac{(22,500)(1,100)}{22,500 + 1,100} = \frac{24,750,000}{23,600} = 1,048.7288$$

$S_1 = 32.38$

20.10 With reference to Problem 20.9, (a) determine the best decision act (A_1: Manufacture the generators, or A_2: Don't manufacture the generators). (b) What is the estimated value of the sample information?

(a)

$$EP(A_1) = k_1 + Nb_1 M_1 = -150,000 + (100)(2)(795.3) = \$9,060$$

$$EP(A_2) = k_2 + Nb_2 M_1 = 0 + (100)(0)(795.3) = 0$$

Therefore, the best act is A_1: Manufacture the generators.

(b) VSI = (posterior expected payoff of best posterior act) − (posterior expected payoff of best prior act)

$= 9,060 - 0 = \$9,060$

20.11 In Problem 20.4 the *EVPI* prior to any sample was determined to be \$7,665. (*a*) How is it possible for the value of sample information determined in Problem 20.10 to exceed the *EVPI*? (*b*) Determine the *EVPI* (posterior) for the generator manufacturing decision. That is, calculate the expected value of perfect information after the sample described in Problem 20.9 has been taken and incorporated into the analysis.

(*a*) *EVPI* is the *expected* value of perfect information. Therefore, it is in fact possible that a particular sample result will have an estimated value which exceeds this long-run average value of perfect information.

(*b*)
$$EVPI \text{ (posterior)} = bS_1 L(D_1) = (200)(32.38)(0.03667) = \$237.47$$

where
$$b = 200$$
$$S_1 = 32.38$$
$$D_1 = \left| \frac{\mu_b - M_1}{S_1} \right| = \left| \frac{750 - 795.3}{32.38} \right| = \frac{45.3}{32.38} = 1.40$$
$$L(D_1) = = 0.03667 \quad \text{(from Appendix 9)}$$

20.12 The office manager in Problems 20.5 to 20.8 decides to maintain a count of the number of blueprint copies required in the department during a 5-week period before purchasing one of the brands of photocopy equipment. He considers these 5 weeks to be a random sample of the 500-week period (again, a simplifying assumption!). The mean number of copies for this sampled period is $\bar{X} = 450$ with the standard deviation $s = 100$. Determine the mean and standard deviation of the posterior distribution.

$$M_1 = \frac{M_0 \sigma_{\bar{x}}^2 + \bar{X} S_0^2}{S_0^2 + \sigma_{\bar{x}}^2} = \frac{(500)(2,000) + (450)(156.25)^2}{(156.25)^2 + 2,000} = \frac{11,986,327}{26,414.062} = 453.8$$

where Est. $\sigma_{\bar{x}}^2 = \frac{s^2}{n} = \frac{10,000}{5} = 2,000$

$$S_1^2 = \frac{S_0^2 \sigma_{\bar{x}}^2}{S_0^2 + \sigma_{\bar{x}}^2} = \frac{(24,414.062)(2,000)}{24,414.062 + 2,000} = 1,848.5655$$

$$S_1 = 42.99$$

20.13 Referring to Problem 20.12, determine (*a*) the best act (A_1: Buy brand *Y* or A_2: Buy brand *Z*) on the basis of the posterior distribution, and (*b*) the estimated value of this sample information.

(*a*)
$$EC(A_1) = k_1 + Nb_1 M_1 = 8,000 + 25(453.8) = \$19,345$$
$$EC(A_2) = k_2 + Nb_2 M_1 = 9,000 + 22.5(453.8) = \$19,210.50$$

Therefore, the best act is A_2: Buy brand *Z*, because this brand has a lower associated expected cost.

(*b*) Since the identification of the best act was not changed as the result of the sample information, the sample had no value. Or using the general formula:

$$VSI = (\text{posterior expected payoff of best posterior act}) - (\text{posterior expected payoff of best prior act})$$
$$= -19,210.50 - (-19,210.50) = \$0$$

THE EXPECTED VALUE OF SAMPLE INFORMATION (*EVSI*) AND THE EXPECTED NET GAIN FROM SAMPLING (*ENGS*)

20.14 Referring to Problems 20.1 and 20.4, determine the expected value of sample information (*EVSI*) associated with a sample of $n = 10$ outlets if Est. $\sigma = 300$.

Given the prior mean $M_0 = 700$, the prior standard deviation $S_0 = 150$, and the estimated standard deviation of sales per outlet $\sigma = 300$,

$$S_*^2 = \frac{S_0^4}{S^2 + \sigma_{\bar{x}}^2} = \frac{(150)^4}{(150)^2 + 8,181} = \frac{506,250,000}{30,681} \cong 16,500$$

$$S_* = 128$$

where Est. $\sigma_{\bar{x}}^2 = \frac{\text{Est. } \sigma^2}{n}\left(\frac{N-n}{N-1}\right) = \frac{90,000}{10}\left(\frac{100-10}{100-1}\right) = (9,000)(0.909) \cong 8,181$

(The finite correction factor is used because $n > 0.05N$. See Section 8.2.)

$$EVSI = bS_*L(D_*) = (200)(128)(0.2339) = \$5,988$$

where $b = |Nb_1 - Nb_2| = |200 - 0| = 200$

$$D_* = \left|\frac{\mu_b - M_0}{S_*}\right| = \left|\frac{750 - 700}{128}\right| = \frac{50}{128} = 0.39$$

$L(D_*) = 0.2339$ (from Appendix 9)

20.15 In Problems 20.6 and 20.7, the prior mean in terms of average copies produced per week was $M_0 = 500$ with $S_0 \cong 156$. Determine the $EVSI$ associated with a sample of $n = 5$ if Est. $\sigma = 250$.

$$S_*^2 = \frac{S_0^4}{S_0^2 + \sigma_{\bar{x}}^2} = \frac{(156)^4}{(156)^2 + 12,500} = \frac{592,240,896}{36,836} = 16,077.77$$

$$S_* = 126.80$$

where Est. $\sigma_{\bar{x}}^2 = \frac{\text{Est. } \sigma^2}{n} = \frac{62,500}{5} = 12,500$

$$EVSI = bS_*L(D_*) = (2.5)(126.80)(0.1223) = \$38.77$$

where. $b = |Nb_1 - Nb_2| = |25 - 22.5| = 2.5$

$$D_* = \left|\frac{\mu_b - M_0}{S_*}\right| = \left|\frac{400 - 500}{126.80}\right| = \frac{100}{126.80} = 0.79$$

$L(D_*) = 0.1223$

In Problem 20.8, the $EVPI$ was determined to be \$61.72. Thus, the expected value of a sample of just 5 weeks would serve to reduce the economic uncertainty associated with this decision situation by \$38.77, on the average.

20.16 With reference to Problem 20.15, suppose the cost of tallying the number of copies currently being produced in the department is evaluated as being \$5.00 per week. Determine the expected net gain associated with a sample of $n = 5$ weeks.

$$ENGS = EVSI - CS = 38.77 - 25.00 = \$13.77$$

Supplementary Problems

DECISION ANALYSIS PRIOR TO ANY SAMPLING

20.17 A distributor has the option of handling a new line of office furniture. Handling this line will require an additional capital expenditure of \$80,000 for new storage facilities. The incremental profit (over variable costs) associated with handling the office furniture is 10 percent of the dollar sales volume. During the period in which the capital expenditure is to be amortized, the owner of the firm estimates that the most likely dollar sales volume for the furniture will be $900,000, and that there is a 50 percent chance the volume will be between $750,000 and $1,050,000.

(a)   Assuming a normal prior probability distribution, determine the mean and standard deviation of this distribution.

(b)   The two possible decision acts are A_1: Distribute and A_2: Don't distribute. Formulate the linear payoff functions associated with these acts and portray these payoff functions on a common graph.

Ans.   (a) $M_0 = \$900,000$ and $S_0 = \$225,000$

20.18   Determine the best act for the decision situation described in Problem 20.17.

Ans. A_1: Distribute, with $EP = \$10,000$

20.19   For Problem 20.17, (a) determine the breakeven value in terms of total dollar sales volume required. (b) Formulate the linear piecewise loss functions for the two possible acts. Portray the loss function of act A_1: Distribute, on a graph with the normal prior probability distribution superimposed on this graph.

Ans.   (a) $X_b = \$800,000$

20.20   For Problems 20.17 to 20.19, (a) what is the maximum average value that any market information can be worth? (b) Suppose the owner of this distributorship chooses the best act as identified by the Bayesian analysis. What is the probability that this decision is correct?

Ans.   (a) $\$4,880.25$,   (b) $P = 0.67$

20.21   The manufacture of a new product will require a capital investment of $100,000. For this product, the variable cost of manufacturing will be $2.00 per unit, the selling price will be $5.00 per unit, and the marketing manager estimates that the most likely average sales level per retail outlet is 80 units. There are 500 retail outlets. Assuming a normal prior probability distribution, determine whether the best act is A_1: Manufacture or A_2: Don't manufacture by determining the expected payoffs associated with these acts.

Ans. A_1: Manufacture, with $EP = \$20,000$

20.22   Continuing with Problem 20.21, the marketing manager states that there is a 70 percent probability that sales of the product will exceed an average of 75 units per outlet. Determine (a) the standard deviation of the normal prior probability distribution and (b) the expected value of perfect information.

Ans.   (a) $S_0 = 9.6$,   (b) $EVPI = \$540$

20.23   The uncertainty inherent in the estimate of the total sales for a new product line is assumed to follow a normal distribution, with $M_0 = 7,000$ units and $S_0 = 1,000$ units. The capital investment required to add the new product line is $80,000, and the markup (over variable cost) for each item sold is $10. Indicate the best decision act for this situation (A_1: Market, or A_2: Don't market) and the associated expected payoff.

Ans. A_2: Don't market, with $EP = 0$

20.24   Refer to Problem 20.23. If the firm conducts a market study to test the product line in a sample region, what is the maximum amount that such additional information could be worth, on the average?

Ans. $\$833$

BAYESIAN POSTERIOR ANALYSIS

20.25   When the prior probability distribution has been determined on a total market basis, the per-outlet values of the prior mean, prior standard deviation, and the breakeven point can be determined simply by dividing each of the respective total-market values by the number of outlets involved. For Problems 20.17 to 20.20, suppose there are a total of 800 outlets nationwide. Determine the per-outlet values of the (a) prior mean, (b) prior standard deviation, and (c) breakeven point.

Ans.   (a) $M_0 = \$1,125$,   ($b$) $S_0 \cong \$281$,   ($c$) $\mu_b = \$1,000$

20.26   In relation to the transformation to per-outlet values carried out in Problem 20.25, suppose a sample of $n = 4$ outlets are randomly selected to test the marketability of the product line. For this sample, the mean sales level for the line of office furniture is $\bar{X} = \$900$ with $s = \$300$. Determine the ($a$) mean and ($b$) standard deviation of the posterior distribution of the estimated level of sales per outlet.

Ans.   (a) $M_1 \cong \$950$,   ($b$) $S_1 \cong \$132$

20.27   (a)   On the basis of the posterior distribution determined in Problem 20.26, determine the best act (A_1: Distribute, or A_2: Don't distribute) for the decision problem described in Problems 20.17 to 20.20.

(b)   What is the estimated value of the sample information obtained in Problem 20.26?

Ans.   (a) A_2: Don't distribute,   (b) $\$4,000$

20.28   Referring to Problems 20.21 and 20.22, suppose the product is offered through 10 randomly selected outlets. The mean sales level of the product in these outlets is $\bar{X} = 60.0$ units with $s = 25.0$ units. Determine the (a) mean and (b) standard deviation of the posterior distribution of the estimated level of sales per outlet.

Ans.   (a) $M_1 = 68.1$,   (b) $S_1 = 6.1$

20.29   Continuing with Problem 20.28, identify (a) the best act and (b) the estimated value of the sample information which was collected.

Ans.   (a) A_1: Manufacture, with $EP = \$2,150$,   ($b$) $\$0$

20.30   (a)   Determine the $EVPI$ for Problems 20.21 and 20.22 after the sample described in Problem 20.28 has been obtained.

(b)   Compare the value of the posterior $EVPI$ determined in (a), above, with the $EVPI$ of $\$540$ determined prior to the sample in Problem 20.22. Explain the meaning of the change which occurred.

Ans.   (a) Posterior $EVPI \cong \$2,694$

THE EXPECTED VALUE OF SAMPLE INFORMATION ($EVSI$) AND THE EXPECTED NET GAIN FROM SAMPLING ($ENGS$)

20.31   In Problems 20.21 and 20.22, the prior probability distribution on a per-outlet basis was determined to be $M_0 = 80.0$ and $S_0 = 9.6$. There are a total of 500 retail outlets. If the product is manufactured, it is estimated that the standard deviation of the sales level within each outlet will be about $\sigma = 30$. Determine the $EVSI$ associated with a sample of $n = 10$ outlets.

Ans. $EVSI = \$96.34$

20.32   Explain the reason for the difference between the $EVSI$ of $\$96.34$ for a sample of $n = 10$ computed in Problem 20.31, and the estimated value of $\$0$ for the sample of $n = 10$ determined in Problem 20.29.

20.33   With respect to the sample being contemplated in Problem 20.31, suppose the cost associated with having each outlet participate in such a study is $\$15$. Determine the $ENGS$ for a sample of $n = 10$.

Ans. $ENGS = -\$53.66$

Chapter 21

Nonparametric Statistical Tests

21.1 SCALES OF MEASUREMENT

Before considering how nonparametric methods differ from the parametric procedures that constitute most of this book, it is useful to define four types of measurement scales in terms of the precision represented by reported values.

In the *nominal scale*, numbers are used only to identify categories. They do not represent any amount or quantity as such.

EXAMPLE 1. If four sales regions are numbered 1 through 4 as general identification numbers only, then the nominal scale is involved, since the numbers simply serve as category names.

In the *ordinal scale*, the numbers represent ranks. The numbers indicate relative magnitude, but the differences between the ranks are not assumed to be equal.

EXAMPLE 2. An investment analyst ranks five stocks from 1 to 5 in terms of appreciation potential. The difference in the appreciation potential between the stocks ranked 1 and 2 generally would not be the same as, say, the difference between the stocks ranked 3 and 4.

In the *interval scale*, measured differences between values are represented. However, the zero point is arbitrary, and not an "absolute" zero. Therefore, the numbers cannot be compared by ratios.

EXAMPLE 3. In either the Fahrenheit or the Celsius temperature scales, a 5° difference, from say 70°F to 75°F is the same amount of difference in temperature as from 80°F to 85°F. However, we cannot say that 60°F is twice as warm as 30°F, because the 0°F point is *not* an absolute zero point (the complete absence of all heat).

In the *ratio scale*, there is a true zero point, and therefore measurements can be compared in the form of ratios.

EXAMPLE 4. Not only is it true that a difference in inventory value of $5,000 is the same amount of difference whether between, say, $50,000 and $55,000 or between $60,000 and $65,000, it is also true that an inventory value of $100,000 is twice as great as an inventory value of $50,000.

21.2 PARAMETRIC VS. NONPARAMETRIC STATISTICAL METHODS

Most of the statistical methods described in this book are called *parametric methods*. The focal point of parametric analysis is some population *parameter* for which the sampling statistic follows a *known distribution* and with measurements being made at the *interval or ratio scale*. When one or more of these requirements or assumptions are not satisfied, then the so-called *nonparametric* methods can be used. An alternative term is *distribution-free methods*, which focuses particularly on the factor that the distribution of the sampling statistic is not known.

If use of a parametric test, such as the *t* test, is warranted, then we would always prefer its use to the nonparametric equivalent. This is so because if we use the same level of significance for both tests, then the power associated with the nonparametric test is always less than the parametric equivalent. (Recall from Section 10.3 that the power of a statistical test is the probability of rejecting a false null hypothesis.) Nonparametric tests often are used in conjunction with small samples, because for such samples the central limit theorem cannot be invoked (see Section 8.2).

Nonparametric tests can be directed toward hypotheses concerning the *form, dispersion,* or *location* (median) of the population. In the majority of the applications, the hypotheses are concerned with the value of a median, the difference between medians, or the differences among several medians. This contrasts with the parametric procedures that are focused principally on population means.

Of the statistical tests already described in this book, the chi-square test covered in Chapter 12 in fact is a nonparametric test. Recall, for example, that the data that are analyzed are at the nominal scale (categorical data). A separate chapter was devoted to the chi-square test because of the extent of its use and the variety of its applications.

21.3 THE RUNS TEST FOR RANDOMNESS

Where a *run* is a series of like observations, the *runs test* is used to test the randomness of a series of observations when each observation can be assigned to one of two categories.

EXAMPLE 5. For a random sample of $n = 10$ individuals, suppose that when they are categorized by sex, the sequence of observations is: M, M, M, M, F, F, F, F, M, M. For these data there are *three* runs, or series of like items.

For numeric data, one way by which the required two-category scheme can be achieved is to classify each observation as being either above or below the median of the group. In general, either too *few* runs or too *many* runs than would be expected by chance would result in rejecting the null hypothesis that the sequence of observations is a random sequence.

The number of runs of like items is determined for the sample data, with the symbol R used to designate the number of observed runs. Where n_1 equals the number of sampled items of one type and n_2 equals the number of sampled items of the second type, the mean and the standard error associated with the sampling distribution of the R test statistic when the sequence is random are

$$\mu_R = \frac{2n_1 n_2}{n_1 + n_2} + 1 \tag{21.1}$$

$$\sigma_R = \sqrt{\frac{2n_1 n_2 (2n_1 n_2 - n_1 - n_2)}{(n_1 + n_2)^2 (n_1 + n_2 - 1)}} \tag{21.2}$$

If either $n_1 > 20$ or $n_2 > 20$, the sampling distribution of r approximates the normal distribution. Therefore under such circumstances the R statistic can be converted to the z test statistic as follows:

$$z = \frac{R - \mu_R}{\sigma_R} \tag{21.3}$$

Where both $n_1 \leq 20$ and $n_2 \leq 20$, tables of critical values for the R test statistic are available in specialized textbooks in nonparametric statistics.

See Problem 21.1 for an application of the runs test for randomness.

21.4 ONE SAMPLE: THE SIGN TEST

The *sign test* can be used to test a null hypothesis concerning the value of the population median. Therefore, it is the nonparametric equivalent to testing a hypothesis concerning the value of the population mean. The values in the random sample are required to be at least at the ordinal scale, with no assumptions required about the form of the population distribution.

The null and alternative hypotheses can designate either a two-tail or a one-tail test. Where Med denotes the population median and Med_0 designates the hypothesized value, the null and alternative hypotheses for a two-tail test are

$$H_0\text{: Med} = \text{Med}_0$$

$$H_1\text{: Med} \neq \text{Med}_0$$

A plus (+) sign is assigned for each observed sample value that is larger than the hypothesized value of the median and a minus sign (−) is assigned for each value that is smaller than the hypothesized value of the median. If a sample value is exactly equal to the hypothesized median, no sign is recorded and the effective sample size thereby is reduced. If the null hypothesis regarding the value of the median is true, the number of plus signs should approximately equal the number of minus signs. Or put another way, the proportion of plus signs (or minus signs) should be about 0.50. Therefore, the null hypothesis tested for a two-tail test is H_0: $\pi = 0.50$, where π is the population proportion of the plus (or the minus) signs. Thus, a hypothesis concerning the value of the median in fact is tested as a hypothesis concerning π. If the sample is small ($n < 30$) the binomial distribution is used to carry out this test, as described in Section 11.4. If the sample is large, the normal distribution can be used, as described in Section 11.5.

See Problem 21.2 for the use of the sign test to test a null hypothesis concerning the population median.

21.5   ONE SAMPLE: THE WILCOXON TEST

Just as is the case for the sign test, the *Wilcoxon test* can be used to test a null hypothesis concerning the value of the population median. Because the Wilcoxon test considers the magnitude of the difference between each sample value and the hypothesized value of the median, it is a more sensitive test than the sign test. On the other hand, because differences are determined, the values must be at least at the interval scale. No assumptions are required about the form of the population distribution.

The null and alternative hypotheses are formulated with respect to the population median as either a one-tail or two-tail test. The difference between each observed value and the hypothesized value of the median is determined, and this difference, with arithmetic sign, is designated d: $d = (X - \mathrm{Med}_0)$. If any difference is equal to zero, the associated observation is dropped from the analysis and the effective sample size is thereby reduced. The absolute values of the differences are then ranked from lowest to highest, with the rank of 1 assigned to the smallest absolute difference. When absolute differences are equal, the mean rank is assigned to the tied values. Finally, the sum of the ranks is obtained separately for the positive and negative differences. The smaller of these two sums is the Wilcoxon T statistic for a two-tail test. In the case of a one-tail test, the smaller sum must be associated with the directionality of the null hypothesis. Appendix 10 identifies the critical values of T according to sample size and level of significance. For rejection of the null hypothesis, the obtained value of T must be *smaller* than the critical value given in the table.

When $n \geq 25$ and the null hypothesis is true, the T statistic is approximately normally distributed. The mean and standard error associated with this sampling distribution are, respectively,

$$\mu_T = \frac{n(n+1)}{4} \tag{21.4}$$

$$\sigma_T = \sqrt{\frac{n(n+1)(2n+1)}{24}} \tag{21.5}$$

Therefore, for a relatively large sample the test can be carried out by using the normal probability distribution and computing the z test statistic as follows:

$$z = \frac{T - \mu_T}{\sigma_T} \tag{21.6}$$

See Problem 21.3 for an application of the Wilcoxon test to test a null hypothesis concerning the population median.

21.6 TWO INDEPENDENT SAMPLES: THE MANN–WHITNEY TEST

The *Mann–Whitney test* can be used to test the null hypothesis that the medians of two populations are equal. It is assumed that the two populations have the same form and dispersion, because such differences also would lead to rejection of the null hypothesis. It is required that the values in the two independent random samples be at least at the ordinal scale.

Two samples are combined into one ordered array, with each sample value identified according to the original sample group. The values are then ranked from lowest to highest, with the rank of 1 assigned to the lowest observed sample value. For equal values the mean rank is assigned to the tied, or equal, values. If the null hypothesis is true, the average of the ranks for each sample group should be approximately equal. The statistic calculated to carry out this test is designated U, and it can be based on the sum of the ranks in either of the two random samples, as follows:

$$U_1 = n_1 n_2 + \frac{n_1(n_1+1)}{2} - R_1 \qquad (21.7)$$

or

$$U_2 = n_1 n_2 + \frac{n_2(n_2+1)}{2} - R_2 \qquad (21.8)$$

where n_1 = size of the first sample

n_2 = size of the second sample

R_1 = sum of the ranks in the first sample

R_2 = sum of the ranks in the second sample

Given that $n_1 > 10$, $n_2 > 10$, and the null hypothesis is true, the sampling distribution of U is approximately normal, with the following parameters:

$$\mu_U = \frac{n_1 n_2}{2} \qquad (21.9)$$

$$\sigma_U = \sqrt{\frac{n_1 n_2 (n_1 + n_2 + 1)}{12}} \qquad (21.10)$$

Therefore, the test statistic for testing the null hypothesis that the medians of two populations are equal is

$$z = \frac{U - \mu_U}{\sigma_U} \qquad (21.11)$$

where U equals U_1 or U_2.

For data situations in which $n_1 < 10$, $n_2 < 10$, or both n_1 and $n_2 < 10$, the normal probability distribution cannot be used for this test. However, special tables for the U statistic are available for such small samples in specialized textbooks on nonparametric statistics.

Problem 21.4 illustrates the use of the Mann–Whitney test.

21.7 PAIRED OBSERVATIONS: THE SIGN TEST

For two samples collected as *paired observations* (see Section 11.3), the *sign test* described in Section 21.4 can be used to test the null hypothesis that the two population medians are equal. The sample values must be at least at the ordinal scale, and no assumptions are required about the forms of the two population distributions.

A plus (+) sign is assigned for each pair of values for which the measurement in the first sample is greater than the measurement in the second sample, and a minus (−) sign is assigned when the opposite condition is true. If a pair of measurements have the same value, these tied values are dropped from the analysis, with the effective sample size thereby being reduced. If the hypothesis that the two populations are at the same level of magnitude is true, the number of plus signs should approximately equal the number of minus signs. Therefore, the null hypothesis tested is H_0: $\pi = 0.50$, where π is the population proportion of the plus (or the minus) signs. If the number of sample pairs is small ($n < 30$), the binomial distribution is used to carry out this test, as described in Section 11.4. If the sample is large ($n > 30$), the normal distribution can be used, as described in Section 11.5. Note that although two samples have been collected, the test is applied to the one set of plus and minus signs that results from the comparison of the pairs of measurements.

Problem 21.5 illustrates use of the sign test for testing the difference between two medians for data that have been collected as paired observations.

21.8 PAIRED OBSERVATIONS: THE WILCOXON TEST

For two samples collected as paired observations, the *Wilcoxon test* described in Section 21.5 can be used to test the null hypothesis that the two population medians are equal. Because the Wilcoxon test considers the magnitude of the difference between the values in each matched pair, and not just the direction, or sign, of the difference, it is a more sensitive test than the sign test. However, the sample values must be at least at the interval scale. No assumptions are required about the forms of the two distributions.

The difference between each pair of values is determined, and this difference, with the associated arithmetic sign, is designated by d. If any difference is equal to zero, this pair of observations is dropped from the analysis, thus reducing the effective sample size. Then the absolute values of the differences are ranked from lowest to highest, with the rank of 1 assigned to the smallest absolute difference. When absolute differences are equal, the mean rank is assigned to the tied values. Finally, the sum of the ranks is obtained separately for the positive and negative differences. The smaller of these two sums is the Wilcoxon T statistic for a two-tail test. For a one-tail test the smaller sum must be associated with the directionality of the null hypothesis, as illustrated in the one-sample application of the Wilcoxon test in Problem 21.3. Appendix 10 identifies the critical value of T according to sample size and level of significance. For rejection of the null hypothesis, the T statistic must be *smaller* than the critical value given in the table.

When $n \geq 25$ and the null hypothesis is true, the T statistic is approximately normally distributed. The formulas for the mean and standard error of the sampling distribution of T and the formula for the z test statistic are given in Section 21.5, on the one-sample application of the Wilcoxon test.

Problem 21.6 illustrates the use of the Wilcoxon test for testing the difference between two medians for data that have been collected as paired observations.

21.9 SEVERAL INDEPENDENT SAMPLES: THE KRUSKAL–WALLIS TEST

The *Kruskal–Wallis test* is used to test the null hypothesis that several populations have the same medians. As such, it is the nonparametric equivalent of the one-factor completely randomized design of the analysis of variance. It is assumed that the several populations have the same form and dispersion for the above hypothesis to be applicable, because differences in form or dispersion would also lead to rejection of the null hypothesis. The values for the several independent random samples are required to be at least at the ordinal scale.

The several samples are first viewed as one array of values, and each value in this combined group is ranked from lowest to highest. For equal values the mean rank is assigned to the tied values. If the null hypothesis is true, the average of the ranks for each sample group should be about equal. The test statistic calculated is designated H and is based on the sum of the ranks in each of the several random samples, as follows:

$$H = \left\{ \left[\frac{12}{N(N+1)} \right] \left[\sum \frac{R_j^2}{n_j} \right] \right\} - 3(N+1) \qquad (21.12)$$

where N = combined sample size of the several samples (note that N does not designate population size in this case)

R_j = sum of the ranks for the jth sample or treatment group

n_j = number of observations in the jth sample

Given that the size of each sample group is at least $n_j \geq 5$ and the null hypothesis is true, the sampling distribution of H is approximately distributed as the χ^2 distribution with $df = K - 1$, where K is the number of treatment, or sample, groups. The χ^2 value that approximates the critical value of the test statistic is always the upper tail value. This test procedure is analogous to the upper tail of the F distribution being used in the analysis of variance.

For tied ranks the test statistic H should be corrected. The corrected value of the test statistic is designated H_c and is computed as follows:

$$H_c = \frac{H}{1 - [\Sigma(t_j^3 - t_j)/(N^3 - N)]} \qquad (21.13)$$

where t_j represents the number of tied scores in the jth sample.

The effect of this correction is to increase the value of the calculated H statistic. Therefore, if the uncorrected value of H leads to the rejection of the null hypothesis, there is no need to correct this value for the effect of tied ranks.

Problem 21.7 illustrates use of the Kruskal–Wallis test for testing the null hypothesis that several populations have the same median.

Solved Problems

THE RUNS TEST FOR RANDOMNESS

21.1   A sample of 36 individuals were interviewed in a market-research survey, with 22 women (W) and 14 men (M) included in the sample. The sampled individuals were obtained in the following order: M, W, W, W, W, M, M, M, W, M, W, W, W, M, M, W, W, W, W, M, W, W, W, M, M, W, W, W, M, W, M, M, W, W, W, M. Use the runs test to test the randomness of this set of observations, using the 5 percent level of significance.

The number of runs in this sample is $R = 17$, as indicated by the underscores above. Where n_1 = the number of women and n_2 = the number of men, we can use the normal distribution to test the null hypothesis that the sampling process was random, because $n_1 > 20$. We compute the mean and the standard error of the sampling distribution of R as follows:

$$\mu_R = \frac{2n_1 n_2}{n_1 + n_2} + 1 = \frac{2(22)(14)}{22 + 14} + 1 = \frac{616}{36} + 1 = 18.1$$

$$\sigma_R = \sqrt{\frac{2n_1 n_2 (2n_1 n_2 - n_1 - n_2)}{(n_1 + n_2)^2 (n_1 + n_2 - 1)}} = \sqrt{\frac{2(22)(14)[2(22)(14) - 22 - 14]}{(22 + 14)^2 (22 + 14 - 1)}}$$

$$= \sqrt{\frac{616(616 - 36)}{36^2 (35)}} = \sqrt{\frac{357,280}{45,360}} = \sqrt{7.8765} = 2.81$$

When the 5 percent level of significance is used, the critical values of the z statistic are $z = \pm 1.96$. The value of the z test statistic for these data is

$$z = \frac{R - \mu_R}{\sigma_R} = \frac{17 - 18.1}{2.81} = \frac{-1.1}{2.81} = -0.39$$

Therefore, using the 5 percent level of significance, we cannot reject the null hypothesis that the sequence of women and men occurred randomly.

ONE SAMPLE: THE SIGN TEST

21.2 It is claimed that the units assembled with a redesigned product assembly system will be greater than with the old system, for which the population median is 80 units per workshift. *Not* giving the benefit of the doubt to the redesigned system, formulate the null hypothesis and test it at the 5 percent level of significance. The sample data are reported in the first part of Table 21.1.

**Table 21.1   Number of Units Assembled with the
Redesigned System**

Sampled workshift	Units assembled (X)	Sign of difference ($X - 80$)
1	75	−
2	85	+
3	92	+
4	80	0
5	94	+
6	90	+
7	91	+
8	76	−
9	88	+
10	82	+
11	96	+
12	83	+

The null and alternative hypotheses are

$$H_0: \text{Med} \le 80$$

$$H_1: \text{Med} > 80$$

Because no assumption is made about the form of the population distribution, a nonparametric test is appropriate. Using the sign test, the null and alternative hypotheses in terms of the proportion of plus signs of the differences, where $d = (X - 80)$, are

$$H_0: \pi \le 0.50$$

$$H_1: \pi > 0.50$$

Referring to Table 21.1, we see that for the fourth sampled workshift the number of units assembled happened to be exactly equal to the hypothesized value of the population median. Therefore, this observation is omitted from the following analysis, resulting in an effective sample size of $n = 11$. Of the 11 signs of the differences reported in Table 21.1, 9 are plus signs. The test is to be carried out using the 5 percent level of significance.

Because the sample size is $n < 30$, the binomial distribution is the appropriate basis for this test. Using the P-value approach to hypothesis testing, as described in Section 10.6, we determine the probability of observing 9 or more plus signs in 11 observations, given that the population proportion of plus signs is 0.50, by reference to Appendix 2 for binomial probabilities:

$$P(X \ge 9 \mid n = 11, \pi = 0.50) = 0.0269 + 0.0054 + 0.0005 = 0.0328$$

Since the P value associated with the sample result is less than 0.05, the null hypothesis is rejected at the 5 percent level of significance for this one-tail test. That is, the probability of observing such a large number (or more) of plus signs when the null hypothesis is true is less than 0.05, and specifically, the probability is 0.0328. Therefore, we accept the alternative hypothesis and conclude that the median output per workshift for the new assembly system is greater than 80 units.

A two-tail sign test is illustrated in Problem 21.5, where the sign test is used in conjunction with paired observations.

ONE SAMPLE: THE WILCOXON TEST

21.3 Use the Wilcoxon test with respect to the null hypothesis and the data in Problem 21.2, above, and compare your solution to the one obtained in Problem 21.2.

$$H_0: \text{Med} \leq 80$$

$$H_1: \text{Med} > 80$$

The sample data are repeated in Table 21.2, which is the worktable for the Wilcoxon test. In this table note that for the fourth sampled workshift the number of units assembled happened to be exactly equal to the hypothesized value of the population median. Therefore, this observation was dropped from the analysis, resulting in an effective sample size of $n = 11$. Also note that the absolute value of d for the first and second sampled workshifts is the same, and therefore the mean rank of 4.5 was assigned to each of these sampled workshifts (in lieu of ranks 4 and 5). The next rank assigned is then rank 6.

Table 21.2 Number of Units Assembled with the Redesigned System and the Determination of Signed Ranks

Sampled workshift	Units assembled (X)	Difference $[d = (X - 80)]$	Rank of $\|d\|$	Signed rank (+)	Signed rank (−)
1	75	−5	4.5		4.5
2	85	5	4.5	4.5	
3	92	12	9	9	
4	80	0			
5	94	14	10	10	
6	90	10	7	7	
7	91	11	8	8	
8	76	−4	3		3
9	88	8	6	6	
10	82	2	1	1	
11	96	16	11	11	
12	83	3	2	2	
Total				58.5	$T = 7.5$

In order to reject the null hypothesis $H_0: \text{Med} \leq 80$ for this one-tail test, it follows that the differences $d = (X - \text{Med}_0)$ must be predominantly positive, for negative differences would represent positive support for the null hypothesis. Therefore, for this upper-tail test the sum of the ranks for the negative differences must be the smaller sum. Referring to Table 21.2, we see that such is in fact the case, and the value of the Wilcoxon test statistic is $T = 7.5$.

The null hypothesis is to be tested at the 5 percent level of significance. Appendix 10 indicates that for $n = 11$ (the effective sample size) the critical value of T for a one-tail test at the 5 percent level of significance is $T = 14$. Because the obtained value of T is smaller than the critical value, the null hypothesis is rejected at the 5 percent level of significance, and we conclude that the median units assembled with the new system is greater than 80 units. This is the same conclusion as with the use of the sign test in the preceding section, in which the P value for the test was $P = 0.0328$. However, reference to Appendix 10 for the Wilcoxon test indicates that for $n = 11$ and $\alpha = 0.025$ the critical value is $T = 11$. Therefore, the null hypothesis could be rejected at the 2.5 percent level of significance with the Wilcoxon test, but not with the sign test. This observation is consistent with our earlier observations that the Wilcoxon test is the more sensitive of the two tests.

TWO INDEPENDENT SAMPLES: THE MANN–WHITNEY TEST

21.4 To evaluate and compare two methods of instruction for industrial apprentices, a training director assigns 15 randomly selected trainees to each of the two methods. Because of normal attrition,

14 apprentices complete the course taught by method 1 and 12 apprentices complete the course taught by method 2. The same achievement test is then given to both trainee groups, as reported in Table 21.3. Test the null hypothesis that the median level of test performance does not differ for the two methods of instruction, using the 5 percent level of significance.

Table 21.3   Achievement Test Scores of Apprentice Trainees Taught by Two Methods of Instruction

Method 1	Method 2
70	86
90	78
82	90
64	82
86	65
77	87
84	80
79	88
82	95
89	85
73	76
81	94
83	
66	

$$H_0: \text{Med}_1 = \text{Med}_2 \qquad H_1: \text{Med}_1 \neq \text{Med}_2$$

Table 21.4 lists the achievement test scores in one array, ranked from lowest to highest score. The ranks are presented in the second column of Table 21.4. Notice the way the rankings are assigned to the tied scores: For the three test scores of 82, for example, the scores are at the rank positions 12, 13, and 14 in the second column of the table. Therefore, the mean of these three ranks, 13, is assigned to these three test scores. Then note that the next rank assigned is 15, not 14, because the assignment of the mean rank of 13 to the three scores of 82 has brought the score of 83 to the 15th position in the array. Similarly, there are two scores of 86; the ranks for these two positions are 18 and 19, and therefore the mean rank of 18.5 is assigned to each of these scores. The next score in the array, 87, is assigned the rank 20. The last two columns in Table 21.4 list the ranks according to the sample, which represents the teaching method used, and the sums of these two columns of ranks are the two values R_1 and R_2, respectively.

Applying formula (*21.7*), we determine the value of U_1 as follows:

$$U_1 = n_1 n_2 + \frac{n_1(n_1+1)}{2} - R_1 = 14(12) + \frac{14(14+1)}{2} - 161$$

$$= 168 + \frac{210}{2} - 161 = 112$$

Because $n_1 > 10$ and $n_2 > 10$, the normal probability distribution can be used to test the null hypothesis that the value of μ_U is the same for both samples. The test is to be carried out at the 5 percent level of significance. Based on formula (*21.9*), the value of μ_U is

$$\mu_U = \frac{n_1 n_2}{2} = \frac{14(12)}{2} = 84$$

Thus the hypotheses are

$$H_0: \mu_U = 84$$

$$H_1: \mu_U \neq 84$$

Table 21.4   Combined Array of the Achievement Test Scores with Associated Ranks

Score*	Rank	Method 1 ranks	Method 2 ranks
64	1	1	
65	2		2
66	3	3	
70	4	4	
73	5	5	
76	6		6
77	7	7	
78	8		8
79	9	9	
80	10		10
81	11	11	
82	13	13	
82	13	13	
82	13		13
83	15	15	
84	16	16	
85	17		17
86	18.5	18.5	
86	18.5		18.5
87	20		20
88	21		21
89	22	22	
90	23.5	23.5	
90	23.5		23.5
94	25		25
95	26		26
Totals		$R_1 = 161.0$	$R_2 = 190.0$

*Scores for method 1 are underscored.

Based on formula (21.10), the standard error of U is

$$\sigma_U = \sqrt{\frac{n_1 n_2 (n_1 + n_2 + 1)}{12}} = \sqrt{\frac{14(12)(14 + 12 + 1)}{12}} = \sqrt{\frac{4,536}{12}} = 19.4$$

Applying formula (21.11), we determine the value of the z test statistic as follows:

$$z = \frac{U_1 - \mu_U}{\sigma_U} = \frac{112 - 84}{19.4} = 1.44$$

Because the critical values of z for a test at the 5 percent level of significance are $z = \pm 1.96$, the null hypothesis that there is no difference between the medians of the two distributions of achievement test scores cannot be rejected.

PAIRED OBSERVATIONS: THE SIGN TEST

21.5   A consumer panel that includes 14 individuals is asked to rate two brands of tea bags according to a point evaluation system based on several criteria. Table 21.5 reports the points assigned and also indicates the sign of the difference for each pair of ratings. Test the null hypothesis that there is no difference in the level of ratings for the two brands of tea bags at the 5 percent level of significance by use of the sign test, formulating the null and alternative hypotheses in terms of the proportion of plus signs.

Table 21.5   Ratings Assigned to Two Brands of Tea Bags by a Consumer Panel

Panel member	Point rating assigned to each brand		Sign of difference
	Brand 1	Brand 2	
1	20	16	+
2	24	26	−
3	28	18	+
4	24	17	+
5	20	20	0
6	29	21	+
7	19	23	−
8	27	22	+
9	20	23	−
10	30	20	+
11	18	18	0
12	28	21	+
13	26	17	+
14	24	26	−

$$H_0: \pi = 0.50$$

$$H_1: \pi \neq 0.50$$

Referring to Table 21.5, we see that two of the consumers rated the two brands of tea bags as equal, and therefore these two panel members are omitted from the following analysis, resulting in an effective sample size of $n = 12$. Of the 12 signs 8 are plus signs.

Because the number of pairs of values is $n < 30$, the binomial distribution is the appropriate basis for this test. As in Problem 21.2, on the one-sample sign test, we use the P-value approach to hypothesis testing. The probability of observing 8 or more plus signs in 12 observations given that the population proportion of plus signs is 0.50 is determined by reference to Appendix 2 for binomial probabilities:

$$P(X \geq 8 \mid n = 12, \pi = 0.50) = 0.1208 + 0.0537 + 0.0161 + 0.0029 + 0.0002$$

$$= 0.1937$$

The probability calculated above is for the one-tail $P(X \geq 8)$. Because a two-tail test is involved in the present application, *this probability must be doubled* to obtain the P value: $P = 2 \times 0.1937 = 0.3874$. That is, the probability is 0.3874 that the observed difference in *either* direction from the expected number of plus signs of $n\pi_0 = 12(0.50) = 6$. This P value is well above the 0.05 probability level, and therefore the null hypothesis that the two brands of tea bags have equal consumer preference clearly cannot be rejected. Note that in Problem 11.8 on p. 191 we obtained the probability for each tail of the binomial distribution separately. In the present case we simply multiply the one tail probability by 2 because the binomial distribution is symmetrical when $\pi = 0.50$

PAIRED OBSERVATIONS: THE WILCOXON TEST

21.6   Apply the Wilcoxon test to the data in Problem 21.5 above, using the 5 percent level of significance.

We repeat the ratings in Table 21.6, determine the values of d, rank the absolute values of d, and finally, we determine the sum of the plus and minus rankings.

Referring to Table 21.6, note that the two panel members who rated the two brands of tea bags equally are dropped from the analysis, because the difference $d = 0$ is not assigned a rank. Therefore, the effective sample size is $n = 12$. Of the 12 absolute values of d several tied pairs result in the assignment of the mean ranks. For instance, there are two absolute differences tied for the positions of rank 1 and rank 2; therefore each of these absolute differences is assigned the mean rank 1.5. The next rank assigned is then rank 3. Because the sum of the ranks for the negative differences is smaller than the sum for the positive differences, this smaller sum is labelled T in Table 21.6.

Table 21.6   Ratings Assigned to Two Brands of Tea Bags by a Consumer Panel and the Determination of Signed Ranks

Panel member	Brand 1 (X_1)	Brand 2 (X_2)	Difference $(d = X_1 - X_2)$	Rank of $|d|$	Signed rank (+)	Signed rank (−)
1	20	16	4	4.5	4.5	
2	24	26	−2	1.5		1.5
3	28	18	10	11.5	11.5	
4	24	17	7	7.5	7.5	
5	20	20	0			
6	29	21	8	9	9	
7	19	23	−4	4.5		4.5
8	27	22	5	6	6	
9	20	23	−3	3		3
10	30	20	10	11.5	11.5	
11	18	18	0			
12	28	21	7	7.5	7.5	
13	26	17	9	10	10	
14	24	26	−2	1.5		1.5
Total					67.5	$T = 10.5$

Appendix 10 indicates that for $n = 12$ (the effective sample size) the critical value of the Wilcoxon T statistic for a two-tail test at the 5 percent level of significance is $T = 14$. Since T is the smaller sum of the signed ranks for a two-tail test, the obtained value of the test statistic as calculated in Table 21.6 is $T = 10.5$. Because this value is less than the critical value, the null hypothesis that there is no difference in the ratings for the two brands of tea bags is rejected.

Note that the null hypothesis of no difference could not be rejected at the 5 percent level when the sign test was applied to these same data in Problem 21.5. Because the Wilcoxon test considers the magnitude of the difference between each matched pair and not just the sign of the difference, the Wilcoxon test for paired observations is a more sensitive test than the sign test.

SEVERAL INDEPENDENT SAMPLES: THE KRUSKAL–WALLIS TEST

21.7   In Problem 13.1 on the one-factor completely randomized design of the analysis of variance we presented achievement test scores for a random sample of $n = 5$ trainees randomly assigned to each of three instructional methods. In those sections the null hypothesis was that the mean levels of achievement for the three different levels of instruction do not differ, and because of the small samples it was necessary to assume that the three populations are normally distributed. Suppose we cannot make this assumption. Apply the Kruskal–Wallis test to test the null hypothesis that the several populations have the same median, using the 5 percent level of significance.

Table 21.7 repeats the data from Table 13.4 and also includes the ranks from lowest to highest when the achievement scores are viewed as one combined group. With reference to the table, we compute the

Table 21.7   Achievement Test Scores of Trainees Under Three Methods of Instruction

Method A_1		Method A_2		Method A_3	
Score	Rank	Score	Rank	Score	Rank
86	12	90	15	82	9.5
79	6	76	5	68	1
81	7.5	88	13	73	4
70	2	82	9.5	71	3
84	11	89	14	81	7.5
Total	$R_1 = 38.5$		$R_2 = 56.5$		$R_3 = 25.0$

value of the test statistic as follows:

$$H = \left\{ \left[\frac{12}{N(N+1)} \right] \left[\sum \frac{R_j^2}{n_j} \right] \right\} - 3(N+1)$$

$$= \left\{ \left[\frac{12}{15(15+1)} \right] \left[\frac{(38.5)^2}{5} + \frac{(56.5)^2}{5} + \frac{(25.0)^2}{5} \right] \right\} - 3(15+1)$$

$$= \frac{12}{240} \left(\frac{5{,}299.5}{5} \right) - 48 = \frac{63{,}594}{1{,}200} - 48 = 52.995 - 48 = 4.995$$

Because there were some ties in the achievement test scores, it is appropriate to correct the value of the test statistic:

$$H_c = \frac{H}{1 - [\sum(t_j^3 - t_j)/(N^3 - N)]}$$

$$= \frac{4.995}{1 - \{[(1^3 - 1) + (1^3 - 1) + (2^3 - 2)]/(15^3 - 15)\}}$$

$$= \frac{4.995}{1 - [(0 + 0 + 6)/(3{,}375 - 15)]} = \frac{4.995}{1 - (6/3{,}360)}$$

$$= \frac{4.995}{1 - 0.00179} = \frac{4.995}{0.99821} = 5.004$$

Thus the corrected value $H_c = 5.004$ is only slightly larger than the uncorrected value of $H = 4.995$, and the ties have a negligible effect on the value of the test statistic. With $df = K - 1 = 3 - 1 = 2$ Appendix 7 indicates that the critical value of the test statistic is $\chi^2 = 5.99$. Because the observed value of the test statistic is less than this critical value, the null hypothesis that the three sample groups were obtained from populations that have the same median cannot be rejected at the 5 percent level of significance. This conclusion is consistent with the conclusion in Problems 13.1 and 13.2 in which the analysis of variance was applied to these data.

Supplementary Problems

THE RUNS TEST FOR RANDOMNESS

21.8   Refer to Table 2.16 on page 26, in which 40 randomly sampled personal loan amounts are reported. The sequence of data collection was according to row in the table (that is, the sequence of observed amounts was $932, $1,000, $356, etc.). The median loan amount for the data in the table is $944.50. Test the randomness of this sequence of loan amounts by classifying each amount as being above or below the median, using the 5 percent level of significance.

Ans. Critical $z = \pm 1.96$, $R = 28$, and $z = 2.56$. The null hypothesis that the sequence is random is rejected. There is an unusually large number of runs.

ONE SAMPLE: THE SIGN TEST

21.9 The following table reports the unit sales of a new tool in a sample of 12 outlets in a given month. The form of the distribution is unknown, and thus a parametric statistical test is not appropriate given the small sample size. Use the sign test with respect to the null hypothesis that the median sales amount in the population is no greater than 10.0 units per outlet, using the 5 percent level of significance.

Tools/outlet	8	18	9	12	10	14	16	7	14	11	10	20

Ans. $P = 0.1719$; the null hypothesis cannot be rejected.

ONE SAMPLE: THE WILCOXON TEST

21.10 Apply the Wilcoxon test with respect to the null hypothesis and the data in Problem 21.9, using the 5 percent level of significance.

Ans. Critical $T = 11$, $T = 10$; the null hypothesis is rejected.

TWO INDEPENDENT SAMPLES: THE MANN–WHITNEY TEST

21.11 A quality department wishes to compare the time required for two alternative systems to diagnose equipment failure. A sample of 30 equipment failures are randomly assigned for diagnosis to the two systems with 14 failures assigned for diagnosis to the first system and 16 assigned for diagnosis to the second system. Table 21.8 reports the total time, in minutes, required to diagnose each failure. No assumption can be made concerning the distribution of the total time required to diagnose such failures. Using the 10 percent

Table 21.8 Time Required to Diagnose Equipment Failures (in Minutes)

System 1	System 2
25	18
29	37
42	40
16	56
31	49
14	28
33	20
45	34
26	39
34	47
30	31
43	65
28	38
19	32
	24
	49

level of significance, test the null hypothesis that the two samples have been obtained from populations that have the same median.

Ans. $R_1 = 173.5$, $R_2 = 291.5$, critical $z = \pm 1.645$, $z = 1.87$; the null hypothesis is rejected at $\alpha = 0.10$.

PAIRED OBSERVATIONS: THE SIGN TEST

21.12   Instead of the experimental design used in Problem 21.11, the same random sample of 10 equipment failures is diagnosed by the two systems. Thus each of the 10 pieces of equipment is subjected to diagnosis twice, and the time required in minutes to diagnose the failure by each system is noted. Table 21.9 indicates the number of minutes required for each diagnosis. Using the 10 percent level of significance, test the null hypothesis that the two samples came from populations that are not different in level. That is, the null hypothesis is that there is no difference in the median amount of time required to diagnose equipment failures by the two methods. Use the sign test for this analysis.

Ans. $P = 0.3438$; the null hypothesis cannot be rejected.

Table 21.9   Time in Minutes Required to Diagnose Equipment Failures by the Paired Observations Design

Sampled equipment	System 1	System 2
1	23	21
2	40	48
3	35	45
4	24	22
5	17	19
6	32	37
7	27	29
8	32	38
9	25	24
10	30	36

PAIRED OBSERVATIONS: THE WILCOXON TEST

21.13   At the 10 percent level of significance apply the Wilcoxon test for matched pairs to the data in Table 21.9, again testing the null hypothesis that there is no difference in the median amount of time required by the two methods. Compare your results with those obtained in Problem 21.12.

Ans.   Critical $T = 11$, $T = 8.0$; the null hypothesis is rejected at $\alpha = 0.10$.

SEVERAL INDEPENDENT SAMPLES: THE KRUSKAL–WALLIS TEST

21.14   In addition to the data reported in Table 21.8, suppose that a third system for diagnosing equipment failure is also evaluated by randomly assigning 12 failures to this system and observing the time required in minutes to diagnose each failure. The required time periods are 21, 36, 34, 19, 46, 25, 38, 31, 20, 26, 30, and 18. Using the 10 percent level of significance, test the null hypothesis that the three samples were obtained from populations with the same median.

Ans. $R_1 = 260.5$, $R_2 = 431.0$, $R_3 = 211.5$, critical $\chi^2 = 4.61$, $H = 5.12$; the null hypothesis is rejected at $\alpha = 0.10$.

Appendix 1

Table of Random Numbers

10097	85017	84532	13618	23157	86952	02438	76520	91499	38631	79430	62421	97959	67422	69992	68479
37542	16719	82789	69041	05545	44109	05403	64894	80336	49172	16332	44670	35089	17691	89246	26940
08422	65842	27672	82186	14871	22115	86529	19645	44104	89232	57327	34679	62235	79655	81336	85157
99019	76875	20684	39187	38976	94324	43204	09376	12550	02844	15026	32439	58537	48274	81330	11100
12807	93640	39160	41453	97312	41548	93137	80157	63606	40387	65406	37920	08709	60623	2237	16505
66065	99478	70086	71265	11742	18226	29004	34072	61196	80240	44177	51171	08723	39323	05798	26457
31060	65119	26486	47353	43361	99436	42753	45571	15474	44910	99321	72173	56239	04595	10836	95270
85269	70322	21592	48233	93806	32584	21828	02051	94557	33663	86347	00926	44915	34823	51770	67897
63573	58133	41278	11697	49540	61777	67954	05325	42481	86430	19102	37420	41976	76559	24358	97344
73796	44655	81255	31133	36768	60452	38537	03529	23523	31379	68588	81675	15694	43438	36879	73208
98520	02295	13487	98662	07092	44673	61303	14905	04493	98086	32533	17767	14523	52494	24826	75246
11805	85035	54881	35587	43310	48897	48493	39808	00549	33185	04805	05431	94598	97654	16232	64051
83452	01197	86935	28021	61570	23350	65710	06288	35963	80951	68953	99634	81949	15307	00406	26898
88685	97907	19078	40646	31352	48625	44369	86507	59808	79752	02529	40200	73742	08391	49140	45427
99594	63268	96905	28797	57048	46359	74294	87517	46058	18633	99970	67348	49329	95236	32537	01390
65481	52841	59684	67411	09243	56092	84369	17468	32179	74029	74717	17674	90446	00597	45240	87379
80124	53722	71399	10916	07959	21225	13018	17727	69234	54178	10805	35635	45266	61406	41941	20117
74350	11434	51908	62171	93732	26958	02400	77402	19565	11664	77602	99817	28573	41430	96382	01758
69916	62375	99292	21177	72721	66995	07289	66252	45155	48324	32135	26803	16213	14938	71961	19476
09893	28337	20923	87929	61020	62841	31374	14225	94864	69074	45753	20505	78317	31994	98145	36168

Appendix 2

Binomial Probabilities*

n	x	.01	.05	.10	.15	.20	.25	p .30	.35	.40	.45	.50
1	0	.9900	.9500	.9000	.8500	.8000	.7500	.7000	.6500	.6000	.5500	.5000
	1	.0100	.0500	.1000	.1500	.2000	.2500	.3000	.3500	.4000	.4500	.5000
2	0	.9801	.9025	.8100	.7225	.6400	.5625	.4900	.4225	.3600	.3025	.2500
	1	.0198	.0950	.1800	.2550	.3200	.3750	.4200	.4550	.4800	.4950	.5000
	2	.0001	.0025	.0100	.0225	.0400	.0625	.0900	.1225	.1600	.2025	.2500
3	0	.9703	.8574	.7290	.6141	.5120	.4219	.3430	.2746	.2160	.1664	.1250
	1	.0294	.1354	.2430	.3251	.3840	.4219	.4410	.4436	.4320	.4084	.3750
	2	.0003	.0071	.0270	.0574	.0960	.1406	.1890	.2389	.2880	.3341	.3750
	3	.0000	.0001	.0010	.0034	.0080	.0156	.0270	.0429	.0640	.0911	.1250
4	0	.9606	.8145	.6561	.5220	.4096	.3164	.2401	.1785	.1296	.0915	.0625
	1	.0388	.1715	.2916	.3685	.4096	.4219	.4116	.3845	.3456	.2995	.2500
	2	.0006	.0135	.0486	.0975	.1536	.2109	.2646	.3105	.3456	.3675	.3750
	3	.0000	.0005	.0036	.0115	.0256	.0469	.0756	.1115	.1536	.2005	.2500
	4	.0000	.0000	.0001	.0005	.0016	.0039	.0081	.0150	.0256	.0410	.0625
5	0	.9510	.7738	.5905	.4437	.3277	.2373	.1681	.1160	.0778	.0503	.0312
	1	.0480	.2036	.3280	.3915	.4096	.3955	.3602	.3124	.2592	.2059	.1562
	2	.0010	.0214	.0729	.1382	.2048	.2637	.3087	.3364	.3456	.3369	.3125
	3	.0000	.0011	.0081	.0244	.0512	.0879	.1323	.1811	.2304	.2757	.3125
	4	.0000	.0000	.0004	.0022	.0064	.0146	.0284	.0488	.0768	.1128	.1562
	5	.0000	.0000	.0000	.0001	.0003	.0010	.0024	.0053	.0102	.0185	.0312
6	0	.9415	.7351	.5314	.3771	.2621	.1780	.1176	.0754	.0467	.0277	.0156
	1	.0571	.2321	.3543	.3993	.3932	.3560	.3025	.2437	.1866	.1359	.0938
	2	.0014	.0305	.0984	.1762	.2458	.2966	.3241	.3280	.3110	.2780	.2344
	3	.0000	.0021	.0146	.0415	.0819	.1318	.1852	.2355	.2765	.3032	.3125
	4	.0000	.0001	.0012	.0055	.0154	.0330	.0595	.0951	.1382	.1861	.2344
	5	.0000	.0000	.0001	.0004	.0015	.0044	.0102	.0205	.0369	.0609	.0938
	6	.0000	.0000	.0000	.0000	.0001	.0002	.0007	.0018	.0041	.0083	.0156
7	0	.9321	.6983	.4783	.3206	.2097	.1335	.0824	.0490	.0280	.0152	.0078
	1	.0659	.2573	.3720	.3960	.3670	.3115	.2471	.1848	.1306	.0872	.0547
	2	.0020	.0406	.1240	.2097	.2753	.3115	.3177	.2985	.2613	.2140	.1641
	3	.0000	.0036	.0230	.0617	.1147	.1730	.2269	.2679	.2903	.2918	.2734
	4	.0000	.0002	.0026	.0109	.0287	.0577	.0972	.1442	.1935	.2388	.2734
	5	.0000	.0000	.0002	.0012	.0043	.0115	.0250	.0466	.0774	.1172	.1641
	6	.0000	.0000	.0000	.0001	.0004	.0013	.0036	.0084	.0172	.0320	.0547
	7	.0000	.0000	.0000	.0000	.0000	.0001	.0002	.0006	.0016	.0037	.0078
8	0	.9227	.6634	.4305	.2725	.1678	.1002	.0576	.0319	.0168	.0084	.0039
	1	.0746	.2793	.3826	.3847	.3355	.2670	.1977	.1373	.0896	.0548	.0312
	2	.0026	.0515	.1488	.2376	.2936	.3115	.2965	.2587	.2090	.1569	.1094
	3	.0001	.0054	.0331	.0839	.1468	.2076	.2541	.2786	.2787	.2568	.2188
	4	.0000	.0004	.0046	.0185	.0459	.0865	.1361	.1875	.2322	.2627	.2734
	5	.0000	.0000	.0004	.0026	.0092	.0231	.0467	.0808	.1239	.1719	.2188
	6	.0000	.0000	.0000	.0002	.0011	.0038	.0100	.0217	.0413	.0703	.1094
	7	.0000	.0000	.0000	.0000	.0001	.0004	.0012	.0033	.0079	.0164	.0312
	8	.0000	.0000	.0000	.0000	.0000	.0000	.0001	.0002	.0007	.0017	.0039
9	0	.9135	.6302	.3874	.2316	.1342	.0751	.0404	.0207	.0101	.0046	.0020
	1	.0830	.2985	.3874	.3679	.3020	.2253	.1556	.1004	.0605	.0339	.0176
	2	.0034	.0629	.1722	.2597	.3020	.3003	.2668	.2162	.1612	.1110	.0703
	3	.0001	.0077	.0446	.1069	.1762	.2336	.2668	.2716	.2508	.2119	.1641
	4	.0000	.0006	.0074	.0283	.0661	.1168	.1715	.2194	.2508	.2600	.2461
	5	.0000	.0000	.0008	.0050	.0165	.0389	.0735	.1181	.1672	.2128	.2461
	6	.0000	.0000	.0001	.0006	.0028	.0087	.0210	.0424	.0743	.1160	.1641
	7	.0000	.0000	.0000	.0000	.0003	.0012	.0039	.0098	.0212	.0407	.0703
	8	.0000	.0000	.0000	.0000	.0000	.0001	.0004	.0013	.0035	.0083	.0176
	9	.0000	.0000	.0000	.0000	.0000	.0000	.0000	.0001	.0003	.0008	.0020
10	0	.9044	.5987	.3487	.1969	.1074	.0563	.0282	.0135	.0060	.0025	.0010
	1	.0914	.3151	.3874	.3474	.2684	.1877	.1211	.0725	.0403	.0207	.0098
	2	.0042	.0746	.1937	.2759	.3020	.2816	.2335	.1757	.1209	.0763	.0439
	3	.0001	.0105	.0574	.1298	.2013	.2503	.2668	.2522	.2150	.1665	.1172
	4	.0000	.0010	.0112	.0401	.0881	.1460	.2001	.2377	.2508	.2384	.2051
	5	.0000	.0001	.0015	.0085	.0264	.0584	.1029	.1536	.2007	.2340	.2461
	6	.0000	.0000	.0001	.0012	.0055	.0162	.0368	.0689	.1115	.1596	.2051
	7	.0000	.0000	.0000	.0001	.0008	.0031	.0090	.0212	.0425	.0746	.1172
	8	.0000	.0000	.0000	.0000	.0001	.0004	.0014	.0043	.0106	.0229	.0439
	9	.0000	.0000	.0000	.0000	.0000	.0000	.0001	.0005	.0016	.0042	.0098
	10	.0000	.0000	.0000	.0000	.0000	.0000	.0000	.0000	.0001	.0003	.0010
11	0	.8953	.5688	.3138	.1673	.0859	.0422	.0198	.0088	.0036	.0014	.0005
	1	.0995	.3293	.3835	.3248	.2362	.1549	.0932	.0518	.0266	.0125	.0054
	2	.0050	.0867	.2131	.2866	.2953	.2581	.1998	.1395	.0887	.0513	.0269
	3	.0002	.0137	.0710	.1517	.2215	.2581	.2568	.2254	.1774	.1259	.0806
	4	.0000	.0014	.0158	.0536	.1107	.1721	.2201	.2428	.2365	.2060	.1611
	5	.0000	.0001	.0025	.0132	.0388	.0803	.1321	.1830	.2207	.2360	.2256
	6	.0000	.0000	.0003	.0023	.0097	.0268	.0566	.0985	.1471	.1931	.2256
	7	.0000	.0000	.0000	.0003	.0017	.0064	.0173	.0379	.0701	.1128	.1611
	8	.0000	.0000	.0000	.0000	.0002	.0011	.0037	.0102	.0234	.0462	.0806
	9	.0000	.0000	.0000	.0000	.0000	.0001	.0005	.0018	.0052	.0126	.0269
	10	.0000	.0000	.0000	.0000	.0000	.0000	.0000	.0002	.0007	.0021	.0054
	11	.0000	.0000	.0000	.0000	.0000	.0000	.0000	.0000	.0000	.0002	.0005
12	0	.8864	.5404	.2824	.1422	.0687	.0317	.0138	.0057	.0022	.0008	.0002
	1	.1074	.3413	.3766	.3012	.2062	.1267	.0712	.0368	.0174	.0075	.0029
	2	.0060	.0988	.2301	.2924	.2835	.2323	.1678	.1088	.0639	.0339	.0161
	3	.0002	.0173	.0852	.1720	.2362	.2581	.2397	.1954	.1419	.0923	.0537
	4	.0000	.0021	.0213	.0683	.1329	.1936	.2311	.2367	.2128	.1700	.1208
	5	.0000	.0002	.0038	.0193	.0532	.1032	.1585	.2039	.2270	.2225	.1934
	6	.0000	.0000	.0005	.0040	.0155	.0401	.0792	.1281	.1766	.2124	.2256
	7	.0000	.0000	.0000	.0006	.0033	.0115	.0291	.0591	.1009	.1489	.1934
	8	.0000	.0000	.0000	.0001	.0005	.0024	.0078	.0199	.0420	.0762	.1208
	9	.0000	.0000	.0000	.0000	.0001	.0004	.0015	.0048	.0125	.0277	.0537
	10	.0000	.0000	.0000	.0000	.0000	.0000	.0002	.0008	.0025	.0068	.0161
	11	.0000	.0000	.0000	.0000	.0000	.0000	.0000	.0001	.0003	.0010	.0029
	12	.0000	.0000	.0000	.0000	.0000	.0000	.0000	.0000	.0000	.0001	.0002
13	0	.8775	.5133	.2542	.1209	.0550	.0238	.0097	.0037	.0013	.0004	.0001
	1	.1152	.3512	.3672	.2774	.1787	.1029	.0540	.0259	.0113	.0045	.0016
	2	.0070	.1109	.2448	.2937	.2680	.2059	.1388	.0836	.0453	.0220	.0095
	3	.0003	.0214	.0997	.1900	.2457	.2517	.2181	.1651	.1107	.0660	.0349
	4	.0000	.0028	.0277	.0838	.1535	.2097	.2337	.2222	.1845	.1350	.0873

*Example: $P(X = 3 \mid n = 5, p = 0.30) = 0.1323$

p

n	x	.01	.05	.10	.15	.20	.25	.30	.35	.40	.45	.50
13	5	.0000	.0003	.0055	.0266	.0691	.1258	.1803	.2154	.2214	.1989	.1571
	6	.0000	.0000	.0008	.0063	.0230	.0559	.1030	.1546	.1968	.2169	.2095
	7	.0000	.0000	.0001	.0011	.0058	.0186	.0442	.0833	.1312	.1775	.2095
	8	.0000	.0000	.0001	.0001	.0011	.0047	.0142	.0336	.0656	.1089	.1571
	9	.0000	.0000	.0000	.0000	.0001	.0009	.0034	.0101	.0243	.0495	.0873
	10	.0000	.0000	.0000	.0000	.0000	.0001	.0006	.0022	.0065	.0162	.0349
	11	.0000	.0000	.0000	.0000	.0000	.0000	.0001	.0003	.0012	.0036	.0095
	12	.0000	.0000	.0000	.0000	.0000	.0000	.0000	.0000	.0001	.0005	.0016
	13	.0000	.0000	.0000	.0000	.0000	.0000	.0000	.0000	.0000	.0000	.0001
14	0	.8687	.4877	.2288	.1028	.0440	.0178	.0068	.0024	.0008	.0002	.0001
	1	.1229	.3593	.3559	.2539	.1539	.0832	.0407	.0181	.0073	.0027	.0009
	2	.0081	.1229	.2570	.2912	.2501	.1802	.1134	.0634	.0317	.0141	.0056
	3	.0003	.0259	.1142	.2056	.2501	.2402	.1943	.1366	.0845	.0462	.0222
	4	.0000	.0037	.0349	.0998	.1720	.2202	.2290	.2022	.1549	.1040	.0611
	5	.0000	.0004	.0078	.0352	.0860	.1468	.1963	.2178	.2066	.1701	.1222
	6	.0000	.0000	.0013	.0093	.0322	.0734	.1262	.1759	.2066	.2088	.1833
	7	.0000	.0000	.0002	.0019	.0092	.0280	.0618	.1082	.1574	.1952	.2095
	8	.0000	.0000	.0000	.0003	.0020	.0082	.0232	.0510	.0918	.1398	.1833
	9	.0000	.0000	.0000	.0000	.0003	.0018	.0066	.0183	.0408	.0762	.1222
	10	.0000	.0000	.0000	.0000	.0000	.0003	.0014	.0049	.0136	.0312	.0611
	11	.0000	.0000	.0000	.0000	.0000	.0000	.0002	.0010	.0033	.0093	.0222
	12	.0000	.0000	.0000	.0000	.0000	.0000	.0000	.0001	.0005	.0019	.0056
	13	.0000	.0000	.0000	.0000	.0000	.0000	.0000	.0000	.0001	.0002	.0009
	14	.0000	.0000	.0000	.0000	.0000	.0000	.0000	.0000	.0000	.0000	.0001
15	0	.8601	.4633	.2059	.0874	.0352	.0134	.0047	.0016	.0005	.0001	.0000
	1	.1303	.3658	.3432	.2312	.1319	.0668	.0305	.0126	.0047	.0016	.0005
	2	.0092	.1348	.2669	.2856	.2309	.1559	.0916	.0476	.0219	.0090	.0032
	3	.0004	.0307	.1285	.2184	.2501	.2252	.1700	.1110	.0634	.0318	.0139
	4	.0000	.0049	.0428	.1156	.1876	.2252	.2186	.1792	.1268	.0780	.0417
	5	.0000	.0006	.0105	.0449	.1032	.1651	.2061	.2123	.1859	.1404	.0916
	6	.0000	.0000	.0019	.0132	.0430	.0917	.1472	.1906	.2066	.1914	.1527
	7	.0000	.0000	.0003	.0030	.0138	.0393	.0811	.1319	.1771	.2013	.1964
	8	.0000	.0000	.0000	.0005	.0035	.0131	.0348	.0710	.1181	.1647	.1964
	9	.0000	.0000	.0000	.0001	.0007	.0034	.0116	.0298	.0612	.1048	.1527
	10	.0000	.0000	.0000	.0000	.0001	.0007	.0030	.0096	.0245	.0515	.0916
	11	.0000	.0000	.0000	.0000	.0000	.0001	.0007	.0024	.0074	.0191	.0417
	12	.0000	.0000	.0000	.0000	.0000	.0000	.0001	.0004	.0016	.0052	.0139
	13	.0000	.0000	.0000	.0000	.0000	.0000	.0000	.0001	.0003	.0010	.0032
	14	.0000	.0000	.0000	.0000	.0000	.0000	.0000	.0000	.0001	.0001	.0005
	15	.0000	.0000	.0000	.0000	.0000	.0000	.0000	.0000	.0000	.0000	.0000
16	0	.8515	.4401	.1853	.0743	.0281	.0100	.0033	.0010	.0003	.0001	.0000
	1	.1376	.3706	.3294	.2097	.1126	.0535	.0228	.0087	.0030	.0009	.0002
	2	.0104	.1463	.2745	.2775	.2111	.1336	.0732	.0353	.0150	.0056	.0018
	3	.0005	.0359	.1423	.2285	.2463	.2079	.1465	.0888	.0468	.0215	.0085
	4	.0000	.0061	.0514	.1311	.2001	.2252	.2040	.1553	.1014	.0572	.0278
	5	.0000	.0008	.0137	.0555	.1201	.1802	.2099	.2008	.1623	.1123	.0667
	6	.0000	.0001	.0028	.0180	.0550	.1101	.1649	.1982	.1983	.1684	.1222

p

n	x	.01	.05	.10	.15	.20	.25	.30	.35	.40	.45	.50
16	7	.0000	.0000	.0004	.0045	.0197	.0524	.1010	.1524	.1889	.1969	.1746
	8	.0000	.0000	.0001	.0009	.0055	.0197	.0487	.0923	.1417	.1812	.1964
	9	.0000	.0000	.0000	.0001	.0012	.0058	.0185	.0442	.0840	.1318	.1746
	10	.0000	.0000	.0000	.0000	.0002	.0014	.0056	.0167	.0392	.0755	.1222
	11	.0000	.0000	.0000	.0000	.0000	.0002	.0013	.0049	.0142	.0337	.0667
	12	.0000	.0000	.0000	.0000	.0000	.0000	.0002	.0011	.0040	.0115	.0278
	13	.0000	.0000	.0000	.0000	.0000	.0000	.0000	.0002	.0008	.0029	.0085
	14	.0000	.0000	.0000	.0000	.0000	.0000	.0000	.0000	.0001	.0005	.0018
	15	.0000	.0000	.0000	.0000	.0000	.0000	.0000	.0000	.0000	.0001	.0002
	16	.0000	.0000	.0000	.0000	.0000	.0000	.0000	.0000	.0000	.0000	.0000
17	0	.8429	.4181	.1668	.0631	.0225	.0075	.0023	.0007	.0002	.0000	.0000
	1	.1447	.3741	.3150	.1893	.0957	.0426	.0169	.0060	.0019	.0005	.0001
	2	.0117	.1575	.2800	.2673	.1914	.1136	.0581	.0260	.0102	.0035	.0010
	3	.0006	.0415	.1556	.2359	.2393	.1893	.1245	.0701	.0341	.0144	.0052
	4	.0000	.0076	.0605	.1457	.2093	.2209	.1868	.1320	.0796	.0411	.0182
	5	.0000	.0010	.0175	.0668	.1361	.1914	.2081	.1849	.1379	.0875	.0472
	6	.0000	.0001	.0039	.0236	.0680	.1276	.1784	.1991	.1839	.1432	.0944
	7	.0000	.0000	.0007	.0065	.0267	.0668	.1201	.1685	.1927	.1841	.1484
	8	.0000	.0000	.0001	.0014	.0084	.0279	.0644	.1134	.1606	.1883	.1855
	9	.0000	.0000	.0000	.0003	.0021	.0093	.0276	.0611	.1070	.1540	.1855
	10	.0000	.0000	.0000	.0000	.0004	.0025	.0095	.0263	.0571	.1008	.1484
	11	.0000	.0000	.0000	.0000	.0001	.0005	.0026	.0090	.0242	.0525	.0944
	12	.0000	.0000	.0000	.0000	.0000	.0001	.0006	.0024	.0081	.0215	.0472
	13	.0000	.0000	.0000	.0000	.0000	.0000	.0001	.0005	.0021	.0068	.0182
	14	.0000	.0000	.0000	.0000	.0000	.0000	.0000	.0001	.0004	.0016	.0052
	15	.0000	.0000	.0000	.0000	.0000	.0000	.0000	.0000	.0001	.0003	.0010
	16	.0000	.0000	.0000	.0000	.0000	.0000	.0000	.0000	.0000	.0000	.0001
	17	.0000	.0000	.0000	.0000	.0000	.0000	.0000	.0000	.0000	.0000	.0000
18	0	.8345	.3972	.1501	.0536	.0180	.0056	.0016	.0004	.0001	.0000	.0000
	1	.1517	.3763	.3002	.1704	.0811	.0338	.0126	.0042	.0012	.0003	.0001
	2	.0130	.1683	.2835	.2556	.1723	.0958	.0458	.0190	.0069	.0022	.0006
	3	.0007	.0473	.1680	.2406	.2297	.1704	.1046	.0547	.0246	.0095	.0031
	4	.0000	.0093	.0700	.1592	.2153	.2130	.1681	.1104	.0614	.0291	.0117
	5	.0000	.0014	.0218	.0787	.1507	.1988	.2017	.1664	.1146	.0666	.0327
	6	.0000	.0002	.0052	.0301	.0816	.1436	.1873	.1941	.1655	.1181	.0708
	7	.0000	.0000	.0010	.0091	.0350	.0820	.1376	.1792	.1892	.1657	.1214
	8	.0000	.0000	.0002	.0022	.0120	.0376	.0811	.1327	.1734	.1864	.1669
	9	.0000	.0000	.0000	.0004	.0033	.0139	.0386	.0794	.1284	.1694	.1855
	10	.0000	.0000	.0000	.0001	.0008	.0042	.0149	.0385	.0771	.1248	.1669
	11	.0000	.0000	.0000	.0000	.0001	.0010	.0046	.0151	.0374	.0742	.1214
	12	.0000	.0000	.0000	.0000	.0000	.0002	.0012	.0047	.0145	.0354	.0708
	13	.0000	.0000	.0000	.0000	.0000	.0000	.0002	.0012	.0045	.0134	.0327
	14	.0000	.0000	.0000	.0000	.0000	.0000	.0000	.0002	.0011	.0039	.0117
	15	.0000	.0000	.0000	.0000	.0000	.0000	.0000	.0000	.0002	.0009	.0031
	16	.0000	.0000	.0000	.0000	.0000	.0000	.0000	.0000	.0000	.0001	.0006
	17	.0000	.0000	.0000	.0000	.0000	.0000	.0000	.0000	.0000	.0000	.0001
	18	.0000	.0000	.0000	.0000	.0000	.0000	.0000	.0000	.0000	.0000	.0000

n	x	.01	.05	.10	.15	.20	.25	.30	.35	.40	.45	.50
19	0	.8262	.3774	.1351	.0456	.0144	.0042	.0011	.0003	.0001	.0000	.0000
	1	.1586	.3774	.2852	.1529	.0685	.0268	.0093	.0029	.0008	.0002	.0000
	2	.0144	.1787	.2852	.2428	.1540	.0803	.0358	.0138	.0046	.0013	.0003
	3	.0008	.0533	.1796	.2428	.2182	.1517	.0869	.0422	.0175	.0062	.0018
	4	.0000	.0112	.0798	.1714	.2182	.2023	.1491	.0909	.0467	.0203	.0074
	5	.0000	.0018	.0266	.0907	.1636	.2023	.1916	.1468	.0933	.0497	.0222
	6	.0000	.0002	.0069	.0374	.0955	.1574	.1916	.1844	.1451	.0949	.0518
	7	.0000	.0000	.0014	.0122	.0443	.0974	.1525	.1844	.1797	.1443	.0961
	8	.0000	.0000	.0002	.0032	.0166	.0487	.0981	.1489	.1797	.1771	.1442
	9	.0000	.0000	.0000	.0007	.0051	.0198	.0514	.0980	.1464	.1771	.1762
	10	.0000	.0000	.0000	.0001	.0013	.0066	.0220	.0528	.0976	.1449	.1762
	11	.0000	.0000	.0000	.0000	.0003	.0018	.0077	.0233	.0532	.0970	.1442
	12	.0000	.0000	.0000	.0000	.0000	.0004	.0022	.0083	.0237	.0529	.0961
	13	.0000	.0000	.0000	.0000	.0000	.0001	.0005	.0024	.0085	.0233	.0518
	14	.0000	.0000	.0000	.0000	.0000	.0000	.0001	.0006	.0024	.0082	.0222
	15	.0000	.0000	.0000	.0000	.0000	.0000	.0000	.0001	.0005	.0022	.0074
	16	.0000	.0000	.0000	.0000	.0000	.0000	.0000	.0000	.0001	.0005	.0018
	17	.0000	.0000	.0000	.0000	.0000	.0000	.0000	.0000	.0000	.0001	.0003
	18	.0000	.0000	.0000	.0000	.0000	.0000	.0000	.0000	.0000	.0000	.0000
	19	.0000	.0000	.0000	.0000	.0000	.0000	.0000	.0000	.0000	.0000	.0000
20	0	.8179	.3585	.1216	.0388	.0115	.0032	.0008	.0002	.0000	.0000	.0000
	1	.1652	.3774	.2702	.1368	.0576	.0211	.0068	.0020	.0005	.0001	.0000
	2	.0159	.1887	.2852	.2293	.1369	.0669	.0278	.0100	.0031	.0008	.0002
	3	.0010	.0596	.1901	.2428	.2054	.1339	.0716	.0323	.0123	.0040	.0011
	4	.0000	.0133	.0898	.1821	.2182	.1897	.1304	.0738	.0350	.0139	.0046
	5	.0000	.0022	.0319	.1028	.1746	.2023	.1789	.1272	.0746	.0365	.0148
	6	.0000	.0003	.0089	.0454	.1091	.1686	.1916	.1712	.1244	.0746	.0370
	7	.0000	.0000	.0020	.0160	.0545	.1124	.1643	.1844	.1659	.1221	.0739
	8	.0000	.0000	.0004	.0046	.0222	.0609	.1144	.1614	.1797	.1623	.1201
	9	.0000	.0000	.0001	.0011	.0074	.0271	.0654	.1158	.1597	.1771	.1602
	10	.0000	.0000	.0000	.0002	.0020	.0099	.0308	.0686	.1171	.1593	.1762
	11	.0000	.0000	.0000	.0000	.0005	.0030	.0120	.0336	.0710	.1185	.1602
	12	.0000	.0000	.0000	.0000	.0001	.0008	.0039	.0136	.0355	.0727	.1201
	13	.0000	.0000	.0000	.0000	.0000	.0002	.0010	.0045	.0146	.0366	.0739
	14	.0000	.0000	.0000	.0000	.0000	.0000	.0002	.0012	.0049	.0150	.0370
	15	.0000	.0000	.0000	.0000	.0000	.0000	.0000	.0003	.0013	.0049	.0148
	16	.0000	.0000	.0000	.0000	.0000	.0000	.0000	.0000	.0003	.0013	.0046
	17	.0000	.0000	.0000	.0000	.0000	.0000	.0000	.0000	.0000	.0002	.0011
	18	.0000	.0000	.0000	.0000	.0000	.0000	.0000	.0000	.0000	.0000	.0002
	19	.0000	.0000	.0000	.0000	.0000	.0000	.0000	.0000	.0000	.0000	.0000
	20	.0000	.0000	.0000	.0000	.0000	.0000	.0000	.0000	.0000	.0000	.0000
25	0	.7778	.2774	.0718	.0172	.0038	.0008	.0001	.0000	.0000	.0000	.0000
	1	.1964	.3650	.1994	.0759	.0236	.0063	.0014	.0003	.0000	.0000	.0000
	2	.0238	.2305	.2659	.1607	.0708	.0251	.0074	.0018	.0004	.0001	.0000
	3	.0018	.0930	.2265	.2174	.1358	.0641	.0243	.0076	.0019	.0004	.0001
	4	.0001	.0269	.1384	.2110	.1867	.1175	.0572	.0224	.0071	.0018	.0004

n	x	.01	.05	.10	.15	.20	.25	.30	.35	.40	.45	.50
25	5	.0000	.0060	.0646	.1564	.1960	.1645	.1030	.0506	.0199	.0063	.0016
	6	.0000	.0010	.0239	.0920	.1633	.1828	.1472	.0908	.0442	.0172	.0053
	7	.0000	.0001	.0072	.0441	.1108	.1654	.1712	.1327	.0800	.0381	.0143
	8	.0000	.0000	.0018	.0175	.0623	.1241	.1651	.1607	.1200	.0701	.0322
	9	.0000	.0000	.0004	.0058	.0294	.0781	.1336	.1635	.1511	.1084	.0609
	10	.0000	.0000	.0000	.0016	.0118	.0417	.0916	.1409	.1612	.1419	.0974
	11	.0000	.0000	.0000	.0004	.0040	.0189	.0536	.1034	.1465	.1583	.1328
	12	.0000	.0000	.0000	.0000	.0012	.0074	.0268	.0650	.1140	.1511	.1550
	13	.0000	.0000	.0000	.0000	.0003	.0025	.0115	.0350	.0760	.1236	.1550
	14	.0000	.0000	.0000	.0000	.0000	.0007	.0042	.0161	.0434	.0867	.1328
	15	.0000	.0000	.0000	.0000	.0000	.0002	.0013	.0064	.0212	.0520	.0974
	16	.0000	.0000	.0000	.0000	.0000	.0000	.0004	.0021	.0088	.0266	.0609
	17	.0000	.0000	.0000	.0000	.0000	.0000	.0001	.0006	.0031	.0115	.0322
	18	.0000	.0000	.0000	.0000	.0000	.0000	.0000	.0001	.0009	.0042	.0143
	19	.0000	.0000	.0000	.0000	.0000	.0000	.0000	.0000	.0002	.0013	.0053
	20	.0000	.0000	.0000	.0000	.0000	.0000	.0000	.0000	.0000	.0001	.0016
	21	.0000	.0000	.0000	.0000	.0000	.0000	.0000	.0000	.0000	.0000	.0004
	22	.0000	.0000	.0000	.0000	.0000	.0000	.0000	.0000	.0000	.0000	.0001
30	0	.7397	.2146	.0424	.0076	.0012	.0002	.0000	.0000	.0000	.0000	.0000
	1	.2242	.3389	.1413	.0404	.0093	.0018	.0003	.0000	.0000	.0000	.0000
	2	.0328	.2586	.2277	.1034	.0337	.0086	.0018	.0003	.0000	.0000	.0000
	3	.0031	.1270	.2361	.1703	.0785	.0269	.0072	.0015	.0003	.0000	.0000
	4	.0002	.0451	.1771	.2028	.1325	.0604	.0208	.0056	.0012	.0002	.0000
	5	.0000	.0124	.1023	.1861	.1723	.1047	.0464	.0157	.0041	.0008	.0001
	6	.0000	.0027	.0474	.1368	.1795	.1455	.0829	.0353	.0115	.0029	.0006
	7	.0000	.0005	.0180	.0828	.1538	.1662	.1219	.0652	.0263	.0081	.0019
	8	.0000	.0001	.0058	.0420	.1106	.1593	.1501	.1009	.0505	.0191	.0055
	9	.0000	.0000	.0016	.0181	.0676	.1298	.1573	.1328	.0823	.0382	.0133
	10	.0000	.0000	.0004	.0067	.0355	.0909	.1416	.1502	.1152	.0656	.0280
	11	.0000	.0000	.0001	.0022	.0161	.0551	.1103	.1471	.1396	.0976	.0509
	12	.0000	.0000	.0000	.0006	.0064	.0291	.0749	.1254	.1474	.1265	.0806
	13	.0000	.0000	.0000	.0001	.0022	.0134	.0444	.0935	.1360	.1433	.1115
	14	.0000	.0000	.0000	.0000	.0007	.0054	.0231	.0611	.1101	.1424	.1354
	15	.0000	.0000	.0000	.0000	.0002	.0019	.0106	.0351	.0783	.1242	.1445
	16	.0000	.0000	.0000	.0000	.0000	.0006	.0042	.0177	.0489	.0953	.1354
	17	.0000	.0000	.0000	.0000	.0000	.0002	.0015	.0079	.0269	.0642	.1115
	18	.0000	.0000	.0000	.0000	.0000	.0000	.0005	.0031	.0129	.0379	.0806
	19	.0000	.0000	.0000	.0000	.0000	.0000	.0001	.0010	.0054	.0196	.0509
	20	.0000	.0000	.0000	.0000	.0000	.0000	.0000	.0003	.0020	.0088	.0280
	21	.0000	.0000	.0000	.0000	.0000	.0000	.0000	.0001	.0006	.0034	.0133
	22	.0000	.0000	.0000	.0000	.0000	.0000	.0000	.0000	.0002	.0012	.0055
	23	.0000	.0000	.0000	.0000	.0000	.0000	.0000	.0000	.0000	.0003	.0019
	24	.0000	.0000	.0000	.0000	.0000	.0000	.0000	.0000	.0000	.0001	.0006
	25	.0000	.0000	.0000	.0000	.0000	.0000	.0000	.0000	.0000	.0000	.0001

Appendix 3

Values of $e^{-\lambda}$

λ	$e^{-\lambda}$	λ	$e^{-\lambda}$
0.0	1.00000	2.5	.08208
0.1	.90484	2.6	.07427
0.2	.81873	2.7	.06721
0.3	.74082	2.8	.06081
0.4	.67032	2.9	.05502
0.5	.60653	3.0	.04979
0.6	.54881	3.2	.04076
0.7	.49659	3.4	.03337
0.8	.44933	3.6	.02732
0.9	.40657	3.8	.02237
1.0	.36788	4.0	.01832
1.1	.33287	4.2	.01500
1.2	.30119	4.4	.01228
1.3	.27253	4.6	.01005
1.4	.24660	4.8	.00823
1.5	.22313	5.0	.00674
1.6	.20190	5.5	.00409
1.7	.18268	6.0	.00248
1.8	.16530	6.5	.00150
1.9	.14957	7.0	.00091
2.0	.13534	7.5	.00055
2.1	.12246	8.0	.00034
2.2	.00180	8.5	.00020
2.3	.10026	9.0	.00012
2.4	.09072	10.0	.00005

Appendix 4

Poisson Probabilities*

X	λ 0.1	0.2	0.3	0.4	0.5	0.6	0.7	0.8	0.9	1.0
0	.9048	.8187	.7408	.6703	.6065	.5488	.4966	.4493	.4066	.3679
1	.0905	.1637	.2222	.2681	.3033	.3293	.3476	.3595	.3659	.3679
2	.0045	.0164	.0333	.0536	.0758	.0988	.1217	.1438	.1647	.1839
3	.0002	.0011	.0033	.0072	.0126	.0198	.0284	.0383	.0494	.0613
4	.0000	.0001	.0002	.0007	.0016	.0030	.0050	.0077	.0111	.0153
5	.0000	.0000	.0000	.0001	.0002	.0004	.0007	.0012	.0020	.0031
6	.0000	.0000	.0000	.0000	.0000	.0000	.0001	.0002	.0003	.0005
7	.0000	.0000	.0000	.0000	.0000	.0000	.0000	.0000	.0000	.0001

X	λ 1.1	1.2	1.3	1.4	1.5	1.6	1.7	1.8	1.9	2.0
0	.3329	.3012	.2725	.2466	.2231	.2019	.1827	.1653	.1496	.1353
1	.3662	.3614	.3543	.3452	.3347	.3230	.3106	.2975	.2842	.2707
2	.2014	.2169	.2303	.2417	.2510	.2584	.2640	.2678	.2700	.2707
3	.0738	.0867	.0998	.1128	.1255	.1378	.1496	.1607	.1710	.1804
4	.0203	.0260	.0324	.0395	.0471	.0551	.0636	.0723	.0812	.0902
5	.0045	.0062	.0084	.0111	.0141	.0176	.0216	.0260	.0309	.0361
6	.0008	.0012	.0018	.0026	.0035	.0047	.0061	.0078	.0098	.0120
7	.0001	.0002	.0003	.0005	.0008	.0011	.0015	.0020	.0027	.0034
8	.0000	.0000	.0001	.0001	.0001	.0002	.0003	.0005	.0006	.0009
9	.0000	.0000	.0000	.0000	.0000	.0000	.0001	.0001	.0001	.0002

X	λ 2.1	2.2	2.3	2.4	2.5	2.6	2.7	2.8	2.9	3.0
0	.1225	.1108	.1003	.0907	.0821	.0743	.0672	.0608	.0550	.0498
1	.2572	.2438	.2306	.2177	.2052	.1931	.1815	.1703	.1396	.1494
2	.2700	.2681	.2652	.2613	.2565	.2510	.2450	.2384	.2314	.2240
3	.1890	.1966	.2033	.2090	.2138	.2176	.2205	.2225	.2237	.2240
4	.0992	.1082	.1169	.1254	.1336	.1414	.1488	.1557	.1622	.1680
5	.0417	.0476	.0538	.0602	.0668	.0735	.0804	.0872	.0940	.1008
6	.0146	.0174	.0206	.0241	.0278	.0319	.0362	.0407	.0455	.0504
7	.0044	.0055	.0068	.0083	.0099	.0118	.0139	.0163	.0188	.0216
8	.0011	.0015	.0019	.0025	.0031	.0038	.0047	.0057	.0068	.0081
9	.0003	.0004	.0005	.0007	.0009	.0011	.0014	.0018	.0022	.0027
10	.0001	.0001	.0001	.0002	.0002	.0003	.0004	.0005	.0006	.0008
11	.0000	.0000	.0000	.0000	.0000	.0001	.0001	.0001	.0002	.0002
12	.0000	.0000	.0000	.0000	.0000	.0000	.0000	.0000	.0000	.0001

X	λ 3.1	3.2	3.3	3.4	3.5	3.6	3.7	3.8	3.9	4.0
0	.0450	.0408	.0369	.0334	.0302	.0273	.0247	.0224	.0202	.0183
1	.1397	.1304	.1217	.1135	.1057	.0984	.0915	.0850	.0789	.0733
2	.2165	.2087	.2008	.1929	.1850	.1771	.1692	.1615	.1539	.1465
3	.2237	.2226	.2209	.2186	.2158	.2125	.2087	.2046	.2001	.1954
4	.1734	.1781	.1823	.1858	.1888	.1912	.1931	.1944	.1951	.1954

*Example: $P(X = 5 \mid \lambda = 2.5) = 0.0668$

X	3.1	3.2	3.3	3.4	λ 3.5	3.6	3.7	3.8	3.9	4.0
5	.1075	.1140	.1203	.1264	.1322	.1377	.1429	.1477	.1522	.1563
6	.0555	.0608	.0662	.0716	.0771	.0826	.0881	.0936	.0989	.1042
7	.0246	.0278	.0312	.0348	.0385	.0425	.0466	.0508	.0551	.0595
8	.0095	.0111	.0129	.0148	.0169	.0191	.0215	.0241	.0269	.0298
9	.0033	.0040	.0047	.0056	.0066	.0076	.0089	.0102	.0116	.0132
10	.0010	.0013	.0016	.0019	.0023	.0028	.0033	.0039	.0045	.0053
11	.0003	.0004	.0005	.0006	.0007	.0009	.0011	.0013	.0016	.0019
12	.0001	.0001	.0001	.0002	.0002	.0003	.0003	.0004	.0005	.0006
13	.0000	.0000	.0000	.0000	.0001	.0001	.0001	.0001	.0002	.0002
14	.0000	.0000	.0000	.0000	.0000	.0000	.0000	.0000	.0000	.0001

X	4.1	4.2	4.3	4.4	λ 4.5	4.6	4.7	4.8	4.9	5.0
0	.0166	.0150	.0136	.0123	.0111	.0101	.0091	.0082	.0074	.0067
1	.0679	.0630	.0583	.0540	.0500	.0462	.0427	.0395	.0365	.0337
2	.1393	.1323	.1254	.1188	.1125	.1063	.1005	.0948	.0894	.0842
3	.1904	.1852	.1798	.1743	.1687	.1631	.1574	.1517	.1460	.1404
4	.1951	.1944	.1933	.1917	.1898	.1875	.1849	.1820	.1789	.1755
5	.1600	.1633	.1662	.1687	.1708	.1725	.1738	.1747	.1753	.1755
6	.1093	.1143	.1191	.1237	.1281	.1323	.1362	.1398	.1432	.1462
7	.0640	.0686	.0732	.0778	.0824	.0869	.0914	.0959	.1002	.1044
8	.0328	.0360	.0393	.0428	.0463	.0500	.0537	.0575	.0614	.0653
9	.0150	.0168	.0188	.0209	.0232	.0255	.0280	.0307	.0334	.0363
10	.0061	.0071	.0081	.0092	.0104	.0118	.0132	.0147	.0164	.0181
11	.0023	.0027	.0032	.0037	.0043	.0049	.0056	.0064	.0073	.0082
12	.0008	.0009	.0011	.0014	.0016	.0019	.0022	.0026	.0030	.0034
13	.0002	.0003	.0004	.0005	.0006	.0007	.0008	.0009	.0011	.0013
14	.0001	.0001	.0001	.0001	.0002	.0002	.0003	.0003	.0004	.0005
15	.0000	.0000	.0000	.0000	.0001	.0001	.0001	.0001	.0001	.0002

X	5.1	5.2	5.3	5.4	λ 5.5	5.6	5.7	5.8	5.9	6.0
0	.0061	.0055	.0050	.0045	.0041	.0037	.0033	.0030	.0027	.0025
1	.0311	.0287	.0265	.0244	.0225	.0207	.0191	.0176	.0162	.0149
2	.0793	.0746	.0701	.0659	.0618	.0580	.0544	.0509	.0477	.0446
3	.1348	.1293	.1239	.1185	.1133	.1082	.1033	.0985	.0938	.0892
4	.1719	.1681	.1641	.1600	.1558	.1515	.1472	.1428	.1383	.1339
5	.1753	.1748	.1740	.1728	.1714	.1697	.1678	.1656	.1632	.1606
6	.1490	.1515	.1537	.1555	.1571	.1584	.1594	.1601	.1605	.1606
7	.1086	.1125	.1163	.1200	.1234	.1267	.1298	.1326	.1353	.1377
8	.0692	.0731	.0771	.0810	.0849	.0887	.0925	.0962	.0998	.1033
9	.0392	.0423	.0454	.0486	.0519	.0552	.0586	.0620	.0654	.0688
10	.0200	.0220	.0241	.0262	.0285	.0309	.0334	.0359	.0386	.0413
11	.0093	.0104	.0116	.0129	.0143	.0157	.0173	.0190	.0207	.0225
12	.0039	.0045	.0051	.0058	.0065	.0073	.0082	.0092	.0102	.0113
13	.0015	.0018	.0021	.0024	.0028	.0032	.0036	.0041	.0046	.0052
14	.0006	.0007	.0008	.0009	.0011	.0013	.0015	.0017	.0019	.0022
15	.0002	.0002	.0003	.0003	.0004	.0005	.0006	.0007	.0008	.0009
16	.0001	.0001	.0001	.0001	.0001	.0002	.0002	.0002	.0003	.0003
17	.0000	.0000	.0000	.0000	.0000	.0001	.0001	.0001	.0001	.0001

X	λ 6.1	6.2	6.3	6.4	6.5	6.6	6.7	6.8	6.9	7.0
0	.0022	.0020	.0018	.0017	.0015	.0014	.0012	.0011	.0010	.0009
1	.0137	.0126	.0116	.0106	.0098	.0090	.0082	.0076	.0070	.0064
2	.0417	.0390	.0364	.0340	.0318	.0296	.0276	.0258	.0240	.0223
3	.0848	.0806	.0765	.0726	.0688	.0652	.0617	.0584	.0552	.0521
4	.1294	.1249	.1205	.1162	.1118	.1076	.1034	.0992	.0952	.0912
5	.1579	.1549	.1519	.1487	.1454	.1420	.1385	.1349	.1314	.1277
6	.1605	.1601	.1595	.1586	.1575	.1562	.1546	.1529	.1511	.1490
7	.1399	.1418	.1435	.1450	.1462	.1472	.1480	.1486	.1489	.1490
8	.1066	.1099	.1130	.1160	.1188	.1215	.1240	.1263	.1284	.1304
9	.0723	.0757	.0791	.0825	.0858	.0891	.0923	.0954	.0985	.1014
10	.0441	.0469	.0498	.0528	.0558	.0558	.0618	.0649	.0679	.0710
11	.0245	.0265	.0285	.0307	.0330	.0353	.0377	.0401	.0426	.0452
12	.0124	.0137	.0150	.0164	.0179	.0194	.0210	.0227	.0245	.0264
13	.0058	.0065	.0073	.0081	.0089	.0098	.0108	.0119	.0130	.0142
14	.0025	.0029	.0033	.0037	.0041	.0046	.0052	.0058	.0064	.0071
15	.0010	.0012	.0014	.0016	.0018	.0020	.0023	.0026	.0029	.0033
16	.0004	.0005	.0005	.0006	.0007	.0008	.0010	.0011	.0013	.0014
17	.0001	.0002	.0002	.0002	.0003	.0003	.0004	.0004	.0005	.0006
18	.0000	.0001	.0001	.0001	.0001	.0001	.0001	.0002	.0002	.0002
19	.0000	.0000	.0000	.0000	.0000	.0000	.0000	.0001	.0001	.0001

X	λ 7.1	7.2	7.3	7.4	7.5	7.6	7.7	7.8	7.9	8.0
0	.0008	.0007	.0007	.0006	.0006	.0005	.0005	.0004	.0004	.0003
1	.0059	.0054	.0049	.0045	.0041	.0038	.0035	.0032	.0029	.0027
2	.0208	.0194	.0180	.0167	.0156	.0145	.0134	.0125	.0116	.0107
3	.0492	.0464	.0438	.0413	.0389	.0366	.0345	.0324	.0305	.0286
4	.0874	.0836	.0799	.0764	.0729	.0696	.0663	.0632	.0602	.0573
5	.1241	.1204	.1167	.1130	.1094	.1057	.1021	.0986	.0951	.0916
6	.1468	.1445	.1420	.1394	.1367	.1339	.1311	.1282	.1252	.1221
7	.1489	.1486	.1481	.1474	.1465	.1454	.1442	.1428	.1413	.1396
8	.1321	.1337	.1351	.1363	.1373	.1382	.1388	.1392	.1395	.1396
9	.1042	.1070	.1096	.1121	.1144	.1167	.1187	.1207	.1224	.1241
10	.0740	.0770	.0800	.0829	.0858	.0887	.0914	.0941	.0967	.0993
11	.0478	.0504	.0531	.0558	.0585	.0613	.0640	.0667	.0695	.0722
12	.0283	.0303	.0323	.0344	.0366	.0388	.0411	.0434	.0457	.0481
13	.0154	.0168	.0181	.0196	.0211	.0227	.0243	.0260	.0278	.0296
14	.0078	.0086	.0095	.0104	.0113	.0123	.0134	.0145	.0157	.0169
15	.0037	.0041	.0046	.0051	.0057	.0062	.0069	.0075	.0083	.0090
16	.0016	.0019	.0021	.0024	.0026	.0030	.0033	.0037	.0041	.0045
17	.0007	.0008	.0009	.0010	.0012	.0013	.0015	.0017	.0019	.0021
18	.0003	.0003	.0004	.0004	.0005	.0006	.0006	.0007	.0008	.0009
19	.0001	.0001	.0001	.0002	.0002	.0002	.0003	.0003	.0003	.0004
20	.0000	.0000	.0001	.0001	.0001	.0001	.0001	.0001	.0001	.0002
21	.0000	.0000	.0000	.0000	.0000	.0000	.0000	.0000	.0001	.0001

X	λ 8.1	8.2	8.3	8.4	8.5	8.6	8.7	8.8	8.9	9.0
0	.0003	.0003	.0002	.0002	.0002	.0002	.0002	.0002	.0001	.0001
1	.0025	.0023	.0021	.0019	.0017	.0016	.0014	.0013	.0012	.0011

X	8.1	8.2	8.3	8.4	λ 8.5	8.6	8.7	8.8	8.9	9.0
2	.0100	.0092	.0086	.0079	.0074	.0068	.0063	.0058	.0054	.0050
3	.0269	.0252	.0237	.0222	.0208	.0195	.0183	.0171	.0160	.0150
4	.0544	.0517	.0491	.0466	.0443	.0420	.0398	.0377	.0357	.0337
5	.0882	.0849	.0816	.0784	.0752	.0722	.0692	.0663	.0635	.0607
6	.1191	.1160	.1128	.1097	.1066	.1034	.1003	.0972	.0941	.0911
7	.1378	.1358	.1338	.1317	.1294	.1271	.1247	.1222	.1197	.1171
8	.1395	.1392	.1388	.1382	.1375	.1366	.1356	.1344	.1332	.1318
9	.1256	.1269	.1280	.1290	.1299	.1306	.1311	.1315	.1317	.1318
10	.1017	.1040	.1063	.1084	.1104	.1123	.1140	.1157	.1172	.1186
11	.0749	.0776	.0802	.0828	.0853	.0878	.0902	.0925	.0948	.0970
12	.0505	.0530	.0555	.0579	.0604	.0629	.0654	.0679	.0703	.0728
13	.0315	.0334	.0354	.0374	.0395	.0416	.0438	.0459	.0481	.0504
14	.0182	.0196	.0210	.0225	.0240	.0256	.0272	.0289	.0306	.0324
15	.0098	.0107	.0116	.0126	.0136	.0147	.0158	.0169	.0182	.0194
16	.0050	.0055	.0060	.0066	.0072	.0079	.0086	.0093	.0101	.0109
17	.0024	.0026	.0029	.0033	.0036	.0040	.0044	.0048	.0053	.0058
18	.0011	.0012	.0014	.0015	.0017	.0019	.0021	.0024	.0026	.0029
19	.0005	.0005	.0006	.0007	.0008	.0009	.0010	.0011	.0012	.0014
20	.0002	.0002	.0002	.0003	.0003	.0004	.0004	.0005	.0005	.0006
21	.0001	.0001	.0001	.0001	.0001	.0002	.0002	.0002	.0002	.0003
22	.0000	.0000	.0000	.0000	.0001	.0001	.0001	.0001	.0001	.0001

X	9.1	9.2	9.3	9.4	λ 9.5	9.6	9.7	9.8	9.9	10.0
0	.0001	.0001	.0001	.0001	.0001	.0001	.0001	.0001	.0001	.0000
1	.0010	.0009	.0009	.0008	.0007	.0007	.0006	.0005	.0005	.0005
2	.0046	.0043	.0040	.0037	.0034	.0031	.0029	.0027	.0025	.0023
3	.0140	.0131	.0123	.0115	.0107	.0100	.0093	.0087	.0081	.0076
4	.0319	.0302	.0285	.0269	.0254	.0240	.0226	.0213	.0201	.0189
5	.0581	.0555	.0530	.0506	.0483	.0460	.0439	.0418	.0398	.0378
6	.0881	.0851	.0822	.0793	.0764	.0736	.0709	.0682	.0656	.0631
7	.1145	.1118	.1091	.1064	.1037	.1010	.0982	.0955	.0928	.0901
8	.1302	.1286	.1269	.1251	.1232	.1212	.1191	.1170	.1148	.1126
9	.1317	.1315	.1311	.1306	.1300	.1293	.1284	.1274	.1263	.1251
10	.1198	.1210	.1219	.1228	.1235	.1241	.1245	.1249	.1250	.1251
11	.0991	.1012	.1031	.1049	.1067	.1083	.1098	.1112	.1125	.1137
12	.0752	.0776	.0779	.0822	.0844	.0866	.0888	.0908	.0928	.0948
13	.0526	.0549	.0572	.0594	.0617	.0640	.0662	.0685	.0707	.0729
14	.0342	.0361	.0380	.0399	.0419	.0439	.0459	.0479	.0500	.0521
15	.0208	.0221	.0235	.0250	.0265	.0281	.0297	.0313	.0330	.0347
16	.0118	.0127	.0137	.0147	.0157	.0168	.0180	.0192	.0204	.0217
17	.0063	.0069	.0075	.0081	.0088	.0095	.0103	.0111	.0119	.0128
18	.0032	.0035	.0039	.0042	.0046	.0051	.0055	.0060	.0065	.0071
19	.0015	.0017	.0019	.0021	.0023	.0026	.0028	.0031	.0034	.0037
20	.0007	.0008	.0009	.0010	.0011	.0012	.0014	.0015	.0017	.0019
21	.0003	.0003	.0004	.0004	.0005	.0006	.0006	.0007	.0008	.0009
22	.0001	.0001	.0002	.0002	.0002	.0002	.0003	.0003	.0004	.0004
23	.0000	.0001	.0001	.0001	.0001	.0001	.0001	.0001	.0002	.0002
24	.0000	.0000	.0000	.0000	.0000	.0000	.0000	.0001	.0001	.0001

Appendix 5

Proportions of Area for the Standard Normal Distribution

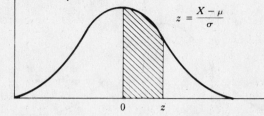

Areas reported below:*

$$z = \frac{X - \mu}{\sigma}$$

z	.00	.01	.02	.03	.04	.05	.06	.07	.08	.09
0.0	.0000	.0040	.0080	.0120	.0160	.0199	.0239	.0279	.0319	.0359
0.1	.0398	.0438	.0478	.0517	.0557	.0596	.0636	.0675	.0714	.0753
0.2	.0793	.0832	.0871	.0910	.0948	.0987	.1026	.1064	.1103	.1141
0.3	.1179	.1217	.1255	.1293	.1331	.1368	.1406	.1443	.1480	.1517
0.4	.1554	.1591	.1628	.1664	.1700	.1736	.1772	.1808	.1844	.1879
0.5	.1915	.1950	.1985	.2019	.2054	.2088	.2123	.2157	.2190	.2224
0.6	.2257	.2291	.2324	.2357	.2389	.2422	.2454	.2486	.2518	.2549
0.7	.2580	2.612	.2642	.2673	.2704	.2734	.2764	.2794	.2823	.2852
0.8	.2881	.2910	.2939	.2967	.2995	.3023	.3051	.3078	.3106	.3133
0.9	.3159	.3186	.3212	.3238	.3264	.3289	.3315	.3340	.3365	.3389
1.0	.3413	.3438	.3461	.3485	.3508	.3531	.3554	.3577	.3599	.3621
1.1	.3643	.3665	.3686	.3708	.3729	.3749	.3770	.3790	.3810	.3830
1.2	.3849	.3869	.3888	.3907	.3925	.3944	.3962	.3980	.3997	.4014
1.3	.4032	.4049	.4066	.4082	.4099	.4115	.4131	.4147	.4162	.4177
1.4	.4192	.4207	.4222	.4236	.4251	.4265	.4279	.4292	.4306	.4319
1.5	.4332	.4345	.4357	.4370	.4382	.4394	.4406	.4418	.4429	.4441
1.6	.4452	.4463	.4474	.4484	.4495	.4505	.4515	.4525	.4535	.4545
1.7	.4554	.4564	.4573	.4582	.4591	.4599	.4608	.4616	.4625	.4633
1.8	.4641	.4649	.4656	.4664	.4671	.4678	.4686	.4693	.4699	.4706
1.9	.4713	.4719	.4726	.4732	.4738	.4744	.4750	.4756	.4761	.4767
2.0	.4772	.4778	.4783	.4788	.4793	.4798	.4803	.4808	.4812	.4817
2.1	.4821	.4826	.4830	.4834	.4838	.4842	.4846	.4850	.4854	.4857
2.2	.4861	.4864	.4868	.4871	.4875	.4878	.4881	.4884	.4887	.4890
2.3	.4893	.4896	.4898	.4901	.4904	.4906	.4909	.4911	.4913	.4916
2.4	.4918	.4920	.4922	.4925	.4927	.4929	.4931	.4932	.4934	.4936
2.5	.4938	.4940	.4941	.4943	.4945	.4946	.4948	.4949	.4951	.4952
2.6	.4953	.4955	.4956	.4957	.4959	.4960	.4961	.4962	.4963	.4964
2.7	.4965	.4966	.4967	.4968	.4969	.4970	.4971	.4972	.4973	.4974
2.8	.4974	.4975	.4976	.4977	.4977	.4978	.4979	.4979	.4980	.4981
2.9	.4981	.4982	.4983	.4983	.4984	.4984	.4985	.4985	.4986	.4986
3.0	.4987									
3.5	.4997									
4.0	.4999									

*Example: For $z = 1.96$, shaded area is 0.4750 out of the total area of 1.0000.

Appendix 6

Proportions of Area for the *t* Distribution

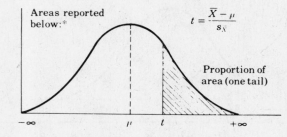

Areas reported below:*

$$t = \frac{\overline{X} - \mu}{s_{\overline{X}}}$$

Proportion of area (one tail)

$-\infty$ μ t $+\infty$

df	0.10	0.05	0.025	0.01	0.005	df	0.10	0.05	0.025	0.01	0.005
1	3.078	6.314	12.706	31.821	63.657	18	1.330	1.734	2.101	2.552	2.878
2	1.886	2.920	4.303	6.965	9.925	19	1.328	1.729	2.093	2.539	2.861
3	1.638	2.353	3.182	4.541	5.841	20	1.325	1.725	2.086	2.528	2.845
4	1.533	2.132	2.776	3.747	4.604	21	1.323	1.721	2.080	2.518	2.831
5	1.476	2.015	2.571	3.365	4.032	22	1.321	1.717	2.074	2.508	2.819
6	1.440	1.943	2.447	3.143	3.707	23	1.319	1.714	2.069	2.500	2.807
7	1.415	1.895	2.365	2.998	3.499	24	1.318	1.711	2.064	2.492	2.797
8	1.397	1.860	2.306	2.896	3.355	25	1.316	1.708	2.060	2.485	2.787
9	1.383	1.833	2.262	2.821	3.250	26	1.315	1.706	2.056	2.479	2.779
10	1.372	1.812	2.228	2.764	3.169	27	1.314	1.703	2.052	2.473	2.771
11	1.363	1.796	2.201	2.718	3.106	28	1.313	1.701	2.048	2.467	2.763
12	1.356	1.782	2.179	2.681	3.055	29	1.311	1.699	2.045	2.462	2.756
13	1.350	1.771	2.160	2.650	3.012	30	1.310	1.697	2.042	2.457	2.750
14	1.345	1.761	2.145	2.624	2.977	40	1.303	1.684	2.021	2.423	2.704
15	1.341	1.753	2.131	2.602	2.947	60	1.296	1.671	2.000	2.390	2.660
16	1.337	1.746	2.120	2.583	2.921	120	1.289	1.658	1.980	2.358	2.617
17	1.333	1.740	2.110	2.567	2.898	∞	1.282	1.645	1.960	2.326	2.576

*Example: For the shaded area to represent 0.05 of the total area of 1.0, value of *t* with 10 degrees of freedom is 1.812.

Source: From Table III of Fisher and Yates, *Statistical Tables for Biological, Agricultural and Medical Research*, 6th ed., 1974, published by Longman Group Ltd., London (previously published by Oliver & Boyd, Edinburgh), by permission of the authors and publishers.

Appendix 7

Proportions of Area for the χ^2 Distribution

Areas reported below:*

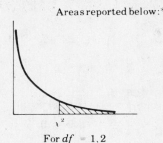

For $df = 1, 2$

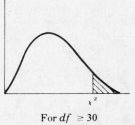

For $df \geq 30$

df	Proportion of area										
	0.995	0.990	0.975	0.950	0.900	0.500	0.100	0.050	0.025	0.010	0.005
1	0.00004	0.00016	0.00098	0.00393	0.0158	0.455	2.71	3.84	5.02	6.63	7.88
2	0.0100	0.0201	0.0506	0.103	0.211	1.386	4.61	5.99	7.38	9.21	10.60
3	0.072	0.115	0.216	0.352	0.584	2.366	6.25	7.81	9.35	11.34	12.84
4	0.207	0.297	0.484	0.711	1.064	3.357	7.78	9.49	11.14	13.28	14.86
5	0.412	0.554	0.831	1.145	1.61	4.251	9.24	11.07	12.83	15.09	16.75
6	0.676	0.872	1.24	1.64	2.20	5.35	10.64	12.59	14.45	16.81	18.55
7	0.989	1.24	1.69	2.17	2.83	6.35	12.02	14.07	16.01	18.48	20.28
8	1.34	1.65	2.18	2.73	3.49	7.34	13.36	15.51	17.53	20.09	21.96
9	1.73	2.09	2.70	3.33	4.17	8.34	14.68	16.92	19.02	21.67	23.59
10	2.16	2.56	3.25	3.94	4.87	9.34	15.99	18.31	20.48	23.21	25.19
11	2.60	3.05	3.82	4.57	5.58	10.34	17.28	19.68	21.92	24.73	26.76
12	3.07	3.57	4.40	5.23	6.30	11.34	18.55	21.03	23.34	26.22	28.30
13	3.57	4.11	5.01	5.89	7.04	12.34	19.81	22.36	24.74	27.69	29.82
14	4.07	4.66	5.63	6.57	7.79	13.34	21.06	23.68	26.12	29.14	31.32
15	4.60	5.23	6.26	7.26	8.55	14.34	22.31	25.00	27.49	30.58	32.80
16	5.14	5.81	6.91	7.96	9.31	15.34	23.54	26.30	28.85	32.00	34.27
17	5.70	6.41	7.56	8.67	10.09	16.34	24.77	27.59	30.19	33.41	35.72
18	6.26	7.01	8.23	9.39	10.86	17.34	25.99	28.87	31.53	34.81	37.16
19	6.84	7.63	8.91	10.12	11.65	18.34	27.20	30.14	32.85	36.19	38.58
20	7.43	8.26	9.59	10.85	12.44	19.34	28.41	31.41	34.17	37.57	40.00
21	8.03	8.90	10.28	11.59	13.24	20.34	29.62	32.67	35.48	38.93	41.40
22	8.64	9.54	10.98	12.34	14.04	21.34	30.81	33.92	36.78	40.29	42.80
23	9.26	10.20	11.69	13.09	14.85	22.34	32.01	35.17	38.08	41.64	44.18
24	9.89	10.86	12.40	13.85	15.66	23.34	33.20	36.42	39.36	42.98	45.56
25	10.52	11.52	13.12	14.61	16.47	24.34	34.38	37.65	40.65	44.31	46.93
26	11.16	12.20	13.84	15.38	17.29	25.34	35.56	38.89	41.92	45.64	48.29
27	11.81	12.83	14.57	16.15	18.11	26.34	36.74	40.11	43.19	46.96	49.64
28	12.46	13.56	15.31	16.93	18.94	27.34	37.92	41.34	44.46	48.28	50.99
29	13.12	14.26	16.05	17.71	19.77	28.34	39.09	42.56	45.72	49.59	52.34
30	13.79	14.95	16.79	18.49	20.60	29.34	40.26	43.77	46.98	50.89	53.67
40	20.71	22.16	24.43	26.51	29.05	39.34	51.81	55.76	59.34	63.69	66.77
50	27.99	29.71	32.36	34.76	37.69	49.33	63.17	67.50	71.42	76.15	79.49
60	35.53	37.43	40.48	43.19	46.46	59.33	74.40	79.08	83.30	88.38	91.95
70	43.28	45.44	48.76	51.74	55.33	69.33	85.53	90.53	95.02	100.4	104.2
80	51.17	53.54	51.17	60.39	64.28	79.33	98.58	101.9	106.6	112.3	116.3
90	59.20	61.75	65.65	69.13	73.29	89.33	107.6	113.1	118.1	124.1	128.3
100	67.33	70.06	74.22	77.93	82.36	99.33	118.5	124.3	129.6	135.8	140.2

Example: For the shaded area to represent 0.05 of the total area of 1.0 under the density function, the value of χ^2 is 18.31 when df 10.

Source: From Table IV of Fisher and Yates, *Statistical Tables for Biological, Agricultural and Medical Research*, 6th ed., 1974, published by Longman Group Ltd., London (previously published by Oliver & Boyd, Edinburgh), by permission of the authors and publishers.

Values of F Exceeded with

df (numerator)

df (denom.)	1	2	3	4	5	6	7	8	9	10	11	12	14	16	20	24	30	40	50	75	100	200	500	∞
1	161 / 4,052	200 / 4,999	216 / 5,403	225 / 5,625	230 / 5,764	234 / 5,859	237 / 5,928	239 / 5,981	241 / 6,022	242 / 6,056	243 / 6,082	244 / 6,106	245 / 6,142	246 / 6,169	248 / 6,208	249 / 6,234	250 / 6,261	251 / 6,286	252 / 6,302	253 / 6,323	253 / 6,334	254 / 6,352	254 / 6,361	254 / 6,366
2	18.51 / 98.49	19.00 / 99.00	19.16 / 99.17	19.25 / 99.25	19.30 / 99.30	19.33 / 99.33	19.36 / 99.36	19.37 / 99.37	19.38 / 99.39	19.39 / 99.40	19.40 / 99.41	19.41 / 99.42	19.42 / 99.43	19.43 / 99.44	19.44 / 99.45	19.45 / 99.46	19.46 / 99.47	19.47 / 99.48	19.47 / 99.48	19.48 / 99.49	19.49 / 99.49	19.49 / 99.49	19.50 / 99.50	19.50 / 99.50
3	10.13 / 34.12	9.55 / 30.82	9.28 / 29.46	9.12 / 28.71	9.01 / 28.24	8.94 / 27.91	8.88 / 27.67	8.84 / 27.49	8.81 / 27.34	8.78 / 27.23	8.76 / 27.13	8.74 / 27.05	8.71 / 26.92	8.69 / 26.83	8.66 / 26.69	8.64 / 26.60	8.62 / 26.50	8.60 / 26.41	8.58 / 26.35	8.57 / 26.27	8.56 / 26.23	8.54 / 26.18	8.54 / 26.14	8.53 / 26.12
4	7.71 / 21.20	6.94 / 18.00	6.59 / 16.69	6.39 / 15.98	6.26 / 15.52	6.16 / 15.21	6.09 / 14.98	6.04 / 14.80	6.00 / 14.66	5.96 / 14.54	5.93 / 14.45	5.91 / 14.37	5.87 / 14.24	5.84 / 14.15	5.80 / 14.02	5.77 / 13.93	5.74 / 13.83	5.71 / 13.74	5.70 / 13.69	5.68 / 13.61	5.66 / 13.57	5.65 / 13.52	5.64 / 13.48	5.63 / 13.46
5	6.61 / 16.26	5.79 / 13.27	5.41 / 12.06	5.19 / 11.39	5.05 / 10.97	4.95 / 10.67	4.88 / 10.45	4.82 / 10.29	4.78 / 10.15	4.74 / 10.05	4.70 / 9.96	4.68 / 9.89	4.64 / 9.77	4.60 / 9.68	4.56 / 9.55	4.53 / 9.47	4.50 / 9.38	4.46 / 9.29	4.44 / 9.24	4.42 / 9.17	4.40 / 9.13	4.38 / 9.07	4.37 / 9.04	4.36 / 9.02
6	5.99 / 13.74	5.14 / 10.92	4.76 / 9.78	4.53 / 9.15	4.39 / 8.75	4.28 / 8.47	4.21 / 8.26	4.15 / 8.10	4.10 / 7.98	4.06 / 7.87	4.03 / 7.79	4.00 / 7.72	3.96 / 7.60	3.92 / 7.52	3.87 / 7.39	3.84 / 7.31	3.81 / 7.23	3.77 / 7.14	3.75 / 7.09	3.72 / 7.02	3.71 / 6.99	3.69 / 6.94	3.68 / 6.90	3.67 / 6.88
7	5.59 / 12.25	4.74 / 9.55	4.34 / 8.45	4.12 / 7.85	3.97 / 7.46	3.87 / 7.19	3.79 / 7.00	3.73 / 6.84	3.68 / 6.71	3.63 / 6.62	3.60 / 6.54	3.57 / 6.47	3.52 / 6.35	3.49 / 6.27	3.44 / 6.15	3.41 / 6.07	3.38 / 5.98	3.34 / 5.90	3.32 / 5.85	3.29 / 5.78	3.28 / 5.75	3.25 / 5.70	3.24 / 5.67	3.23 / 5.65
8	5.32 / 11.26	4.46 / 8.65	4.07 / 7.59	3.84 / 7.01	3.69 / 6.63	3.58 / 6.37	3.50 / 6.19	3.44 / 6.03	3.39 / 5.91	3.34 / 5.82	3.31 / 5.74	3.28 / 5.67	3.23 / 5.56	3.20 / 5.48	3.15 / 5.36	3.12 / 5.28	3.08 / 5.20	3.05 / 5.11	3.03 / 5.06	3.00 / 5.00	2.98 / 4.96	2.96 / 4.91	2.94 / 4.88	2.93 / 4.86
9	5.12 / 10.56	4.26 / 8.02	3.86 / 6.99	3.63 / 6.42	3.48 / 6.06	3.37 / 5.80	3.29 / 5.62	3.23 / 5.47	3.18 / 5.35	3.13 / 5.26	3.10 / 5.18	3.07 / 5.11	3.02 / 5.00	2.98 / 4.92	2.93 / 4.80	2.90 / 4.73	2.86 / 4.64	2.82 / 4.56	2.80 / 4.51	2.77 / 4.45	2.76 / 4.41	2.73 / 4.36	2.72 / 4.33	2.71 / 4.31
10	4.96 / 10.04	4.10 / 7.56	3.71 / 6.55	3.48 / 5.99	3.33 / 5.64	3.22 / 5.39	3.14 / 5.21	3.07 / 5.06	3.02 / 4.95	2.97 / 4.85	2.94 / 4.78	2.91 / 4.71	2.86 / 4.60	2.82 / 4.52	2.77 / 4.41	2.74 / 4.33	2.70 / 4.25	2.67 / 4.17	2.64 / 4.12	2.61 / 4.05	2.59 / 4.01	2.56 / 3.96	2.55 / 3.93	2.54 / 3.91
11	4.84 / 9.65	3.98 / 7.20	3.59 / 6.22	3.36 / 5.67	3.20 / 5.32	3.09 / 5.07	3.01 / 4.88	2.95 / 4.74	2.90 / 4.63	2.86 / 4.54	2.82 / 4.46	2.79 / 4.40	2.74 / 4.29	2.70 / 4.21	2.65 / 4.10	2.61 / 4.02	2.57 / 3.94	2.53 / 3.86	2.50 / 3.80	2.47 / 3.74	2.45 / 3.70	2.42 / 3.66	2.41 / 3.62	2.40 / 3.60
12	4.75 / 9.33	3.88 / 6.93	3.49 / 5.95	3.26 / 5.41	3.11 / 5.06	3.00 / 4.82	2.92 / 4.65	2.85 / 4.50	2.80 / 4.39	2.76 / 4.30	2.72 / 4.22	2.69 / 4.16	2.64 / 4.05	2.60 / 3.98	2.54 / 3.86	2.50 / 3.78	2.46 / 3.70	2.42 / 3.61	2.40 / 3.56	2.36 / 3.49	2.35 / 3.46	2.32 / 3.41	2.31 / 3.38	2.30 / 3.36
13	4.67 / 9.07	3.80 / 6.70	3.41 / 5.74	3.18 / 5.20	3.02 / 4.86	2.92 / 4.62	2.84 / 4.44	2.77 / 4.30	2.72 / 4.19	2.67 / 4.10	2.63 / 4.02	2.60 / 3.96	2.55 / 3.85	2.51 / 3.78	2.46 / 3.67	2.42 / 3.59	2.38 / 3.51	2.34 / 3.42	2.32 / 3.37	2.28 / 3.30	2.26 / 3.27	2.24 / 3.21	2.22 / 3.18	2.21 / 3.16
14	4.60 / 8.86	3.74 / 6.51	3.34 / 5.56	3.11 / 5.03	2.96 / 4.69	2.85 / 4.46	2.77 / 4.28	2.70 / 4.14	2.65 / 4.03	2.60 / 3.94	2.56 / 3.86	2.53 / 3.80	2.48 / 3.70	2.44 / 3.62	2.39 / 3.51	2.35 / 3.43	2.31 / 3.34	2.27 / 3.26	2.24 / 3.21	2.21 / 3.14	2.19 / 3.11	2.16 / 3.06	2.14 / 3.02	2.13 / 3.00
15	4.54 / 8.68	3.68 / 6.36	3.29 / 5.42	3.06 / 4.89	2.90 / 4.56	2.79 / 4.32	2.70 / 4.14	2.64 / 4.00	2.59 / 3.89	2.55 / 3.80	2.51 / 3.73	2.48 / 3.67	2.43 / 3.56	2.39 / 3.48	2.33 / 3.36	2.29 / 3.29	2.25 / 3.20	2.21 / 3.12	2.18 / 3.07	2.15 / 3.00	2.12 / 2.97	2.10 / 2.92	2.08 / 2.89	2.07 / 2.87
16	4.49 / 8.53	3.63 / 6.23	3.24 / 5.29	3.01 / 4.77	2.85 / 4.44	2.74 / 4.20	2.66 / 4.03	2.59 / 3.89	2.54 / 3.78	2.49 / 3.69	2.45 / 3.61	2.42 / 3.55	2.37 / 3.45	2.33 / 3.37	2.28 / 3.25	2.24 / 3.18	2.20 / 3.10	2.16 / 3.01	2.13 / 2.96	2.09 / 2.89	2.07 / 2.86	2.04 / 2.80	2.02 / 2.77	2.01 / 2.75
17	4.45 / 8.40	3.59 / 6.11	3.20 / 5.18	2.96 / 4.67	2.81 / 4.34	2.70 / 4.10	2.62 / 3.93	2.55 / 3.79	2.50 / 3.68	2.45 / 3.59	2.41 / 3.52	2.38 / 3.45	2.33 / 3.35	2.29 / 3.27	2.23 / 3.16	2.19 / 3.08	2.15 / 3.00	2.11 / 2.92	2.08 / 2.86	2.04 / 2.79	2.02 / 2.76	1.99 / 2.70	1.97 / 2.67	1.96 / 2.65
18	4.41 / 8.28	3.55 / 6.01	3.16 / 5.09	2.93 / 4.58	2.77 / 4.25	2.66 / 4.01	2.58 / 3.85	2.51 / 3.71	2.46 / 3.60	2.41 / 3.51	2.37 / 3.44	2.34 / 3.37	2.29 / 3.27	2.25 / 3.19	2.19 / 3.07	2.15 / 3.00	2.11 / 2.91	2.07 / 2.83	2.04 / 2.78	2.00 / 2.71	1.98 / 2.68	1.95 / 2.62	1.93 / 2.59	1.92 / 2.57

Probabilities of 5 and 1 Percent

Each cell lists the 5 % value (upper) and 1 % value (lower).

df																								
19	4.38/8.18	3.52/5.93	3.13/5.01	2.90/4.50	2.74/4.17	2.63/3.94	2.55/3.77	2.48/3.63	2.43/3.52	2.38/3.43	2.34/3.36	2.31/3.30	2.26/3.19	2.21/3.12	2.15/3.00	2.11/2.92	2.07/2.84	2.02/2.76	2.00/2.70	1.96/2.63	1.94/2.60	1.91/2.54	1.90/2.51	1.88/2.49
20	4.35/8.10	3.49/5.85	3.10/4.94	2.87/4.43	2.71/4.10	2.60/3.87	2.52/3.71	2.45/3.56	2.40/3.45	2.35/3.37	2.31/3.30	2.28/3.23	2.23/3.13	2.18/3.05	2.12/2.94	2.08/2.86	2.04/2.77	1.99/2.69	1.96/2.63	1.92/2.56	1.90/2.53	1.87/2.47	1.85/2.44	1.84/2.42
21	4.32/8.02	3.47/5.78	3.07/4.87	2.84/4.37	2.68/4.04	2.57/3.81	2.49/3.65	2.42/3.51	2.37/3.40	2.32/3.31	2.28/3.24	2.25/3.17	2.20/3.07	2.15/2.99	2.09/2.88	2.05/2.80	2.00/2.72	1.96/2.63	1.93/2.58	1.89/2.51	1.87/2.47	1.84/2.42	1.82/2.38	1.81/2.36
22	4.30/7.94	3.44/5.72	3.05/4.82	2.82/4.31	2.66/3.99	2.55/3.76	2.47/3.59	2.40/3.45	2.35/3.35	2.30/3.26	2.26/3.18	2.23/3.12	2.18/3.02	2.13/2.94	2.07/2.83	2.03/2.75	1.98/2.67	1.93/2.58	1.91/2.53	1.87/2.46	1.84/2.42	1.81/2.37	1.80/2.33	1.78/2.31
23	4.28/7.88	3.42/5.66	3.03/4.76	2.80/4.26	2.64/3.94	2.53/3.71	2.45/3.54	2.38/3.41	2.32/3.30	2.28/3.21	2.24/3.14	2.20/3.07	2.14/2.97	2.10/2.89	2.04/2.78	2.00/2.70	1.96/2.62	1.91/2.53	1.88/2.48	1.84/2.41	1.82/2.37	1.79/2.32	1.77/2.28	1.76/2.26
24	4.26/7.82	3.40/5.61	3.01/4.72	2.78/4.22	2.62/3.90	2.51/3.67	2.43/3.50	2.36/3.36	2.30/3.25	2.26/3.17	2.22/3.09	2.18/3.03	2.13/2.93	2.09/2.85	2.02/2.74	1.98/2.66	1.94/2.58	1.89/2.49	1.86/2.44	1.82/2.36	1.80/2.33	1.76/2.27	1.74/2.23	1.73/2.21
25	4.24/7.77	3.38/5.57	2.99/4.68	2.76/4.18	2.60/3.86	2.49/3.63	2.41/3.46	2.34/3.32	2.28/3.21	2.24/3.13	2.20/3.05	2.16/2.99	2.11/2.89	2.06/2.81	2.00/2.70	1.96/2.62	1.92/2.54	1.87/2.45	1.84/2.40	1.80/2.32	1.77/2.29	1.74/2.23	1.72/2.19	1.71/2.17
26	4.22/7.72	3.37/5.53	2.98/4.64	2.74/4.14	2.59/3.82	2.47/3.59	2.39/3.42	2.32/3.29	2.27/3.17	2.22/3.09	2.18/3.02	2.15/2.96	2.10/2.86	2.05/2.77	1.99/2.66	1.95/2.58	1.90/2.50	1.85/2.41	1.82/2.36	1.78/2.28	1.76/2.25	1.72/2.19	1.70/2.15	1.69/2.13
27	4.21/7.68	3.35/5.49	2.96/4.60	2.73/4.11	2.57/3.79	2.46/3.56	2.37/3.39	2.30/3.26	2.25/3.14	2.20/3.06	2.16/2.98	2.13/2.93	2.08/2.83	2.03/2.74	1.97/2.63	1.93/2.55	1.88/2.47	1.84/2.38	1.80/2.33	1.76/2.25	1.74/2.21	1.71/2.16	1.68/2.12	1.67/2.10
28	4.20/7.64	3.34/5.45	2.95/4.57	2.71/4.07	2.56/3.76	2.44/3.53	2.36/3.36	2.29/3.23	2.24/3.11	2.19/3.03	2.15/2.95	2.12/2.90	2.06/2.80	2.02/2.71	1.96/2.60	1.91/2.52	1.87/2.44	1.81/2.35	1.78/2.30	1.75/2.22	1.72/2.18	1.69/2.13	1.67/2.09	1.65/2.06
29	4.18/7.60	3.33/5.42	2.93/4.54	2.70/4.04	2.54/3.73	2.43/3.50	2.35/3.33	2.28/3.20	2.22/3.08	2.18/3.00	2.14/2.92	2.10/2.87	2.05/2.77	2.00/2.68	1.94/2.57	1.90/2.49	1.85/2.41	1.80/2.32	1.77/2.27	1.73/2.19	1.71/2.15	1.68/2.10	1.65/2.06	1.64/2.03
30	4.17/7.56	3.32/5.39	2.92/4.51	2.69/4.02	2.53/3.70	2.42/3.47	2.34/3.30	2.27/3.17	2.21/3.06	2.16/2.98	2.12/2.90	2.09/2.84	2.04/2.74	1.99/2.66	1.93/2.55	1.89/2.47	1.84/2.38	1.79/2.29	1.76/2.24	1.72/2.16	1.69/2.13	1.66/2.07	1.64/2.03	1.62/2.01
32	4.15/7.50	3.30/5.34	2.90/4.46	2.67/3.97	2.51/3.66	2.40/3.42	2.32/3.25	2.25/3.12	2.19/3.01	2.14/2.94	2.10/2.86	2.07/2.80	2.02/2.70	1.97/2.62	1.91/2.51	1.86/2.42	1.82/2.34	1.76/2.25	1.74/2.20	1.69/2.12	1.67/2.08	1.64/2.02	1.61/1.98	1.59/1.96
34	4.13/7.44	3.28/5.29	2.88/4.42	2.65/3.93	2.49/3.61	2.38/3.38	2.30/3.21	2.23/3.08	2.17/2.97	2.12/2.89	2.08/2.82	2.05/2.76	2.00/2.66	1.95/2.58	1.89/2.47	1.84/2.38	1.80/2.30	1.74/2.21	1.71/2.15	1.67/2.08	1.64/2.04	1.61/1.98	1.59/1.94	1.57/1.91
36	4.11/7.39	3.26/5.25	2.86/4.38	2.63/3.89	2.48/3.58	2.36/3.35	2.28/3.18	2.21/3.04	2.15/2.94	2.10/2.86	2.06/2.78	2.03/2.72	1.98/2.62	1.93/2.54	1.87/2.43	1.82/2.35	1.78/2.26	1.72/2.17	1.69/2.12	1.65/2.04	1.62/2.00	1.59/1.94	1.56/1.90	1.55/1.87
38	4.10/7.35	3.25/5.21	2.85/4.34	2.62/3.86	2.46/3.54	2.35/3.32	2.26/3.15	2.19/3.02	2.14/2.91	2.09/2.82	2.05/2.75	2.02/2.69	1.96/2.59	1.92/2.51	1.85/2.40	1.80/2.32	1.76/2.22	1.71/2.14	1.67/2.08	1.63/2.00	1.60/1.97	1.57/1.90	1.54/1.86	1.53/1.84
40	4.07/7.31	3.23/5.18	2.84/4.31	2.61/3.83	2.45/3.51	2.34/3.29	2.25/3.12	2.18/2.99	2.12/2.88	2.07/2.80	2.04/2.73	2.00/2.66	1.95/2.56	1.90/2.49	1.84/2.37	1.79/2.29	1.74/2.20	1.69/2.11	1.66/2.05	1.61/1.97	1.59/1.94	1.55/1.88	1.53/1.84	1.51/1.81

(continued)

df (numerator)

df (denom.)	1	2	3	4	5	6	7	8	9	10	11	12	14	16	20	24	30	40	50	75	100	200	500	∞
42	4.07 / 7.27	3.22 / 5.15	2.83 / 4.29	2.59 / 3.80	2.44 / 3.49	2.32 / 3.26	2.24 / 3.10	2.17 / 2.96	2.11 / 2.86	2.06 / 2.77	2.02 / 2.70	1.99 / 2.64	1.94 / 2.54	1.89 / 2.46	1.82 / 2.35	1.78 / 2.26	1.73 / 2.17	1.68 / 2.08	1.64 / 2.02	1.60 / 1.94	1.57 / 1.91	1.54 / 1.85	1.51 / 1.80	1.49 / 1.78
44	4.06 / 7.24	3.21 / 5.12	2.82 / 4.26	2.58 / 3.78	2.43 / 3.46	2.31 / 3.24	2.23 / 3.07	2.16 / 2.94	2.10 / 2.84	2.05 / 2.75	2.01 / 2.68	1.98 / 2.62	1.92 / 2.52	1.88 / 2.44	1.81 / 2.32	1.76 / 2.24	1.72 / 2.15	1.66 / 2.06	1.63 / 2.00	1.58 / 1.92	1.56 / 1.88	1.52 / 1.82	1.50 / 1.78	1.48 / 1.75
46	4.05 / 7.21	3.20 / 5.10	2.81 / 4.24	2.57 / 3.76	2.42 / 3.44	2.30 / 3.22	2.22 / 3.05	2.14 / 2.92	2.09 / 2.82	2.04 / 2.73	2.00 / 2.66	1.97 / 2.60	1.91 / 2.50	1.87 / 2.42	1.80 / 2.30	1.75 / 2.22	1.71 / 2.13	1.65 / 2.04	1.62 / 1.98	1.57 / 1.90	1.54 / 1.86	1.51 / 1.80	1.48 / 1.76	1.46 / 1.72
48	4.04 / 7.19	3.19 / 5.08	2.80 / 4.22	2.56 / 3.74	2.41 / 3.42	2.30 / 3.20	2.21 / 3.04	2.14 / 2.90	2.08 / 2.80	2.03 / 2.71	1.99 / 2.64	1.96 / 2.58	1.90 / 2.48	1.86 / 2.40	1.79 / 2.28	1.74 / 2.20	1.70 / 2.11	1.64 / 2.02	1.61 / 1.96	1.56 / 1.88	1.53 / 1.84	1.50 / 1.78	1.47 / 1.73	1.45 / 1.70
50	4.03 / 7.17	3.18 / 5.06	2.79 / 4.20	2.56 / 3.72	2.40 / 3.41	2.29 / 3.18	2.20 / 3.02	2.13 / 2.88	2.07 / 2.78	2.02 / 2.70	1.98 / 2.62	1.95 / 2.56	1.90 / 2.46	1.85 / 2.39	1.78 / 2.26	1.74 / 2.18	1.69 / 2.10	1.63 / 2.00	1.60 / 1.94	1.55 / 1.86	1.52 / 1.82	1.48 / 1.76	1.46 / 1.71	1.44 / 1.68
55	4.02 / 7.12	3.17 / 5.01	2.78 / 4.16	2.54 / 3.68	2.38 / 3.37	2.27 / 3.15	2.18 / 2.98	2.11 / 2.85	2.05 / 2.75	2.00 / 2.66	1.97 / 2.59	1.93 / 2.53	1.88 / 2.43	1.83 / 2.35	1.76 / 2.23	1.72 / 2.15	1.67 / 2.06	1.61 / 1.96	1.58 / 1.90	1.52 / 1.82	1.50 / 1.78	1.46 / 1.71	1.43 / 1.66	1.41 / 1.64
60	4.00 / 7.08	3.15 / 4.98	2.76 / 4.13	2.52 / 3.65	2.37 / 3.34	2.25 / 3.12	2.17 / 2.95	2.10 / 2.82	2.04 / 2.72	1.99 / 2.63	1.95 / 2.56	1.92 / 2.50	1.86 / 2.40	1.81 / 2.32	1.75 / 2.20	1.70 / 2.12	1.65 / 2.03	1.59 / 1.93	1.56 / 1.87	1.50 / 1.79	1.48 / 1.74	1.44 / 1.68	1.41 / 1.63	1.39 / 1.60
65	3.99 / 7.04	3.14 / 4.95	2.75 / 4.10	2.51 / 3.62	2.36 / 3.31	2.24 / 3.09	2.15 / 2.93	2.08 / 2.79	2.02 / 2.70	1.98 / 2.61	1.94 / 2.54	1.90 / 2.47	1.85 / 2.37	1.80 / 2.30	1.73 / 2.18	1.68 / 2.09	1.63 / 2.00	1.57 / 1.90	1.54 / 1.84	1.49 / 1.76	1.46 / 1.71	1.42 / 1.64	1.39 / 1.60	1.37 / 1.56
70	3.98 / 7.01	3.13 / 4.92	2.74 / 4.08	2.50 / 3.60	2.35 / 3.29	2.23 / 3.07	2.14 / 2.91	2.07 / 2.77	2.01 / 2.67	1.97 / 2.59	1.93 / 2.51	1.89 / 2.45	1.84 / 2.35	1.79 / 2.28	1.72 / 2.15	1.67 / 2.07	1.62 / 1.98	1.56 / 1.88	1.53 / 1.82	1.47 / 1.74	1.45 / 1.69	1.40 / 1.62	1.37 / 1.56	1.35 / 1.53
80	3.96 / 6.96	3.11 / 4.88	2.72 / 4.04	2.48 / 3.56	2.33 / 3.25	2.21 / 3.04	2.12 / 2.87	2.05 / 2.74	1.99 / 2.64	1.95 / 2.55	1.91 / 2.48	1.88 / 2.41	1.82 / 2.32	1.77 / 2.24	1.70 / 2.11	1.65 / 2.03	1.60 / 1.94	1.54 / 1.84	1.51 / 1.78	1.45 / 1.70	1.42 / 1.65	1.38 / 1.57	1.35 / 1.52	1.32 / 1.49
100	3.94 / 6.90	3.09 / 4.82	2.70 / 3.98	2.46 / 3.51	2.30 / 3.20	2.19 / 2.99	2.10 / 2.82	2.03 / 2.69	1.97 / 2.59	1.92 / 2.51	1.88 / 2.43	1.85 / 2.36	1.79 / 2.26	1.75 / 2.19	1.68 / 2.06	1.63 / 1.98	1.57 / 1.89	1.51 / 1.79	1.48 / 1.73	1.42 / 1.64	1.39 / 1.59	1.34 / 1.51	1.30 / 1.46	1.28 / 1.43
125	3.92 / 6.84	3.07 / 4.78	2.68 / 3.94	2.44 / 3.47	2.29 / 3.17	2.17 / 2.95	2.08 / 2.79	2.01 / 2.65	1.95 / 2.56	1.90 / 2.47	1.86 / 2.40	1.83 / 2.33	1.77 / 2.23	1.72 / 2.15	1.65 / 2.03	1.60 / 1.94	1.55 / 1.85	1.49 / 1.75	1.45 / 1.68	1.39 / 1.59	1.36 / 1.54	1.31 / 1.46	1.27 / 1.40	1.25 / 1.37
150	3.91 / 6.81	3.06 / 4.75	2.67 / 3.91	2.43 / 3.44	2.27 / 3.14	2.16 / 2.92	2.07 / 2.76	2.00 / 2.62	1.94 / 2.53	1.89 / 2.44	1.85 / 2.37	1.82 / 2.30	1.76 / 2.20	1.71 / 2.12	1.64 / 2.00	1.59 / 1.91	1.54 / 1.83	1.47 / 1.72	1.44 / 1.66	1.37 / 1.56	1.34 / 1.51	1.29 / 1.43	1.25 / 1.37	1.22 / 1.33
200	3.89 / 6.76	3.04 / 4.71	2.65 / 3.88	2.41 / 3.41	2.26 / 3.11	2.14 / 2.90	2.05 / 2.73	1.98 / 2.60	1.92 / 2.50	1.87 / 2.41	1.83 / 2.34	1.80 / 2.28	1.74 / 2.17	1.69 / 2.09	1.62 / 1.97	1.57 / 1.88	1.52 / 1.79	1.45 / 1.69	1.42 / 1.62	1.35 / 1.53	1.32 / 1.48	1.26 / 1.39	1.22 / 1.33	1.19 / 1.28
400	3.86 / 6.70	3.02 / 4.66	2.62 / 3.83	2.39 / 3.36	2.23 / 3.06	2.12 / 2.85	2.03 / 2.69	1.96 / 2.55	1.90 / 2.46	1.85 / 2.37	1.81 / 2.29	1.78 / 2.23	1.72 / 2.12	1.67 / 2.04	1.60 / 1.92	1.54 / 1.84	1.49 / 1.74	1.42 / 1.64	1.38 / 1.57	1.32 / 1.47	1.28 / 1.42	1.22 / 1.32	1.16 / 1.24	1.13 / 1.19
1000	3.85 / 6.66	3.00 / 4.62	2.61 / 3.80	2.38 / 3.34	2.22 / 3.04	2.10 / 2.82	2.02 / 2.66	1.95 / 2.53	1.89 / 2.43	1.84 / 2.34	1.80 / 2.26	1.76 / 2.20	1.70 / 2.09	1.65 / 2.01	1.58 / 1.89	1.53 / 1.81	1.47 / 1.71	1.41 / 1.61	1.36 / 1.54	1.30 / 1.44	1.26 / 1.38	1.19 / 1.28	1.13 / 1.19	1.08 / 1.11
∞	3.84 / 6.64	2.99 / 4.60	2.60 / 3.78	2.37 / 3.32	2.21 / 3.02	2.09 / 2.80	2.01 / 2.64	1.94 / 2.51	1.88 / 2.41	1.83 / 2.32	1.79 / 2.24	1.75 / 2.18	1.69 / 2.07	1.64 / 1.99	1.57 / 1.87	1.52 / 1.79	1.46 / 1.69	1.40 / 1.59	1.35 / 1.52	1.28 / 1.41	1.24 / 1.36	1.17 / 1.25	1.11 / 1.15	1.00 / 1.00

df (denominator)

Source: Reprinted by permission from *Statistical Methods*, 6th ed., by George W. Snedecor and William G. Cochran, © 1967, by the Iowa State University Press, Ames, Iowa.

Appendix 9

Unit Normal Loss Function

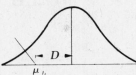

D	.00	.01	.02	.03	.04	.05	.06	.07	.08	.09
.0	.3989	.3940	.3890	.3841	.3793	.3744	.3697	.3649	.3602	.3556
.1	.3509	.3464	.3418	.3373	.3328	.3284	.3240	.3197	.3154	.3111
.2	.3069	.3027	.2986	.2944	.2904	.2863	.2824	.2784	.2745	.2706
.3	.2668	.2630	.2592	.2555	.2518	.2481	.2445	.2409	.2374	.2339
.4	.2304	.2270	.2236	.2203	.2169	.2137	.2104	.2072	.2040	.2009
.5	.1978	.1947	.1917	.1887	.1857	.1828	.1799	.1771	.1742	.1714
.6	.1687	.1659	.1633	.1606	.1580	.1554	.1528	.1503	.1478	.1453
.7	.1429	.1405	.1381	.1358	.1334	.1312	.1289	.1267	.1245	.1223
.8	.1202	.1181	.1160	.1140	.1120	.1100	.1080	.1061	.1042	.1023
.9	.1004	.09860	.09680	.09503	.09328	.09156	.08986	.08819	.08654	.08491
1.0	.08332	.08174	.08019	.07866	.07716	.07568	.07422	.07279	.07138	.06999
1.1	.06862	.06727	.06595	.06465	.06336	.06210	.06086	.05964	.05844	.05726
1.2	.05610	.05496	.05384	.05274	.05165	.05059	.04954	.04851	.04750	.04650
1.3	.04553	.04457	.04363	.04270	.04179	.04090	.04002	.03916	.03831	.03748
1.4	.03667	.03587	.03508	.03431	.03356	.03281	.03208	.03137	.03067	.02998
1.5	.02931	.02865	.02800	.02736	.02674	.02612	.02552	.02494	.02436	.02380
1.6	.02324	.02270	.02217	.02165	.02114	.02064	.02015	.01967	.01920	.01874
1.7	.01829	.01785	.01742	.01699	.01658	.01617	.01578	.01539	.01501	.01464
1.8	.01428	.01392	.01357	.01323	.01290	.01257	.01226	.01195	.01164	.01134
1.9	.01105	.01077	.01049	.01022	$.0^2 9957$	$.0^2 9698$	$.0^2 9445$	$.0^2 9198$	$.0^2 8957$	$.0^2 8721$
2.0	$.0^2 8491$	$.0^2 8266$	$.0^2 8046$	$.0^2 7832$	$.0^2 7623$	$.0^2 7418$	$.0^2 7219$	$.0^2 7024$	$.0^2 6835$	$.0^2 6649$
2.1	$.0^2 6468$	$.0^2 6292$	$.0^2 6120$	$.0^2 5952$	$.0^2 5788$	$.0^2 5628$	$.0^2 5472$	$.0^2 5320$	$.0^2 5172$	$.0^2 5028$
2.2	$.0^2 4887$	$.0^2 4750$	$.0^2 4616$	$.0^2 4486$	$.0^2 4358$	$.0^2 4235$	$.0^2 4114$	$.0^2 3996$	$.0^2 3882$	$.0^2 3770$
2.3	$.0^2 3662$	$.0^2 3556$	$.0^2 3453$	$.0^2 3352$	$.0^2 3255$	$.0^2 3159$	$.0^2 3067$	$.0^2 2977$	$.0^2 2889$	$.0^2 2804$
2.4	$.0^2 2720$	$.0^2 2640$	$.0^2 2561$	$.0^2 2484$	$.0^2 2410$	$.0^2 2337$	$.0^2 2267$	$.0^2 2199$	$.0^2 2132$	$.0^2 2067$
2.5	$.0^2 2004$	$.0^2 1943$	$.0^2 1883$	$.0^2 1826$	$.0^2 1769$	$.0^2 1715$	$.0^2 1662$	$.0^2 1610$	$.0^2 1560$	$.0^2 1511$
2.6	$.0^2 1464$	$.0^2 1418$	$.0^2 1373$	$.0^2 1330$	$.0^2 1288$	$.0^2 1247$	$.0^2 1207$	$.0^2 1169$	$.0^2 1132$	$.0^2 1095$
2.7	$.0^2 1060$	$.0^2 1026$	$.0^3 9928$	$.0^3 9607$	$.0^3 9295$	$.0^3 8992$	$.0^3 8699$	$.0^3 8414$	$.0^3 8138$	$.0^3 7870$
2.8	$.0^3 7611$	$.0^3 7359$	$.0^3 7115$	$.0^3 6879$	$.0^3 6650$	$.0^3 6428$	$.0^3 6213$	$.0^3 6004$	$.0^3 5802$	$.0^3 5606$
2.9	$.0^3 5417$	$.0^3 5233$	$.0^3 5055$	$.0^3 4883$	$.0^3 4716$	$.0^3 4555$	$.0^3 4398$	$.0^3 4247$	$.0^3 4101$	$.0^3 3959$
3.0	$.0^3 3822$	$.0^3 3689$	$.0^3 3560$	$.0^3 3436$	$.0^3 3316$	$.0^3 3199$	$.0^3 3087$	$.0^3 2978$	$.0^3 2873$	$.0^3 2771$
3.1	$.0^3 2673$	$.0^3 2577$	$.0^3 2485$	$.0^3 2396$	$.0^3 2311$	$.0^3 2227$	$.0^3 2147$	$.0^3 2070$	$.0^3 1995$	$.0^3 1922$
3.2	$.0^3 1852$	$.0^3 1785$	$.0^3 1720$	$.0^3 1657$	$.0^3 1596$	$.0^3 1537$	$.0^3 1480$	$.0^3 1426$	$.0^3 1373$	$.0^3 1322$
3.3	$.0^3 1273$	$.0^3 1225$	$.0^3 1179$	$.0^3 1135$	$.0^3 1093$	$.0^3 1051$	$.0^3 1012$	$.0^4 9734$	$.0^4 9365$	$.0^4 9009$
3.4	$.0^4 8666$	$.0^4 8335$	$.0^4 8016$	$.0^4 7709$	$.0^4 7413$	$.0^4 7127$	$.0^4 6852$	$.0^4 6587$	$.0^4 6331$	$.0^4 6085$
3.5	$.0^4 5848$	$.0^4 5620$	$.0^4 5400$	$.0^4 5188$	$.0^4 4984$	$.0^4 4788$	$.0^4 4599$	$.0^4 4417$	$.0^4 4242$	$.0^4 4073$
3.6	$.0^4 3911$	$.0^4 3755$	$.0^4 3605$	$.0^4 3460$	$.0^4 3321$	$.0^4 3188$	$.0^4 3059$	$.0^4 2935$	$.0^4 2816$	$.0^4 2702$
3.7	$.0^4 2592$	$.0^4 2486$	$.0^4 2385$	$.0^4 2287$	$.0^4 2193$	$.0^4 2103$	$.0^4 2016$	$.0^4 1933$	$.0^4 1853$	$.0^4 1776$
3.8	$.0^4 1702$	$.0^4 1632$	$.0^4 1563$	$.0^4 1498$	$.0^4 1435$	$.0^4 1375$	$.0^4 1317$	$.0^4 1262$	$.0^4 1208$	$.0^4 1157$
3.9	$.0^4 1108$	$.0^4 1061$	$.0^4 1016$	$.0^5 9723$	$.0^5 9307$	$.0^5 8908$	$.0^5 8525$	$.0^5 8158$	$.0^5 7806$	$.0^5 7469$
4.0	$.0^5 7145$	$.0^5 6835$	$.0^5 6538$	$.0^5 6253$	$.0^5 5980$	$.0^5 5718$	$.0^5 5468$	$.0^5 5227$	$.0^5 4997$	$.0^5 4777$
4.1	$.0^5 4566$	$.0^5 4364$	$.0^5 4170$	$.0^5 3985$	$.0^5 3807$	$.0^5 3637$	$.0^5 3475$	$.0^5 3319$	$.0^5 3170$	$.0^5 3027$
4.2	$.0^5 2891$	$.0^5 2760$	$.0^5 2635$	$.0^5 2516$	$.0^5 2402$	$.0^5 2292$	$.0^5 2188$	$.0^5 2088$	$.0^5 1992$	$.0^5 1901$
4.3	$.0^5 1814$	$.0^5 1730$	$.0^5 1650$	$.0^5 1574$	$.0^5 1501$	$.0^5 1431$	$.0^5 1365$	$.0^5 1301$	$.0^5 1241$	$.0^5 1183$
4.4	$.0^5 1127$	$.0^5 1074$	$.0^5 1024$	$.0^6 9756$	$.0^6 9296$	$.0^6 8857$	$.0^6 8437$	$.0^6 8037$	$.0^6 7655$	$.0^6 7290$
4.5	$.0^6 6942$	$.0^6 6610$	$.0^6 6294$	$.0^6 5992$	$.0^6 5704$	$.0^6 5429$	$.0^6 5167$	$.0^6 4917$	$.0^6 4679$	$.0^6 4452$
4.6	$.0^6 4236$	$.0^6 4029$	$.0^6 3833$	$.0^6 3645$	$.0^6 3467$	$.0^6 3297$	$.0^6 3135$	$.0^6 2981$	$.0^6 2834$	$.0^6 2694$
4.7	$.0^6 2560$	$.0^6 2433$	$.0^6 2313$	$.0^6 2197$	$.0^6 2088$	$.0^6 1984$	$.0^6 1884$	$.0^6 1790$	$.0^6 1700$	$.0^6 1615$
4.8	$.0^6 1533$	$.0^6 1456$	$.0^6 1382$	$.0^6 1312$	$.0^6 1246$	$.0^6 1182$	$.0^6 1122$	$.0^6 1065$	$.0^6 1011$	$.0^7 9588$
4.9	$.0^7 9096$	$.0^7 8629$	$.0^7 8185$	$.0^7 7763$	$.0^7 7362$	$.0^7 6982$	$.0^7 6620$	$.0^7 6276$	$.0^7 5950$	$.0^7 5640$

*The small numbers which appear as superscripts indicate the number of zeros immediately following the decimal point. For example, $.0^2 9957$ is the value .009957.

Source: Reproduced by permission of the copyright holders, The President and Fellows of Harvard College, from the Unit Normal Loss Integral table which appears as Table IV in *Introduction to Statistics for Business Decisions* by Robert Schlaifer, published by McGraw-Hill Book Company, New York, 1961.

Appendix 10

Critical Values of *T* in the Wilcoxon Test

1-sided	2-sided	n = 5	n = 6	n = 7	n = 8	n = 9	n = 10
P = 0.05	P = 0.10	1	2	4	6	8	11
P = 0.025	P = 0.05		1	2	4	6	8
P = 0.01	P = 0.02			0	2	3	5
P = 0.005	P = 0.01				0	2	3

1-sided	2-sided	n = 11	n = 12	n = 13	n = 14	n = 15	n = 16
P = 0.05	P = 0.10	14	17	21	26	30	36
P = 0.025	P = 0.05	11	14	17	21	25	30
P = 0.01	P = 0.02	7	10	13	16	20	24
P = 0.005	P = 0.01	5	7	10	13	16	19

1-sided	2-sided	n = 17	n = 18	n = 19	n = 20	n = 21	n = 22
P = 0.05	P = 0.10	41	47	54	60	68	75
P = 0.025	P = 0.05	35	40	46	52	59	66
P = 0.01	P = 0.02	28	33	38	43	49	56
P = 0.005	P = 0.01	23	28	32	37	43	49

1-sided	2-sided	n = 23	n = 24	n = 25	n = 26	n = 27	n = 28
P = 0.05	P = 0.10	83	92	101	110	120	130
P = 0.025	P = 0.05	73	81	90	98	107	117
P = 0.01	P = 0.02	62	69	77	85	93	102
P = 0.005	P = 0.01	55	68	68	75	84	92

1-sided	2-sided	n = 29	n = 30	n = 31	n = 32	n = 33	n = 34
P = 0.05	P = 0.10	141	152	163	175	188	201
P = 0,025	P = 0.05	127	137	148	159	171	183
P = 0.01	P = 0.02	111	120	130	141	151	162
P = 0.005	P = 0.01	100	109	118	128	138	149

1-sided	2-sided	n = 35	n = 36	n = 37	n = 38	n = 39	
P = 0.05	P = 0.10	214	228	242	256	271	
P = 0.025	P = 0.05	195	208	222	235	250	
P = 0.01	P = 0.02	174	186	198	211	224	
P = 0.005	P = 0.01	160	171	183	195	208	

1-sided	2-sided	n = 40	n = 41	n = 42	n = 43	n = 44	n = 45
P = 0.05	P = 0.10	287	303	319	336	353	371
P = 0.025	P = 0.05	264	279	295	311	327	344
P = 0.01	P = 0.02	238	252	267	281	297	313
P = 0.005	P = 0.01	221	234	248	262	277	292

1-sided	2-sided	$n = 46$	$n = 47$	$n = 48$	$n = 49$	$n = 50$
$P = 0.05$	$P = 0.10$	389	408	427	446	466
$P = 0.025$	$P = 0.05$	361	379	397	415	434
$P = 0.01$	$P = 0.02$	329	345	362	380	398
$P = 0.005$	$P = 0.01$	307	323	339	356	373

(*Source*: From F. Wilcoxon and R. A. Wilcox, Some Rapid Approximate Statistical Procedures, 1964, Reproduced with permission of Lederle Laboratories Division, American Cyanamid Company.)

Index

A priori probability, 65
Addition, rules of, 67–68
Aggregate index number, 291
Aggregate price index, 291–292
Alternative hypothesis, 157
Amount certain, 308
Analysis of variance, 222–227
  basic rationale, 222–223
  one-factor completely randomized design, 223–224
  randomized block design, 224–225
  in regression analysis, 264–265
  two-factor completely randomized design, 225–226
  two-way analysis, 224
"And under" type of frequency distribution, 11
ANOVA (*see* Analysis of Variance)
Arithmetic average, 30
Arithmetic mean, 30, 32–33
Autocorrelation, 266
Average, 30
Average deviation, 46, 50–51

Backward stepwise regression analysis, 263
Bar chart, 11–12
Base period, shifting of, 292
Bayes' theorem, 71–72, 324–325
Bayesian criterion, 307
Bayesian decision analysis, 1, 302–309, 322–331, 338–349
  compared with classical decision procedures, 348–349
Bayesian posterior analysis, 325–327, 344–346
Bernoulli process, 93
Bimodal distribution, 31
Binomial probabilities, 93–95
  normal approximation of, 114–115
  Poisson approximation of, 97–98
  table of, 373–375
Bivariate normal distribution, 246
Boundaries of a class in a frequency distribution, 7
Breakeven point, 340–341
Business indicators, 279–280
Business statistics, definition of, 1

Census, 1
Central limit theorem, 129–130
Central tendency, 30
Chance events in decision tree analysis, 308
Chi-square distribution:
  for contingency table tests, 202–204
  for estimating the population variance, 148–149
  for goodness of fit tests, 200–202

Chi-square distribution (*continued*)
  table of values for, 383
  for testing the difference between two proportions, 205–206
  for testing the differences among several proportions, 206–207
  for testing the value of the population proportion, 204–205
  for testing the value of the population variance, 185–186
Chi-square test, 200–207
Class boundaries, 7
Class interval, 7–8
Class limits:
  exact, 7
  stated, 7
Class midpoint, 7
Classical approach to probability, 65
Classical statistics, 1
  contrasted to Bayesian decision analysis, 348–349
Cluster sample, 2
Coefficient:
  of correlation, 247–249
  of determination, 247
  of multiple correlation, 265
  of multiple determination, 265
  of partial correlation, 266
  of partial determination, 266
  of variation, 119
Coinciding indicators, 280
Colinearity, 266
Combinations, 74–75
Component bar chart, 11–12
Computational formulas, 47
Computer applications:
  for analysis of variance, 227, 229, 234, 238–239
  for chi-square test, 207, 218
  for confidence intervals, 149, 153–154
  for contingency table test, 207, 218
  for correlation analysis, 250, 256–258
  for descriptive measures for data sets, 35, 41, 52, 61
  for discrete probability distributions, 98, 106
  for estimating the mean, 133, 140–141
  for forming frequency distributions, 13, 24–25
  for generating random numbers, 3, 5
  for hypotheses concerning differences between means, 187, 196
  for hypotheses concerning the mean, 165, 173–174
  for multiple correlation analysis, 267–273
  for multiple regression analysis, 267–273

Computer applications (*continued*)
 for normally distributed variables, 116–117, 123–124
 for regression analysis, 250, 256–258
 for seasonal analysis, 281, 287–288
 for trend analysis, 281, 288–289
Conditional mean, in regression analysis, 245–246
Conditional probability, 68–69
Confidence interval approach to hypothesis testing, 164–165
Confidence intervals:
 for difference between means, 144–145
 for difference between proportions, 147–148
 for mean, 131–133
 for proportion, 146–147
 for standard deviation, 148–149
 summary table for the mean, 133
 for variance, 148–149
Consumer Price Index (CPI), 293
Contingency table, 72
Contingency table test, 202–204
Continuity correction factor (*see* Correction for continuity)
Continuous random variable, 91, 110
Continuous variable, 2
Correction for continuity, 115
Correlation analysis, 246–249
Correlation coefficient (*see* Coefficient of correlation)
Covariance, 249
CPI, 293
Credibility interval, 339
Critical value, 158, 160
Cumulative frequency distribution, 10–11
Curvilinear relationship, 242
Cycle chart, 277
Cyclical fluctuations, 275
Cyclical forecasting, 279–280
Cyclical turning points, 279
Cyclical variations, analysis of, 277

Deciles, 32, 34–35
Decision points, 308
Decision tree analysis, 307–308
Deductive reasoning, 131
Deflation of time series values, 293
Degree of belief, 65
Degree of confidence, 131
Degrees of freedom (df), 133
 for contingency table tests, 203
 for F distribution, 186
 for goodness of fit tests, 201
Density function, 110
 for normal distribution, 111
Dependent events, 68–71
Dependent variable, 242

Descriptive statistics, 1
Deseasonalized data, 278
Deviation formulas, 47
Difference between means:
 confidence intervals for, 144–145
 testing hypotheses for, 178–182
Difference between proportions:
 confidence interval for, 147–148
 testing of by use of chi-square distribution, 205–206
 testing of by use of normal distribution, 184–185
Differences among several means, testing for, 222–227
Differences among several proportions, testing for, 206–207
Discrete random variable, 91
Discrete variable, 1
Distribution-free methods, 357
Double exponential smoothing, 280
Dummy variable, 263–264

$e^{-\lambda}$ values, table of, 376
Economic consequences, in decision analysis, 304
Empirical approach to probability, 65
ENGS, 329–331, 347–348
EOL, 307, 342–344
EP, 306
EPPI, 322
Estimation (*see* Confidence intervals)
Events:
 conditional, 68–69
 dependent, 68–71
 independent, 68–71
 mutually exclusive, 67–68
 nonexclusive, 67–68
 in payoff table, 302
EVPI, 322, 342–344
EVSI, 327–329, 346–347
Exact class limits, 7
Expected frequency (f_e), 200
Expected loss, 307
Expected net gain from sampling (*ENGS*), 329–331, 347–348
Expected opportunity loss (*EOL*), 307, 342–344
Expected payoff criterion, 306–307
Expected payoff with perfect information (*EPPI*), 322
Expected utility criterion, 308–309
Expected value of discrete random variable, 91–92
Expected value of perfect information (*EVPI*), 322–323, 342–344
Expected value of sample information (*EVSI*), 327–329, 346–347
Experiment, 2
Exponential probability distribution, 116–117
Exponential smoothing, 280
Exponential trend curve, 276

F distribution, 186–187, 222
    tables of areas for, 384–386
F ratio, 222–223
Factorial design, 226–227
Finite correction factor, 129
Fitted value, 244
Fixed-effects model, 226
Folding back, in decision tree analysis, 308
Forecasting, 278–281
Forward stepwise regression analysis, 263
Frequency curve, 9–10
Frequency distribution, 7
Frequency polygon, 9

Gompertz curve, 277
Goodness of fit tests, 200–202
Greek letters, use of as symbols, 30
Grouped data, 7
    formulas for, 33–35, 50–52

Histogram, 8
Homogeneity of variance, 222
Homoscedasticity, 242
Hypergeometric distribution, 95–96
Hypothesis testing:
    for difference between means, 178–182
    for difference between medians, 360–361
    for difference between proportions, 184–185,
        205–206
    for difference between variances, 186–187
    for differences among several means, 222–223
    for differences among several medians, 361–362
    for differences among several proportions, 206
    for mean, 157–165
    for median, 358–359
    for proportion, 182–184, 204–205
    for standard deviation, 185–186
    summary table for the mean, 165
    for variance, 185–186

Incomplete block design, 226
Independence, test for categorical data, 202–204
Independent events, 68–71
Independent variable, 242
Index numbers, 291–293
Index of Industrial Production, 293
Indicator variables, 263–264
Inductive reasoning, 131
Inferential statistics, 1
Interaction, in analysis of variance, 224
Intersection of two events, 69
Interval of class, 7–8
Interval scale, 357

Joint probability table, 72

Kruskal–Wallis test, 361–362
Kurtosis, 9–10

Lagging indicators, 280
Laspeyres' index, 292
Latin square design, 226
Leading indicators, 279
Least squares criterion, 243
    use of in regression analysis, 243–244
Leptokurtic frequency curve, 9–10
Level of significance, 157
Line graph, 12
Linear payoff function, 340–342
Linear piecewise loss function, 342–344
Linear regression analysis, 242–246
Link relatives, 292
Location, measures of, 30
Loss function, 342

Mann–Whitney test, 360
Marginal probability, 72
Matched pairs, 180
Maximax criterion, 304–305
Maximin criterion, 304
Mean, 30
Mean square, 222
Mean square error (*MSE*):
    in analysis of variance, 222
    in regression analysis, 265
Median, 31, 33–34, 358–359
Mesokurtic frequency curve, 9–10
Midpoint of class, 7
Minimax regret criterion, 305–306
Mode, 31, 34
Modified ranges, 45–46, 50
MSE, 222, 265
MSTR, 222
Multicolinearity, 266
Multimodal distribution, 31
Multiple correlation analysis, 265–267
Multiple regression analysis, 242
Multiplication, rules of, 69–71
Mutually exclusive events, 67–68

Negative skewness, 9–10
Net overall terminal expected payoff (*NOTEP*), 330
Net regression coefficient, 262
Nominal scale, 357
Nonexclusive events, 67–68
Nonparametric statistical methods, 357–358
Nonparametric statistical tests, 357–362
Normal approximation:
    of binomial probabilities, 114–115
    of Poisson probabilities, 115–116
Normal prior probability distribution, 338–340

Normal probability distribution, 110–114
  in Bayesian analysis, 338–349
  table of area for, 381
NOTEP, 330
Null hypothesis, 157

Observed frequency (f_o), 200
OC curve, 161
Odds ratio, 66–67
Ogive, 10
Ogive curve, 10
One-factor analysis of variance, 223–224
One-tail test, 159
Operating characteristic (*OC*) curve, 161
Opportunity loss, 305
Opportunity loss function, 342
Optimum sample size in Bayesian analysis, 329–331
Ordinal scale, 357
OTEP, 330
Overall terminal expected payoff (*OTEP*), 330

P-value approach to hypothesis testing, 164
Paasche's index, 292
Paired observations, 180–182
Parameter, 1, 30, 128
Parametric statistical methods, 357–358
Partial correlation coefficient, 266
Partial regression coefficient, 262
Payoff table, 302–303
Payoffs, 302
Pearson's coefficient of skewness, 49
Percentage pie chart, 12–13
Percentiles, 11, 32, 34–45
  for normally distributed variables, 113–114
Permutations, 73–74
Personalistic approach to probability, 65–66
Pie chart, 12–13
Platykurtic frequency curve, 9–10
Point estimator, 128
Poisson approximation of binomial probabilities, 97–98
Poisson probability distribution, 96–98
  normal approximation of, 115–116
  table of values for, 377–380
Poisson process, 96
Population parameter, 1, 30, 128
Positive skewness, 9–10
Posterior analysis, 325–327, 345–346
Posterior expected payoff, 345–346
Posterior mean, 344–346
Posterior probability distribution, 323
Posterior variance, 345–346
Power, in hypothesis testing, 162
Power curve, 162
Prediction interval, 246
  in multiple regression analysis, 263

Preposterior analysis, 327–329, 346–347
Price index, 291
Price relative, 291
Prior mean, 338–339
Prior probabilities, 303
Prior probability distribution, 323
Prior standard deviation, 339–340
Probability, 65–66
  a priori, 65
  classical, 65
  conditional, 68–69
  empirical, 65
  personalistic, 65–66
  relative frequency, 65
  subjective, 65–66
Probability curve, 110
Probability density function, 110
  for normal distribution, 111
Probability distribution, 91
Probability sample, 2
Producer Price Indexes, 293
Purchasing power of dollar, 293

Quantity index, 291
Quantity relative, 291
Quartiles, 32, 34–35

Random-effects model, 226
Random numbers, 3
  table of, 372
Random sample, 2
Random variable:
  continuous, 91
  discrete, 91
Randomized block design, 224–225
Range, 45, 50
Ratio scale, 357
Ratio-to-moving-average method, 278
Reciprocal property of F distribution, 186
Reference contract, 308–309
Regression analysis, 242–246
  multiple, 262–265
Regression equation, 243–244
  multiple, 262
Regret, in decision analysis, 305
Relative frequency approach to probability, 65
Relative frequency distribution, 11
Replication, 224
Residual, 244, 264
Residual plot, 244, 264
Risk averter, 309
Risk neutral, 309
Risk seeker, 309
Roman letters, use of as symbols, 30
Runs test for randomness, 358

Sample:
  cluster, 3
  probability, 2
  random, 2
  scientific, 2
  simple random, 2
  stratified, 3
  systematic, 2
Sample size:
  in Bayesian analysis, 329–331
  for estimating mean, 132
  for estimating proportion, 146–147
  for testing mean, 163
  for testing proportion, 184
Sample statistic, 1, 30
Sampling distribution, 128
Sampling without replacement, 95
Scatter plot, 242
Scientific sample, 2
Seasonal adjustments, 278
Seasonal variations, 275
  measurement of, 277–278
Seasonally adjusted data, 278
Sign test:
  for one sample, 358–359
  for paired observations, 360–361
Simple correlation analysis, 246–249
Simple index number, 291
Simple random sample, 2
Simple regression analysis, 242–246
Single exponential smoothing, 280
Skewness, 9–10, 32
  coefficient of, 49–50
Slope of regression line, 243
  inferences concerning, 245
Splicing of index numbers, 293
Standard deviation, 46–49, 51–52
Standard error:
  of conditional mean, 245
  of difference between means, 144
  of difference between proportions, 147
  of estimate, 244–245
  of forecast, 246
  of mean, 129
  of mean difference, 181
  of proportion, 146
Standard normal distribution, 111
Stated class limits, 7
Statistics, 1
  descriptive, 1
  inferential, 1
Stepwise regression analysis, 263
Stratified sample, 3
Student's t distribution, 132–133
  table of areas, 382
Subjective approach to probability, 65–66

Survey, 2
Symmetrical distribution, 9–10
Systematic sample, 2

t distribution (see Student's t distribution)
T statistic in Wilcoxon test, 359, 361
  table of critical values for, 388–389
Test statistic, 158
Testing a hypothesis (see Hypothesis testing)
Time series, 275
Time series analysis, 275–281
Time series model, 275–276
Trend, 275
Trend analysis, 276–277
Trend line, 276
Triple exponential smoothing, 280
Two-factor analysis of variance, 225–226
Two-tail test, 158
Two-way analysis of variance, 224
Type I error, 157, 161
Type II error, 157, 161–163

Unbiased estimator, 128
Ungrouped data, 7
Unimodal distribution, 31
Union of two events, 67
Unit normal loss function, 343
  table of values for, 387
Utility, 308
Utility function, 309

Value index, 291
Value relative, 291
Value of sample information (VSI), 326–327,
  345–346
Variability, measures of, 45
Variables:
  continuous, 2
  discrete, 1
  random, 91
Variance, 46–49, 51–52
Variation, coefficient of, 49
Venn diagram, 66
VSI, 326–327, 345–346

Weighted aggregate of prices, 292
Weighted average, 30–31
Weighted mean, 30–31
Wholesale price index, 293
Wilcoxon test:
  for one sample, 359
  for paired observations, 361
  table of critical values for, 388–389

z distribution, table of areas for, 381
z value, 111